Fundamentals of Adhesion

Fundamentals of Adhesion

Edited by

Lieng-Huang Lee

Xerox Corporation
Webster, New York

Plenum Press • New York and London

Library of Congress Cataloging-in-Publication Data

Fundamentals of adhesion / edited by Lieng-Huang Lee.
p. cm.
Includes bibliographical references and indexes.
ISBN 0-306-43470-9
1. Adhesion. I. Lee, Lieng-Huang, date.
QC183.F88 1991
541.3'3--dc20 90-15509
CIP

ISBN 0-306-43470-9

A Division of Plenum Publishing Corporation
233 Spring Street, New York, N.Y. 10013

Printed in the United States of America

Contributors

John E. E. Baglin • IBM Almaden Research Center, San Jose, California 95120–6099

Robert E. Baier • Industry/University Center for Biosurfaces, State University of New York at Buffalo, Buffalo, New York 14214

Amitava Banerjea • NASA Lewis Research Center, Cleveland, Ohio 44135 and Physics Department, Kent State University, Kent, Ohio 44242

F. Brochard-Wyart • Laboratoire de Physique de la Matière Condensée, Collège de France, F-75231 Paris Cedex 05, France and Structure et Réactivité aux Interfaces, Université Pierre et Marie Curie, F-75231 Paris Cedex 05, France

Manoj K. Chaudhury • Dow Corning Corporation, Midland, Michigan 46868

P. G. de Gennes • Laboratoire de Physique de la Matière Condensée, Collège de France, F-75231 Paris Cedex 05, France

F. Faupel • IBM Thomas J. Watson Research Center, Yorktown Heights, New York 10598. *Present address*: Institut für Metall-physik der Universität Göttingen, D-3400 Göttingen, Federal Republic of Germany

John Ferrante • NASA Lewis Research Center, Cleveland, Ohio 44135

Robert J. Good • Department of Chemical Engineering, State University of New York at Buffalo, Buffalo, New York 14260

W. Gutowski • Division of Building, Construction and Engineering, Commonwealth Scientific and Industrial Research Organization (CSIRO), Melbourne, Australia

Richard Haight • IBM Thomas J. Watson Research Center, Yorktown Heights, New York 10598

Dan A. Hays • Webster Research Center, Xerox Corporation, Webster, New York 14580

Paul S. Ho • IBM Thomas J. Watson Research Center, Yorktown Heights, New York 10598

Hatsuo Ishida • Department of Macromolecular Science, Case Western Reserve University, Cleveland, Ohio 44106

Lieng-Huang Lee • Webster Research Center, Xerox Corporation, Webster, New York 14580

Anne E. Meyer • Industry/University Center for Biosurfaces, State University of New York at Buffalo, Buffalo, New York 14214

James D. Miller • Department of Macromolecular Science, Case Western Reserve University, Cleveland, Ohio 44106. *Present address*: Amoco Performance Products, Inc., Bound Brook, New Jersey 08805

Edwin P. Plueddemann • Dow Corning Corporation, Midland, Michigan 48640

B. D. Silverman • IBM Thomas J. Watson Center, Yorktown Heights, New York 10598

John R. Smith • Physics Department, General Motors Research Laboratories, Warren, Michigan 48090

Carel J. van Oss • Department of Chemical Engineering and Department of Microbiology, State University of New York at Buffalo, Buffalo, New York 14214

Robert C. White • IBM Thomas J. Watson Research Center, Yorktown Heights, New York 10598. *Present address*: Department of Electrical Engineering, Columbia University, New York, New York 10027

Richard P. Wool • Department of Materials Science and Engineering, University of Illinois, Urbana, Illinois 61801

Preface

For several years, I have been responsible for organizing and teaching in the fall a short course on "Fundamentals of Adhesion: Theory, Practice, and Applications" at the State University of New York at New Paltz. Every spring I would try to assemble the most pertinent subjects and line up several capable lecturers for the course. However, there has always been one thing missing—an authoritative book that covers most aspects of adhesion and adhesive bonding. Such a book would be used by the participants as a main reference throughout the course and kept as a sourcebook after the course had been completed. On the other hand, this book could not pretend to be one of those "All you want to know about" volumes, simply because adhesion is an interdisciplinary and ever-growing field.

For the same reason, it would be very difficult for a single individual, especially me, to undertake the task of writing such a book. Thus, I relied on the principle that one leaves the truly monumental jobs to experts, and I finally succeeded in asking several leading scientists in the field of adhesion to write separate chapters for this collection. Some chapters emphasize theoretical concepts and others experimental techniques. In the humble beginning, we planned to include only twelve chapters. However, we soon realized that such a plan would leave too much ground uncovered, and we resolved to increase the coverage. After the book had evolved into thirty chapters, we started to feel that perhaps our mission had been accomplished. Originally we had thought to publish the entire project under the title *Fundamentals of Adhesion;* then it was deemed necessary to split a portion of it off into the second volume, *Adhesive Bonding*. Since these two volumes include chapters by many authors, some overlap is inevitable.

In this first volume we focus on adhesion with or without the use of an adhesive, while in the second we shall concentrate on bonding with the aid of adhesives. In this volume our major concern is the mechanisms of bond formation, while in the second volume our scope will be broadened to both bond-forming and bond-breaking processes. With both of these volumes, we hope that our readers will become familiar with both the science and technology of adhesion.

In this volume, we shall stress the fundamentals of adhesion. Chapter 1 is an overview of the chemistry and physics of adhesion. Chapter 2 summarizes the thermodynamics of adhesion. Chapters 3 and 4 cover the theories of adhesive forces across interfaces. Chapter 5 discusses briefly the dynamics of wetting. Chapters 6 and 7 investigate the diffusion mechanism of adhesion chiefly on the basis of the reptation theory. Chapter 8 deals with the controversial topic of electronic adhesion. Chapters 9 and 10 touch upon the phenomena of chemical interactions at the interface. Chapters 11 and 12 explore the theories of metal adhesion. Chapters 13 and 14 pave the way for us to understand thin-film adhesion related to electronic devices. Finally, Chapter 15 introduces us to the related growing field of bioadhesion, which none of us should ignore.

For better readability, all of the chapters have been refereed and separate nomenclature lists are included at the end of most of the chapters. I should like to take this opportunity to thank the many referees for their efforts, and especially Professor Robert Good for his patience in refereeing more than one chapter for this volume. I also wish to thank all of the authors for their fine cooperation. Finally, I sincerely appreciate the support of the Webster Research Center of Xerox Corporation in helping me complete this task, and the help of Mr. F. G. Belli and Ms. E. Jonas of the Technical Information Center in preparing the indexes.

Lieng-Huang Lee

August, 1990

Contents

1. The Chemistry and Physics of Solid Adhesion

Lieng-Huang Lee

6. Kinetics of Polymer–Polymer Interdiffusion

F. Brochard-Wyart

7. Welding, Tack, and Green Strength of Polymers

Richard P. Wool

13. Interface Design for Thin Film Adhesion

John E. E. Baglin

14. Chemistry, Microstructure, and Adhesion of Metal–Polymer Interfaces

Paul S. Ho, Richard Haight, Robert C. White, B. D. Silverman, and F. Faupel

15. Aspects of Bioadhesion

Robert E. Baier and Anne E. Meyer

1

The Chemistry and Physics of Solid Adhesion

Lieng-Huang Lee

> When a plate of gold shall be bonded with a plate of silver or joined thereto,
> it is necessary to beware of three things, of dust, of wind and of moisture:
> for if any come between the gold and silver they may not be joined together . . .
>
> *De Proprietatibus Rerum (The Properties of Things) . . . 1250 AD*

1. INTRODUCTION

This review about bond formation resembles a jigsaw puzzle because we cannot see the overall picture until we put most of the pieces together. There have been many questions about solid adhesion. However, it is surprising to learn how little we do know about this subject matter.

Do we know for sure how things stick together? How does nature hold things together? Can things be bonded together forever? How can we learn from nature? Are there any magical criteria for achieving good bonding of solids? What is the natural glue or superglue that binds solids together without any externally applied adhesive? How can one make some seemingly ungluable materials adhere together? Why, in some instances, does there appear to be absolutely no adhesion between some solids? In this review, we may not be able to provide our readers with satisfactory answers to some of the above questions. However, our readers will soon find out how extensively we have advanced in searching for those clues and how well we have attempted to tie those clues together. It is now up to our readers to question, dissect, and formulate their own answers.

In this guided tour, we first start with a minimum coverage of four forces of nature without invoking much of atomic and nuclear physics. From the inner core of nuclei, we shall show the first superglue of nature—gluons, which bind all things from here to eternity. Then we revisit some basic concepts related to primary and secondary intermolecular forces to pave a way for detailed discussions by other authors. Of utmost importance are the molecular interactions which are in between primary and secondary forces. As it turns out in recent years both physicists and chemists have paid a great deal of

Lieng-Huang Lee • Webster Research Center, Xerox Corporation, Webster, New York 14580.

attention to molecular interactions, but under different names. The physicists prefer to call them "donor–acceptor" interactions, while chemists call them "acid–base" interactions. We shall discuss them from both points of view.

To understand molecular interactions, we need to have some knowledge about orbital perturbation, interaction energy, natural bond order (NBO), frontier orbital, etc. Since electrons are involved in every respect, electrons seem to act as a glue for matter. Consequently, we should learn how physicists describe electrons with density of states, work function, barrier height, and Fermi level, etc.

In a recent fascinating book [*Solids and Surfaces—A Chemist's View of Bonding in Extended Structures,* VCH Publishers, Weinheim and New York (1988)], Professor Roald Hoffmann has carefully discussed some of these basic concepts for the chemists' sake. Interestingly, he has also sown the idea that chemistry and physics should meet in the solid state. Thus, we shall introduce some of his sound ideas to explain molecular interactions and adhesion, especially on solid surfaces.

After a brief exposure to solid state physics, we shall seek a soft landing on some down-to-earth problems about how things stick together. For this purpose, we chose metals as the starting point partly because some of them have been discovered and utilized since prehistoric time and partly because metals have been used in bonding or joining with other materials.

As a part of the grand tour, we start to search for the clues about how metals stick to other solids, such as metals, semiconductors, salts, oxides, ceramics, superconductors, glass, diamond, polymers, and rubbers. Along the way, we shall also reveal some untold stories about how chemists invent ways to bond metals to some ungluable materials. We are hoping that by the end of this exercise we can learn some guiding principles about how solids stick (or do not stick) together. After we excite (or charge up) our readers with enough curiosities, we shall let other authors follow-up in feeding them with proper theories or new experimental skills to solve the mystery of adhesion.

2. THE FOUR FORCES OF NATURE

According to the Gauge theories of forces,[(1–3)] all matter is made up of quarks and leptons, which interact with one another presumably through four forces: gravity, electromagnetism, the weak force, and the strong force. Gravity and electromagnetism, with an unlimited range, are the most familiar to ordinary people. The electromagnetic force binds nuclei and electrons, which are one kind of leptons, into atoms, and also holds atoms in molecules. The weak force, which is less familiar, is responsible for certain kinds of nuclear decay, while the strong force binds quarks together tightly to form protons and neutrons and also holds protons and neutrons in place within nuclei. However, both the weak and strong forces operate only over a very short range, no larger than the radius of an atomic nucleus. At a distance of 10^{-13} cm (=1 fm), the strengths of these four forces (Table 1)[(3)] can be ranked in a decreasing order: strong force > electromagnetism ≫ weak force ≫ gravity.

Beside these four forces, there were speculations about the existence of the fifth force prior to the research of gravity[(3)] under the thick ice of Greenland. However, before the end of 1988, scientists concluded[(4)] that more studies are needed before they can be sure about the fifth force. They might have found a non-Newtonian component[5] of gravity instead of the fifth force.

2.1. Carriers of Four Forces

To understand how nature holds matter together it is necessary to learn how physicists theorize the particle carriers[3] of these four forces (Table 1). For gravity,[6] the carrier is assumed to be graviton, which has a zero rest-mass and a zero electric charge but a spin of two. For electromagnetism, the carrier is photon which has a zero rest-mass, a zero electric charge, and a spin of one. For the weak force, the carriers are intermediate bosons, W^+, W^-, and Z^0, all of which have a rest-mass (Table 1). Finally for the strong force, the carriers are gluons which, like photons, have a zero rest-mass, a zero electric charge, and a spin of one. In addition, all gluons are assumed to carry color charges.

A simple description of these four forces also shows the division between the classical field theory and quantum field theory. The classical field theory maintains that energy and momentum are carried by a field, while the quantum field theory assumes that both energy and momentum exist in discrete units, such as quanta, which can also be described as particles.

Thus, in the quantum electrodynamic theory (QED), electromagnetic forces are said to occur through particles, such as photons. Forces such as gravity and electromagnetism, transmitted by massless particles, have apparent finite ranges, and those forces decrease in strength inversely proportional to the square of the interparticle distance. QED is an extremely successful theory in that its predictions have been verified experimentally to great precision, in part because it is characterized by a dimensionless "coupling constant" (or fine-structure constant), which is derived from the well-known Coulomb's law.

In the relativistic quantum theory, it is customary to measure the product of charges

TABLE 1

Four Forces of Nature and Related Carriers[a]

Force	Range	Strength at 10^{-13} centimeter in comparison with strong force	Carrier	Mass at rest (Ge V/c^2)	Spin	Electric charge	Remarks
Gravity	Infinite	10^{-38}	Graviton	0	2	0	Conjectured
Electromagnetism	Infinite	10^{-2}	Photon	0	1	0	Observed directly
Weak	Less than 10^{-16} cm	10^{-13}	Intermediate Bosons: W^+	81	1	+1	Observed directly
			W^-	81	1	−1	Observed directly
			Z^0	93	1	0	Observed directly
Strong	Less than 10^{-13} cm	1	Gluons	0	1	0	Permanently confined

[a] From C. Quigg, *Sci. Am.* **252**(4), 84 (April, 1985) (reproduced with permission).

(in esu) in units of $\hbar c$, where $\hbar$ is h (Planck constant)/2π, and c is the velocity of light. The charges are also quantized as integral multiples of the proton charge, e. Hence, Coulomb's law[7] can also be expressed as

$$F/\hbar c = \alpha QQ'/r^2 \tag{1}$$

where Q and Q' are integers, and α is the dimensionless coupling constant

$$\alpha = e^2/\hbar c \tag{2}$$

and it has an experimental value close to 1/137. It is small enough to allow straightforward perturbation expansions in powers of α.

The usefulness of Coulomb's law at the molecular level will be conveniently demonstrated later in the SI unit. The factor $4\pi\epsilon_0$ (where ϵ_0 is the permittivity in the free space) will be added to the denominator.

Since this section is about forces, we need to explain the meaning of spin. For the even-integer spin, as in the case of graviton, there is only an attractive force, while for the odd-integer spin, as in the case of photon and gluon, there are both attractive and repulsive forces involved. In the following subsection we shall describe briefly the nature of gluon, which will lead us to the concept of how matter is bound together even at the nucleon level.

2.2. Gluons and Adhesion of Nuclei

The name gluon,[8] created by physicists, is presumably by coupling "glue" and "on." By definition, gluons are those particles (see Table 1) binding quarks together and quarks have been postulated to be constituents of protons, neutrons, and many related particles. A quark[9,10] has the unique property of color charge, and the quark field consists of three-component vectors in a "color space"; those components correspond to three different "colors": "red," "white," and "blue," respectively. Of course, none of these colors is seen visually. Instead, the details of these components can only be described by quantum chromodynamics (QCD) and will not be delineated here.

Interestingly, when a quark is shaken, it can release any one of the eight gluons. On the other hand, a quark can also absorb a gluon. In this respect, a gluon resembles a photon; these two particles are both massless and travel at the speed of light. Unlike a photon, which is electrically neutral, a gluon has a color charge. As a result, the glue that is supposed to stick quarks together can also adhere to itself because two gluons can be held together by the exchange of other gluons to form glueballs.[8] However, the lifetime of a glueball produced through a nuclear collision is extremely short: on the order of 10^{-25} second. Thus, it is not possible to observe glueballs directly. However, it is likely that two glueballs[8] have already been detected indirectly at SLAC using the Crystal Ball detector.

In principle, all nuclear physical phenomena can be described by QCD, and most atomic, molecular, and solid state (or condensed matter) physics are expressed in terms of QED because in the latter cases electrically charged particles are generally involved. Before we leave this subject of QCD, we should point out that both gluons and quarks are confined and cannot be identified or observed as separate particles. Since a proton consists of three quarks and gluons, which are confined, a proton has an unusually long lifetime of about 10^{31} years, or its decay rate[11] is calculated to be one decay per proton in 10^{31}

years. In terms of adhesion, we should say that gluons are one of the outstanding superglues, and gluons yield an adhesive bonding of immeasurable durability. Since the life of our universe has been estimated to be only 10^{10} years, this adhesive bond should outlast our universe.

If there is any significance in pointing out gluons, it is the interparticle distance. The closer the particles, the stronger the force, or the energy. We shall reiterate this fundamental relationship at both atomic and molecular levels.

2.3. Nuclear Equation of State

Recently, the relationship between interparticle distance and energy in a nucleus has been expressed by Rose *et al.*[12] in terms of a nuclear equation of state. This finding is derived from the scaling relations for solids discovered by Ferrante *et al.*[13] in 1979. In the case of solids, the total binding energy $E(a)$ per particle can be written as

$$E(a) = \Delta E E^* (a^*) \tag{3}$$

Here ΔE is the equilibrium binding energy and a is a coordinate describing the separation between the constituents (e.g., the distance between the two nuclei of a diatomic molecule). The scaled coordinate a^* is expressed as

$$a^* = (a - a_0)/l \tag{4}$$

where a_0 is the equilibrium spacing and l is the scaling length. Thus, $E^*(a^*)$ describes the shape of the universal bind-energy relation.

For nucleons, the $\bar{V}_{14}$ potential is averaged from the spin–isospin-dependent *s*- and *p*-wave components to represent nuclear matter. In the region of strong bonding, the $\bar{V}_{14}$ potential is virtually indistinguishable from the universal binding-energy relation of solids as shown in Figure 1. Consequently, it has been established that the surface and cohesive

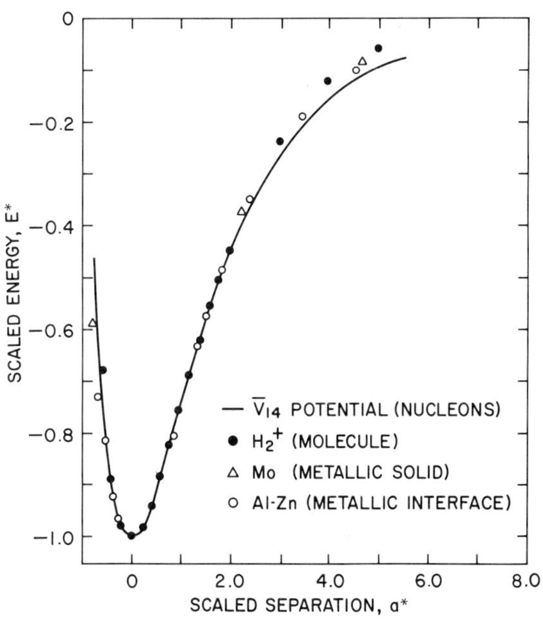

FIGURE 1. Comparison of a scaled two-nucleon ($\bar{V}_{14}$) with scaled binding energy relations for the molecule $H_2{}^+$, the bulk metal Mo, and the biometallic interface Al–Zn. [After J. H. Rose, J. P. Vary, and J. R. Smith, *Phys. Rev. Lett.* **53**(4), 344 (1984) (reproduced with permission).]

energies of nuclei can be related to the nucleon separation as in the case of metals and the electron–hole liquids.

3. CHEMICAL BONDING AND INTERMOLECULAR FORCES

Now let us examine how some of the four forces operate in atoms, molecules, and solids. First let us discuss briefly some aspects of chemical bonding and intermolecular forces. Detailed discussions of the subject are given by Gutowski[(14)] and Good *et al.*[(15)]

The short-range forces involved in chemical bonding are quantum mechanical in nature and generally attractive. The exception is the repulsive forces resulting from the exchange interactions due to Pauli exclusion principle. However, these repulsive forces can be balanced by the attractive forces at very short distances.

On the other hand, the long-range attractive interactions can be generated through (1) Coulombic or electrostatic forces and (2) Lifshitz–van der Waals forces. The electrostatic forces arise from the Coulomb interactions between charges, permanent dipoles, and quadrupoles. In contrast, the Lifshitz–van der Waals forces include dispersion and polarization forces that arise from dipole moments induced in atoms and molecules by the electric fields of nearby charges and permanent dipoles. Now, let us discuss briefly these forces in the following subsections.

3.1. Chemical Forces and Covalent Bonding

The forces that bind atoms together tightly within the molecules are covalent forces, which are of short range over the order of an interatomic separation of 0.1–0.2 nm (or 1–2 Å). The strengths of common covalent bonds[(16)] are mainly in the range of 100–300 kT per bond, or 200–800 kJ mol^{-1} (at 298 K, 1 kJ $mol^{-1} \simeq$ 0.4 kT per bond). In general, the bond strength decreases with increasing bond length.

The stabilization of a covalent bond can be derived from the delocalization[(17)] of the motion of one or more electrons in a molecule as demonstrated in a square well model. The delocalization can be visualized as a transfer of fluid from one "box" to the other so as to drop the Fermi level E_F (∞). This drop in E_F brings forth a decrease in the total energy and a stabilization of the structure. A normal covalent bond is formed by the delocalization of one or more pairs of electrons over two atomic centers. Further delocalization over more than two atomic centers can lead to subsequent delocalization, as in the case of some organic reactions.

3.2. Coulomb Force and Ionic Bond

In addition to covalent bond, there are ionic and partial ionic bonds. The attraction between charges of opposite signs is the important underlying force in partial ionic and ionic bonds. This force, Coulomb force, is also the major component of intermolecular forces. The Coulomb force between two charged atoms, or ions, can even be stronger than the force of covalent bond. This force is expressed in terms of electrostatic force F_{es} in the SI unit:[(16)]

$$F_{es} = \frac{Q_1 Q_2}{4\pi\epsilon_0 \epsilon r^2} = \frac{Z_1 Z_2 e^2}{4\pi\epsilon_0 \epsilon r^2} \tag{5}$$

where Q_1 and Q_2 are two point charges; Z_1 and Z_2 are two respective ionic valencies; e is the elementary electronic charge ($e = 1.602 \times 10^{-19}$ C); ϵ_0 is the permittivity of free space ($\epsilon_0 = 8.9 \times 10^{-12}$ fd/m or $C^2N^{-1}m^{-2}$); ϵ is the relative permittivity or dielectric constant of the medium and r is the distance between two ionic charges. The conversion between the SI unit and the cgs unit has been explained by Good.(15)

The corresponding Coulomb energy E_{es} is expressed as

$$E_{es} = -\frac{Q_1Q_2}{4\pi\epsilon_0\epsilon r} = -\frac{Z_1Z_2e^2}{4\pi\epsilon_0\epsilon r} \tag{6}$$

For example, for two isolated monovalent ions (e.g., Na^+ and Cl^-) in contact, r is the sum of two ionic radii (0.276 nm), and the Coulombic energy is -8.4×10^{-19} J. In terms of kT at 300 K, this energy is approximately 200 kT per ion pair in vacuum. Though the Coulomb energy is strong and rather unlimited in range, the above energy per ion pair will fall below 1 kT beyond 56 nm.

3.3. Electronegativity and Partial Ionic Bond

Since not all covalent bonds are perfectly covalent, most of them possess some ionic character, which can be described by either electronegativity or ionicity. By the original definition, electronegativity χ(18) is the power of an atom to attract electrons to itself. Recently, a modified definition(19) stated that electronegativity is the ability of an atom or a group of atoms to attract or hold electrons to itself when combining with other atoms or groups.

Further confusion has been created by the existence of at least seven different electronegativity scales. Among them, the more commonly used are the Pauling's scale, χ^P,(20) and the Mulliken's scale, χ^M.(21) The χ^P scale is a macroscopic concept based on the heat of formation of an ionic compound. On the other hand, χ^M is a microscopic concept, based on atomic spectra,

$$\chi^M = (I + A)/2 \tag{7}$$

where I is the ionization potential and A the electron affinity. By definition, ionization potential is the energy difference between the top of the valence band and the vacuum level, while electron affinity is that between the bottom of the conduction band and the vacuum level. One more distinction between χ^P and χ^M rests on the basis of the bonded state. The χ^P scale describes the power of an atom to attract electrons as it is modified in the bonded condition, while χ^M is based on the ionization potentials and the electron affinities of valence states of unbonded atoms. In general, χ^M is preferred, but χ^P values are more readily available. We shall discuss this later in conjunction with the work function of metal.

Ionicity was originally defined by Pauling(20) as

$$f_i = 1 - \exp\left[-\frac{1}{4}(\chi_A - \chi_B)^2\right] \tag{8}$$

where χ_A and χ_B are electronegativities of atoms A and B, respectively.

Later, Phillips(22) defined f_i for A^NB^{8-N} semiconductor compounds as

$$f_i = C^2/(E_h^2 + C^2) \tag{9}$$

where E_h is the covalent component of the average energy gap E_g^{Av} and C the ionic component. Thus, the average energy gap is related to these two components as

$$(E_g^{Av})^2 = E_h^2 + C^2 \tag{10}$$

Basically, E_h is similar to the resonance integral β of the Hückel theory, while C is similar to the Pauling's electronegativity difference $\Delta\chi$. In comparing these two f_i values (Table 2), the latter due to Phillips is about 10 times more accurate for the A^NB^{8-N} compounds than the former due to Pauling. The f_i value due to Phillips has been more commonly used in describing the ionicity of a semiconductor compound. In Table 2, the Phillips's ionicities of semiconductor compounds are arranged in an increasing order.

3.4. Lifshitz–van der Waals Forces Between Microscopic Bodies

After the above brief discussion of primary forces, we should briefly examine secondary forces, such as Lifshitz–van der Waals (LW) forces, between microscopic bodies and between macroscopic bodies. Detailed discussion of these forces will be presented by Gutowski(14) and Good *et al.*(15)

3.4.1. Dispersion Forces

LW forces consist of dispersion and polarization forces. Dispersion forces, which make up at least one-third of the total LW forces, like gravitation forces, act between all atoms and molecules including totally neutral ones. The main features of dispersion forces may be summarized as follows:(16)

- The dispersion forces are long range and can be effective from large distances ($>$10 nm) down to interatomic distances.
- The dispersion forces may be repulsive or attractive, and generally those forces between two molecules or large particles do not follow a simple power law.
- These forces can also align or orient molecules, though this orientation effect is usually weak.
- The dispersion interaction of two bodies is affected by the presence of other bodies nearby. This is also known as the nonadditivity of an interaction.

The perturbation responsible for dispersion forces is the instantaneous Coulomb interaction between electrons and nuclei in the two molecules. London(23) derived an approximate expression for the dispersion energy E_D for the two attracting spherical molecules in terms of their polarizabilities α and ionization potentials $I(\simeq h\nu_0)$. Here ν_0 is the frequency of the zero-point motion of the electron. The rapidly varying dipole produced by the zero-point motion gives rise to an electric field, and this field polarizes the other atom in which a dipole is induced in phase with the first; this, in turn, interacts with the instantaneous dipole. $(E_D)_{1,1}$ between two similar molecules is given in Eq. (1) in Table 3, while $(E_D)_{1,2}$ between two dissimilar molecules is given in Eq. (2). It should be noted that the energy E_D is inversely proportional to the sixth powder of the interatomic distance. Moreover, London's dispersion energy is present in all parts of atoms or molecules.

TABLE 2
Comparison between Phillip's and Pauling's Partial Ionicities for $A^N B^{8-N}$ Compounds[a,b]

Crystal	Type*	f_i Phillips	f_i Pauling (1932)
C	D	0	0
Si	D	0	0
Ge	D	0	0
Sn	D	0	0
SiC	Zb, W	0.177	0.11
BN	Zb	0.256	0.22
GaSb	Zb	0.261	0.02
AlAs	Zb	0.274	0.06
AlP	Zb	0.307	0.01
GaAs	Zb	0.310	0.04
InAs	Zb	0.357	0.02
GaP	Zb	0.374	0.06
InP	Zb	0.421	0.04
AlSb	Zb	0.426	0.04
AlN	W	0.449	0.43
ZnTe	Zb	0.546	0.06
ZnO	W	0.616	0.59
ZnS	Zb, W	0.623	0.18
CdTe	Zb	0.675	0.04
ZnSe	Zb	0.676	0.15
CdS	Zb, W	0.685	0.18
CdSe	W	0.699	0.15
CaS	R	0.902	0.43
SrO	R	0.926	0.79
RbF	R	0.960	0.92

[a] D denotes diamond, Zb zinc blend, W wurtzite, and R rock salt.
[b] From J. C. Phillips, *Rev. Mod. Phys.* **42**(3), 317 (1970).

3.4.2. Retardation Forces

Later, Casimir and Polder[(24,25)] proposed that the London theory must be modified to account for the interaction between molecules at large distances, e.g., greater than 100 Å. At long distances, the retardation effect occurs because a finite time is required for an electromagnetic wave to be propagated through the medium in which the particles are situated. If the time is long, the electron may be in a less favorable position for attraction. The retardation energy $(E_R)_{1,1}$ for two similar molecules is given in Eq. (3) in Table 3, and that $(E_R)_{1,2}$ for two dissimilar molecules in Eq. (4). It should be noted that the dependence of the energy on distance is now r^{-7}, i.e., there is an even more rapid drop-off of energy with distance when the "retardation" effect of the London forces comes into play.

TABLE 3
Van der Waals Interaction Energies for Microscopic Bodies
[in Vacuum ($\epsilon = 1$) and SI unit][a,b,c]

Dispersion Energy, E_D

Similar: $$(E_D)_{1,1} = -\frac{3}{4}\left[\frac{1}{(4\pi\,\epsilon_0)^2}\right]\left(\frac{I\,\alpha^2}{r^6}\right) \quad (1)$$

Dissimilar: $$(E_D)_{1,2} = -\frac{3}{2}\left[\frac{1}{(4\pi\,\epsilon_0)^2}\right]\left(\frac{I_1 I_2}{I_1 + I_2}\right)\left(\frac{\alpha_1\alpha_2}{r^6}\right) \quad (2)$$

Retardation Energy, E_R

Similar: $$(E_R)_{1,1} = -\frac{23}{8\pi^2}\left[\frac{1}{(4\pi\,\epsilon_0)^2}\right]\left(\frac{\alpha_2}{r^6}\right)\left(\frac{h\,c}{r}\right) \quad (3)$$

Dissimilar: $$(E_R)_{1,2} = -\frac{23}{8\pi^2}\left[\frac{1}{(4\pi\,\epsilon_0)^2}\right]\left(\frac{\alpha_1\alpha_2}{r^6}\right)\left(\frac{h\,c}{r}\right) \quad (4)$$

Orientation Energy, E_O
(Keesom)

Dipole–dipole (freely rotating)

$$\left(kT > \frac{\mu_1\mu_2}{4\,\pi\,\epsilon_0\,r^3}\right) \qquad E_O = -\frac{2}{3}\left[\frac{1}{(4\pi\,\epsilon_0)^2}\right]\left(\frac{\mu_1^2\,\mu_2^2}{k\,T\,r^6}\right) \quad (5)$$

Induction Energy, E_I
(Debye)

Dipole–induced dipole (rotating): $$(E_I)_{p,p} = -\left[\frac{1}{(4\pi\,\epsilon_0)^2}\right]\left(\frac{\alpha_1\mu_2^2 + \alpha_2\mu_1^2}{r^6}\right) \quad (6)$$

Dipole–nonpolar (rotating): $$(E_I)_{n,p} = -\left[\frac{1}{(4\pi\,\epsilon_0)^2}\right]\left(\frac{\alpha_2\mu_1^2}{r^6}\right) \quad (7)$$

[a] From J. N. Israelachvili, *Intermolecular and Surface Forces,* Academic Press, New York (1985) (adopted with permission).
[b] $E_{SI} = E_{cgs} \times 1/(4\pi\epsilon_0)^2$.
[c] Nomenclature:

αc	Polarizability	I	Ionization constant
c	Light velocity	k	Boltzmann constant
ϵ_0	Permittivity in free space	r	Separation distance
ϵ	Permittivity in the medium	T	Absolute temperature
h	Planck constant	μ	Dipole moment

3.4.3. Orientation Forces

Besides the most basic nonpolar interaction, dispersion forces, there are polarization or polar interactions between molecules of counterbodies, such as the dipole–dipole interaction (Keesom) and the dipole-induced dipole interaction (Debye). The essential difference between dispersion and polarization forces is that, while the virtual transitions responsible for the former involve simultaneous excitation of both molecules, those for the latter only involve a passive partner.

The Keesom interaction[26] (orientation) between two molecules with permanent dipoles is temperature-dependent and the attraction energy E_O in vacuum is given in Eq. (5) in Table 3.

3.4.4. Induction Forces

Debye[27] argued that if the attraction was simply due to the Keesom effect, then the energy should be greatly reduced at high temperatures. Since the experimental results were contrary to the prediction, he concluded that an additional attractive effect should be involved. Thus he showed that another polar interaction should be induced between a permanent dipole and an induced dipole. The energy $(E_{\mathrm{I}})_{\mathrm{p,p}}$ for this type of interaction is given in Eq. (6) in Table 3, while that energy $(E_{\mathrm{I}})_{\mathrm{n,p}}$ between a dipole and a nonpolar molecule is given in Eq. (7). It is important to note that for Eqs. (5)–(7), the molecules are not fixed but freely rotating.

3.5. *Nonretarded Lifshitz–van der Waals Forces Between Macroscopic Bodies*

In reality, the contacts between all solids are made between macroscopic bodies. In the following, we shall demonstrate how we can use the above microscopic relations to calculate LW forces in macroscopic bodies.

3.5.1. Hamaker Constant

Hamaker[28] developed equations for calculating the interactions between systems by assuming the additivity of dispersion forces, and in all his equations, a constant, A_{H}, is involved. This constant depends on the nature of the two interacting materials. Originally, this constant[28] A_{H} or $A_{1,2}$ was defined only for the dispersion forces:

$$A_{1,2} = r^2 N_1 N_2 B_{1,2} \tag{13}$$

where N_1 is the number of molecules per unit volume in material 1, N_2 in material 2, and $B_{1,2}$ is the coefficient of r^{-6} in the expression for the dispersion energy $(E_{\mathrm{D}})_{1,2}$ between an isolated pair of molecules 1 and 2, separated by a distance r. Thus, $E_{\mathrm{D}} \sim -B_{1,2}r^{-6}$.

However, later, as the vdW theory was further developed, it was realized[29] that the Hamaker constant should contain all other components of vdW interaction energies:

$$A_{1,2} = (A_{\mathrm{D}})_{1,2} + (A_{\mathrm{O}})_{1,2} + (A_{\mathrm{I}})_{1,2} + \text{coupling terms} \tag{14}$$

The details of this extension are discussed by Good *et al.*[15]

With the aid of the Hamaker constant, we are now able to calculate the nonretarded forces (or energies) between macroscopic bodies (Figure 2). The three typical cases[30] are: (1) between two thick parallel plates, (2) between a plate and a sphere, and (3) between two spheres. The vdW forces can be expressed in terms of two factors: one of them depends only on the geometry of the system (e.g., R, the radius and Z, the distance between the two counterbodies), and the other, the Hamaker constant. For macroscopic bodies in contact, the nonretarded energy of interaction is, to a first approximation, made up of the sum of all interacting bodies between all the atoms in the counterbodies. If the distance Z between the two bodies is very small in comparison with other dimensions of the system, the summation procedure can yield results for the above three cases as shown in Table 4.

What happens if retardation takes effect when $Z > 100$ Å? Let us use the example of

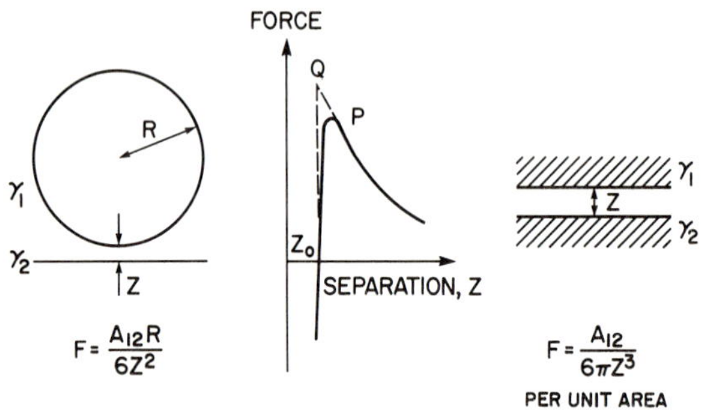

FIGURE 2. Nonretarded van der Waals forces between macroscopic bodies. A_{12} is the Hamaker constant and Z is the separation distance. [After D. Tabor, *J. Colloid Interface Sci.* **58**(1), 2 (1977) (reproduced with permission).]

the interaction between two thick plates. The vdW energy should decay as the square of the separation at a rather close range. When retardation sets in, the decay is hastened as predicted as the cube of Z when $Z > 100$ Å. Actually, this type of experiment has also been used to verify the validity of Hamaker's reasoning.

3.5.2. Lifshitz–van der Waals Constant

The second novel approach, developed by Dzyaloshinskii, Lifshitz, and Pitaevskii,(31) in calculating van der Waals forces is based on the electromagnetic radiation field in each of the two bodies. The expression for the two infinite parallel plates is very complicated. However, under certain conditions, it can be simplified as follows:

$$F^{\mathrm{LW}} = \frac{\hbar\bar{\omega}}{8\pi^2 Z^3} \tag{15}$$

where F^{LW} is the Lifshitz–van der Waals component of the attractive force, and $\bar{\omega}$ the averaged angular frequency:

$$\bar{\omega} = \int_0^\infty \left(\frac{\epsilon_1(i\xi) - 1}{\epsilon_1(i\xi) + 1}\right)\left(\frac{\epsilon_2(i\xi) - 1}{\epsilon_2(i\xi) + 1}\right) d\xi \tag{16}$$

In Eq. (16), the subscripts 1 and 2 refer to two interacting bodies; $\epsilon(i\xi)$ is related to the imaginary part; $\epsilon''(\omega)$, of the complex dielectric constant, by the Kronig–Kramers relation,(32) i.e., by

$$\epsilon(i\xi) - 1 = (2/\pi)\int_0^\infty \left(\frac{\omega\epsilon''(\omega)}{\omega^2 + \xi^2}\right) d\omega \tag{17}$$

Thus, if $\epsilon''(\omega)$ is known as a function of frequency, $\bar{\omega}$ can be calculated from Eqs. (16) and (17). The value of $\bar{\omega}$ is very large for bodies with large ϵ'', i.e., strong absorptions, at large angular frequencies (far-UV region).

Krupp,(33) in using the Lifshitz–van der Waals constant $\hbar\bar{\omega}$ to calculate the Hamaker constant, obtained

$$A_{\mathrm{H}} = \frac{3\hbar\bar{\omega}}{4\pi} \tag{18}$$

TABLE 4
Nonretarded van der Waals Forces and Energies for Macroscopic Bodies[a]

	Force	Energy
Two thick plates	$A/6\pi Z^3$	$-A/12\pi Z^2$ (per unit area)
Plate–sphere	$AR/6Z^2$	$-AR/6Z$
Two spheres	$AR/12Z^2$	$-AR/12Z$

[a] From M. van den Tempel, *Adv. Colloid Interface Sci.* **3**, 137(1972).

From the $\hbar\bar{\omega}$ data of Krupp for identical materials, we have calculated both $\hbar\bar{\omega}$ and A_H for materials interacting with polystyrene (PS) (Table 5).

3.5.3. Adhesion Forces Derived from van der Waals Interactions

In the case of pure solid-to-solid adhesion, we do not consider using an adhesive to adhere the solids together. However, the principle involved can also be applied to the case that an adhesive is actually being applied. The data in Table 5 can be used to calculate the adhesion force derived solely from vdW interactions. Thus, the force between an undeformable sphere and a plate in Table 4 can be expressed as

$$F^{\mathrm{LW}} = \frac{A_{12}R}{6Z^2} = \frac{\hbar\bar{\omega}R}{8\pi Z^2} \tag{19}$$

If the two bodies come into atomic approach, the distance Z becomes the equilibrium separation Z_0 of the atoms.

TABLE 5
Calculated Lifshitz–van der Waals Constants and Hamaker Constants for Polystyrene Interacting with Other Materials

	$\hbar\bar{\omega}$ (eV)	$A_{12} \times 10^{20}$ (J)
PS–Graphite	4.59	17.0
PS–Au	4.52	17.0
PS–Ag	4.23	16.3
PS–Ge	3.67	13.8
PS–Cu	3.57	13.8
PS–Diamond	3.57	13.8
PS–Si	3.34	13.2
PS–Si + Oxide (130 Å)	3.01	11.3
PS–Al_2O_3	2.64	10.1
PS–CdS	2.62	10.1
PS–KCl	1.66	6.3
PS–H_2O	1.40	5.3
PS–SiO_2 (glass)	1.24	5.0

3.5.4. Work of Adhesion—Lifshitz–van der Waals Component

The thermodynamic maximum work of adhesion $W_{(A)}$ can be obtained through Dupree's equation[34]

$$W_A = F_1 + F_2 - F_{1,2} \tag{20}$$

where F_1 and F_2 are surface free energies of body 1 and body 2, respectively, and $F_{1,2,}$ is the interfacial free energy. In reality, γ_1 and γ_2, surface tensions of the two bodies, and $\gamma_{1,2}$ interfacial tension, are used instead. This maximum work of adhesion is actually only the Lifshitz–van der Waals component of the total W_A, and it is expressed as W_A^{LW}.

If we assume that a sphere and a plate are separated from an equilibrium Z_o to infinity, then the work done to create two new surfaces instead of the interface is equal to W_A^{LW}. Hence,

$$W_A^{LW} = \int_{Z_0}^{\infty} f dZ = \gamma_1 + \gamma_2 - \gamma_{1,2} = \frac{A_{1,2}}{12\pi Z_0^2} \tag{21}$$

From Eq. (19)

$$F_0^{LW} = 2\pi R(\sigma_1 + \sigma_2 - \sigma_{1,2}) \tag{22}$$

If both the sphere and the plate are of the same material, $\gamma_1 = \gamma_2$ and $\gamma_{1,2} = 0$, then

$$F_0^{LW} = 4\pi R\gamma \tag{23}$$

This equation was first derived by Bradley,[35] not by the thermodynamic relation but by the pairwise summation. Here, we need to emphasize that the considerations for W_A are still based on the equilibrium condition, which is rarely attainable in the actual solid contact. Further discussion about the thermodynamic of adhesion is given by Gutowski.[14]

In the following section, we shall consider the effect of deformation on the adhesive forces between solids.

3.5.5. Effect of Deformation on Adhesive Forces Derived from the Lifshitz–van der Waals Interactions

This study related to deformation has been reviewed by Tabor.[36] Derjaguin[37] first proposed that when two solid bodies were brought together by an attractive force, deformation could take place at the point of contact. The radius of contact a_0 is related to the modulus of the material and the surface force.

In the case that the elastic contact is a purely Hertzian contact,[38] the surface force is not taken into account; then the contact radius between a sphere and a plate can be expressed as

$$a_0 = \left(\frac{3RF}{4E^*}\right)^{1/3} \tag{24}$$

where

$$\frac{1}{E^*} = \frac{(1 - \nu_1^2)}{E_1} + \frac{(1 - \nu_2^2)}{E_2} \tag{25}$$

R is the radius of the sphere, F the applied load, or force, E_1 and E_2 are Young's moduli of the two elastic bodies, and v is Poisson's ratio of the sphere.

When the surface force is taken into account, the calculated a_0 and F, the pull off force, are different between those obtained by Derjaguin–Muller–Toporov (DMT)[39] and those by Johnson–Kendall–Roberts (JKR)[40]:

	DMT	JKR
Pull–off force F_0^{LW}	$4\pi R\gamma$	$3\pi R\gamma$
Contact radius a_0	$\left[\frac{3\pi R^2\gamma}{2}\left(\frac{1-v^2}{E}\right)\right]^{1/3}$	$\left[9\pi R^2\gamma\left(\frac{1-v^2}{E}\right)\right]^{1/3}$

Generally, the pull-off force as calculated by the JKR method is smaller than that by DMT, and the radius of contact by DMT is about twice that as derived by JKR. Moreover, the pull-off force is identical with or without adhesion as calculated by DMT.

In the case of JKR, the results have been verified with the contact of an optically smooth rubber sphere in contact with a smooth glass plate. The experiments agree with their equilibrium theory of adhesion. The assumption of the theory is that the forces of attraction at the contact area are identical to those outside that area. That area is reached when the total energy of the system, including the stored elastic energy, the mechanical energy in the applied load, and the lost surface energy, is at equilibrium.

3.5.5.1. Equilibrium Theory of Adhesion. Similar to the above conditions, when one rubber sphere of radius R and Young's modulus E is touching a rigid plate, Johnson, Kendall, and Roberts[40] have found that the contact radius a_0[41] is larger than that obtained from the Hertz theory:

$$a_0^3 = \frac{3}{4}\left(\frac{1-v^2}{E}\right) R \left\{F + 3\pi R\gamma + \left[(3\pi R\gamma)^2 + 6\pi R\gamma F\right]^{1/2}\right\} \tag{26}$$

where F is the load or force, and γ is the interfacial energy of the contact. Quantity γ can be defined as the work done against the surface attraction in breaking a unit area of the solid–solid interface. On the other hand, γ is also the work done against the surface attraction when forming a unit area of solid–solid contact. Thus, the process is reversible in theory. However, in practice, the forming and breaking of a bond is rarely reversible. Therefore, energy can be dissipated easily during the process and this raises the problem which will be considered as adhesive hysteresis.[42]

3.5.5.2. Disagreements between DMT and JKR Theories. Since the publication of Tabor's paper in 1977, there have been several communications[43,44] between Tabor and Derjaguin about the DMT and JKR theories. Finally, in 1983 a paper by Muller *et al.*[45] appeared to have settled the issue. In that paper they concluded that when the particles are sufficiently hard and small, and/or a material is of low specific energy of adhesion, the DMT approach holds; otherwise, at high values of the parameter, the JKR theory is valid. Originally, JKR dealt with a soft material with a modulus of 10^5–10^6 dyn/cm.

Muller *et al.*[45] have introduced a new parameter, μ, as a guide to determining the regime where either of the two theories applies:

$$\mu = \frac{32}{3\pi}\left[\frac{2R(\Delta\gamma)^2}{\pi E^* Z_0^3}\right]^{1/3} \tag{27}$$

where Z_0 is the interatomic spacing. A low value of μ (<1) corresponds to the regime where the DMT model applies, and a high value of μ (>1) to that where the JKR model applies. As μ increases, the behavior changes so that the pull-off force occurs at a finite area of contact as predicted by the the JKR model. The dividing line appears to be at μ = 0.8.

3.5.5.3. Recent Work on the Effect of Deformation. Though the JKR model can adequately describe the case when $\mu > 1$, a recent work by Rimai *et al.*[(46)] raises a new question about the JKR model. Their experiments involve the use of spheroidal gold particles (about 3 μm) deposited on both hard and soft planar, electronically conducting substrates. The particles appear to embed into the softer substrate, but not into the harder one. The deformation has been found to be plastic rather than elastic, thus the surface tension effect is more significant. The radius of contact can be estimated better with the combination of Krupp's model[(33)] (based on the vdW attraction) and the JKR model (based on tensile response due to surface tension or surface energy) than with the JKR model alone.

4. MOLECULAR INTERACTIONS

The total forces of adhesion between two solids are not derived solely from Lifshitz–van der Waals (LW) and Coulombic attraction. Besides these forces, there are other short-range (<0.3 nm) forces resulting from molecular interactions. One of the major molecular interactions has been identified as the donor–acceptor or acid–base interaction. Later, we shall mention briefly the background of this interaction. First, let us discuss molecular interactions between simple molecules.

4.1. Morse Interaction Curve

The interaction between two hydrogen atoms[(47)] is the simplest case of interaction. However, there has been much theoretical and experimental work dealing with this problem. In this section we merely intend to point out the experimental result on the energy of the interaction of two 1(s) hydrogen atoms as a function of internuclear separation.

The Morse equation is given as

$$\Phi_{\text{Morse}} = 109.46\,[-2\exp(-X) + \exp(-2X)] \quad \text{kcal/mol} \tag{28}$$

where

$$X = 1.0298\,(r_{ab}/a_0) - 1.401$$

Φ is the interaction energy, and r_{ab} is the interatomic distance. These curves are simple and can be compared with the experimental result by Rydberg[(48)] and Beutler.[(49)]

The general shape of these curves for nitrogen (Figure 3) is similar to the one reported for the nucleon–nucleon interaction (Figure 1) and those for the metal–metal adhesion to be discussed by Banerjea, Ferrante, and Smith.[50]

4.2. Perturbation Theory

Hudson and Klopman[51] proposed an equation to describe the effect of orbital perturbation of two molecules on chemical reactivity. Their equation for the interaction energy ΔE can be simplified by introducing into the second term the interaction between the HOMO (highest occupied molecular orbital) of a nucleophile (or base) and the LUMO (lowest unoccupied molecular orbital) of an electrophile (or acid):

$$\Delta E = -\frac{Q_{\mathrm{Nu}}Q_{\mathrm{El}}}{\epsilon R} + \frac{2(C_{\mathrm{Nu}}C_{\mathrm{El}}\beta)^2}{E_{\mathrm{HOMO}} - E_{\mathrm{LUMO}}} \tag{29}$$

(The Coulombic term) (The frontier orbital term)

where Q_{Nu} and Q_{El} are the total charges for the nucleophile and electrophile, respectively; C_{Nu} and C_{El} are the coefficients of the atomic orbital Nu and El, respectively; β is the resonance integral, ϵ is the permittivity, and R is the distance between Nu and El.

This simplified equation indicates that the molecular interaction generally consists of at least two terms: the Coulombic term and the frontier orbital term. For the electrostatic interaction, the first term dominates, while for the donor–acceptor (EDA) interaction, the second term dominates.

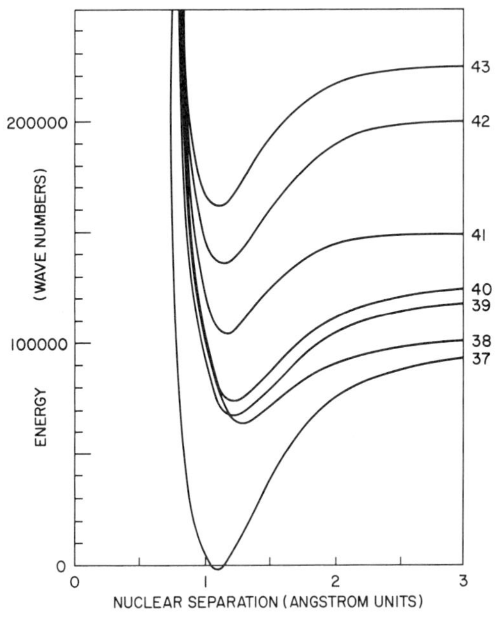

FIGURE 3. Potential energy curves for nitrogen. [After P. M. Morse, *Phys. Rev.* **34,** 57 (1929) (reproduced with permission).]

In the following sections, we shall discuss briefly the two current theories about molecular interactions. The first is that of Kitaura and Morokuma[52]; the second is the natural bond orbital (NBO) theory. The first emphasizes the electrostatic interaction, while the second emphasizes the orbital interaction.

4.3. Kitaura–Morokuma's Decompsition of Interaction Energies

According to Kitaura and Morokuma,[52–54] the molecular interaction energy can be decomposed into the following components:

$$\Delta E_{\text{int}} = \Delta E_{\text{es}} + \Delta E_{\text{po}} + \Delta E_{\text{ex}} + \Delta E_{\text{ct}} + \Delta E_{\text{mix}} \tag{30}$$

where the suffix es denotes electrostatic, po polarization, ex exchange, ct charge transfer, and mix denotes coupling terms of higher order, in some cases including the dispersion component of LW forces.

The above four types of interaction have been illustrated schematically by Isaacs[55] (Figure 4). The dominance of any of the four interactions depends upon the overlapping of atomic orbitals:

Case 1. Without overlap of atomic orbitals

$$\Delta E_{\text{int}} = \underset{\substack{\text{(First}\\\text{order)}}}{\Delta E_{\text{es}}} + \underset{\substack{\text{(Second}\\\text{order)}}}{\Delta E_{\text{po}}}$$

Case 2. With overlap of atomic orbitals

$$\Delta E_{\text{int}} = \underbrace{\Delta E_{\text{es}} + \Delta E_{\text{po}}}_{\text{(First order)}} + \underbrace{\Delta E_{\text{ind}} + \Delta E_{\text{dis}} + \Delta E_{\text{ex-po}}}_{\text{(Second order)}}$$

In both cases, the interaction energy involves primarily the electrostatic energy and the LW energies, e.g., polarization (induction), dispersion, exchange–polarization, etc. Since the orbital overlap is directly determined by the interatomic distance, that distance becomes the critical factor in deciding the dominance of some of the components of the interaction energy.

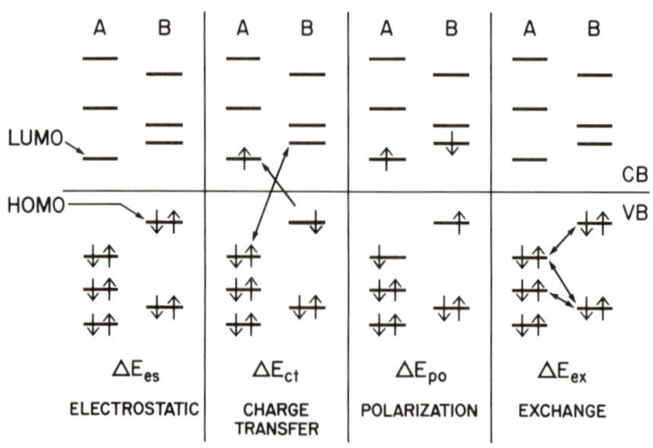

FIGURE 4. Molecular orbital interactions between two close-shell molecules.

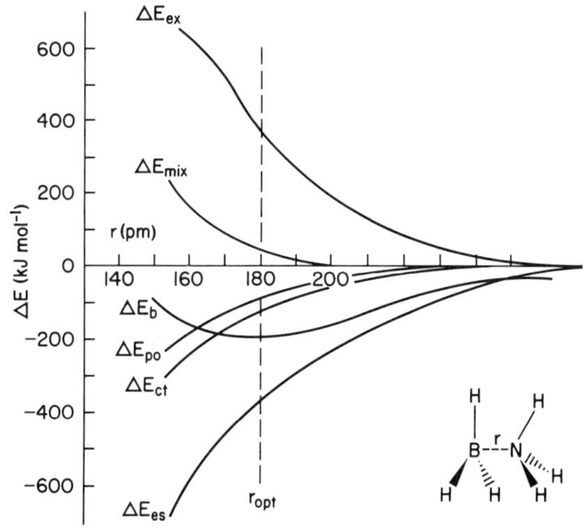

FIGURE 5. Interatomic distances between borane and ammonia. [After N. S. Isaacs, *Physical Organic Chemistry*, p. 59, Longman Scientific and Technical; Wiley, New York (1987).]

The effect of the interatomic distance between BH_3 (borane) and NH_3 on various energy components is illustrated by Isaacs[(55)] on the basis of KM's data (Figure 5). In this case, the ΔE_{es} value is the most dominant. The value of ΔE_{ex} begins to compete strongly at a separation under 2 Å (or 200 pm) and rises steeply, but is balanced by the growing polarization (ΔE_{po}) and charge-transfer component (ΔE_{ct}). A minimum total energy is reached at 1.8 Å (or 180 pm) at which ΔE_{int} or the binding energy is −187 kcal (or −44.7 kcal/mol). In general, for the borane–amine interaction, the various components decrease with a decrease in distance according to the order: $\Delta E_{es} > \Delta E_{ct}$ and ΔE_{po}.

Though the KM method, based on the *ab initio* self-consistent-field (SCF) theory, has enjoyed wide acceptance for many years, recently the critique[(56)] has claimed that the KM method has consistently led to a much smaller estimate of charge transfer. This problem should serve to bring forth the NBO model in the following subsection.

4.4. Natural Bond Orbital (NBO) Donor–Acceptor Approach

The molecular interaction of special interest is the "intermediate" interaction between the "chemical" reaction of the covalent type and the "physical" interaction of the van der Waals type. Since the "intermediate" interaction is in the "gray" area, sometimes it is difficult to differentiate whether it is an acid–base interaction or a donor–acceptor interaction.[(57)] Amid all these confusions in the past 15 years, the "intermediate" bonding has also been arbitrarily classified as "van der Waals bonding."

The KM treatment of this "intermediate" interaction is based on the electrostatic overlapping charge distribution. Thus, it favors the "acid–base" viewpoint, which will be enumerated by Good *et al.*[(15)] On the other hand, an independent method, the natural bond orbital (NBO), considers both bonding and antibonding orbitals. Consequently, it is synonymous with the "donor–acceptor" concept. A comprehensive review of the NBO method has recently been given by Reed, Curtiss, and Weinhold,[(56)] but we only attempt to summarize several key concepts related to this method on the basis of their review.

Originally, the NBO method was developed by Foster and Weinhold[58] as an analysis for studying hybridization and covalency effects in polyatomic wave functions. NBO deals with the fundamental chemical concepts of localized bonds and lone pairs of electrons in the structure. According to the NBO method, a localized σ_{AB} bond can be expressed in terms of the orthonormal hybrids h_A, h_B [natural hybrid orbitals (NHOs)]:

$$\sigma_{AB} = c_A h_A + c_B h_B \tag{31}$$

where c_A and c_B are the polarization coefficients for A and B, respectively, that describe the relative polarity of the bond orbital toward A and B. In turn, the NHOs consist of a set of effective valence-shell atomic orbitals (NAOs). The filled NBOs for σ_{AB} of the natural Lewis structure are suitable to describe the covalency of a molecule.

On the other hand, the antibonds are formed from the unoccupied orbitals. As the counterpart of σ_{AB}, σ^*_{AB} is expressed as

$$\sigma^*_{AB} = c_B h_A - c_A h_B \tag{32}$$

This departure from the idealized Lewis structure is a small correction of the covalency.

As a result, the total energy E can be decomposed into

$$E = E_{\sigma\sigma} + E_{\sigma\sigma^*} \tag{33a}$$

$$= E_{\text{Lewis}} + E_{\text{non-Lewis}} \tag{33b}$$

The symbols σ and σ^* are used in the generic sense referring to filled and unfilled orbitals of the Lewis structure. In fact, σ may be a core orbital (c), a lone pair (n), a σ bond or a π bond, etc., while σ^* may be a σ or π antibond (σ^* or π^*), or an extravalence shell Rydberg (r) orbital, etc.

The donor–acceptor interaction as shown in Figure 6 is an example of the $\sigma \rightarrow \sigma^*$ interactions between filled (donor) and unfilled (acceptor) orbitals. This type of interaction has also been called the "charge-transfer" or the generalized "acid–base" interaction. That is why it is rather difficult to sort out the true nature of an interaction merely from these names.

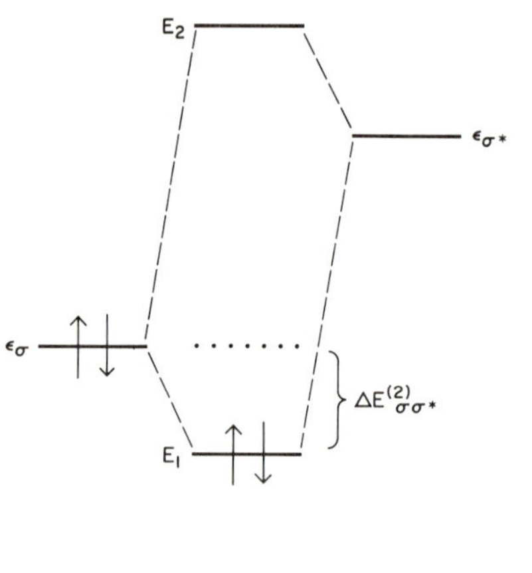

FIGURE 6. Perturbative donor–acceptor interaction, involving a filled orbital σ and an unfilled orbital σ^*. [After A. E. Reed, L. A. Curtiss, and J. Weinhold, *Chem. Rev.* **88,** 899 (1988) (reproduced with permission).]

In the case of complex formation, the complexation energy ΔE_{com} can be divided into two components:

$$\Delta E_{com} = \Delta E_{CT} + \Delta E_{NCT} \tag{34}$$

where CT denotes charge transfer and NCT no charge transfer. For example, in the case of water dimer formation through hydrogen bonding, the respective values are −4.1, −6.5, and +2.4 kcal/mol. The no-charge-transfer appears to dominate the complexation.

In general, the charge-transfer not only results in an increase in binding energy but allows a significant amount of exclusion repulsion to be overcome, allowing molecules to approach closer and penetrate significantly into the van der Waals contact distance. It is noteworthy that the vdW penetration distance d_p is actually a criterion for a significant donor–acceptor ("chemical" overlap) character of the bonding:

$$d_p = R_{vdW} - R_{eq} \tag{35}$$

where R_{vdW} is the vdW radius and R_{eq} the equilibrium radius. For a donor–acceptor interaction to occur, d_p should be appreciably positive, or $d_p \geqq 0.1$ Å. Since the empirical vdW radius is related to effective gas-phase collision diameters under ambient conditions (1 kT ≃ 1 kcal/mol), one may assume that the donor–acceptor interaction is significant if the complexation energy is 1 kcal/mol or higher.

In summary, the NBO analysis stresses the importance of quantum-mechanical orbital interaction and exchange effects in the van der Waals regime. Since these exchange effects are distinguishable from classical electrostatic effects, hydrogen bonding can be treated more like a donor–acceptor interaction. Of course, customarily hydrogen bonding has already been treated as an acid–base interaction. Since the charge-transfer component of the total energy as calculated by the NBO model is higher than that by the KM method, the NBO analysis appears to be able to differentiate between hydrogen-bonded and non-hydrogen-bonded species.

4.5. Acid–Base Interaction

Now that we have read the other side of the story about the donor–acceptor interaction, let us examine the concept of the acid–base interaction. Presumably, both of the above two models can be used to analyze the acid–base interaction. However, in retrospect the KM method appears to be more fitting.

The acid–base interactions[59] encompass not only hydrogen bonding and the electron pair donor–acceptor interactions, but also the electrophile–nucleophile interactions in organic chemistry.[60] A general equation can describe an acid–base interaction:

$$\underset{\text{acid}}{\mathrm{A}} + \underset{\text{base}}{:\mathrm{B}} \rightarrow \underset{\text{acid–base complex}}{\mathrm{A:B}}$$

As discussed in the above sections, both covalent and noncovalent (or ionic) factors are involved in these interactions.

4.5.1. Drago's Equation

On the same basis, Drago *et al.*[(61)] introduced four parameters for the calculation of the enthalpy ΔH^{a-b} of an acid–base interaction:

$$-\Delta H^{a-b} = E_A E_B + C_A C_B \tag{36}$$

where E_A and E_B are the susceptibilities of the acid (A) and base (B), respectively, to undergo the electrostatic interaction, while C_A and C_B are those to form covalent bonds.

The four parameters of 31 acids and 43 bases are given elsewhere.[(61)] Several examples of them are listed in Table 6.

4.5.2. Work of Adhesion—The Acid–Base Component

From the enthalpy of the acid–base interaction, ΔH^{a-b}, one can readily obtain the work of adhesion W_A^{a-b} resulting from the interaction:

$$W_{int}^{a-b} = W_A^{a-b} \tag{37}$$

However, in the case of a solid–solid interaction, the surface area involved can be very localized. Thus, we need to determine the surface area fraction or the population of the interaction,[(62)] n^{a-b}, in terms of moles per unit area. Then the actual W_A^{a-b} is given by

$$W_A^{a-b} = -f\, n^{a-b}\, \Delta H^{a-b} \tag{38}$$

where f is an enthalpy–to–free-energy correction factor, which has recently been found to be significantly below unity. (Refer to J. C. Berg and M. D. Vrbanac in a paper presented to the 64th Colloid and Suface Science Symposium, June 18–20, 1990.)

4.5.3. Total Work of Adhesion

If a solid surface involves only Lifshitz–van der Waals and the acid–base interactions, the total W_A is the sum of the two components:

$$W_A = W_A^{LW} + W_A^{a-b} \tag{39}$$

In the above discussion of the KM method of the decomposition of interaction energy, it is possible that a fraction of the LW component has also been taken into account in the interaction energy. Thus, the W_A^{a-b} term is likely to include a fraction of W_A^{LW}.

TABLE 6
Drago's Parameters for Acids and Bases[a]

Acids	C_A	E_A	Bases	C_B	E_B
Iodine	1.00	1.00	Pyridine	6.40	1.17
Phenol	0.44	4.33	Ammonia	3.46	1.36
Boron trifluoride	3.08	7.96	Ethyl acetate	1.74	0.98
Sulfur dioxide	0.81	0.92	Acetone	2.33	0.99
Chloroform	0.15	3.31	Benzene	0.71	0.49
Water	2.45	0.33	Triethyl amine	11.09	0.99

[a] From R. S. Drago *et al.*, *J. Am. Chem. Soc.* **93**(24), 6014 (1971).

Fowkes[62] has demonstrated that the Lifshitz–van der Waals component W_A^{LW} can be obtained from the dispersion component of the $W_{1,2}$:

$$W_A^{LW} = W_{1,2}^d \tag{40}$$

where

$$W_{1,2}^d = (W_{1,1}^d \cdot W_{2,2}^d)^{1/2} \tag{41}$$

$$= (\gamma_1^d \cdot \gamma_2^d)^{1/2} \tag{42}$$

Thus, W_A^{LW} can be derived from the geometric mean of γ^d. However, W_A^{a-b} cannot be estimated from the geometric mean of γ^a and γ^b.

Hence, the total W_A in Eq. (39) can be expressed as

$$W_A = W_A^d + W_A^{a-b}$$

$$= (\gamma_1^d \cdot \gamma_2^d)^{1/2} + W_A^{a-b} \tag{43}$$

5. ROLE OF DIFFUSION IN ADHESION

In the above discussion of van der Waals and molecular interaction, the chief requirement for solid bodies is to approach to a molecular distance. If the intermolecular separation is long range, both LW and Coulombic forces are operative. If the separation is less than 3 Å, both LW forces and molecular interactions can function to achieve adhesion. One of the means in achieving a close contact is wetting as discussed by de Gennes.[63,64] The other mean is diffusion. Under proper conditions all solids, e.g., metals,[65] semiconductors,[66] polymers,[67] can diffuse across the interfaces to achieve a close contact. Thus, diffusion is another important process in achieving adhesion. In this section we shall briefly examine several common relations pertaining to diffusion.

5.1. Fick's Laws of Diffusion

At the steady state of diffusion, the concentration remains unchanged and does not vary with time. Then according to Fick's first law,[67] the flux in the x-direction (F_x) is proportional to the concentration gradient ($\delta c/\delta x$):

$$F_x = -D\,(\delta c/\delta x) \tag{44}$$

where D is the diffusion constant.

If the concentration varies with time, the diffusion is not at the steady state. Then Fick's second law of diffusion should be used to determine the diffusion constant:

$$\left(\frac{\partial c}{\partial t}\right) = D\left(\frac{\partial^2 c}{\partial x^2} + \frac{\partial^2 c}{\partial y^2} + \frac{\partial^2 c}{\partial z^2}\right) \tag{45}$$

When diffusion is limited to the x-direction, Eq. (45) can be simplified to

$$\frac{\partial c}{\partial t} = D\,\frac{\partial^2 c}{\partial x^2} \tag{46}$$

5.2. Activation Energy of Diffusion

The activation rate theory of Glasstone, Laidler, and Eyring[68] has been applied to analyze the diffusion process of a solid. Thus, the diffusion constant D can be expressed as

$$D = \lambda^2 \frac{kT}{h} \cdot \frac{F^{\ddagger}}{F} e^{-\epsilon_0/kT} \tag{47}$$

where λ is the elementary jump length, k Boltzmann's constant, h the Planck constant, $F^{\ddagger}$ the standard free energy of activation per mole, F the free energy at the normal state, and ϵ_0 the activation energy per molecule at 0 K.

The activation energy of diffusion Q can also be expressed in terms of kilocalorie per mole as in the following equation:

$$D = D_0 e^{-Q/RT} \tag{48}$$

where D_0 is the frequency factor. Since $R = N_A k$ ($N_A = 6.02 \times 10^{23}$/mol), the activation energy can be readily converted from the molar basis into the molecular basis. In the literature, the activation energy on the molecular basis is generally expressed in terms of electron volt (1 eV $= 1.6 \times 10^{-19}$ J $= 3.8 \times 10^{-23}$ kcal). Thus, 1 eV per molecule is equivalent to 22.8 kcal per mole.

In general, Q depends primarily on two factors: (1) the strength of the atomic bond of the solid, and (2) the atomic mechanism for diffusion. When bond strengths increase as in the covalently bonded metals, such as Si and Ge, Q increases accordingly. Indirectly, Q can be correlated with the melting temperature of a solid, T_m, which is a reflection of bond strength (Table 7). The diffusion data in Table 7 have been well verified. Generally, when a radiotracer diffuses into the same metal, the activation energy of self-diffusion appears to increase with the increase in melting temperature.

Diffusion in amorphous materials is also dependent on the atomic bonding. In the case of polymers, strong covalent bonds exist within the chains but weak van der Waals bonds influence intermolecular interactions. Thus, diffusion takes place generally by the motion of molecular chains. Regarding the subject of polymer diffusion, one should refer to the review by Brochard-Wyart[69] and Chalykh *et al.*[70]

5.3. Diffusion and Adhesion of Polymers

Diffusion as one of the adhesion mechanisms for polymers was first proposed by Voyuskii.[71] Originally for the diffusion mechanism, Vasenin[72] developed the kinetic aspects of adhesion. However, in light of the reptation theory of de Gennes,[73] the diffusion kinetics have been rewritten. A general discussion of the effect of diffusion on adhesion is discussed by Wool.[74] The phenomena of tack, green strength, and welding are better understood on the basis of the reptation theory.

Polymer adhesion can be adversely affected by the diffusion of water into the matrix. A review by Comyn[67] pointed out the detrimental effect of water in joint durability. Generally, the diffusion of plasticizer or solvent toward the interface can adversely affect the bond strength.

TABLE 7

Radiotracer Self-Diffusion Data for Pure Metals[a,b]

Metal	Tracer[(1)]	Crystal[(2)] form	Metal[(2)] form and purity	mp[(1)](°C)	Temp[(2)] range(°C)	Form[(1)] of analysis	Activation[(1)] energy Q (kcal/mol)	Frequency[(1)] factor D_0 (cm²/sec)	Remarks
K	K^{42}	BCC	S (99.7)	63.7	−52–61	SS	9.36	0.16	
Na	Na^{22}	BCC	P (99.99)	97.8	0–98	SS	10.09	0.145	
In	In^{114}	Face-tetragonal	S⊥C (99.99)	156	44–144	SS	18.7	3.7	
Li	Li^{6}	BCC	P (99.98)	181	35–178	SS	12.60	0.14	
Se	Se^{75}	Hexagonal	P	217	35–140	—	11.70	1.4×10^{-4}	
Sn	Sn^{113}	Tetragonal	S⊥C (99.999)	231.9	160–226	SS	25.1	10.7	
Cd	Cd^{115}	HCP	S (99.95)	321	110–283	RA	19.3	0.14	
Pb	Pb^{204}	FCC	S (99.999)	327	150–320	SS	25.52	0.887	
Zn	Zn^{65}	HCP	S⊥C (99.999)	419	240–418	SS	23.0	0.18	
Mg	Mg^{28}	HCP	S‖C	650	467–635	SS	32.2	1.0	
Al	Al^{27}	FCC	S	660	450–650	SS	34.0	1.71	
Ge	Ge^{71}	Diamond cubic	S	937	766–928	SS	68.5	7.8	Covalent
Ag	Ag^{110}	FCC	S (99.999)	960.8	640–955	SS	45.2	0.67	
Au	Au^{198}	FCC	S (99.97)	1063	850–1050	SS	42.26	0.107	
Cu	Cu^{67}	FCC	S (99.999)	1083	698–1061	SS	50.5	0.78	
Si	Si^{31}	Diamond cubic	S (99.99999)	1410	1225–1400	SS	110.0	1800	Covalent
Ni	Ni^{63}	FCC	P (99.95)	1453	1042–1404	SS	68	1.9	
Co	Co^{60}	HCP	P (99.9)	1495	1100–1405	SS	67.7	0.83	
α-Fe	Fe^{55}	BCC	P (99.92)	1536	809–889	—	60.3	5.4	
Pd	Pd^{103}	FCC	S (99.999)	1552	1060–1500	SS	63.6	0.205	
β-Ti	Ti^{44}	HCP	P (99.95)	1668	900–1540	SS	60.0	1.09	
Pt	Pt^{195}	FCC	P (99.99)	1769	1325–1600	SS	68.2	0.33	
Cr	Cr^{51}	BCC	P (99.98)	1875	1030–1545	SS	73.7	0.2	
V	V^{48}	BCC	S,P (99.9)	1900	1360–1830	SS	94.14	214.0	
Mo	Mo^{99}	BCC	P	2610	1850–2350	SS	96.9	0.5	
W	W^{185}	BCC	P (99.99)	3410	1800–2403	SS	140.3	1.88	

[a] From (1) R. C. West (ed.), *Handbook of Chemistry and Physics*, CRC, Boca Raton, Florida (1989–1900) (adapted with permission); and (2) C. R. Barrett, W. D. Nix, and A. S. Tetelman, *The Principle of Engineering Materials*, p. 540, Prentice-Hall, N.J. (1973).

[b] Crystal forms: S, single crytsal; P, polycrystalline; ⊥c, perpendicular to c-direction. Analysis methods: SS, serial sectioning; RA, residual analysis.

5.4. Diffusion of Solid into Polymers

On the basis of Buche–Cashin–Debye's equation,[75] Lee[76] assumed that polymer adhesion can be subdivided into rubbery-polymer–rubber polymer adhesion (R–R adhesion), rubbery polymer–glassy polymer adhesion (R–G adhesion), and rubbery polymer–non-polymer adhesion (R–S adhesion). According to the original calculation, the glassy polymer does not diffuse readily at a viscosity greater than 10^{13} poise. However, at that time, we did not explore the role of a solid on adhesion if it diffuses into a glassy polymer.

Recently Faupel *et al.*[77] studied the diffusion of Cu into a cured polyimide below the glass transition temperature. They used low-energy ion sputtering in conjunction with a radiotracer technique. The initial decay was shown to be non-Fickian due to the formation of Cu cluster, while the tail of the decay appeared to be Fickian due to the diffusion of single Cu atoms. The results suggest that the diffusion into a glassy polymer seems to be controlled by a free-volume mechanism below the glass transition temperature.

6. PHYSICOCHEMICAL BONDING ON SOLID SURFACES

Solid–solid adhesion[78] is complex. Sometimes it involves more than LW forces and molecular interactions. There are cases in which true physicochemical or chemical bonds are formed at the interface. Plueddemann[79] and Miller and Ishida[80] will discuss various interactions at interfaces. In the case of thin film technology, layers of film, metals, or nonmetals are deposited in vacuum. The interactions of materials in the vacuum environments can readily involve physicochemical or chemical reactions. Later in this paper, we shall enumerate those reactions with metals. In this volume, Baglin[81] and Ho *et al.*[82] will discuss the adhesion of thin films in more details.

To understand physicochemical interaction, we need to investigate basic principles related to reactions and bonding on solid surfaces. Recently Hoffmann[83,84] treated this subject matter chiefly from the frontier orbital approach. We shall first briefly review some basic concepts regarding electronic structures of solids before we discuss some of Hoffmann's new findings.

6.1. Density of States (DOS)

When a large number of free electrons exist in a solid, their energies can be expressed in terms of density of states $N(E)$. This quantity is defined as the number of quantum states per unit crystal volume between the energy levels E and $E + \mathrm{d}E$, while $N(E)$ is expressed as kinetic energy because the number of quantum states increases with increase in energy according to the following relation:

$$N(E) = \frac{\pi}{4}\left(\frac{8m}{h^2}\right)^{3/2} E^{1/2} \tag{49}$$

Thus, $N(E)$ increases parabolically with energy E[85] as shown in Figure 7.

The distribution of the energies of free electrons is described by the Fermi function

$$P(E) = \frac{1}{\exp[(E - E_F)kT] + 1} \tag{50}$$

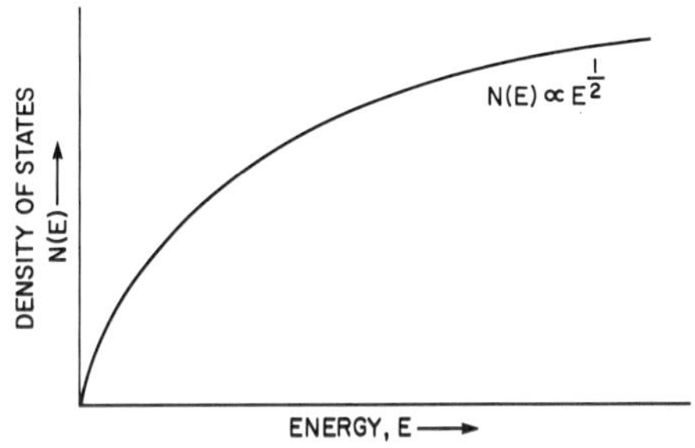

FIGURE 7. Density of states $N(E)$ as a function of energy level. [After C. R. Barrett, W. D. Nix, and A. S. Tetelman, *The Principles of Engineering Materials,* p. 363, Prentice-Hall, New Jersey (1973) (reproduced with permission).]

where $P(E)$ is the probability that an energy level E is fully occupied and E_F is the Fermi energy, which will be discussed in Section 6.3.

6.2. Electron Work Function

The potential energy[86] of the free electrons in a one-dimensional metal is illustrated in Figure 8. A finite potential energy barrier, W, exists at each end of the solid. In order for an electron to escape from a metal, it must have a kinetic energy to overcome the barrier, W. Since the free electrons[86] in the metal have kinetic energies ranging from near zero at the bottom of the potential energy well to E_F at the Fermi level, the critical work that must be supplied to remove a free electron from the solid at 0 K is called the work function ϕ:

$$\phi = W - E_F \tag{51}$$

This is the most important static property of a metal. We shall discuss the effect of work function on adhesion and contact charging later.

The photoelectric effect (or photoemission) has been one of the convenient ways besides contact potential to determine the work function according to the following equation:

$$\phi = hc/\lambda_c \tag{52}$$

where λ_c is the longest photon wavelength to cause electrons to be emitted from the surface of a metal. Though for metals, the photoelectric threshold is equal to the work

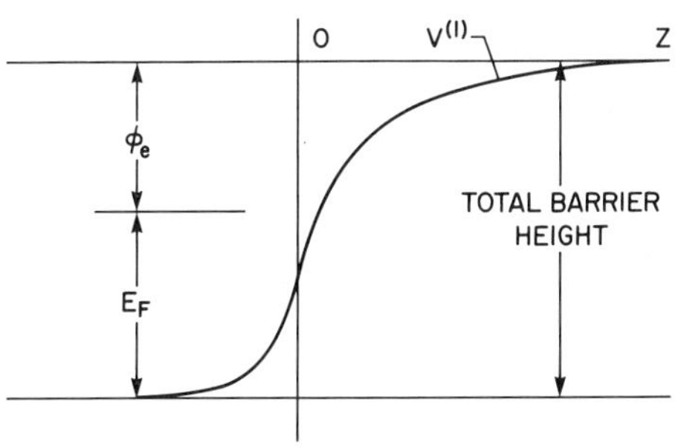

FIGURE 8. Relation between the electron work function ϕ_e, the Fermi energy E_F, and the effective one-electron potential energy for a state at the top of the Fermi distribution $V^{(1)}$. [After J. R. Smith, *Phys. Rev.* **181**(2), 522 (1969) (reproduced with permission).]

function, for semiconductors, this is not generally true. For semiconductors, the photoelectric threshold is equal to the energy difference between the vacuum level and the top of the valence band (or ionization potential, I).

In Table 8, we compare Pauling's[87] and Mulliken's[88] electronegativity values and work functions[89] of polycrystalline metals. In general, the work function increases with the increase in Pauling's electronegativity value χ^P. However, there is no simple correlation between them. On the other hand, the work functions of the bulk elements are very close to Mulliken's values, χ^M, for free atoms.

At this point, we should remember that the electronic properties of atoms need not be identical to those of elemental solids. In atoms, the charging effects are large and nonlinear, while in solids they are linear. It it is also known that in a free atom the wave functions of valence electrons differ from those of an atom in a solid. The different energy level of a free atom and a solid is illustrated in Figure 9.

In one of the earlier sections, the electronegativity (atomic) as defined by Mulliken[21] is given as $\chi^M = \frac{1}{2}(I + A)$, which is also the negative of chemical potential, μ. Thus, it is interesting to see the close correlation between χ^M and the work function of a bulk metal as shown in Table 8.

Michaelson[88] has also shown that χ^M and ϕ give all linear fits for subgroups of elements. The basis for these linear fits can be illustrated in Figure 9 by the relation between the electron transition energy in an atom (χ^M) and the electron transition energy in an elemental bulk metal (ϕ). The theoretical work by Steiner and Gyftopoulos[90] has shown that the work function ϕ for a metal surface is equal to the neutral electronegativity of the surface atoms. Thus

$$\Phi = \frac{I + A}{2} \sim \chi^M \tag{53}$$

6.3. Fermi Energy

Besides the work function, Fermi energy is another important property of metals. Assuming that a cubic metal sample with sides of length L contains N free electrons, then at 0 K the electrons will occupy the lowest possible energy levels according to the Pauli exclusion principle. When all of the N free electrons have occupied the lowest possible levels, the energy of the most energetic electron is called the Fermi energy, E_F. In other

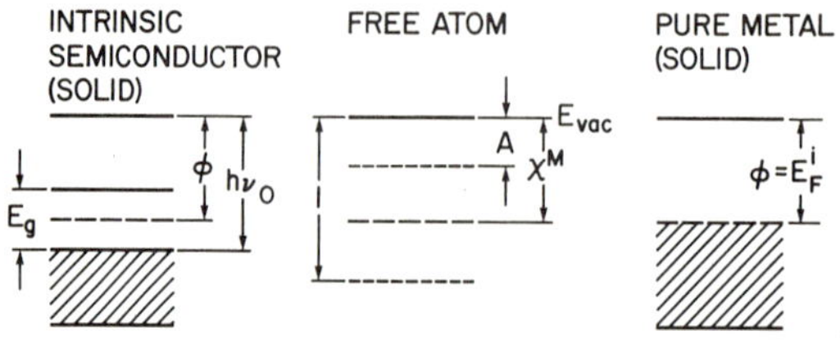

FIGURE 9. Energy level diagram showing quantities discussed in the text for intrinsic semiconductor (solid), free atom, and pure metal (solid). [After H. B. Michaelson, *IBM J. Res. Dev.* **22**(1), 72 (1978) (reproduced with permission).]

TABLE 8

Comparison of Pauling's and Mulliken's Electronegativities and Work Functions of Polycrystalline Metals

Metal	Electronegativity χ		Work function[c] ϕ	
	Pauling's[a]	Mulliken's[b]	Method of determination[e]	Preferred value
Se	2.45	5.9	P	5.9
Pt	2.2	5.6	P	5.65
Ni	1.8	4.4	P	5.15
Pd	2.2	4.5	P	5.12
Au	2.4	5.8	P	5.1
Ge	1.9	4.6	CPD	5.0
C	2.6	6.3	CPD	5.0
Co	1.8	4.3	P	5.0
Be	1.5	4.7	P	4.98
Te	2.3	5.5	P	4.95
Si	1.9	4.8	CPD	4.85 (n)
Os	2.2	4.9	T	4.83
Cu	1.9–2.0 (1^+, 2^+)	4.5	P	4.65
Mo	1.8	4.1	P	4.60
W	1.70	4.3	CPD	4.55
Fe	1.8–1.9 (2^+, 3^+)	4.1	P	4.50
Cr	1.6	3.7	P	4.50
Hg	1.9	4.9	P	4.49
Sn	1.8–1.9 (2^+, 4^+)	4.3	CPD	(4.42)[d]
Ti	1.8	3.5, 4.1	P	(4.33)
Zn	1.5	4.3	P	4.33
Al	1.5	3.2	P	4.28
Ag	1.9	4.5	P	4.26
Ga	1.6	3.2	CPD	4.2
In	1.7	3.0	P	4.12
Rb	0.8	2.4	P	3.98
As	2.0	5.3	P	3.75
Mg	1.2	3.8	P	(3.66)
Li	1.0	3.0	F	2.90
Na	0.9	2.9	P	2.75
K	0.8	2.4	P	2.30
Cs	0.7	2.2	P	2.14

[a] From J. A. Dean (ed.), Lange's *Handbook of Chemistry,* 12th ed. McGraw-Hill, New York pp. 3–11 (1978).
[b] From H. B. Michaelson, *IBM J. Res. Dev.* **22**(1), 72 (1978).
[c] From H. B. Michaelson, in: *Handbook of Chemistry and Physics* (R. C. West, ed.), p. E-93, CRC Press, Florida (1989–1990).
[d] Those values in parentheses are not well substantiated.
[e] Methods used for the determination of the work function are: P, photoelectric; CPD, contact potential difference; T, thermionic, F, field emission.

words, the Fermi level is the highest occupied molecular orbital (HOMO) of the metal. The Fermi energy of a metal can also be calculated from the following equation:[(85)]

$$E_F = \frac{h^2}{8m}\left(\frac{3\eta}{\pi}\right)^{2/3} \tag{54}$$

where $\eta = N/L^3$ is the number of free electrons per unit crystal volume and m is the mass of the electron. Fermi energies of several metals[(85)] are listed in Table 9. By adding work functions (from Table 8) to Fermi energies, we can obtain the barrier heights (W) for several metals (Table 9).

In recent years, Anderson[(91)] has determined Fermi energies for many metals. On the basis of those values, Hoffmann[(83,84)] has prepared a plot (Figure 10) to compare the Fermi energies of several transition metals. From this plot, we can estimate the reactivities of transition metals relating to one another or to other molecules. In general, the Fermi levels in those transition metals fall in the d band. The filling of the d band increases toward the right side of the series. Over the greater part of the series, the Fermi level falls. For example, the E_F vlaue of Ni (or Cu) is below that of Cr. Thus a molecule may react with Cr, but not necessarily with Ni (or Cu). The Fermi level of transition metals will help explain some of the observations made by Buckley about metal–ceramic interactions to be discussed in one of the forthcoming sections.

6.4. *Frontier Orbital Concept*

In Section 4.2, we briefly mentioned the perturbation of orbitals with the simplified equation for the interaction energy derived by Klopman and Hudson.[(51)] When two reacting molecules start to approach each other, their orbitals perturb and interact. As a result, two transition orbitals are formed, i.e., bonding and antibonding. The formation of the former is accompanied by the evolution of heat, and the reaction is exothermic. On the

TABLE 9
Fermi Energies and Barrier Heights for Several Monovalent Metals[a]

Metal	Crystal structure	Lattice parameter a (Å)	η (cm^{-3}) (number of free electrons per unit volume)	F_F (eV) (measured from the bottom of the free electron energy band)	ϕ (eV) (work function)	W (eV) (barrier height)
Cs	BCC	6.05	$2/a^3 = 9.03 \times 10^{21}$	1.53	2.14	3.67
Rb	BCC	5.62	$2/a^3 = 1.13 \times 10^{22}$	1.82	2.16	3.98
K	BCC	5.33	$2/a^3 = 1.32 \times 10^{22}$	2.14	2.30	4.44
Na	BCC	4.28	$2/a^3 = 2.55 \times 10^{22}$	3.16	2.75	5.91
Li	BCC	3.50	$2/a^3 = 4.66 \times 10^{22}$	4.72	2.90	7.62
Ag	FCC	4.07	$4/a^3 = 5.96 \times 10^{22}$	5.51	4.26	9.77
Cu	FCC	3.60	$4/a^3 = 8.52 \times 10^{22}$	7.04	4.65	11.69

[a] Data in columns 1–5 from C. R. Barrett, W. D. Nix, and A. S. Tetelman, *The Principles of Engineering Materials*, p. 365, Prentice-Hall, New Jersey (1973).

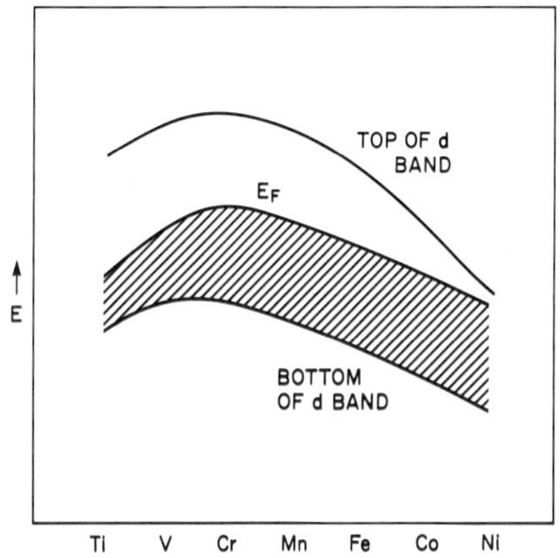

FIGURE 10. Fermi levels of transition metals. [After R. Hoffmann, *Rev. Mod. Phys.* **60**(3), 601 (1988) (reproduced with permission).]

other hand, the formation of the latter requires the absorption of heat, and the reaction is endothermic because two electrons must go into the antibond.

In the case of combining unfilled (unoccupied) orbitals of both molecules, there is no energy required. We shall illustrate later the role of unfilled orbitals on solid surfaces. On the other hand, the interactions between an unfilled orbital and a filled (occupied) orbital have an energy-lowering effect.

Fukui(92) has shown that the most important interaction between two molecules is that between the HOMO of A-molecule and the LUMO of B-molecule because it will result in the largest drop in energy. In one of the earlier sections, this type of interaction has been discussed as the donor–acceptor or the acid–base interaction. These orbitals are the frontier orbitals. The other interactions between the occupied orbitals and unoccupied orbitals are comparatively secondary and affect only the energetics of transition states.

In terms of the electronic structure in solids, the HOMO is equivalent to the top of the valence band, the LUMO to the bottom of the conduction band. Thus, in terms of energy, the energy of the HOMO is E_v, and that of the LUMO is E_c. The Fermi energy E_F for an intrinsic semiconductor is

$$E_F = \frac{1}{2}(E_{\text{HOMO}} + E_{\text{LUMO}}) = \frac{1}{2}(E_c + E_v) \tag{55}$$

Furthermore, as stated in the Koopmans theorem,(93) the frontier orbital energies are given by

$$-E_{\text{HOMO}} = I \tag{56}$$

and

$$-E_{\text{LUMO}} = A \tag{57}$$

where I is the ionization potential (or energy) and A the electron affinity.

7. INTERACTIONS ON SOLID SURFACES

7.1. Interactions Between Two Discrete Molecules

In 1964, Blyholder[94] first applied the orbital concept for the study of solid surfaces. Recently, Hoffmann[83,84] employed the frontier orbital concept[95] much more extensively and also very successfully. He categorizes the solid interactions according to the number of electrons exchanging intermolecularly:

1. two-electron interaction involving an electron pair,
2. four-electron exchange,
3. no-electron exchange.

In the diagram (Figure 11a), a is shown in interactions ① and ②, b in interaction ③, and c in interaction ④. Interactions ① and ② result in a true covalent or dative bonding; thus its overall effect is stabilization of the system owing to the lowering of energy. If it is a complex formation, the bond is dative in nature involving a donor and an acceptor of the electron. This is what we discussed in earlier sections. Interaction ③ is the exchange that is repulsive in nature. When the overlap of orbitals takes place, the antibonding level rises and the bonding level decreases. Thus, the energy for going up is more than that for coming down. However, the total energy[84] of the overlapped orbitals is higher than that of separately isolated orbitals.

Finally, interaction ④ which involves no electron from each of the unoccupied orbitals produces no net gain or loss in energy. In other words, this is the effortless combination of orbitals for a pair of molecules.

7.2. Interactions Between a Molecule and a Solid

If one of the pair is not a discrete molecule but a solid (Figure 11b), Hoffmann[84] points out a new concern, that interaction ③ may prevent interactions ① and ② from happening. Why? The reason is that the antibonding level of the molecule may rise above the Fermi level of the solid. Then, unexpectedly, the electrons may flow from the antibonding level to the Fermi level of the solid, and consequently the stabilization effect starts to diminish.

For a metal surface which is nearly a continuum, one more interaction can be realized, i.e., interaction ⑤. Here, the Fermi level is at the top of the valence band (or the

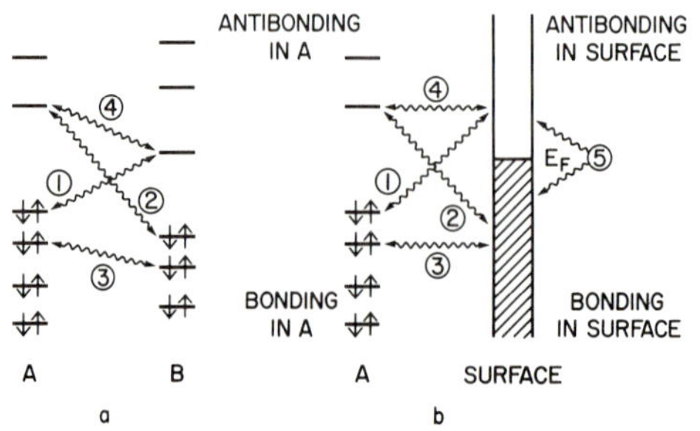

FIGURE 11. Comparison of orbital interactions: (a) between two discrete molecules and (b) between a molecule and a solid surface. [After R. Hoffmann, *Rev. Mod. Phys.* **60**(3), 601 (1988) (reproduced with permission).]

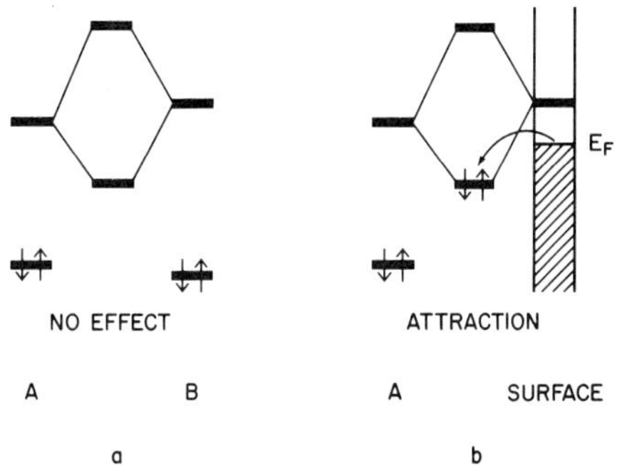

FIGURE 12. Comparison of interaction (4): (a) between two discrete molecules and (b) between a molecule and a solid surface. [After R. Hoffmann, *Rev. Mod. Phys.* **60**(3), 601 (1988) (reproduced with permission).]

occupied level), and electrons (or holes) will flow in the surface and in the bulk beneath it in an attempt to balance the interactions. This particular interaction has a sort of house-cleaning effect. Thus, all four first-order interactions (① → ④) will move the metal's level up and down, but interaction ⑤ influences only the second-order energetics and bonding resulting from the shift of electron density around the Fermi level.

In Figure 12, interaction ④ is compared between (a) two discrete molecules and (b) a molecule and a solid surface. As mentioned previously, for two discrete molecules, indeed interaction ④ produces no net change in energy. However, in the case between a discrete molecule and a surface, the situation is different. In the solid, there is a continuum of levels, the bonding of the interacting levels may fall below the Fermi level of the solid (Figure 12b). After becoming filled, it will promote the fragment A-surface bonding. Moreover, the surface may supply the electrons to fill that level. Therefore interaction ④ may produce a change in energy between a discrete molecule and a solid surface, but not between two discrete molecules.

7.3. Chemisorption

What happens if there is chemisorption on the solid surface? The net result is that the metal–adsorbate bond is strengthened at the expense of the bonding of the adsorbate within the bulk. This somewhat resembles the push-and-pull mechanism in organic chemistry, and there appears to be an overall conservation of energy. This interesting aspect of chemisorption has been applied to surface interactions between gas molecules and solids. We believe that some of those aspects may be related to solid adhesion. However, we do not intend to elaborate the subject in this section.

8. HARD–SOFT ACID–BASE (HSAB) PRINCIPLE

8.1. The HSAB Principle in Solution

Before we leave the subject of frontier orbital concept, let us remind ourselves once more about the above interactions ① and ②. Since interactions ① and ② are nothing but the donor–acceptor (or acid–base) interactions, Pearson's hard–soft acid–base (HSAB) principle[96,97] should apply. In terms of frontier orbital,[97] the HSAB principle can be stated as follows:

- Hard electrophiles (acids) have a high-energy LUMO and usually have a positive charge.
- Soft electrophiles have a low-energy LUMO but do not necessarily have a positive charge.
- Hard nucleophiles (bases) have a low-energy HOMO and usually have a negative charge.
- Soft nucleophiles have a high-energy HOMO but do not necessarily have a negative charge.
- A hard–hard reaction is fast because of a large Coulombic attraction. Thus, the reaction is controlled by the first term of the perturbation energy in Eq. (29).
- A soft–soft reaction is fast because of a large interaction between the HOMO of a nucleophile and the LUMO of an electrophile. Thus, the reaction is controlled by the second term in Eq. (29).

8.2. Extension of the HSAB Principle to the Solid State

Recently Lee[98,99] extended the HSAB principle to solid adhesion and surface interactions by considering the electronic structure of the solid state. This extension is one of the important bridges between chemistry and physics in the solid state. In chemistry, an important term, the absolute or chemical hardness, η, has been introduced by Pearson[97] to be

$$\eta = -\frac{1}{2}(E_{\mathrm{HOMO}} - E_{\mathrm{LUMO}}) \tag{58}$$

as shown in Figure 13.

On the other hand, in physics, the average energy gap, E_g^{Av} as defined by Phillips,[22] has been used to determine the value of gap separation between E_{HOMO} and E_{LUMO}. Thus,

$$E_g^{\mathrm{Av}} = -(E_{\mathrm{HOMO}} - E_{\mathrm{LUMO}}) \tag{59}$$

Therefore, if we combine these two entirely different terms, we obtain

$$\eta = \frac{1}{2}E_g^{\mathrm{Av}} \tag{60}$$

It is surprising to find that the chemical hardness is directly related to the energy gap. However, Lee[98,99] has pointed out that this quantity E_g is not the minimum energy gap, E_g^{o}, reported in the literature. The average energy gap E_g^{Av} is generally larger than the minimum energy gap. Furthermore, E_g^{Av} is affected by both the covalent and ionic factors.

On the basis of this extension, we can classify the following materials according to their chemical hardnesses:[96,97]

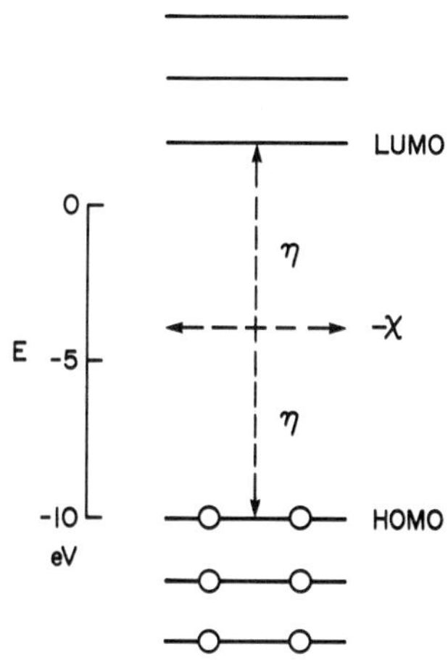

FIGURE 13. Chemical hardness (η) and orbital energy diagram for a molecule. HOMO, highest occupied MO; LUMO, lowest occupied MO. [After R. G. Pearson, *J. Chem. Educ.* **64**(7), 563 (1987) (reproduced with permission).]

- All metals are soft, and they are mostly acids.
- All semimetals are soft.
- All semiconductors are rather soft, and most of them are bases.
- Most insulators, including polymers, are hard.

This type of classification can assist us in predicting surface interactions and adhesion between solids. We shall discuss various interactions between solids in the forthcoming sections, partly according to the extended HSAB principle.

9. METAL-TO-METAL CONTACT

A review of problems related to interfacial bonding and adhesion of solid contact has been presented by Buckley *et al.*[(100)] In that paper the main subject is metal-to-metal contact that is frequently encountered in friction and wear of materials. A real contact involves the interactions between many asperities on each counterface. The deformation and adhesion of these asperities can directly affect the outcome of friction and wear. A study involving the interaction between a single asperity and a counterface can be carried out with a sharp metal point and a flat surface, e.g., the pin-and-disk device. The force required to separate these contacts in ultrahigh vacuum after a known load is applied can be accurately measured.

Some of the experimental results obtained (not under vacuum) by Pashley *et al.*[(101,102)] indicate that at very low applied loads the contact is generally controlled by surface forces derived from Lifshitz–van der Waals interactions discussed in Section 3. The surface forces can be strong enough to cause plastic flow and deformation. When the real surface of metal is taken into account, there are always some oxides formed. The presence of an oxide layer tends to lower the surface energy and affect adhesion, friction, and wear. For the metal-to-metal single asperity contacts, surface energy can be an important factor. For example, in the case of an elastic contact, the adhesion force F may vary between 1.5

$\pi R\Delta\gamma$ and $2\pi R\Delta\gamma$,[101] while for a plastic contact, it should be $2\pi R\Delta\gamma$ or slightly less (where $\Delta\gamma = \gamma_1 + \gamma_2 - \gamma_{12}$).

9.1. Metal–Metal Adhesion

For a clean metal surface, as in the case of an ultravacuum environment, metal-to-metal contact can lead to adhesion. This is especially true when both metals are the same. In the case of dissimilar metals, Keller[103] has shown that there are some exceptions depending upon the free energy of atomic bond formation, ΔG_f. Those that adhere to form some intermediate phase at the contact have negative ΔG_f, while those that do not adhere have positive ΔG_f (as shown in Table 10).

In the following section, we shall discuss a universal relationship governing the metal–metal adhesion. We hope that this theoretical development can provide us with a general understanding about the metal–metal adhesion.

9.2. Metal–Metal Adhesion Theory

In Section 2.3, we briefly mentioned the work by Ferrante *et al.*[104,105] on the universality of adhesive energy E_{ad} versus the separation a between the surface of metals. They have found that most metals when adhered follow the universal relationship between the scaled adhesive energy, $E^*_{ad}(a)$, and the scaled separation,

$$a^*E_{ad}(a) = \Delta E \cdot E^*_{ad}(a^*) \tag{61}$$

where $E_{ad}(a) = [E(a) - E(\infty)]/2A$; A is the cross-sectional area, $a^* = (a - a_m)/l$ and

$$l_{\text{(scaling length)}} = \left\{ \frac{\Delta E}{[\mathrm{d}^2 E_{ad}(a)/\mathrm{d}a^2]a_m} \right\}^{1/2}$$

ΔE is the equilibrium binding energy or the magnitude of $E_{ad}(a)$ at the respective minima of the curves, $E^*_{ad}(a^*)$ is a universal energy relation, a_m is the equilibrium separation, and a is the atomic separation from the equilibrium value a_m.

When two metals are identical, ΔE becomes the surface energy of the metal.

9.3. Effect of Critical Separation on Metal–Metal Adhesion

It is easy for clean metals to cling together under high vacuum. However, for contaminated metals or rough surfaces, adhesion drops off considerably. There appears to be a critical distance necessary for metal to achieve adhesion.

Ferrante and Smith[106] have stated that it is the kinetic energy that initiates the bonding, but it is the exchange-correlation energy that determines the strength of an adhesive bond at the bimetallic interface. The range of strong bonding is about 2 Å (or 0.2 nm) for the Al (111)–Mg (011) contact. It should be noted that the electrostatic energy is repulsive at small separations but attractive at large distances. The dominant repulsive term is the kinetic energy at small separation. Indeed, the decomposition of energies for the bimetallic adhesion is similar to that by the KM method. The critical separation discussed here is also similar to that (1.8 Å) for the acid–base complex BH_3–NH_3.[55]

TABLE 10
Adhesion of Metallic Pairs[a]

Adhesion observed ($\Delta G_f < 0$)	Adhesion not observed ($\Delta G_f > 0$)
Fe–Al	Cu–Mo
Cu–Ag	Ag–Mo
Ni–Cu	Ag–Fe
Ni–Mo	Ag–Ni

[a] From D. V. Keller, *Wear* **6**, 353 (1963).

Interestingly, the critical distance for an electrophile–nucleophile interaction[107] has also been found to be 2.3–2.7 Å. In brief, the critical separation should be 2 to 3 Å.

9.4. Effect of the Acid–Base Interaction on Metal–Metal Adhesion

Is the metal bonding another form of the A–B interaction? Yes, if the atomic separation is around <3 Å as discussed in the preceding section. Since most metals are also classified as "soft"[96] according to the absolute chemical hardness scale, the metal–metal adhesion can be directly derived from the soft–acid soft–base interaction, which is predominantly controlled by their frontier orbitals. In other words, a good contact between metals in a clean environment can involve both long-range and short-range interactions.

The acid–base interaction between intermetallics containing d metals has been demonstrated by Brewer.[108] For example, hafmium (Hf) with the unoccupied d orbitals has been shown to be a Lewis acid, on opposing a Lewis base, such as a platinum group metal containing nonbonding electron pairs. At low concentrations of niobium (Nb), palladium (Pd) acts as a strong base, while at high concentrations of Nb, rhodium (Rh) becomes a strong base. In general, transition metals possess the acid–base character because the d electrons, unlike s–p electrons, tend to assume a much more localized electronic behavior. In one of the later sections, we shall discuss some special cases of d metals.

Recently, the chemical interaction at the Cr/Cu interface has been modeled by Cain *et al.*[109] as an acid–base interaction. The modeling is made on the basis of the tight binding band of the extended Hückel method. Metals have been treated as amphoteric materials, which may be directly related to the band structure and the Fermi level of the metal. Cu can be considered to be more basic than Cr because it has more filled than empty orbitals. On the other hand, the Fermi level of Cr splits the 3d bands giving Cr as many unfilled orbitals as filled ones. Thus, Cr can act as either a Lewis acid or a Lewis base depending upon the opposing metal. For the Cu–Cr interaction, Cr becomes a Lewis acid, or an electron acceptor, and the bonding with Cu takes place at the bridge sites.

9.5. Effect of Crystal Structure on Metal–Metal Adhesion

Besides surface energy and the acid–base nature of metals, their crystal structures can also influence adhesion. In the case of Cu single crystals,[110] the high atomic density

(111) plane exhibits the lowest coefficient of adhesion, i.e., adhesion bond force divided by the applied load, among the three orientations of Cu examined (Table 11). The (111) plane has the highest modulus of elasticity and is the least resistant to deformation under a given load. Thus, the real area of contact is less than those of the other less dense planes.

Crystal structure also affects the adhesion between dissimilar metals. In the case of Au–Cu adhesion, the adhesion forces (in mg) for three Cu crystal planes are 390 for Cu (110), 185 for Cu (100), and 80 for Cu (111). In this case, surface energies of the three planes also decrease in the same order.

9.6. Effect of Contact Electrification on Metal–Metal Adhesion

At the metal-to-metal contact, electrons (or charges) can transfer. Does the transfer affect adhesion? When two dissimilar metals make a contact (Figure 14) and an equilibrium is reached, a contact potential[111] V_c is established as

$$V_c = (\phi_1 - \phi_2)e \tag{62}$$

where ϕ is the work function of the metal. Since ϕ ranges between 3 and 5 eV (see Table 7 in Section 6.2) V_c varies from 0 to 2 V. If the average distance over which V_c varies is assumed to be d and the electric field at the interface to be V_c/d, then the electrostatic pressure calculated by Hays[112] is

$$P_A = \epsilon_0 V_c^2/(ed^2) \tag{63}$$

Assuming $k = 1$, $V_c = 1$ V, and $d = 5$ Å or 0.5 nm, then P_A is only 20 MPa.

Since the Lifshitz–van der Waals adhesive pressure (or force) under the same condition is approximately 100 MPa, the effect of the electronic transfer is far less (or $\frac{1}{5}$) than that of LW forces. Thus in the case of metal-to-metal contact, depending on the distance, the two primary effects are the surface forces and the Coulombic interaction at long range and the acid–base interaction at short range. The effect of contact electrification is secondary at best.

9.7. Effect of Diffusion on Metal–Metal Adhesion

Metals can be bonded together through diffusion in the absence of a chemical reaction. In practice, a diffusion bonding process[113] has been established upon interdiffusing of surface atoms of similar or dissimilar metals in the solid state at one-half to two-thirds of T_m below the melting point of the basic metal. Moreover, diffusion bonding can also be carried out in the liquid state. If, in the solid state, both temperature and pressure are applied to achieve the optimum diffusion, a good vacuum ($<10^{-3}$ torr) or oxygen-free inert environment is best suited for the process. Moisture-free argon or helium is commonly employed at <-70 °F dew point. A reducing atmosphere of hydrogen is sometimes utilized. For hard metals, a good surface finish (no rougher than 16 rms) is preferred.

Sometimes, an intermediate metal film, instead of an oxide, is introduced to enhance the diffusion. For example, recently a gold–cadmium (Au–Cd) alloy has been reported by Decroupet *et al.*[114] to enhance the diffusion bonding of gold. A high diffusion rate at low temperature (about 700 °C) has been obtained with the use of a Au–Cd alloy.

TABLE 11
Young's Moduli and Adhesion Coefficients of Single-Crystal and Polycrystalline Copper (99.999%)[a]

Copper form and orientation	Young's modulus ($\times 10^{11}$ dyn/cm^2)	Adhesion coefficient[b]
Single crystal (100) matched planes and directions	6.67	1.02
Single crystal (110) matched planes and directions	13.1	0.61
Single crystal (111) matched planes and directions	19.4	0.30
Polycrystal	12.0	1.00

[a] From D. H. Buckley, J. Ferrante, M. D. Pashley, and J. R. Smith, Solids in contact, *Mater. Sci. Eng.* **83**, 177 (1986).
[b] Load, 50 gf; pressure, 10^{-9} Pa.

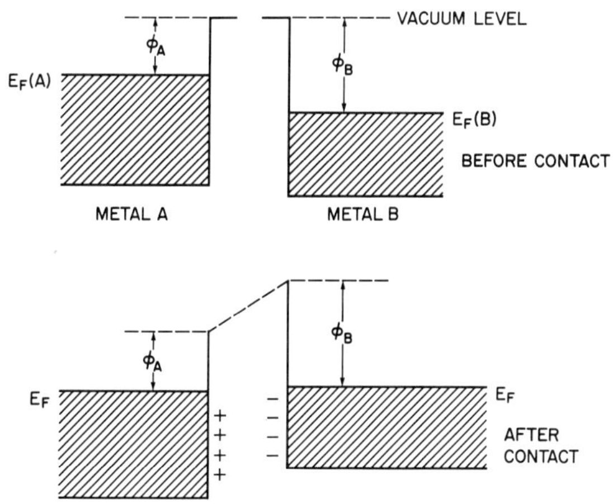

FIGURE 14. Metal-to-metal contact. [After C. R. Barrett, W. D. Nix, and A. S. Tetelman, *The Principles of Engineering Materials,* p. 376, Prentice-Hall, New Jersey (1973) (reproduced with permission).]

10. METAL–SEMICONDUCTOR CONTACT

The interface between metal and semiconductor has been studied intensively.(115) In this paper, we only intend to summarize those results related to interactions and adhesion. With regard to interactions, there have been numerous reports(115) about interdiffusion(116) at the interface within 1000 Å. Actually, this region should be properly called *interphase;* the actual interface(117) may be only around 10 Å (Figure 15). For example, at the transition metal–silicon interface,(118) there are the silicide formation and interfacial mixing at a considerable depth (about 100 Å or more).

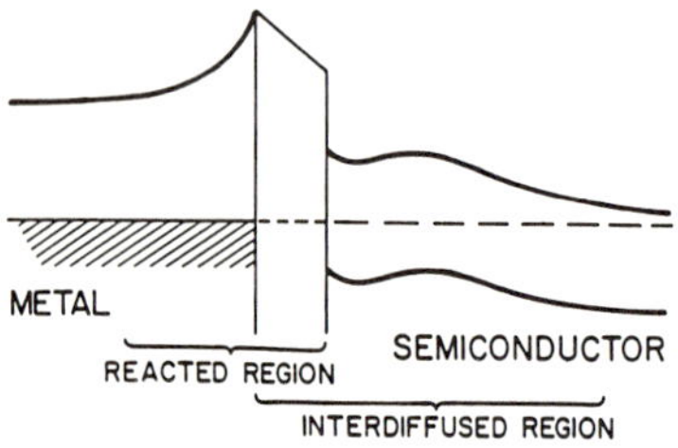

FIGURE 15. Schematic diagram of the extended metal–semiconductor interface. [After L. J. Brillson *et al.*, *J. Vac. Sci. Technol.* **20,** 661 (1981) (reproduced with permission).]

10.1. Ionicity of Semiconductors

In 1969, Kurtin, McGill, and Mead[119] revealed that the sensitivity of a semiconductor's Schottky barrier height to different metals increased with the ionicity of the semiconductor. Hence, an index of the interface behavior S is defined as

$$\phi_{SB} = S\,(\phi_m - \phi_{SC}) + K \tag{64}$$

where ϕ_{SB} is Schottky barrier height, ϕ_m and ϕ_{SC} are work functions of the metal and semiconductor, respectively, while K is a constant. Since ϕ_m is closely related to χ_m, a plot of ϕ_{SB} and χ_m gives a linear slope of S as shown in the inset in Figure 16. The master curve in Figure 16 is a well-known plot obtained by Kurtin *et al.*[119] on the relation between S and the ionicity of a semiconductor, or the electronegativity difference $\Delta\chi^P$ between anion and cation. The increase of S with $\Delta\chi$ has been interpreted as evidence for the increase in intrinsic surface states. There is a distinct difference between covalent and ionic semiconductors. For covalent semiconductors, Mead[120] concluded that the surface

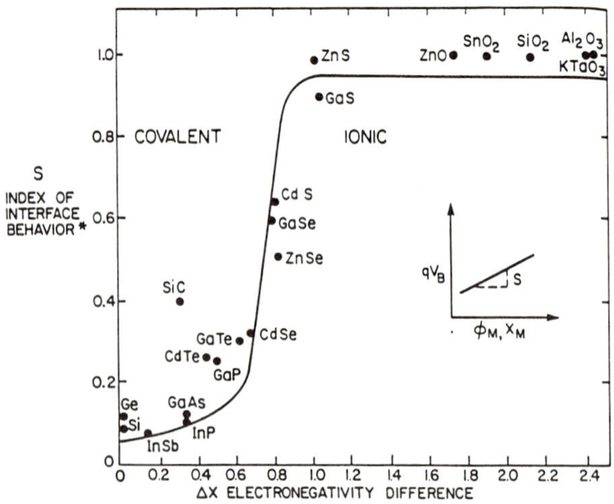

FIGURE 16. Index of interface behavior S versus electronegativity difference. [After S. Kurtin, T. C. McGill, and C. A. Mead, *Phys. Rev. Lett.* **22,** 1433 (1970) (reproduced with permission).]

lattice disruption is large, in contrast to ionic semiconductors. The latter group has lower surface states due to the more stable Coulombic crystal bonding and less disruption of lattice potential. Duke *et al.*[121] have also shown that the atoms on the surface of covalent semiconductors move more readily toward surface reconstruction than those of ionic semiconductors. This distinction can also affect surface charge density, which will be discussed in a later section.

10.2. Metal–Semiconductor Adhesion

Since most semiconductors are "soft" bases, they can readily react with metals, which are supposed to be "soft" acids. Kinetically, a "soft" acid–"soft" base reaction can take place at a good rate. When gold contacts silicon, adhesion between Au and Si[122] is achieved, and even after separation Au tends to adhere to the Si surface owing to the lower cohesive energy of Au. The bonding between Au and Si[116] is well studied, especially related to the diffusional intermixing at the interface. However, no well-defined compound has been found. In the case of Ag/Si, there has also been no reaction. The inertness of Au and Ag will be discussed later.

On the other hand, for d metals, such as Ni, Pd, and Pt,[116] silicide-like compounds have been detected at the interface. Presumably, the chemically driven process is caused by the energy gained through the hybridization of d electrons of the metal with sp electrons of Si.

In the case of the Au–Ge contact,[122] Au does not transfer to Ge because Ge is cohesively weaker than both Au and Si. As a result, Ge is transferred to the Au surface. Since the Au–Ge interfacial energy is stronger than the cohesive energy of Ge, the fracture at separation occurs in the bulk of Ge.

Au and Au alloys are known to interact readily with GaAs. However, recently the adhesion has been shown by Weizer *et al.*[123] to be also a dissociative diffusion process. The adhesion is greatly enhanced if the samples are annealed in a vacuum environment. At high temperatures, e.g., 850 °C in the presence of H_2, Yamada *et al.*[124] have shown that the single-crystal GaAs can adhere to itself. This may be due to the surface cleaning of the crystal by H_2 and the self-diffusion of GaAs.

Generally, a metal and a semiconductor can react chemically or interdiffuse[125] to form new interfacial phases. Both chemical reaction and interdiffusion should promote adhesion at the interface and also influence Schottky barrier height.

10.3. Effect of Chemical Reaction on Metal–Semiconductor Interface

The interactions between metals and semiconductors have been summarized by Brillson[126] to be related to the heat of reaction ΔH_R at the interface. In Table 12, the reactivities of metal–semiconductor pairs are listed according to their ΔH_R values. The critical heat of reaction, ΔH_R^C, is 0.5 eV/metal atom. Similar to the cases of metal–metal interactions studied by Keller,[97] the negative heat of reaction ($\Delta H_R < H_R^C$) is an important thermodynamic requirement for the interaction to take place at the interface.

The general trends for metal–semiconductor reactions can be illustrated with GaAs (Reference 691 in our Reference 115): (1) for unreactive noble metals, e.g., Au, Ag, and Cu, the predominant process is the outdiffusion of semiconductor; (2) for near-noble

TABLE 12
Reactivities and Heats of Reaction of Metal–Semiconductor Pairs[a]

Reactive ($\Delta H_R < \Delta H_R^C$)	Unreactive ($\Delta H_R > \Delta H_R^C$)
Al–$CdS_{(11\bar{2}0)}$	Au–$CdS_{(11\bar{2}0)}$
Al–$CdSe_{(11\bar{2}0)}$	Au–$CdS_{(11\bar{2}0)}$ (intrinsic)
Cu–$CdS_{(11\bar{1}0)}$	Au–$CdS_{(11\bar{1}0)}$
Cu–$CdSe_{(11\bar{2}0)}$	Au–$CdSe_{(11\bar{2}0)}$
Al–$GaAs_{(110)}$	Ag–$CdS_{(10\bar{1}0)}$
Cu–$GaAs_{(110)}$	Ag–$CdSe_{(11\bar{2}0)}$
	Ag–$ZnS_{(110)}$

[a] From L. J. Brillson, *J. Vac. Sci. Technol.* **15**(4), 1378(1978).

metals, e.g., Ni, Pd, Pt, Rh, Ir, and Os, stable compounds can be formed (for example, Ni and GaAs can yield NiAs and β-NiGa); (3) for early transition metals containing d electrons, e.g., Cr, Mo, W, V, Nb, Ta, Ti, Zr, and Hf, they are generally reactive; (4) for Group III metals, e.g., Al, Ga, and In, they can form compounds but are less reactive than transition metals.

One chemical trend[115] is clear. Proceeding from Group IV to Group III–V to Group II–VI to insulator–metal interfaces, the metal bonding to anions becomes increasingly significant. Consequently, the more covalent semiconductors, as in Group IV and Group III–V compounds, have a greater tendency to form metal–anion and metal–cation phases (especially for a relatively unreactive metal such as Au), while the more ionic II–VI compounds and insulators form predominantly metal–anion compounds and localized distribution of dissociated cations.

In terms of the acid–base interaction, metals behave predominantly as "soft acids" and occasionally as "soft bases" to covalent semiconductors. To ionic semiconductors, metals act as soft acids exclusively. In the former case, metals are amphoteric, while in the latter, metals are monopolar. We shall differentiate between covalent and ionic semiconductors in terms of surface density of states in the following section.

10.4. Effect of Contact Electrification on Metal–Semiconductor Adhesion

10.4.1. Surface Charge Density

Cowley and Sze[127] derived an expression for the calculation of surface charge density for the metal–semiconductor (n-type) contact. If it is assumed that the charge exchange occurs at a gap of ≦ 4 Å with the surfaces and at equilibrium, then the surface charge density σ can be expressed as

$$\sigma = eD_S \left[\frac{\phi_m - E_g^0 - \chi + \phi_0}{(1 + e^2 Z D_S/\epsilon)} \right] \tag{65}$$

where e is the electronic charge ($e = 1.6 \times 10^{-19}$ C), ϕ_m the work function of metal, χ the electronegativity of the semiconductor, E_g^0 the minimum band gap, and ϕ_o is the distance between the top edge of the valence band and the Fermi level which measures the energy range within which the surface states are filled (Figure 17). The work function of the semiconductor ϕ_{SC} is given by

$$\phi_{SC} = E_g^0 + \chi - \phi_0 \tag{66}$$

10.4.2. Large D_S for Covalent Semiconductors

Krupp[128] has analyzed the two conditions for both covalent and ionic semiconductors. The covalent semiconductors which generally have an index of interface behavior S less than 0.2 possess a very large density of surface states in the range of $D_S \geqq 10^{14}$ cm^{-2} (eV)$^{-1}$. For example, Si, Ge, GaAs, and GaP are covalent semiconductors. For these semiconductors, Eq. (65) can be simplified to

$$\begin{aligned} \sigma &\simeq \frac{\epsilon_0}{eZ}(\phi_m - \phi_{SC}) \\ &\simeq -\frac{\epsilon_0}{eZ}\Delta\phi \end{aligned} \tag{67}$$

where $\Delta\phi = -(\phi_m - \phi_{SC})$. Experimentally, ϕ_o is approximately $\frac{2}{3}E_g^0$, ϕ_m is between 3 and 4.7 eV, and ϕ_{SC} ranges from 4.0 (GaP) to 4.8 eV (CdS). The E_g^0 values are 2.2 eV (GaP) and 2.4 eV (CdS). Under these conditions, surface charge density $\sigma \leqq 10^{13}$ e/cm^2 at the metal–semiconductor contact. Typically, σ ought to be 10^{12} e/cm^2. Assuming a crystalline surface containing 10^{15} atoms/cm^2, there are approximately 0.1% surface atoms actually charged.

10.4.3. Small D_S for Ionic Semiconductors

In contrast, the ionic semiconductors with an interface index $S \geqq 0.8$, such as ZnS, SiO_2, and Al_2O_3, have comparatively small density of surface states [$D_S \leqq 10^{12}$ cm^{-2} (eV)$^{-1}$]. Then

$$\frac{e^2Z}{\epsilon_0}D_S << 1$$

For these semiconductors, Eq. (65) becomes

$$\begin{aligned} \sigma &\simeq eD_s(\phi_m - \phi_{sc}) \\ &\simeq -eD_s\Delta\phi \end{aligned} \tag{68}$$

Typically, σ ought to be about $\pm 10^{10}$ e/cm^2, and the surface charged atoms are only 0.001%. This is about two orders of magnitude smaller than that at the interface containing high density of states. In this case, the surface charge density is dependent on both the density of surface states and the difference in work functions. In summary, we compare the contact electrification of metal with both covalent and ionic semiconductors in Table 13.

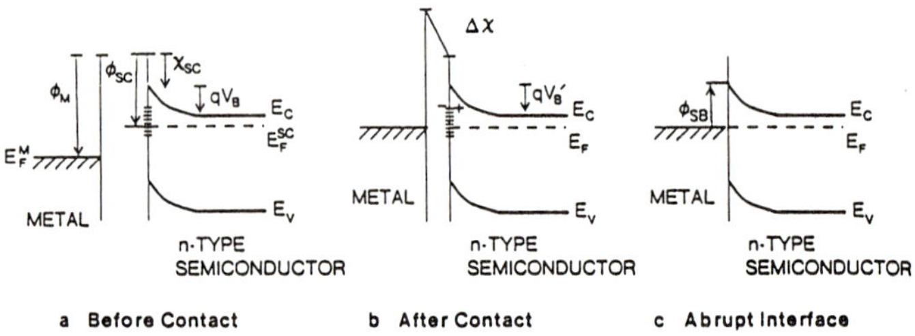

FIGURE 17. Scheme of metal–semiconductor (with high density of states) contact. [After L. J. Brillson, *Surf. Sci. Rep.* **2,** 123 (1982) (reproduced with permission).]

10.4.4. Lifshitz–van der Waals Forces versus Electrostatic Attraction

The above comparison between covalent and ionic semiconductors suggests that LW forces are more important for the adhesion at the metal–semiconductor interface than the electrostatic attraction. The actual contact charges determined by Dervos and Truscott[(129)] for the three metal–semiconductor pairs are:

Metal–semiconductor	10^{10} e/cm^2
Au–Si	1.5
Cu–Si	0.4
Al–Si	0.2

These results are in agreement with the above estimates. In other words, the surface charge densities at the metal–semiconductor interface are much less than those at the metal–metal interfaces. Thus, the electronic adhesion mechanism between metal and semiconductor cannot contribute much to the overall adhesion under those conditions.

TABLE 13
Contact Electrification at n-Type Semiconductor–Metal Interfaces[a]

	Covalent	Ionic
Semiconductors	Si, Ge, Sn, GaAs, GaP	ZnS, SiO_2, $SrTiO_3$, Al_2O_3, SnO_2, AlN
S, interface index (of ionicity)	<0.2	≧0.8
Density of surface states, D_S [cm^2 (eV)$^{-1}$]	≧10^{14}	≦10^{12}
Surface charge denisty, σ	≦10^{13}	≦10^{10}
Equation for σ	$-(\epsilon_0/eZ)\ \Delta\phi$	$-eD_S\ \Delta\phi$

[a] Adapted from H. Krupp, in *Static Electrification,* No. 11, p. 1, Institute of Physics, Bristol (1972).
[b] Z is gap distance, ϵ_0 permittivity of the free space, e electronic charge, and $\Delta\phi = -(\phi_m - \phi_{SC})$.

Since LW interaction plays a major role in the metal–semiconductor adhesion, accurate surface energy data of semiconductors are needed for the calculation of the work of adhesion. For example, the surface energies of Si, GaAs, and GaP measured by Messmer and Billeto[130] with a modified spark discharge method should be valuable for adhesion studies:

Semiconductor	Surface Energy (mJ/M_2)
Si (111)	1140
Si (110)	1900
GaAs (110)	860
GaP (110)	1900

Since these values were obtained by the cleavage of crystals with a minimal plastic deformation, presumably they should be close to intrinsic surface energies.

11. METAL–SALT AND METAL–OXIDE ADHESION

In ordinary situations, metals are in contact with other insulators such as salts and oxides. Most insulators should be considered as "hard" according to the chemical hardness scale, partly due to their ionicities and high energy gaps.[98,99]

In the studies of metallization of oxides for microelectronic thin films (<1 μm), Mattox[131] has reviewed factors affecting adhesion. One of the factors that appears to affect the adhesion between metals and oxides is the heat of formation ΔH_f of the oxide as shown in Table 14. Generally, the more reactive the metal with O_2, the more negative the ΔH_f value. From these data, we can see why Ag and Au are generally less reactive than other metals. For oxides and salts, reactive metals also favor adhesion.

When Weaver[132] studied the adhesion between metal thin films and alkali halides (e.g., NaCl, KCl, and KBr), he concluded that the shear forces are linearly proportional to the Lifshitz–van der Waals energies of metals.

TABLE 14
Heats of Formation of Metal Oxides

Metal oxide	ΔH_f (kcal/mol)
Ta_2O_3	−500
Al_2O_3	−399
V_2O_5	−290
Cr_2O_3	−270
TiO_2	−280
WO_3	−200
MoO_3	−180
Cu_2O	−40
Ag_2O	−7
Au_2O_3	+19

The metal–oxide thick film (>1 μm) adhesion can be affected by environmental factors as in the case of Al_2O_3.[133] The water at the interface can convert Al_2O_3 into a hydroxide $Al(OH)_3$, bayerite, which eventually weakens the bond by changing the morphologies and bond strengths.[134] In the case of TiO_2, though it is more stable than Al_2O_3, under severe environmental conditions, the oxide can undergo a polymorphic transformation leading to bond strength degradation.

Johnson and Pepper[135] have used the self-consistent-field X-alpha scattered-wave cluster molecular-orbital models to determine the interfacial strengths between metals and sapphire. Their results show that for transition metals chemical bonds are established between d electrons and the nonbonding 2p orbital electrons of the oxygen anions on the Al_2O_3 surface. This could be unique evidence for the acid–base interaction between metal and oxide. The decrease of contact shear strength through the series of Fe, Ni, Cu, and Ag has been explained by an increasing number of occupied metal–sapphire antibonding MO.

Kim *et al.*[136] have studied the adhesion of Ti thin films to several oxide substrates. Ti is deposited with an electron beam onto single-crystal sapphire (Al_2O_3), magnesia (MgO), quartz (SiO_2), and fused silica (SiO_2) substrates. ESCA spectra show that Ti reacts with Al_2O_3 on sapphire and SiO_2 on the fused silica through oxygen. The reactions have been confirmed with X-ray induced O (KVV) Auger spectra. However, they have been unable to correlate their results with adhesive strengths, e.g., peel strengths, because those strengths could be affected by other factors, such as deformation.

Varchenya *et al.*[137] have measured adhesive strengths of juvenile contacts between metallic films and SiO_2, alkali metal halides and thallium halides. Their calculations are based on the Griffith's criteria, and they intend to correlate adhesive strengths with various surface forces. As expected, the adhesive strengths that measure practical adhesion correlate better with the breaking stresses than with London's dispersion energies. Theoretically, adhesive strengths cannot be equated to the thermodynamic work of adhesion as discussed in an earlier section.

12. METAL–CERAMIC ADHESION

Since in recent years ceramics have been widely used for high-strength materials and superconductors, the adhesion between ceramics and metals has become a timely research topic. Metal–ceramic bonding can be achieved through Lifshitz–van der Waals interactions or molecular interactions.[138] Murr[139] determined interfacial energetics for metal–ceramic systems. For example, the work of adhesion between Ni(S) and ThO_2(S) is 1100 mJ/m^2 at 1200 °C, while that between Ni(l) and ThO_2(S) at 1455 °C is only 515 mJ/m^2; presumably, this decrease in surface energy is due to an increase in temperature.

Wettability of ceramics by metals has been studied for achieving better adhesion. Naka *et al.*[140] have found that Al definitely wets Si_3N_4, and Cu wets better than Ag. The wetting of Si_3N_4 ceramics may be related to the ease of nitride formation at the interface. In the case of ZrO_2, Nikolopoulos *et al.*[141] have reported that the work of adhesion follows a decreasing order of Ni > Co > Cu. For SiC, Shimbo *et al.*[142] have shown a similar trend for the work of adhesion: Al > Cu > Ag, and Al_4C_3 has been identified at the Al/SiC interface. In the case of Sn–Al alloys studied by Li *et al.*[143] the wetting of sapphire is favored by the high concentration of Al, thus W_A increases with the addition of Al to Sn.

Klomp and Vrugt[(144)] have applied the electromotive force measurements between metals and ceramics and concluded that the interactions can be both physical and physicochemical in nature. Strong adhesive bonding has been observed for metal to oxide and nonoxide ceramics. The metals involved are mostly transition metals containing d electrons. Buckley[(138)] has indicated that a correlation exists between the bond strength and the percent d-valence bond saturation for transition metals. The greater the degree of d-bond saturation, the less the adhesive bond strength. This finding reinforces the results obtained by Brewer[(108)] on the acid–base interaction of d metals. When metals are in contact with ferrites[(138)] (e.g., Ni-Zn and MnOZn ferrites), the adhesion can also be attributed to the presence of d-band electrons in metals.

When oxide ceramics[(138)] are in contact with mechanically softer materials, such as most metals, the distinctive difference in the elastic and plastic responses of the two materials can cause considerable deformation of the softer material. Consequently, adhesion at the junctions (or real contacts) is achieved. When metals contact oxide ceramics,[(138)] surface energetics play an important role. With metals that form stable oxides such as Cu, Ni, Rh, Co, and Be, adhesion occurs to the oxygen ions in the outermost atomic layer of sapphire.

For nonoxide ceramics, Buckley[(138)] has shown that the freshly formed SiC is highly reactive with metals. On the basis of the HSAB principle, SiC acts as a "soft" base and metals act as "soft" acids. Thus, a soft acid–soft base interaction takes place readily. In the case of Fe–SiC, the metal is transferred to the SiC surface after the sliding experiment. For ceramics like Si_3N_4, Morita *et al.*[(145)] have found that silicate and silica formed at the interface can enhance wettability and adhesion between Si_3N_4 and Al. Interdiffusion has been shown to play a role in adhesion between metals and high-strength ceramics, such as Si_3N_r and SiC. According to Iino *et al.*,[(146)] a new technique involves the use of an interdiffusing layer to achieve metal–ceramic bonding. For example, the Ni and Nb interdiffusing layer are used for these ceramics. The interdiffusing layer is composed of an interface metal (foil or thin film) and an adjacent metal (bulk, plate, sheet, or thin film). During the bonding process, the adjacent metal diffuses through the metal to the metal–ceramic interface and enhances the adhesion.

Diffusion has also been reported by Baumgartl *et al.*[(147)] to be an important process in achieving the adhesion between metal and enamel. The interface is approximately 5-μm thick and the diffusion process can be followed with X-ray and Auger microanalysis of the tapered section.

Two new methods have been mentioned regarding the enhancement of metal–ceramic adhesion. Iino *et al.*[(148)] have applied an electric discharge machined surface to join ceramic and metal. Iino and Elssner[(149)] first reported the addition of a small amount of Y (0.05 to 1 wt%) in Ni to increase the toughness of the adhesive bond for Al_2O_3/Ni. Under that condition, only the active metal near the interface (about 20 μm deep) appears to be affected by the improved bonding. Now that surface can be treated by electric discharge machined,[(149)] i.e., the electrode element (e.g., Y) transferred to the machined surface uniformly by several micrometers. The joint, known as YNi(Y), actually is tough and contains some oxygen in the form of Y_2O_3.

The application of laser treatment (LT) to enhance between Cu films and sapphire substrates through a sequential sputtering deposition process has recently been developed by Pedraza *et al.*[150] The enhancement is not attributed to the formation of an oxide at the interface but to the surface energy gradient resulting from the thermal gradient in the melt. The joint strength can be increased by two to three times with the use of LT, and the strong bond is maintained throughout successive melting. These enhancements are presumably due to a better contact of the liquid Cu with sapphire at high temperatures under the laser irradiation.

Recently, Miyoshi[151] has found that the metal–ceramic adhesion is predominantly affected by the hardness of the metal and less by the surface energy of the metal. However, his finding has not been further verified.

13. METAL–SUPERCONDUCTOR ADHESION

The discovery of high T_c superconductors by Bednorz and Müller[152] has opened up a new era for ceramic materials. The addition of yttrium to the oxide has been discovered by Chu *et al.*[153] and Wu *et al.*[154] to raise considerably the value of T_c to 90 K. Recently, bulk superconductors have been processed into $YBa_2Cu_3\ O_{7-x}$ films.[155–159] However, the adhesion between those superconductor films and metals has presented a problem.[159]

Recently Cheung and Ruckenstein[160] have observed the interactions between the Y–Ba–Cu superconductors and substrates. A reactivity scale has been established to be in the following decreasing order:

$$\mathrm{Nb} > \mathrm{Si} > \mathrm{ZrO_2} \geqq \mathrm{Al_2O_3} > \mathrm{SrTiO_3} > \mathrm{MgO} \geqq \mathrm{Cu} > \mathrm{Ag}$$

Moreover, the reactivity of the Y–Ba–Cu oxide is mostly controlled by Ba^{21} ions. The reactions of the oxide with metals vary. That with Nb degrades the sample, and no reaction takes place with Ag. The reaction with Cu results in the formation of CuO and Cu_2O accompanying the degradation of the 123 phase of the superconductor.

14. METAL–GLASS ADHESION

Similar to ceramics, glasses are rather ionic. Thus, their adhesion to metals is generally similar to that between ceramics and metals. Klomp[161] concluded that there appears to be no reactions at metal–glass interfaces. To achieve good LW bonding, selected metals can adhere to glasses and ceramics if the components are in intimate contact and heated to about $0.9T_m$ (T_m is the melting temperature). However, the bonding pressure is as much as 1.5 MN/m^2 for nonoxygen-active metals but up to 10 MN/m^2 for oxygen-active metals. This difference in pressure signifies the need for glass to achieve an intimate contact with the genuine metal surface. The bond strength generally depends on the strength of the metal, but can be as high as that of alumina. Furthermore, the bond strengths vary linearly with the melting temperatures of metals. Benjamin and Weaver[162] studied the adhesion of thin metal films to glasses. For nonoxidizable metals, such as Au, the adhesion is governed by LW interactions. On the other hand, films with many oxidizable metals show an increase in bond strength with time, and the final strength is generally one or two orders of magnitude greater than that of a gold film. Thus, oxygen is the major ingredient in achieving the increase in adhesive strength with time. It appears that the adhesion for oxygen-active metals and glass generally involves three steps: (1) LW interaction, (2) oxide formation, and (3) chemical bonding.

The role of oxide has been further shown by Weaver[(132)] to be important for the metal-to-glass adhesion. Those metals, e.g., Au and Ag, which cannot form oxides readily, generally adhere poorly to glass. On the other hand, those metals, e.g., Mg, Zr, Mo, Ti, Fe, Cr, Ni, and Cu, which have strong affinity for oxygen, adhere very well to glass. The first three metals form nonremovable bonds immediately after the deposition. Films of other metals, except Ni, produce nonremovable bonds after a short period of contact. It is important to note that all these metals contain d electrons. Other metals, e.g., Zn, Cd, Sn, and Pb, have good affinity for oxygen but are poor in film-forming due to aggregation of metal. Thus, the adhesion to glass is also poor. Al and Mn have other problems. Al forms oxides so readily that the oxides trap gases and the adhesion cannot increase with time. In the case of Mn, its film undergoes a phase transformation and the resultant film is weak and easy to crack.

An interesting kinetic study of adhesion of glass to a solid has been carried out by Hrma.[(163)] He considered three different single, rate-controlling processes: glass flow, interfacial reaction, and diffusion into the glass. Pask[(164)] carefully studied the interfaces and bonding between glass and metal. He theorized the existence of an oxide layer —M—M—M—O—Si—O—M—O—Si—O—Na— to be the key to the metal-glass bonding. He called this oxide layer "-ous oxide" layer that could be as thin as one monomolecular layer. Both metal and glass have to be in equilibrium with the oxide layer before bonding can be successfully achieved. However, this work, which was done before modern analytical techniques were available, has not been verified with new surface tools.

15. METAL–DIAMOND ADHESION

Diamond is the mechanically hardest material. It is extremely wear-resistant and has a low friction coefficient. The crystallographic network of diamond[(165)] consists exclusively of covalently bonded, aliphatic sp^3 hybridized atoms spaced tetrahedrally by a uniform distance of 1.545 Å between atoms. The tetrahedrons are connected to one another at their tips to establish a unique crystal lattice. This structure accounts for many of its excellent mechanical and physical properties.

One of the future technologies is to coat diamond[(166)] on metals, glass, or semiconductors by vapor deposition to form wear-resistant and corrosion-resistant films. Thus, the adhesion of diamond to other materials, such as metals, deserves special attention.

Adhesion of Ti to diamond under high vacuum has been observed by Buckley.[(122)] After separation, the softer material, Ti, is transferred to the diamond surface. Interestingly, the adhesion of Cu to diamond has been shown by Pepper[(166)] to be affected by annealing. Without annealing, there is no adhesion. After vacuum annealing at 900 °C, some occupied and unoccupied surface states appear with the presence of some electrical conductivity, and adhesion between diamond and metal results. This adhesion is manifested by the increase in the interfacial friction, which can be reflected in the increase in adhesive strength. With the increase in the annealing temperature, the adhesive strength increases accordingly. The adhesion has been explained in terms of the minimum energy gap and surface states.

Pepper[(167)] has also studied the transformation of the diamond (110) face, and his results show that the annealing operation at 900 °C presumably removes the adsorbed hydrogen on the surface. Thus, the electrical conductivity increases somewhat after the annealing. The role of hydrogen is further established by the re-adsorption. After re-adsorption, the conductivity indeed diminishes.

16. METAL–POLYMER ADHESION

Metals and polymers require strong bonding under two entirely different applications: (1) to form structural components and (2) to metallize polymeric substrates. For structural components, polymers serve as adhesives or binders for metals adherends. For metallization, a metal film is either deposited from a solution or in vacuum on a polymer substrate. The adhesive bonding of structural components has been discussed elsewhere.[(168)] In this section, we shall devote our discussions chiefly to the metal–polymer adhesion under the metallization situation to form electronic components, e.g., packaging materials.[(169)]

16.1. Mechanisms of Metal–Polymer Adhesion

Mittal[(170)] reviewed various mechanisms on the adhesion aspects of metallization:

- mechanical or interlocking,
- weak boundary layer,
- chemical,
- electrostatic.

In this section, we shall briefly demonstrate the more relevant mechanisms to be:

- physisorption based on Lifshitz–van der Waals forces,
- molecular interactions, e.g., acid–base (or donor–acceptor) interactions,
- chemical interactions, and, to a lesser extent,
- electrostatic attraction.

Most of these mechanisms have been discussed in several earlier sections of this chapter.

16.2. Classification of Polymers According to their Wettabilities

In comparison with metals and metal oxides, all polymers are low surface energy materials, and the surface energies range between 10 and 50 mJ/m^2. For the convenience of this discussion, we arbitrarily divide polymers according to their wettabilities, which are indirectly determined as Zisman's critical surface tensions, γ_c.[(171)] For example, polymers[(172,173)] may be arranged into three classes:

1. Low wettability (30 mJ/m^2 > γ_c > 10 mJ/m^2):
   Most fluoropolymers, polysiloxanes and polyolefins, etc.

2. Medium wettability (40 mJ/m^2 > γ_c > 30 mJ/m^2):
   Most vinylpolymers, e.g., poly(vinyl acetate), poly(styrene), poly(vinyl chloride), poly(ethyl acrylate), poly(methyl methacrylate), etc.
3. High wettability (γ_c > 40 mJ/m^2):
   Most condensation polymers, e.g., poly(carbonate), polyesters, nylons, epoxy resins, etc.

We can now separately discuss metal–polymer adhesion according to the above three classes. Rubbers are also of either low or medium wettability,(174) and we shall discuss rubber–metal adhesion separately in the last section partly because the applications of rubbers are not for metallization under vacuum conditions.

16.3. Metal Adhesion to Low Wettability Polymers

16.3.1. Fluoropolymers

16.3.1.1. Thick Films. The adhesion between metals and fluoropolymers as thick films is generally poor. Thus, several conventional surface modification techniques(175) have been used:

1. Etching by sodium naphthalene:
   For polytetrafluoroethylene PTFE (γ_c = 18–18.5 mJ/m^2) only the surface region is attacked by sodium naphthalene, as indicated by the color change. For the treated fluorinated ethylene–propylene copolymer FEP, Dwight and Riggs(176) have shown with ESCA the presence of unsaturation, >C=O and —COOH groups. Andrews and Kinloch(177) reported the increase in wettability by etching. For polychlorotrifluoroethylene PCTFE (γ_c = 31 mJ/m^2) and FEP,(178) the etching resulted in an increase in the butt joint strength.
2. Mechanical and morphological treatments:
   The surfaces of fluoropolymers can be abraded(179,180) to increase mechanical interlocking. After PTFE is abraded, it can be bonded to itself and to Al. The improved adhesive strength is attributed to surface de-contamination and mechanical activation through some mechanochemical process which, in turn, can generate free radicals on the PTFE surface.

An early work by Schonhorn and Ryan(181) has claimed that the FEP melt against high energy surfaces, e.g., Au or Al, is nucleated to form a transcrystalline interface of higher γ_c ($\simeq$40.4 mJ/m^2) than that of regular PTFE ($\simeq$18.5 mJ/m^2). This change in wettability has been postulated to arise from an increase in the surface density of PTFE. The polymer surface is no longer representative, with respect to density, of an amorphous state in the bulk, which yields a lower surface energy than that of a crystalline phase.

16.3.1.2. Thin Films. Chemical modification of the FEP surface by several metals during the vacuum deposition presents some new possibilities in improving the adhesion between fluoropolymers and metals. Roberts *et al.*(182) have found that by depositing first Al and then Au, the adhesion between the fluoropolymer and Au, after Al is washed off, can be improved substantially. Similar improvements(183) in adhesive strength can be realized by depositing Al and Ti, respectively, on Teflon FEP. However, in those cases, the F atoms are presumably extracted from the polymer and a Ti—F or Al—F bond is formed, instead. Later, Vogel and Schonhorn[184] proposed a scheme on the basis of their work with ESCA to account for the surface crosslinking of PTFE:

$$\begin{matrix}\text{Fluorocarbon}\\ \text{polymers}\end{matrix} \xrightarrow[\text{metals}]{\text{evaporated}} \begin{matrix}\text{Metal}\\ \text{fluoride}\\ +\\ \text{unsaturated}\\ \text{polymer}\end{matrix} \left(\begin{matrix}\text{organometallic}\\ \text{complex}\end{matrix}\right) \rightarrow \text{Crosslinking}$$

Vilenskii *et al.*[185] have applied glow discharge to improve the adhesion between PTFE and metals. The formation of $RO_2\cdot$ has been shown to lead to a better adhesion of PTFE to Al. Michael and Stulik[186] have reported unique core formation after the PTFE surface is bombarded with KeV Xe atoms. The modification of surface topography of PTFE has also been observed by Chang *et al.*[187,188] during the presputtering. The presputtered surface demonstrates strong adhesion to Cu partly due to chemical interaction detected separately with ESCA. The importance of both physical and chemical factors has been established by Haag and Suhr[189] through their experiments with the pretreatment of PTFE by reactive gases, such as O_2 and CF_4/O_2. The deposition of Pd, Pt, Au, and Cu layers by the PECVD method results in a strong Cu–PTFE bond after the pretreatment.

The effect of irradiation of PTFE film on adhesion has been studied. For electron irradiation, Yamada *et al.*[190] have shown that the adhesive strength between the irradiated PTFE and metal films increases, and the increase is greater for the Ni film than the Au film. Wheeler and Pepper[191] have reported the effect of X-ray flux on the surface of PTFE. The heavily crosslinked or branched structure has been identified after the irradiation. The enhancement of adhesion between PTFE and Au has been achieved by Werner *et al.*[192] and Jacobson *et al.*[193] with the use of high-energy ion irradiation.

The chemical interaction between PTFE and metal surfaces has been further confirmed by Buckley[194] with the use of a field ion micrograph (FIM). For example, the reaction between PTFE and the tungsten-tip has been postulated to be through the bonding of CF_2 to the W surface. Other polymers, such as polyimide, have the same tendency to react with the metal surface. However, FIM shows no reaction between PTFE and In.

Indeed, the type of metal does affect the adhesive strength between metal and PTFE. Vogel and Schonhorn[184] have found that the tensile shear strengths of the evaporated metal films follow a decreasing order:

$$\text{Ti} > \text{Fe} > \text{Ni} > \text{Al} > \text{Au} \gg \text{Cu}$$

This result appears to correlate with free energies ΔG_f and heats of formation ΔH_f of fluorine compounds[195] (as shown in Table 15). The more negative the ΔH_f (or ΔG_f), the more rective the metal.

16.3.2. Polyolefins

16.3.2.1. Thick Films. The adhesion between metals and polyolefins is generally poor because of low wettability and the lack of reactive group or oxygen on the surface to achieve acid–base interaction[196,197] or chemical bonding. For thick films, several common methods have been used to enhance adhesion.[175]

TABLE 15
Free Energies and Heats of Formation of Metallic Fluorides[a]

Compound	Free energy of formation, ΔG_f (kcal/mol)	Heat of formation, ΔH_f (kcal/mol)
TiF_4	−403.7	−394.2
FeF_3	−256.0	−249.0
NiF_2	−160.9	−155.7
AlF_3	−360.7	−356.0
AuF_3	(unstable)	(unstable)
CuF	N/A	+44.0
CuF_2	N/A	−160.0

[a] From V. A. Karagounis, unpublished results.

Mechanical factors(198,199) have been shown to be significant for the adhesion of low-density polyethylene PE (γ_c = 31 mJ/m^2) to Al. The polymer enters the porous anionic film and creates a mechanical keying action. For electroplating,(200) the polymer (e.g. polypropylene, PP, γ_c = 29 mJ/m^2) surface is usually etched for metal (e.g., Cu) coating. However, mechanical interlocking is not enough for achieving optimum adhesion; an additional chemical reaction(201) between stannous ions and Cu to generate polar groups on the PP surface is needed to produce a better result.

Two commonly used methods for improving the adhesion between metals and polyolefins are corona discharge treatment (CDT)(202) and flame treatment (FT).(203) Though CDT is an old technique, the mechanisms have only been elucidated in the last several years. Briggs and Kendall(204) have identified the enolic–OH group on the surface of corona-treated LDPE. Carley and Kitze(205) have found rather stable peroxides of the forms of RO_2R and RO_3R on the CDT–PE surfaces. Gerenser *et al.*(206) have identified more species with ESCA on the CDT–PE surfaces. Functional groups, such as hydroxy, epoxy, hydroperoxy, carboxylic acid, and carbonyl, are all present as revealed by the derivatization technique. For the sample treated with CDT of an energy greater than 4000 J/m^2, water washing can only remove 10% of the incorporated oxygen. Moreover, homogeneous oxidation reaches approximately 50 Å below the polymer surface.

The flame treatment (FT) of PE has been studied by Briggs *et al.*(203) with ESCA. The FT method can oxidize PE surface to a depth of 40 to 90 Å comparable to the CDT–PE surface. A normal FT treatment of LDPE is equivalent to a level of the wet treatment by chromic acid. The oxidized surface can remain unchanged for as long as 12 months. The oxidized products on the surface are —OH, —C═O and —COOH, etc. Garbassi *et al.*(207) have studied the FT-polypropylene with XPS. According to them the first step is to produce hydroperoxide and later to generate other groups, e.g., —OH, —COOH, —C═O, —C—O—C, RO_2, etc. Their results indicate that the surface polarity is relatively more sensitive to the presence of the —COOH group.

For thick films, the presence of an acidic functional group on the polyolefin surface is known to enhance adhesion, presumably through the acid–base interactions. Lewis and Forrestal(208) reported an example of grafting of bis(2-chloroethyl)vinyl phosphonate to

PP to enhance adhesion to metals, e.g., Al, steel, and Cu. The chromic acid treatment of PE surface has been studied by Briggs[209] with the derivatization techniques. A Cr complex has been postulated to lead to the appearance of the carboxylic group. The $KMnO_4/H_2SO_4$ treatment of PE has been shown by Golander[210] to enhance adhesion. Both sulfonate and vinylidene groups have been identified with ESCA.

Schultz *et al.*[211] have shown that grafting of a small amount of acrylic acid (about 1% by wt) to PE can increase its adhesion to Al by about 10 times. The oriented carboxylic group at the interface tends to form a chemical linkage with Al_2O_3 on the Al surface.

16.3.2.2. Thin Films. The principles involved in the adhesion enhancement of thick-film polyolefins can be readily extended to that of thin films. For example, the adhesion of Al/PE has been studied in parallel to that of Al/polyacrylic acid (PAA).[212] Reactions have been postulated by DeKoven *et al.*[212] to occur between both C—O and C—C linkages and Al oxide to form carbide-like species.

Surface treatments of thin-film polyolefins with ion beam, electron beam, and plasma bombardment can enhance adhesion to metals. Bodö and Sundgren[213,214] applied ion bombardment to promote adhesion between Ti and PE. Ion bombardment appears to remove impurities on the PE surface and form the Ti–C interfacial layer. Without bombardment various metals[215] yield different adhesive strengths: high for Ti, Ni, and Cr but low for Al, Cu, Ag, and Au films. Similar to PTFE, the adhesive strengths,[184] in terms of tensile strengths, vary with metals according to the following decreasing order:

$$\text{Fe} \simeq \text{Ni} > \text{Ti} \gg \text{Al} \gg \text{Au} > \text{Cu}$$

When the HDPE films are *in situ* Ar-ion bombarded, the adhesion to all metal films, except Ag and Au, increases more rapidly than the cohesive strength of HDPE. The ion bombardment presumably creates an activated and crosslinked PE surface as revealed by XPS.

The surface modification with plasma can also be applied to PE for enhancing its adhesion to metals. The adhesive strength of the vacuum-deposited Ag on PE has been shown by Gerenser[216] to increase in the following order: untreated < Ar-plasma < O_2-plasma < N_2-plasma treated PE. Chemical interactions have been manifested by identifying the Ag—O—C and Ag—N—C species with XPS. These species presumably act as nucleation and chemical reaction sites. The focus of failure upon peeling changes from interfacial adhesive failure for the untreated and the Ar-plasma treated PE to cohesive failure for the O_2-plasma and N_2-plasma treated PE.

An interesting approach by Bekiarov *et al.*[217] is to evaporate a PE thin film on a metal substrate. In this manner, wetting is preferred because of depositing a low wettability polymer on a high surface energy metal. Moreover, the application of UV irradiation during the deposition tends to maintain a high adhesive strength of the deposited film.

16.3.3. Polysiloxanes

Poly(dimethyl siloxane), PDMS (γ_c = 24 mJ/m^2), is another low wettability polymer. Bodö and Sundgren[218] have studied the effect of surface modification of PDMS on its adhesion to Ti. Two pretreatment methods have been used: (1) Ar^+ bombardment and

(2) O_2-plasma treatment. Both methods lead to crosslinking of the surface and an increase in the polymer strength. The XPS results indicate the presence of Ti—C and Ti—O bonds, respectively. The ion bombardment of the polymer film also tends to affect the crystal growth of Ti from a three-dimensional to a two-dimensional mode.

16.4. Metal Adhesion to Medium Wettability Polymers

Polymers with a critical surface tension between 30 and 40 mJ/m^2 may be considered to be of medium wettability. In this group, there are mostly vinyl polymers[172] as shown in Table 16.

We shall discuss their adhesion to metals in two categories as in the above section: thick films and thin films.

16.4.1. Thick Films

One of the metal-electroplated plastics is the ABS (acrylonitrile–butadiene–styrene) copolymer.[219] Morphologically, the copolymer is a two-phase material. During the solution etching of the copolymer, the grafted rubber particles are dissolved out to leave spherical holes behind. The activation and chemical deposition of the metal takes place predominantly inside the wall of those holes. The filling of those holes with metal is the major step of electroplating. Thus, the mechanism involved appears to be mechanical interlocking. However, a chemical activation step is necessary to catalyze the deposition of Ag^+, Pd^{2+}, or Sn^{2+} on the inner wall surface. The final step is the electrodeposition of Cu or Ni metal. Other polymers, e.g., modified polypropylene (PP) and modified poly(phenylene oxide) (PPO), have also been electroplated. The electroplating of PP has been briefly discussed in the above section.

16.4.2. Thin Films

The ABS polymer can also be electroplated without electroless plating. A Cu thin film has been reported by Tsukada and Hosokawa[220] to deposit on the ABS polymer by

TABLE 16
Medium Wettability Polymers[a]

Polymer	Abbreviation	γ_c (mJ/m^2)
Poly(vinyl methyl ether)	PVME	29–30
Poly(propylene oxide)	PPO	32
Poly(ethyl methacrylate)	PEMA	33
Poly(ethyl acrylate)	PEA	35
Poly(styrene)	PS	33–36
Poly(vinyl acetate)	PVAc	37
Poly(vinyl alcohol)	PVA	37
Poly(methyl methacrylate)	PMMA	39
Poly(vinyl chloride)	PVC	39
Poly(vinylidene chloride)	$PVCl_2$	40

[a] From L. H. Lee, *J. Appl. Polym. Sci.* **12**, 719(1968).

an internal-magnet magnetron sputtering system. Oxygen-plasma treatment of the polymer surface improves the adhesion, presumably owing to the change of topography. However, ESCA has also detected some chemical bonding at the interface.

Burkstrand[221–223] has established that the adhesion between styrene and metals can be achieved by oxidation of the polymer. His results indicate that a metal complex of the M—O—C type is responsible for the increase in adhesive strengths to Cu,[221], Ni, or Cr.[222,223] This is one of the early works with XPS (or ESCA) confirming chemical interactions to be important to adhesion at the metal–polymer interface.

Later, Burkstrand[224,225] concluded that if a polymer already contains an oxygen it is not necessary to oxidize the polymer and form the complex. For example, in the case of polyvinyl alcohol (PVA), a similar complex to that in the oxidized polystyrene has been identified at the interface. His work has been extended to include other oxygen-containing polymers,[226] e.g., poly(ethylene oxide) (PEO), poly(vinyl methyl ether) (PVME), poly(vinyl acetate) (PVAc), and poly(methyl methacrylate) (PMMA). In all cases, a metal-chelate type complex is essential in achieving higher adhesive strength than that of the unoxidized polystyrene. Among metals, Au film, which never forms a complex at the interface, has been confirmed[227] to produce no adhesion to these polymers.

Burkstrand[228,229] has further established that the interfacial species are different when Ar^+ sputter-deposited "hot" atoms hit the polymer surface. The XPS data suggest that the new complexes on the PVA and PVME surface are bound more strongly to the polymer. However, the results for the PMMA and PVAc surfaces are more difficult to interpret.

The complexity of a sputtered surface has been clarified by Grant *et al.*[230] with both Ar^+ and O_2. They have also examined with XPS and IRRAS (infrared reflection absorption spectroscopy) two other polymers besides polystyrene, i.e., polypropylene and poly(ethylene terephthalate) (PET). Oxygen or argon sputter-etching of these polymers causes surface oxidation and crosslinking. For polystyrene and polypropylene, the predominant new functionalities generated are C═O and C—O groups. For PET, the major result is crosslinking. Even in the presence of O_2 plasma, there is only a trace of oxidation. Finally, sputtering with "hot" atoms appears to cause some chemical interactions at the interface between polystyrene and Ni as examined by Buckstrand.[228,229]

16.5. *Metal Adhesion to High Wettability Polymers*

Most condensation polymers and some vinyl polymers with a critical surface tension >40 mJ/m^2 can be considered to be of high wettability. These polymers are arranged in Table 17.[172]

In general, the adhesion of these polymers to most metals is good, though their surface energies are still much lower than those of oxides and metals. The wettability is still a problem if a metal is deposited on a solid polymer. On the other hand, if a molten polymer is coated on a solid metal, a spontaneous wetting should result. In the following, we shall discuss their adhesion to metals under both thick-film and thin-film conditions.

16.5.1. Thick Films

In this section, we briefly discuss the adhesion of metals to thick films of PET and PAN. As in the case of surface modification of polyolefins, corona discharge treatment

TABLE 17
High Wettability Polymers[a]

Polymer	Abbreviation	γ_c (mJ/m^2)
Poly(carbonate)	PC	42
Poly(6-aminocaproic acid)	Nylon 6	42
Poly(ethylene terephthalate)	PET	43
Poly(acrylonitrile)	PAN	44
Poly(hexamethylene adipamide)	Nylon 6, 6	46
Epoxy resins	Epoxide	43

[a] From L. H. Lee, *J. Appl. Polym. Sci.* **12**, 719(1968).

(CDT) has been a common practice for improving the adhesion of PET. Several studies in recent years have helped elucidate the mechanisms of adhesion enhancement. Ledercq *et al.*(231) have observed the effect of corona treatment on chemical and physical changes on the PET surface. The chemical change is determined as the increase in polarity by the adsorption of radioactive calcium ions and by the contact angle on the surface. An increase of functional groups has been estimated to be approximately 2×10^{13} sites/cm^2. Briggs *et al.*(232) have identified with XPS the reactive groups of —OH and —COOH on the PET surface. Pochan *et al.*(233) have demonstrated that about 4% of O_2 reacts on the PET surface with dry-air corona and 75% of the oxidation products are hydroperoxide, epoxy, hydroxyl, carboxylic, and isolated carbonyl species. A physical change in morphology has also been detected by Cross *et al.*(234) by sputter etching of the PET surface. The above results are similar to those discussed in one of the previous sections on CDT–PE. Both chemical and physical factors serve to enhance the adhesion of metals to CDT–PET.

Besides surface treatment, the heat setting temperature of the PET film has been found by Puydt *et al.*(235) to affect the adhesion between PET and Al. The heat setting conditions appear to have affected the grain size of Al. Large grains without orientation on the Al layer give poor adhesion.

For the adhesion between Ni and polyacrylonitrile, Boiziau *et al.*(236) concluded that the monomer, acrylonitrile, VCN, can graft onto Ni to form PAN. This graft polymerization assumes a form of the acid–base interaction between the monomer and metal surface. Monomer is presumably adsorbed on the metal surface by anchoring both the double bond and the nucleophile (C≡N) and then the polymerization takes place by stabilizing interfacial bonds.

Polyimides are versatile high-termperature polymers which have been used for both thick-film and thin-film applications. For thick-film coating, the internal reflectance IR spectroscopy has been used by Leidheiser and Deck(237) to study the interaction between Co and PI. The difference spectrum obtained from IR indicates that a chemical reaction has taken place at the interface. The emission Mössbauer technique has been applied by Vértes *et al.*(238) to determine the type of interaction occurring at the interface. The results are in accord with the IR findings, and PI has been found to bond to the cobalt substrate by means of carboxylate groups with Co^{2+} ions in the oxide film at the interface.

16.5.2. Thin Films

Yamada *et al.*[190] have indicated that adhesive strengths between Ni and polymers increase with the increase in critical surface tensions of polymers. Thus, the polymers, e.g., Nylon 6,6, discussed in this section are generally better in adhesion to Au and Ni than low wettability polymers, e.g., PTFE and UHMWPE (ultrahigh molecular weight PE) discussed previously.

In the following section, we shall discuss chiefly the adhesion of metals to polyimide (PI) thin films. Chou and Tang[239] have employed XPS to study the interactions between metals and PI during the metallization in UHV. The results indicate that Cr and Ni react with the pendant oxygen on the PI surface but Cu and Ag do not. A thermodynamic model has been proposed to predict the interaction based on partial molar energy. However, that model[240] alone is insufficient to explain the experimental results without a detailed quantum-mechanical treatment.

The same thermodynamic model has further been challenged by Anderson *et al.*[241] as too simple to account for actual interactions at the metal–PI interface. The Co–PI interface has been studied by them with high-resolution XPS between 20 and 400 °C. Below 5 Å, a vigorous but selective reaction takes place. Then a Co overlayer containing polymer fragment is formed. The stepwise heating of PI with a 40-Å Co overlayer enhances the reaction but somehow degrades the product.

Among metal–PI reactions, the interaction involving Cr[242–245] is the most well studied. Chromium is one of the more reactive transition metals, and the chemical interaction has been identified by Ho *et al.*[82,245] to be the formation of a charge-transfer complex between Cr and not the carbonyl group but the π system of the PMDA (pyromellytic dianhydride–oxydianiline) unit of PI. Recently Lee[98,99] has postulated that interaction to be an acid–base interaction.

The reactions between PI and several reactive metals besides Cr have been studied. For Al, Bartha *et al.*[246] have applied PES to identify the formation of an intermixed layer resulting from the reaction between Al and the carbonyl group. By using XPS, Atanasoska *et al.*[247] have shown that Al indeed reacts at the carbonyl sites. The adatoms of Al first transfer charges to carbonyl carbon atoms via oxygen. This resonance hybrid of C—O—Al weakens, but does not break, the carbonyl bonds. With increasing Al coverage, finally stronger Al—O bonds result by forming a complex.

Cerium has been found by Clabes[248] to be too reactive to form a selective Ce–PI complex. The unreacted metal is then nucleated at surface defects to grow into huge planar islands. Titanium is another reactive metal which has been studied by Ohuchi and Freilich[249] with XPS and PES. They have found that at low coverages the PI surface reacts strongly in the form of charge transfer through carbonyl groups. At high coverages, a Ti—C bond formation has been identified. This Ti—C bond has also been observed by Bödo *et al.*[250] in their study related to the ion bombardment of the PI surface for the adhesion to Ti. The unreacted metal grows three dimensionally into islands. With the Ar^+ bombardment on the PI surface prior to deposition, a preferential bond breakage of the imide group and the formation of graphite-like carbons result. Since the number of the nucleation site is so large, the Ti film growth is comparable to a calculated layer-by-layer type growth.

The reactivity of germanium in the bond formation with PI has been reported by

Atanasoska *et al.*[251] Ge reacts first with the —C=O groups of PI below 0.5 Å, then yields an imide–benzene ring complex after increased deposition. These rings are vulnerable to Ge attack and consequently chemical bonds of Ge—C type are formed. The later stage of growth of Ge is also layer-by-layer like.

Following the general trend in metal reactivities, Cu[252] and Au[253] are also unreactive with PI. However, as a component part for microelectronics, Cu is more attractive partially due to its lower cost than most metals. Thus, there has been a great deal of work toward achieving the adhesion between Cu and PI.

Since Cu does not react with PI, recently Cu has been found by Faupel *et al.*[77] to diffuse into PI presumably producing no enhancement in adhesion. However, there have been several attempts to attain the adhesion between Cu and PI. One of the methods involves the oxygen-reactive ion-beam bombardment of the polymer surface.[254] The adhesive strength can be increased by a factor of 25. A second method is to apply a thin Cr interlayer to adhere Cu and PI together.[255]

The above difficult bonding situation can be completely reversed, if PI is deposited on Cu from solution. First of all, wetting is favored by placing a lower surface tension liquid on a high surface energy solid.[256] Chambers *et al.*[257] have shown that an oxide layer of CuO is formed at the interface about one hour after the coating. CuO can enhance the acid–base interaction by raising the acid strength of Cu. Recently, Cain *et al.*[258] have applied the tight binding calculations of the extended Hückel type to establish relative acid strengths of Cu compounds in a decreasing order:

$$CuF_2 > CuO > CuF \simeq Cu_2O > Cu$$

Thus, the oxidation of Cu is necessary to increase its acid strength resulting in a higher adhesive strength between PI/Cu(O). For the same reason, the argon ion bombardment of the Cu surface has been shown by Paik *et al.*[259] to produce an oxide (perhaps Cu_2O) at the interface. The oxygen-rich region grows beyond 50 Å before the PI deposition and approximately 2000 Å afterward.

The interfacial reaction between poly(amide-acid), PI precursor, and Cu has been studied by Burrell *et al.*[260] The intermixing between Cu and the polymer on the Cu surface has been detected to be approximately 2000 Å. A copper carboxylate is formed at the acid site in the precursor. Salaneck *et al.*[261] have carried out a model compound study on the basis of phthalimide on Cu. According to their results, the model compound lies flat on the polycrystalline Cu surface. Charge transfer from Cu to the adsorbate is also observed.

Finally, how about the Au–PI adhesion? Meyer *et al.*[253] have detected no Au–PI interaction at less than 1 Å. Above 1 Å, the growth of Au cluster occurs instead. Since Au is the most electronegative metal, Au starts to inject charges into PI. This may suggest some role of contact electrification for thin-film adhesion.

At this point we should summarize the discussion on the metal–polyimide interactions. First, the reactivities of metals with PI appear to follow a decreasing order:

$$Ce > Ti > Cr > Co > Ni \gg Cu > Au$$

The adhesive strengths determined by Kim *et al.*[262] in the form of a 90° peel test show that the adhesive strengths of PI to metals are generally very high in comparison with those of metals to PI. Among Cr, Ni, Cu, and Au, Cr has the highest adhesive strength in both cases. Near the interface of PI/Cu (or Ni), a respective oxide is identified: Cu_2O for PI/Cu and NiO for PI/Ni. The formation of the oxide is presumably derived from the complex of the polyamic acid and the respective metal. In terms of the acid–base interaction concept, an oxide is more acidic (or harder) than the bulk metal, and it can favor the AB interaction. In terms of surface energetics, an oxide has a lower surface energy than the respective metal, and it is more stable than the corresponding metal.

16.6. Metal–Rubber Adhesion

Like vinyl polymers, most rubbers are of low to medium wettability (Table 18),[(263)] thus they are difficult to adhere to metals. To achieve metal–rubber adhesion, the rubbers have to be modified through chemical reactions with halogens, or cyclization, or copolymerization with polar monomers, e.g., those containing cyano- or carboxylic acid groups. For example, after poly(1,3-butadiene) is transformed into poly(2-chloro-1,3-butadiene), the wettability increases from 31 to 38 mJ/m^2.

16.6.1. Natural Rubber–Steel Adhesion

During the last forty-five years, the adhesion between natural rubber (NR) and brass-plated steel for the steel cord tire application has been well studied. Without[(264,265)] the plated brass, there is no adhesion between NR and steel (Table 19). Interestingly, sulfur, the crosslinking (or vulcanizing) agent in the NR, actually also functions as the crossbridging agent between NR and brass to form the Cu_xS/ZnS interphase. In the model (Figure 18), the brass develops a strong adhesive bond to NR through the *in situ* growth of a thin (≦50 nm) copper sulfide film on the brass surface before crosslinking sets in. Since this sulfide film is porous, the rubber chains become entangled throughout this film.

TABLE 18
Wettabilities of Rubbers[a]

Rubbers	Abbreviation	γ_c (mJ/m^2)
Isobutene–isoprene copolymer	Butyl	27
Ethylene–propylene rubber	EP	28
Poly(isoprene), *trans*	PIP (trans)	30
Poly(1,3-butadiene), *trans*	PB (trans)	31
Poly(isoprene), *cis*	PIP (cis)	31
Poly(1,3-butadiene), *cis*	PB (cis)	32
Styrene–butadiene copolymer (25 : 75)	SBR	33
Cyclized poly(isoprene)	—	34
Poly(epichlorohydrin)	—	35
Rubber hydrochloride	—	36
Chlorosulfonated polyethylene	CSPE	37
Butadiene–acrylonitrile	BN	37
Chlorinated poly(isoprene)	—	37
Chloroprene	—	38

[a] From L. H. Lee, *J. Polym. Sci. A-2*, **5**, 1103(1967).

TABLE 19
Adhesion of Various Materials to Natural Rubber[a,b]

Material	Adhesion level[d]	Interface	Remarks
Iron, steel	0	—	No adhesion
Copper	0	Excess Cu_2S	Some adhesion if undercured
Copper-plated steel[e]	700–900	—	Good adhesion if thickness < 50 nm
Cu_2S-coated steel[f]	700–880	—	Good adhesion with fresh Cu_2S
Zinc	100–200	Some ZnS	Poor adhesion
Copper-plated zinc[e]	700–800	Cu_2S	Good adhesion if thickness < 50 nm
Brass 70/30	700–1200	Cu_2S and some ZnS	Good adhesion depends on surface preparation

[a] From W. J. van Ooij, *Rubber Chem. Technol.* **52**, 605(1979).
[b] Compound of type A; cured 25 min at 150 °C.
[c] Sheets of 0.5 mm thickness.
[d] N/64 mm².
[e] Electroless immersion plating.
[f] Copper-plated steel sulfidized by S in paraffin.

16.6.2. Natural Rubber–Nickel Alloy Adhesion

In the last several years, a new coating containing one of Ni–Cu–Zn alloys[(266)] has been developed to replace the brass coating. These ternary alloys are less sensitive to dezincification than ordinary brass. Furthermore, these alloys do not form a conducting oxide layer and, consequently, the formation of the Cu_xS layer can be suppressed. The formulation of these alloys can affect adhesion. For example, the optimum Co content is approximately between 0.05 and 0.2 phr (part per hundred of the metal). In the rubber formulation, 4 phr of the insoluble sulfur gives a good overall performance.

In brief, the ternary alloy-coated cord can equal and, in the case of salt water

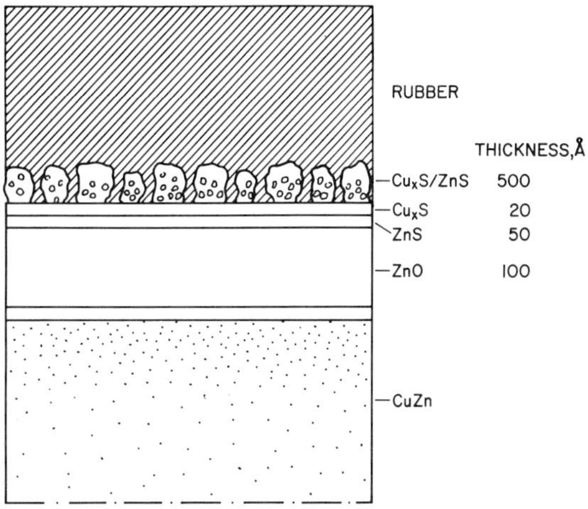

FIGURE 18. Schematic of interfacial sulfide in natural rubber–brass bonding showing mechanical interlocking. [After W. J. van Ooij, *Rubber Chem. Technol.* **52,** 605 (1979) (reproduced with permission).]

immersion, surpass the adhesion of low copper content brass-coated cord. Therefore, the use of ternary alloys should be one of the important developments for the future tire cord technology.

16.6.3. Polybutadiene–Cobalt Interaction

The interaction between polybutadiene (PB) and cobalt has been studied by Leidheiser *et al.*[267] with the ^{57}Co emission Mössbauer technique. The results indicate that the interfacial contact is through a layer of oxide and hydrous oxide, and the oxide on Co does not thicken during the baking process. On the steel side, the thermal treatment at 200 °C causes a significant amount of conversion of Fe^{3+} to Fe^{2+} at the oxide coating interface. The latter part of PB/Fe reaction has been verified by Watts *et al.*[268] In this case, PB acts like a reducing agent.

16.7. Effect of Contact Electrification on Metal–Polymer Adhesion

In the above discussions about the adhesion of metals to various polymers, we have seen the evidence of the acid–base (or donor–acceptor) interaction and chemical bonding at the interface. In the case of Au/PI contact, we also saw the evidence of the charge (or electron) injection into PI. Hence, it is natural at this point to ask whether triboelectrification (or electronic adhesion) plays any role in adhesion between metals and polymers, which are mostly insulators.

This subject about the role of electrostatics in adhesion will be discussed by Hays.[112] Let us briefly summarize some major concepts in this section.

Similar to the discussion in Section 8.3 about the effect of contact electrification on metal–semiconductor adhesion, the role of contact electrification on metal–polymer adhesion is only a secondary effect. As discussed by Bauser *et al.*[269] and Morris,[270] if the contact between a polymer and a metal is essentially real,[271] a metal may not acquire an appreciable surface charge at least for the following two reasons:

1. There are not enough energy levels at appropriate energy to accommodate surface charges of a density of 10^{10} e/cm^2 or more on various kinds of polymers.
2. The time required to transfer charges into the bulk of a polymer is by orders of magnitude larger than the usual contact times.

Thus, according to Krupp[128] the contact scheme as shown in Figure 19 has to be modified in two ways: (1) introducing surface states, and (2) introducing bulk defects within the energy gap.

The surface states may be of two types:

- intrinsic, e.g., functional groups on the polymer surface, such as ester or carboxyl groups protruding on the surface;
- extrinsic, e.g., catalyst residual, antioxidants, or products formed from the reaction with the environment, such as O_2, O_3, etc.

The review by Hays[112] indicates that if the electrostatic contribution is significant, the surface charge density should be greater than 1600 nC/cm^2, which is equivalent to a density of surface states D_s equal to 10^{13} electrons/cm^2. Since the surface atoms have

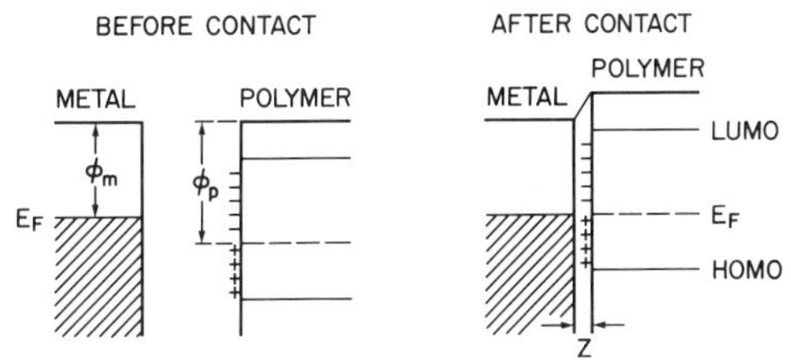

FIGURE 19. Contact between a metal and a polymer.

been estimated to be 10^{15} atoms/cm^2, this is equivalent to only 1% of the surface being charged. In reality, the highest density of surface states for the covalent semiconductors has been estimated to be even less: only about 10^{12} electrons/cm^2. Thus, for polymers, which are similar to covalent semiconductors, the electrostatic contribution is small in comparison to the LW forces because of low surface charge density.

Medley[(272)] has reported that the maximum charge exchange density measured upon separation for a mercury–polymer contact in vacuum is only 160 nC/cm^2. The contact between a polymer and a liquid metal like mercury should be rather real. Even under this ideal situation, the charge density is still an order of magnitude smaller than that to be significant in adhesion. However, one may argue that the actual charge in the electrical double layer (EDI)[(273)] can be much higher before the separation. Recent work by Lowell[(274)] indicates that the back-flow of electrons only takes place for a very high charge density at the interface.

Roberts[(275)] has studied the role of surface charge contribution to rubber adhesion and concluded that the electrostatic contribution is less than 10%. Krupp[(33)] also reached a similar conclusion.

Possart *et al.*[(276)] have attempted to determine the actual charge density and the work of the electrical double layer without breaking the bond. Recently, they have shown[(276–278)] that the electrostatic component of W_{el} due to EDL is only 1.71×10^{-3} J/m^2. Here W_{el} is the interaction energy between the space charges in the undisturbed EDL.

In comparison to the actual peeling work, the peeling of LDPE (low-density polyethylene) from the aluminum foil yields a specific peel work of 1 J/m^2, which is about 600 times that of the W_{el}. Thus, the specific peel work is much larger than the stored electrostatic energy due to the presence of the EDL. This is an indication that the electrostatic component is only a fraction of the total adhesive strength. Thus, their result still cannot confirm Deryagin's original hypothesis on the dominant role of EDL in adhesion.

17. POLYMER–SOLID ADHESION

Besides metals, polymer can form an adhesive bond with other insulating solids. When a polymer contacts another insulator, it can form a joint depending upon the distance of the contact. When a polymer is used to adhere two solids (similar or dissimilar) together, it becomes an adhesive. We shall separate these two situations and briefly discuss them below.

17.1. Polymer–Solid Contact

When a polymer contacts another polymer or insulator, there will be some electron exchange at molecular distances (<3 Å) according to the scheme shown in Figure 20. The direction of the electron flow depends on work functions of polymer counterbodies.[279] As a rule, the electron will flow from one solid of low work function to that of high work function. For this reason, a tribo series[280] has been established for polymers and other insulators (Table 20). This series is an empirical guide to the direction of the flow of electrons. After a polymer donates an electron, it becomes positively charged, while after a polymer accepts an electron, it is negatively charged. The polymers on top of the series have low work functions, while those in the bottom have high work functions. Thus, knowledge of the work function of a polymer (or an insulator) is very valuable in predicting the outcome of static charging.

In Table 21, we also note that the dielectric constant K correlates well with work function ϕ of dielectrics. Gallo and Lama[281] have established the following relation between the two parameters:

$$\phi = \left(\frac{e^2}{4X_0}\right)\left(\frac{K+7}{K+1}\right) \tag{69}$$

where ϕ is the work function of a plane dielectric, K the dielectric constant, X_0 the maximum excursion distance of an electron from the surface, and e is electronic charge.

In an earlier section we already discussed the relationship between χ^M (Mulliken's electronegativity) and the work function. Thus, all three of these properties are a measure of the chemical potential of a polymer (or a solid). In other words, the direction of electron flow is controlled by the chemical potentials between a polymer (or a material) and a metal.

Besides electron flow, the polymer and the solid can both deform to form a joint. The effect of LW forces upon the adhesion of the deformed junctions has been discussed earlier. It was concluded that van der Waals forces play a dominant role in polymer–solid adhesion without an adhesive. Recent work of Possart[277,278] has confirmed that the effect of EDL plays only a minor role in adhesion even for the metal–insulator contact. Since the surface charge density is generally low for insulators, the chance of achieving electronic adhesion is even less for the polymer–insulator contact.

Surface modification of polymers, e.g., polyolefins, PTFE, etc., is designed to increase surface functionalities, or donor–acceptor sites. This increase of functionalities

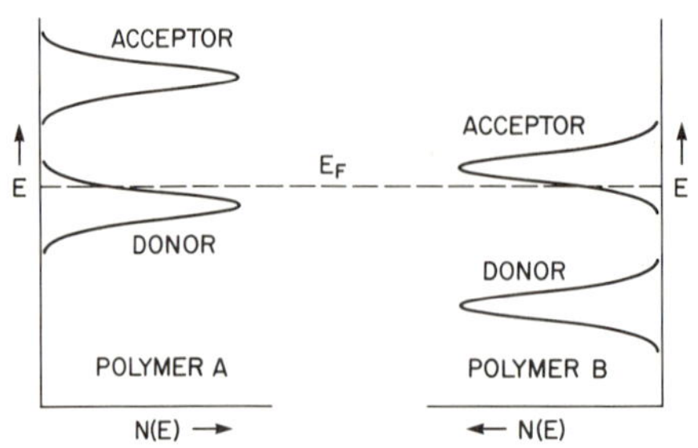

FIGURE 20. Energy diagram for polymer-to-polymer contact—the spreading distribution of donor level (HOMO) and acceptor level (LUMO).

TABLE 20
Triboelectricity Series of Polymers and Other Materials[a]

Positive
Silicone elastomer with silica filler
Borosilicate glass, fire-polished
Window glass
Aniline–formol resin, acid-catalyzed
Polyformaldehyde
Polymethylmethacrylate
Ethylcellulcse
Polyamide 11
Polyamide 6-6
Rock salt (NaCl)
Melamine formol
Wool, knitted
Silica, fire-polished
Silk, woven
Polyethylene glycol succinate
Cellulose acetate
Polyethylene glycol adipate
Polydiallyl phthalate
Cellulose (regenerated) sponge
Cotton, woven
Polyurethane elastomer
Styrene–acrylonitrile copolymer
Styrene–butadiene copolymer
Polystyrene
Polyisobutylene
Polyurethane flexible sponge
Borosilicate glass, ground surface
Polyethylene glycol terephthalate
Polyvinyl butyral
Formo-phenolic, hardened
Epoxide resin
Polychlorobutadiene
Butadiene–acrylonitrile copolymer
Natural rubber
Polyacrylonitrile
Sulfur
Polyethylene
Polydiphenylol-propane carbonate
Chlorinated polyether
Polyvinylchloride with 25% DOP
Polyvinylchloride without plasticizer
Polytrifluorochloroethylene
Polytetrafluoroethylene
Negative

[a] From J. Henniker, *Nature* **196**, 474 (1962) (reproduced with permission).

TABLE 21
Work Functions and Dielectric Constants of Polymers[a]

Polymers	Work function, ϕ (eV)	K
Poly(tetrafluoroethylene)	5.75	2.1
Poly(vinyl chloride)	5.13	3–4
Poly(sulfone)	4.95	3.0
Poly(styrene)	4.90	2.6
Poly(ethylene)	4.90	2.3
Poly(carbonate)	4.80	3.17
Poly(methyl methacrylate)	4.68	3.5–4.5
Poly(vinyl acetate)	4.38	5.5
Nylon 6,6	4.3	3.9–7.6
Poly(ethylene oxide)	3.95	6.0

[a] The values of ϕ are the unpublished data of S. Strella while the values of K are from work by Gallo and Lama.(281)

can affect the apparent work function of the related surface. As a result, the surface charge density can be increased. Hays has studied the UV irradiation of polystyrene(282) and the oxidation of polyethylene(283) and found that the surface charge density increases with the extent of irradiation or oxidation. He has also postulated(112) that if the density of an acceptor is much greater than that of a donor (10^{17} to $10^{18}/m^2$ or 0.01 to 0.1 of a monolayer), the charge transfer density can be 10^{15} to $10^{17}/m^2$ for the acceptor–donor energy level difference of 0.5 to 1 eV and a separation of 5 to 10 Å. Assuming the dielectric constant to be 10 for this hypothetical case, the adhesive pressure for the higher surface charge density can be comparable to the LW contribution.

Coincidentally, this type of surface modification has also been used to enhance adhesion, presumably for a different reason, e.g., to promote the acid–base (or donor–acceptor) interaction. The difficult question is whether the enhancement is actually due to the electronic mechanism. It appears that if the modification results in 0.01 to 0.1 of monolayer of acceptor (or *vice versa*), a good adhesion can be achieved on either count, electronic or the acid–base interaction. Otherwise, the LW interaction predominates.

17.2. Polymers as Adhesives

This review does not deal with polymers used directly as adhesives. Fortunately, there have been many books covering this general subject,(284–296) and the readers should seek appropriate references. However, in a separate volume entitled *Adhesive Bonding*,(297) we shall bring the subject matter up to date and in more detail. In that book, Lee(298) reviews polymer adhesion mechanisms. Krieger(299) examines flow properties of adhesives. Good and Gupta(300) study interfacial, rheological, and thermal control mechanisms in polymer adhesion, while Gupta(301) also discusses extension rheometry of polymer melts. Chu(302) deals with dynamic mechanical properties of pressure-sensitive adhesives. Regarding surface characterization, Davis(303) introduces various techniques, and Filbey and Wightman(304) apply them to the study of polymer/metal adhesion.

For structural adhesives, Clearfield, McNamara, and Davis[(305)] investigate adherend surface preparation, and Minford[(306)] presents his study on durability of adhesive bonded structure. In addition, Krieger[(307)] focuses on the principle and practice in testing structural adhesives for properties necessary for stress analysis.

For adhesive technology, it is as important to understand mechanisms of bond-breaking as those about bond-forming discussed in this volume. Previously, Greenwood and Johnson[(308)] discussed the adhesion of viscoelastic solids. Maugis[(309)] sums up his extensive studies on adherence and fracture mechanics, while Andrews[(310)] focuses on the application of fracture mechanics to adhesion. For those interested in engineering aspects, Reddy and Roy[(311)] introduce the concepts of finite-element analysis of adhesive joints. For those more interested in the physical phenomena of fracture, Dickinson[(312)] demonstrates the versatile fracto-emission technique for the study of adhesive joints. Finally, Rose[(313)] briefly introduces the ultrasonic nondestructive test (NDT) for the inspection of adhesive bonds and composite materials.

18. DISCUSSION

After the above review, we can now see more clearly some of the following clues about various types of solid adhesion:

1. *Forces of Nature and Adhesion.* At the atomic and molecular levels, one of the most important forces controlling adhesion and bonding is the electromagnetic force. The fundamental relationship about the electrostatic force between two point charges is described by Coulomb's equation. Unlike van der Waals forces, the Coulombic interaction can take place within a nuclei as well as beyond several hundred angstroms. For a monovalent ion pair, the attractive energy, which varies with r^{-1}, is about 200 kT at 300 K for a distance of 2.76 Å but falls off to 1 kT beyond 560 Å.

2. *Intermolecular Distance.* The distance between two discrete molecules or a discrete molecule and a solid surface is the dominant controlling factor for adhesion. In one extreme, if the distance is less than 10^{-13} cm, as in the case within a nucleus, then gluon, the superglue, can generate the strong force to bind proton and neutron together within a nucleus. Of course, this is not the situation that we are facing for normal adhesion problems. However, nature has set an unsurpassable example for us to appreciate the importance of the separation distance.

3. *Lifshitz–van der Waals Interactions.* Generally, the intermolecular distance is the key for achieving various types of interaction at interfaces. If the intermolecular distance is long range, nonretarded LW and Coulombic interactions should control adhesion. If the distance is beyond 100 Å, besides the Coulombic energies, the retarded LW energy should be taken into account. For microscopic systems, the nonretarded LW energies are inversely proportional to 10^{-6} of the separation distance r, while the retarded LW energies are inversely proportional to 10^{-7} of r. For macroscopic bodies, the effect of separation distance on LW energies is less critical; the nonretarded LW energies vary with the distance of Z or Z^2 depending on the geometry of the counterbodies.

4. *Molecular Interactions.* If the intermolecular distance is short range (less than 3 Å), a stronger molecular interaction can also take place between a donor and an acceptor (or acid and base). Here, we can apply the concepts related to reactivities of two molecules. Klopman and Hudson considered the orbital perturbation for their calculation of the interaction energy from two major terms: the ionic or Coulombic and the covalent or frontier orbital. The contribution of each of these two terms varies with the structure of the two counterbodies. For the second term, the frontier orbitals involve both a donor and an acceptor. The interaction may be called either the donor–acceptor interaction according to the NBO theory or the acid–base interaction according to Kitaura and Morokuma's (KM) approach. Both theories differ in detail according to their respective emphases.

According to the KM approach, the interaction energy (or binding energy) should consist of primarily the electrostatic component, ΔE_{es}, and the charge-transfer component, ΔE_{ct}, and secondarily the polarization component, ΔE_{po}, and the exchange component, ΔE_{ex}. The relative importance of the electrostatic and charge-transfer components depends on the ionicity of the molecules.

5. *Chemical Reactions at Interfaces.* If the interaction is stronger yet, e.g., within 1–2 Å, a chemical reaction can take place at the interface to form a chemical or covalent bond. In this review, there have been many actual examples of chemical interactions between a metal and a solid. For chemical interactions, several important thermodynamic considerations should be taken into account. For example, the metal reactivity is related to the heat of the formation of the corresponding oxide. The more negative the heat of formation of the oxide, the more reactive the metal.

6. *Covalency versus Ionicity.* Ionicity affects both the chemistry and physics of the solid state. In this review, one may find several ways to define ionicity, e.g., electronegativity, work function, dialectric constant, Phillips's partial ionicity, etc. Though each of them serves some specific purpose, they all are somehow related.

It appears that in many instances, such as the interaction energy between two molecules, it is necessary to consider both covalency and ionicity together. In the case of the enthalpy of the acid–base interaction, Drago's constant consists of both covalent and ionic factors. In the case of the average energy gap E_g^{Av} of a solid, Phillips has also proposed considering both factors, e.g., covalent and ionic. Thus, a partial ionicity has been determined for many solids, especially semiconductors.

7. *Electrical Double Layer.* If a solid interface contains two opposing charges, a double layer is formed. Deryaguin proposed that the electrical double layer (EDL) is responsible in adhering two surfaces together. Since most solids are not fully ionic and the ionic sites on the surface are only a fraction, the effect of EDL on the overall adhesion is secondary at best.

In this review, we only mentioned briefly contact electrification pertaining to the metal–metal contact, metal–semiconductor contact, metal–polymer contact, and polymer–solid contact. In all four cases, the extent of charge exchange is limited by low surface charge density. Thus, the contribution of charge exchange to the overall adhesion has been demonstrated to be only a fraction, e.g., 10% or less. Consequently, the original electronic mechanism on the basis of the EDL cannot account for all adhesion aspects of solids.

8. Electrons as Glue. Besides electrical charges, what else can hold solids together? The adhesiveless or glueless examples in this review suggest that electrons may have functioned as a glue at close molecular distances. When two discrete molecules or a discrete molecule and a solid are brought together, orbitals perturb and electrons interact; consequently, a bond is formed. Electrons not only bind but also enter electrostatic interaction (like those between charges), charge transfer, exchange, and polarization. Thus, if we assume electrons to be a glue, we need not have to treat charges separately as if they were different from electrons.

For electrons to function as a glue, there should be some basic rules guiding the success of the outcome. Though electrons are basically fleeting, they should flow according to a proper potential (chemical) difference. Thus, the work function and Fermi level can foretell the direction of the flow. Since not all electrons are equally energetic, the density of states (DOS) can further characterize the type of interacting bodies. For example, most ionic semiconductors are of low density of surface states (or low surface charge density), while covalent semiconductors are of high density of states. As a result, metal–ionic semiconductor interactions appear to be affected by the work function of the metal, while metal–covalent semiconductor interactions appear to be less dependent on work function. Interestingly, the same relationship also applies to contact charging, which can be a separate subject on its own.

9. Impurity and Contamination. Why do some solids appear to show no adhesion to each other? Besides surface asperities, impurities and contaminations, which were mentioned as early as 1250 AD (see the beginning of this chapter), are the worst enemies of adhesion. Both of them can keep real surfaces apart. As a result, the intermolecular distance can be much greater than 100 Å, and at such a distance the Lifshitz–van der Waals interactions are weak, and molecular interaction cannot be realized.

Recently, Smith and Cianciolo[(314)] have found that impurities on the solid surface can adversely affect the adhesive binding energy ΔE. At low temperatures, if it is energetically more favorable for impurities to segregate on the surface, then ΔE will decrease. However, the converse is also true; if the surface segregation is endothermic, ΔE should increase. For the latter case, the impurities tend to segregate at the grain boundary. In either case, impurities are detrimental to adhesion.

10. Avalanche on Solid Surfaces. Recent work on the effect of impurities suggests that it is of the utmost importance to keep solid surfaces clean. If the intermolecular distance falls below a critical distance, Smith *et al.*[(315)] have observed that an avalanche takes place on the surfaces as shown in Figure 21. This surface collapse resembles "cold-welding" of metal surfaces under an ultrahigh vacuum or in outer space. This type of avalanche is good for adhesion but bad for friction and wear. The avalanche can cause an instantaneous seizure on solid surfaces. After a surface collapses, it requires energy to break the junctions apart. Thus, high friction and high wear result because friction is a measure of resistance to motion, and wear is simply the fracture of unwanted junctions.

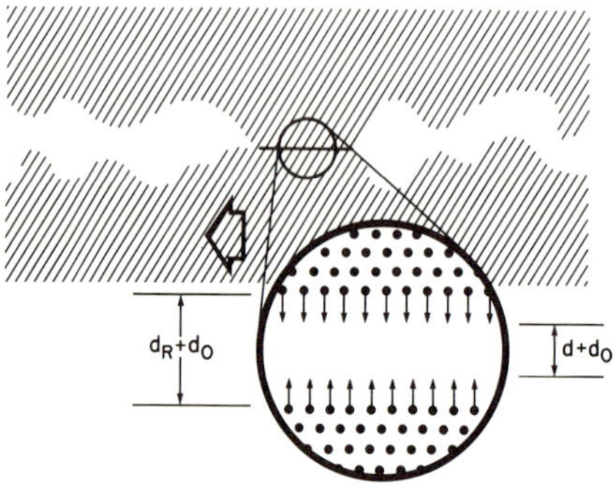

FIGURE 21. Cross section of sliding surface. Inset: Region that has avalanched together, with arrows showing the displacements of the atoms in the surface layers due to avalanche. Also shown are the rigid interfacial separation $d_R + d_0$ and the relaxed interfacial separation $d + d_0$, where d_0 is the bulk interplanar spacing. [After J. R. Smith, B. Bozzolo, A. Banerjea, and J. Ferrante, *Phys. Rev. Lett.* **63**(2), 1269 (1989) (reproduced with permission).]

11. Solid Surfaces versus Discrete Molecules. Hoffmann has demonstrated with the frontier orbital method that the orbital interactions between two discrete molecules are somewhat different from those between a molecule and a solid surface. A solid is like a reservoir which can balance itself by moving electrons above and below the Fermi level (sea). This adjustment is like a cleaning act that discrete molecules cannot perform. This insight about solid surfaces should provide a better understanding about chemisorption and catalytic reactions besides adhesion. Solid surfaces are not a simple matter and require different considerations.

12. HSAB Principle of Solids. Recently, on the basis of the frontier orbital concept, Lee has attempted to relate chemical hardness η to the average energy gap E_g^{Av} of a solid. As a result, the hard–soft acid–base (HSAB) principle of solutions has been extended to solid interactions. Since chemical hardness is derived from Mulliken's electronegativity, not from Pauling's electronegativity, both covalent and ionic factors have been equally considered.

With this extension of the HSAB principle, chemical reactivities of various solids can be correlated with the barrier of electrons crossing the energy gap. Consequently, chemical hardness becomes a bridge linking chemistry and physics in the solid state.

By the HSAB terminology, all metals are soft; all semiconductors and semimetals are also relatively soft, and all insulators including polymers are generally hard. Then for achieving good reactivities or adhesion at the interface, it is necessary to match a "hard" acid with a "hard" base or a "soft" acid with a "soft" base, provided that the two counterbodies are within a molecular distance.

In a separate paper,[316] we shall demonstrate that instead of μ and η, the work function, ϕ, and the average energy gap, E_g^{Av}, can be used to estimate the number of electrons, ΔN, transferred between systems A and B for an acid–base interaction, thus

$$\Delta N \approx \Delta\phi / \Sigma E_g^{Av} \tag{70}$$

13. Metal Adhesion. Interestingly, when metals adhere or react with other molecules or solids, they generally follow a similar pattern despite the nature of the counterbody. Though all metals are "soft," not all metals are equally reactive. In fact, noble and near-noble metals are relatively unreactive. As expected, most transition metals are rather reactive, and their reactivities are closely related to the availability of d electrons and their Fermi levels. In general, the reactions of metals leading to bonding may undergo several steps: physisorption ⇒ chemisorption ⇒ complex formation ⇒ chemical reaction. The bond strength should increase according to the same order.

14. Surface Modification of Polymers. Unlike metals, polymers are generally "hard" and of low surface energy. Basically, it is difficult for solid polymers to stick to other solids, e.g., metals. Thus, surface modification has often been applied to alter surface characteristics and enhance adhesion. In this review, we have cited many examples of chemical etching, ion-sputtering, corona-, flame-, and plasma-treatments for various polymers. Effective surface modification methods can cause unadherable or ungluable solids to adhere to each other. Thus, the chemistry involved in surface modification of polymers as well as other solids is exciting and deserves continuing exploration.

19. SUMMARY

In this review, we have examined both chemical and physical factors affecting solid adhesion. It appears that, through the electromagnetic force, all matter can be bonded or glued together with or without an adhesive, if conditions are right. Depending upon intermolecular distances, either long-range or short-range interactions can take place at the interface to enhance adhesion. For long-range interactions, Lifshitz–van der Waals and Coulombic forces (or energies) dominate; for short range (<3 Å), molecular interactions can supersede LW forces. Under thermodynamically and kinetically favorable conditions, chemical reactions can also occur at the interface within 1–2 Å to form chemical bonds. Thus, to achieve good adhesion between solids, the intermolecular distance must be maintained at a minimum through wetting or diffusion. The presence of impurities or contaminants tends to prevent solid adhesion. Similarly, the presence of asperties also hinders good solid contact. Only clean and smooth surfaces can attain optimum adhesion.

It is a useful concept to regard electrons as a glue, as suggested by Ferrante *et al.* Though electrons play an important role in solid adhesion, we should not equate what we described in this review to the electronic mechanism proposed by Deryagin. His theory is based on the electrical double layer (EDL) that has been demonstrated by others to influence only a small fraction of the total adhesion.

It is important to note that this review concerns chiefly the bond-forming process. Thus, the clues we found may be applicable to bond formation, but may not apply to bond breaking. However, some effective surface modification methods appear to improve both the work of adhesion and adhesive strengths. Finally, the chemistry related to surface modification has been demonstrated to play an important role in enhancing adhesion.

APPENDIX

A. Fundamental Physical Constants[a]

Quantity	Symbol	Value	Units	Relative uncertainty (ppm)
Speed of light in vacuum	c	299792458	$m\ s^{-1}$	(exact)
Permeability of vacuum	μ_0	$4\pi\ 4\times\ 10^{-7}$	$N\ A^{-2}$	(exact)
		$= 12.566370614\ldots$	$10^{-7}\ N\ A^{-2}$	
Permittivity of vacuum	ϵ_0	$1/\mu_0 c^2$	$10^{-12}\ F\ m^{-1}$	(exact)
		$= 8.854187817\ldots$		
Newtonian constant of gravitation	G	6.67259(85)	$10^{-11}\ m^3\ kg^{-1}\ s^{-2}$	128
Planck constant	h	6.6260755(40)	$10^{-34}\ J\ s$	0.60
$h/2\pi$	$\hbar$	1.05457266(63)	$10^{-34}\ J\ s$	0.60
Elementary charge	e	1.60217733(49)	$10^{-19}\ C$	0.30
Magnetic flux quantum, $h/2l$	ϕ_0	2.06783461(61)	10^{-35} Wb	0.30
Electron mass	m_e	9.1093897(54)	10^{-31} kg	0.59
Proton mass	m_p	1.6726231(10)	10^{-27} kg	0.59
Proton–electron mass ratio	m_p/m_e	1836.152701(37)		0.020
Fine-structure constant, $\mu_0 ce^2/2h$	α	7.29735308(33)	10^{-3}	0.045
Inverse fine-structure constant	α^{-1}	137.0359895(61)		0.045
Rydberg constant, $1/2m,\ c\alpha^2/2h$	R_∞	10973731.534(13)	m^{-1}	0.0012
Avogadro constant	N_A, L	6.0221367(36)	$10^{23}\ mol^{-1}$	0.59
Faraday constant, $N_A l$	F	96485.309(29)	$C\ mol^{-1}$	0.30
Molar gas constant	R	8.314510(70)	$J\ mol^{-1}\ K^{-1}$	8.4
Boltzmann constant, R/N_A	k	1.380658(12)	$10^{-23}\ J\ K^{-1}$	8.5
Stafan–Boltzmann constant, $(\pi^2/60)k^4/h^3l^2$	σ	5.67051(19)	$10^{-8}\ W\ m^{-2}\ K^{-4}$	34

B. Non-SI Units Used with SI[b]

Quantity	Symbol	Value	Units	Relative uncertainty (ppm)
Electron volt, (e/C)J = {e}J	eV	1.60217733(49)	10^{-19} J	0.30
(Unified) atomic mass unit, 1 μ = m_μ = 1/12 m(^{12}C)	u	1.6605402(10)	10^{-27} kg	0.59

[a] From B. W. Petley, *Sci. Prog.* (*Oxford*) **73**, 231 (1989) (reproduced with permission), also see Ref. 317.

[b] A selection of the values recommended for the fundamental physical constants by Cohen and Taylor, 1987, under the auspices of the CODATA Task Group on Fundamental Physical Constants. The figures in parentheses are the estimated standard deviation uncertainties in the last digit quoted. Note that the cull covariance matrix must be used in evaluating the uncertainties of qualities computed from them, since many of the uncertainties are correlated.

ACKNOWLEDGMENT

The author would like to thank Professor David Tabor of the Department of Physics, Carendish Laboratory, Cambridge University for his valuable comments on the original manuscript.

NOMENCLATURE

A	Electron affinity
$A_{1,2}, A_{H}$	Hamaker constant
a	Lattice parameter
a	Coordinate describing the separation between constituents
a^*	Scaled coordinate
a_0	Equilibrium spacing
a_0	Contact radius
a_m	Equilibrium separation
$B_{1,2}$	Coefficient of dispersion energy
C	Ionic component of the average energy gap
$C_{A,B}$	Covalent interaction susceptibility
C_{Nu}, C_{El}	Coefficient of the atomic orbital of the nucleophile or the electrophile
c	Concentration
c	Velocity of light
c_A, c_B	Polarization coefficients for A and B, respectively
D	Diffusion constant
D_0	Frequency factor
d	Distance between two counterbodies
d_p	vdW penetration distance
E	Electrophile
E	Young's modulus
e	Elementary charge
$E(a)$	Total binding energy
$E_{A,B}$	Electrostatic interaction
E_c	Energy of the conduction band
E_D	Dispersion energy
E_F	Fermi energy
E_g^p	Minimum energy gap
E_g^{Av}	Average energy gap
E_h	Covalent component of the average energy gap
E_I	Induction energy
E_O	Orientation energy
E_R	Retardation energy
E_V	Energy of the valence band
ΔE	Equilibrium binding energy
ΔE_{com}	Complexation energy
ΔE_{CT}	Charge transfer component of the complexation energy
ΔE_{int}	Interaction energy
ΔE_{NCT}	Noncharge transfer component of the complexation energy
ΔE_{ct}	Charge transfer component of the interaction energy
ΔE_{es}	Electrostatic component of the interaction energy
ΔE_{ex}	Exchange component of the interaction energy
ΔE_{po}	Polarization component of the interaction energy

F	Force or applied load
F	Surface free energy
F	Free energy at normal state
F_{es}	Electrostatic energy
F^{LW}	Lifshitz–van der Waals component of the attractive force
f	Enthalpy–to–free-energy correction factor for W_A^{a-b}
f_i	Ionicity or partial ionic character
ΔG_f	Free energy of atomic bond formation
h	Planck constant
h_A, h_B	Orthonormal hybrid
$\hbar$	$h/2\pi$
ΔH^{a-b}	Enthalpy of the acid–base interaction
ΔH_f	Heat of formation
ΔH_r	Heat of reaction
ΔH_R^C	Critical heat of reaction
I	Ionization potential
K	Dielectric constant
k	A constant
l	Length scale
m	Mass of electron
N	Nucleophile
N	Number of free electrons
N	Number of molecules per unit volume
N_A	Avogadro's constant
N_E	Density of states
n^{a-b}	Surface population of acid–base sites
P_A	Electrostatic pressure
P_E	Probability that an energy level E is fully occupied
Q	Activation energy of diffusion
Q or q	Point charge
R	Gas constant
R	Radius of a sphere
R_{eq}	Equilibrium radius
R_{vdW}	van der Waals radius
r	Interparticle (interatomic) distance
S	Index of interface behavior
T	Temperature
T_m	Melting temperature
V_c	Contact potential
$\bar{V}_{14}$	Nuclear potential
W	Barrier energy
W_A	Work of adhesion
W_A^{LW}	Lifshitz–van der Waals component of the work of adhesion
W_A^{a-b}	Acid–base component of the work of adhesion
W_{el}	Work of electrical double layer
X_0	Minimum excursion distance of an electron from the surface

Z	Ionic valency
Z	Distance between two counterbodies
Z_0	Equilibrium distance between two counterbodies
α	Polarizability
α	Dimensionless coupling constant (or fine-structure constant)
β	Resonance integral
γ	Surface tension or surface energy
$\gamma_{1,2}$	Interfacial tension
γ_c	Zisman's critical surface tension
γ^d	Dispersion component of surface tension
ϵ_0	Activation energy of diffusion per molecule at 0 K
ϵ_0	Permittivity in the free space (vacuum)
ϵ	Dielectric permittivity
η	Chemical hardness
η	Number of free electrons per unit crystal volume
κ	Boltzmann's constant
λ	Elementary jump length
λ_c	Longest photon wavelength
μ	Dipole moment
ν_0	Frequency of zero-point motion of electron (or natural frequency)
ν	Poisson's ratio
ξ	Frequency
σ	Surface charge density
σ_{AB}	Localized sigma bond
σ^*_{AB}	Localized sigma antibond
ϕ_{Morse}	Morse interaction energy
ϕ	Work function
ϕ_m	Work function of metal
ϕ_{sc}	Work function of semiconductor
ϕ_0	Distance between the top edge of the valence band and the Fermi level
ϕ_{SB}	Schottky barrier height
χ^M	Mulliken's electronegativity
χ^P	Pauling's electronegativity
ω	Average angular frequency

REFERENCES

1. S. Weinberg. Unified theories of elementary-particle interaction, *Sci. Am.* **231**(1), 50 (July, 1974).
2. G. 't Hooft, Gauge theories of the forces between elementary particles, *Sci. Am.* **242**(6), 104 (June, 1980).
3. C. Quigg, Elementary particles and forces, *Sci. Am.* **252**(4), 84 (April, 1986).
4. R. J. Hughes, T. Goldman, and M. M. Nieto, "Gravity in Greenland," in: Proceedings of the Division of Particles Fields Meeting, American Physical Society, Storrs, Connecticut (August, 1988).
5. M. E. Ander, M. A. Zumberge, T. Lautzenhiser, R. L. Parker, C. L. V. Aiken, M. R. Gorman, M. M. Nieto, A. P. R. Cooper, J. F. Ferguson, E. Fisher, G. A. McMechan, G. Sasagawa, J. M. Stevenson, G.

Backus, A. D. Chave, J. Greer, P. Hammer, B. L. Hansen, J. A. Hildebrand, J. R. Kelty, C. Sidles, and J. Wirtz, Test of Newton's inverse-square law in the Greenland ice cap, *Phys. Rev. Lett.* **62**(9), 985 (1989).
6. T. Goldman, R. J. Hughes, and M. M. Nieto, Gravity and antimatter, *Sci. Am.* **258**(3), 48 (March, 1988).
7. H. Georgi and S. L. Glashow, Unified theory of elementary-particle forces, *Phys. Today,* 30 (Sept. 1980).
8. K. Ishikawa, Glueballs, *Sci. Am.* **247**(5), 142 (Nov. 1982); also S. U. Chung (ed.), *Glueballs, Hybrids and Exotic Hadrons,* No. 185, Proceedings of a workshop held at Brookhaven National Laboratory, Upton, New York, August 29–September 1, 1988, American Institute of Physics (1989).
9. N. Isgur and G. Karl, Hadron spectroscopy and quarks, *Phys. Today,* 36 (Nov. 1983).
10. H. Harari, The structure of quarks and leptons, *Sci. Am.* **248**(4), 56 (1983).
11. J. M. LoSecco, F. Reines, and D. Sinclair, The search for proton decay, *Sci. Am.* **252**(6), 54 (June, 1985).
12. J. H. Rose, J. P. Vary, and J. R. Smith. Nuclear equation of state from scaling relations for solids, *Phys. Rev. Lett.* **53**(4), 344 (July, 1984).
13. J. Ferrante and J. R. Smith, Theory of metallic adhesion, *Phys. Rev. B* **19**(8), 3911 (1979).
14. W. Gutowski, Thermodynamics of adhesion, in: *Fundamentals of Adhesion* (L. H. Lee, ed.), Chapter 2, Plenum Press, New York (1990), this volume.
15. R. J. Good and M. K. Chaudhury, Theory of adhesive forces across interfaces: (1) The Lifshitz–van der Waals component of interaction and adhesion, in: *Fundamentals of Adhesion* (L. H. Lee, ed.), Chapter 3, Plenum Press, New York (1990), this volume.
16. J. N. Israelachvili, *Intermolecular and Surface Forces,* Academic Press, New York (1985).
17. S. Nordholm, Delocalization—the key concept of covalent bonding, *Chem. Educ.* **65**(7), 581 (1988).
18. L. Pauling, The nature of chemical bond IV. The energy of single bonds and the relative electronegativity of atoms, *J. Am. Chem. Soc.* **54,** 3570 (1932).
19. S. G. Bratsch, Revised Mulliken electronegativities II. Applications and limitations, *J. Chem. Educ.* **65**(3), 223 (1988).
20. L. Pauling, *The Nature of Chemical Bond,* Cornell University Press, Ithaca, New York (1966).
21. R. S. Mulliken, Electronic structures of molecules, *J. Chem. Phys.* **2,** 782 (1934).
22. J. C. Phillips, Ionicity of chemical bond in crystals, *Rev. Mod. Phys.* **42**(3), 317 (1970).
23. F. London, The general theory of molecular forces, *J. Chem. Soc., Faraday Trans.,* **33,** 8 (1937).
24. H. B. C. Casimir and D. Polder, The influence of retardation on the London-van der Waals forces, *Phys. Rev.* **73**(4), 360 (1948).
25. L. Spruch, Retarded, or Casimir, long-range potentials, *Phys. Today,* 37 (Nov. 1986).
26. W. H. Keesom, *Phys. Z.* **22,** 126 (1921); **23,** 225 (1922).
27. P. J. W. Debye, *Phys. Z.* **21,** 178 (1920); **22,** 302 (1921).
28. H. C. Hamaker, *Physica* **4,** 1058 (1937).
29. C. J. van Oss, M. K. Chaudhury, and R. J. Good, Interfacial Lifshitz–van der Waals and polar interactions in macroscopic systems, *Chem. Rev.* **88,** 927 (1988).
30. M. van den Tempel, Interaction forces between condensed bodies in contact, *Adv. Colloid Interface Sci.* **3,** 137 (1972).
31. I. E. Dzyaloshinskii, E. M. Lifshitz, and L. P. Pitaevskii, The general theory of van der Waals forces, *Adv. Phys.* **10,** 165 (1961).
32. E. M. Lifshitz, *Zh. Eksp. Teor. Fiz.* **29,** 94 (1955).
33. H. Krupp, Particle adhesion, theory and experiment, *Adv. Colloid Interface Sci.* **3,** 137 (1972).
34. A. Dupree, *Théorie Méchanique de La Chaleur,* p. 393, Gauthier-Villars, Paris (1869).
35. R. S. Bradley, *Philos. Mag.* **13,** 583 (1932).
36. D. Tabor, Surface forces and surface interactions, *J. Colloid Interface Sci.* **58**(1), 2 (1977).
37. B. V. Derjaguin, *Koll. Z.* **69,** 155 (1934).
38. H. Hertz, *Miscellaneous Papers,* Macmillan, London (1896).
39. B. V. Derjaguin, V. M. Muller, and Yu, P. Toporov, Effect of contact deformations on the adhesion of particles, *J. Colloid Interface Sci.* **53,** 314 (1975).
40. K. L. Johnson, K. Kendall, and A. D. Roberts, Surface energy and the contact of elastic solids, *Proc. R. Soc. London, Ser. A,* **324,** 301 (1971).
41. K. Kendall, The adhesion and surface energy of elastic solids, *J. Phys. D* **4,** 1186 (1971).
42. K. Kendall, Sticky solids, *Contemp. Phys.* **21**(3), 277 (1988).
43. B. V. Derjaguin, V. M. Muller, and Yu P. Toporov, On the role of molecular forces in contact deformations

(critical remarks concerning Dr. Tabor's paper), *J. Colloid Interface Sci.* **67**(2), 378 (1978); *J. Colloid Interface Sci.* **67**(2), 380 (1978).

44. V. M. Muller, V. S. Yushchenko, and B. V. Derjaguin, General theoretical consideration of the influence of surface forces of contact deformations and the reciprocal adhesion of elastic spherical particles, *J. Colloid Interface Sci.* **92**(1), 92 (1983).
45. V. M. Muller, B. V Derjaguin, and Yu P. Toporov, On two methods of calculation of the force of sticking of an elastic sphere to a rigid plane, *Colloids and Surfaces* **7,** 251 (1983).
46. D. S. Rimai, L. P. DeMejo, and R. C. Bowen, Observations of adhesion-induced deformations between spheroidal gold particles and conducting substrates, *J. Appl. Phys.* **65**(2), 755 (1989); "Adhesion-induced deformations of polymeric substrates: Particle size dependence of the contact area, *J. Appl. Phys.* **66**(8), 3574 (1989).
47. P. Morse, Diatomic molecules according to the wave mechanics, II. Vibrational levels. *Phys. Rev.* **34,** 57 (1929).
48. R. Rydberg, *Z. Phys.* **73,** 376 (1931).
49. H. Beutler, *Z. Phys. Chem.* **B27,** 287 (1934).
50. A. Banerjea, J. Ferrante, and J. R. Smith, Adhesion at metal interfaces, in: *Fundamentals of Adhesion* (L. H. Lee, ed.), Chapter 11, Plenum Press, New York (1990), this volume.
51. G. Klopman and R. F. Hudson, Polyelectronic perturbation of chemical reactivity, *Theor. Chim. Acta (Berl.)* **8,** 165 (1967).
52. K. Kitaura and K. Morokuma, A new energy decomposition scheme for molecular interactions within the Hartree–Fock approximation, *Int. J. Quantum Chem.* **10,** 325 (1976).
53. K. Morokuma, Why do molecules interact? The origin of electron donor–acceptor complexes, hydrogen bonding, and proton affinity, *Acc. Chem. Res.* **10,** 294 (1977).
54. K. Morokuma and K. Kitaura, Variational approach (SCF ab-initio calculations) to the study of molecular interactions: the origin of molecular interactions, in: *Molecular Interactions* (H. Ratajczak and W. J. Orville-Thomas, eds.), Wiley, New York (1980).
55. N. S. Isaacs, *Physical Organic Chemistry,* Longman Scientific and Technical; Wiley, Harlow, Essex, England; New York (1987).
56. A. E. Reed, L. A. Curtiss, and F. Weinhold, Intermolecular interactions from a natural bond orbital, donor–acceptor viewpoint, *Chem. Rev.* **88,** 899 (1988).
57. V. Gutmann, *The Donor–Acceptor Approach to Molecular Interactions,* Plenum Press, New York (1978).
58. J. P. Foster and F. Weinhold, Neutral hybrid orbitals, *J. Am. Chem. Soc.* **102,** 7211 (1980).
59. W. B. Jensen, *The Lewis Acid–Base Concepts: An Overview,* Wiley–Interscience, New York (1980).
60. W. B. Jensen, The Lewis acid–base definitions: a status report, *Chem. Rev.* **78**(1), 1 (1978).
61. R. S. Drago, G. C. Vogel, and T. E. Needham, A four-parameter equation for predicting enthalpies of adduct formation, *J. Am. Chem. Soc.* **93**(24), 6014 (1971).
62. F. M. Fowkes, Role of acid–base interfacial bonding in adhesion, *J. Adhes. Sci. Technol.* **1**(1), 7 (1987).
63. P. G. de Gennes, Wetting, statics and dynamics, *Rev. Mod. Phys.* **57**(3), Part 1, 827 (1985).
64. P. G. de Gennes, The dynamics of wetting, in: *Fundamentals of Adhesion* (L. H. Lee, ed.), Chapter 5, Plenum Press, New York (1990), this volume.
65. J. M. Poate, K. N. Tu, and J. W. Mayer (eds.), *Thin Films—Interdiffusion and Reactions,* Wiley, New York (1978).
66. B. I. Boltaks, *Diffusion in Semiconductors* (J. I. Carasso, transl. and H. J. Goldsmid, ed.), Academic Press, New York (1963).
67. J. Comyn (ed.), *Polymer Permeability,* Elsevier Applied Science, London and New York (1985).
68. S. Glasstone, K. J. Laidler, and H. Eyring, *The Theory of Rate Processes,* McGraw-Hill, New York (1941).
69. F. Brochard-Wyart, Kinetics of polymer–polymer interdiffusion, in: *Fundamentals of Adhesion* (L. H. Lee, ed.), Chapter 6, Plenum Press, New York (1990), this volume.
70. A. E. Chalykh and V. B. Zlobin, Modern ideas about diffusion in polymer systems, *Russ Chem. Rev.* **57**(6), 504 (1988).
71. S. S. Voyutskii, *Autohesion and Adhesion of High Polymers* (transl. S. Kaganoff), Interscience, New York (1963).
72. R. M. Vasenin, Adhesion of high polymers, *Adhes. Age* **8,** 18 (May 1965); **8,** 30 (June, 1965).

73. P. G. de Gennes, Reptation of a polymer chain in the presence of fixed obstacles, *J. Chem. Phys.* **55**(2), 572 (1971).
74. R. P. Wool, Welding, tack, and green strength of polymers, in: *Fundamentals of Adhesion* (L. H. Lee, ed.), Chapter 7, Plenum Press, New York (1990), this volume.
75. F. Bueche, W. M. Cashin, and P. Debye, The measurements of self-diffusion in solid polymers, *J. Chem. Phys.* **20**(12), 1956 (1952).
76. L. H. Lee, Adhesion of high polymers. I. Influence of diffusion, adsorption, and physical state on polymer adhesion, *J. Polym. Sci. A-2* **5,** 751 (1967).
77. F. Faupel, D. Gupta, B. D. Silverman, and P. S. Ho, Direct measurements of Cu diffusion into a polyimide below the glass transition temperature, *Appl. Phys. Lett.* **55**(4), 357 (1989).
78. K. Kendall, Theoretical aspects of solid-solid adhesion, *Sci. Prog. (Oxford)* **72,** 155 (1988).
79. E. P. Plueddemann, Adhesion through silane coupling agents, in: *Fundamentals of Adhesion* (L. H. Lee, ed.), Chapter 9, Plenum Press, New York (1990), this volume.
80. J. D. Miller and H. Ishida, Adhesive–adherend interface and interphase, in: *Fundamentals of Adhesion* (L. H. Lee, ed.), Chapter 10, Plenum Press, New York (1990), this volume.
81. J. E. E. Baglin, Interface design for thin film adhesion, in: *Fundamentals of Adhesion* (L. H. Lee, ed.), Chapter 13, Plenum Press, New York (1990), this volume.
82. P. S. Ho, R. A. Haight, R. C. White, B. D. Silverman, and F. Faupel, Chemistry, microstructure, and adhesion of metal–polymer interfaces, in: *Fundamentals of Adhesion* (L. H. Lee, ed.), Chapter 14, Plenum Press, New York (1990), this volume.
83. R. Hoffmann, How chemistry and physics meet in the solid state, *Angew. Chem.*, Int. Ed. Engl. **26,** 846 (1987); A chemical and theoretical way to look at bonding on surfaces, *Rev. Mod. Phys.* **60**(3), 601 (1988).
84. R. Hoffmann, *Solids and Surfaces—A Chemist's View of Bonding in Extended Structures,* VCH Publishers, Weinheim and New York (1988).
85. C. R. Barrett, W. D. Nix, and A. S. Tetelman, *The Principle of Engineering Materials,* p. 375, Prentice-Hall, New Jersey (1973).
86. J. R. Smith, Self-consistent many-electron theory of electron work functions and surface potential characteristics for selected metals, *Phys. Rev.* **181**(2), 522 (1969).
87. J. A. Dean, Electronegativities of the elements, in: *Lange's Handbook of Chemistry,* 12th ed. pp. 3–11, McGraw-Hill, New York (1978).
88. H. B. Michaelson, Relation between an atomic electronegativity scale and the work function, *IBM J. Res. Dev.* **22**(1), 72 (1978).
89. H. B. Michaelson, Electron work functions of the elements, in: *Handbook of Chemistry and Physics* (R. C. West, ed.), pp. 2–93, CRC Press, Florida (1989–1990).
90. D. Steiner and E. P. Gyftopoulos, *Proceedings of the 27th Annual Conference on Physical Electronics,* p. 160, Massachusetts Institute of Technology, Cambridge (March 1967).
91. O. K. Anderson, in: *The Electronic Structure of Complex Systems* (P. Phariseau and W. M. Temmerman, eds.), Plenum Press, New York (1984).
92. K. Fukui, Role of frontier orbital in chemical reactions, *Science* **218**(4574), 747 (Nov. 1982).
93. T. Koopmans, Ordering of wave functions and eigenenergies to the individual electrons of an atom, *Physica* **1,** 104 (1934).
94. G. Blyholder, Molecular orbital view of chemisorbed carbon monoxide, *J. Phys. Chem.* **68**(10), 2772 (1964).
95. I. Fleming, *Frontier Orbitals and Organic Chemical Reactions,* Wiley, London and New York (1976).
96. R. G. Pearson, Hard and soft acids and bases, *J. Am. Chem. Soc.* **85**(22), 3533 (1963).
97. R. G. Pearson, Recent advances in the concept of hard and soft acids and bases, *J. Chem. Educ.* **64**(7), 561 (1987).
98. L. H. Lee, Molecular dynamics and energetics of metal–polymer friction and wear. I. Applications of the hard–soft acid–base (HSAB) principle to solid adhesion and surface trobointeractions, in: *Proceedings 5th International Congress on Tribology* (K. Holmberg and I. Nieminen, eds.), Vol. 3, p. 308, Finish Society for Tribology, Espoo, Finland (1989).
99. L. H. Lee, Hard–soft acid–base (HSAB) principle for solid adhesion and surface interactions, in: *Fundamentals of Adhesion* (L. H. Lee, ed.), Chapter 12, Plenum Press, New York (1990), this volume.

100. D. H. Buckley, J. Ferrante, M. D. Pashley, and J. R. Smith, Solids in contact, in: *Panel Report on Interfacial Bonding and Adhesion,* Chapter 1, *Mater. Sci. Eng.* **83,** 169 (1986).
101. M. D. Pashley, J. B. Pethica and D. Tabor, Adhesion and micromechanical properties of metal surfaces, *Wear* **100,** 7 (1984).
102. M. D. Pashley and J. B. Pethica, The role of surface forces in metal–metal contacts, *J. Vac. Sci. Technol.* **A-3**(3), 757 (1985).
103. D. V. Keller, Adhesion between solid metals, *Wear* **6,** 353 (1963).
104. J. Ferrante, G. H. Bozzolo, C. W. Finley, and A. Banerjea, Interfacial adhesion, theory and experiment, *Mater. Res. Soc. Symp. Proc.* **119,** 3 (1988).
105. J. H. Rose, J. R. Smith, and J. Ferrante, Universal features of bonding in metals, *Phys. Rev. B* **28**(4), 1835 (1983).
106. J. Ferrante and J. R. Smith, Theory of bimetallic interface, *Phys. Rev. B* **31**(6), 3427 (1985).
107. F. M. Menger, Nucleophilicity and distance, in: *Nucleophilicity* (J. M. Harris and S. P. McManus, eds.), Chapter 14, Chemistry Symposium Series No. 209, American Chemical Society, Washington, D.C. (1987).
108. L. Brewer, A bonding model for strong generalized Lewis acid–base interactions in intermetallics, *Pure Appl. Chem.* **60**(3), 281 (1988).
109. S. R. Cain, L. J. Matienzo, and F. Emmi, Modeling of chemical interactions at Cr/Cu interfaces, *J. Phys. Chem. Solids* **50**(1), 87 (1989).
110. D. H. Buckley, *Surface Effects in Adhesion, Friction, Wear and Lubrication,* Elsevier, Amsterdam (1981).
111. R. M. Rose, L. A. Shepard, and J. Wulff, *The Structure and Properties of Materials,* Vol. IV, *Electronic Properties,* p. 120, Wiley, New York (1966).
112. D. Hays, Role of electrostatics in adhesion, in: *Fundamentals of Adhesion* (L. H. Lee, ed.), Chapter 8, Plenum Press, New York (1990), this volume.
113. G. V. Alm, Diffusion bonding—methods and applications. Part 1—Terminology, *Adhes. Age* **13,** 28 (July, 1970).
114. D. Decroupet, S. Mathot, and G. Demortier, Diffusion bonding of gold using microscopic Au–Cd alloys, *J. Mater. Sci. Lett.* **8,** 849 (1989).
115. L. J. Brillson, The structure and properties of metal–semiconductor interfaces, *Surf. Sci. Rep.* **2,** 123 (1982).
116. G. Margaritondo, L. J. Brillson, V. Brusic, J. R. Chelikowsky, R. W. Grant, and G. W. Rubloff, Semiconductor interfaces, in: *Panel Report on Interfacial Bonding and Adhesion,* Chapter 6, *Mater. Sci. Eng.* **83,** 169 (1986).
117. M. L. Cohen, Pseudopotential calculations for ideal interface, in: *Surfaces and Interfaces in Ceramic and Ceramic–Metal Systems* (J. A. Pask and A. Evans, eds.), Plenum Press, New York (1988).
118. J. C. Phillips, Chemistry and structure of transition metal silicide–silicon interfaces, in Proc. Symp. on *Thin Film Phenomena—Interface and Interactions* (J. E. E. Baglin and J. M. Poate, eds.), 78-2, Electrochemical Society, Princeton (1978).
119. S. Kurtin, T. C. McGill, and C. A. Mead, Fundamental transition in the electronic nature of solids, *Phys. Rev. Lett.* **22,** 1433 (1970).
120. C. A. Mead, Surface states on semiconductor crystals; barrier on the Cd(Se:S) system, *Appl. Phys. Lett.* **6,** 103 (1965).
121. C. B. Duke, A. R. Lubinsky, B. W. Lee, and P. Mark, Atomic geometry of cleavage surfaces of tetrahedrally coordinated compound semiconductors, *J. Vac. Sci. Technol.* **13,** 761 (1976).
122. D. H. Buckley, The role of material properties in adhesion, *NASA Technical Memo* 83792 (1984).
123. V. G. Weizer and N. S. Fatemi, The interaction of gold with gallium arsenide, *J. Appl. Phys.* **64**(9), 4618 (1988).
124. A. Yamada, M. Oasa, H. Nagabuchi, and M. Kawashima, Direct adhesion of single-crystal GaAs wafer, *Mater. Lett.* **6**(5,6), 167 (1988).
125. L. J. Brillson and G. Margaritondo, Adsorption and Schottky barrier formation on compound semiconductor surfaces, in: *The Chemical Physics of Solid Surfaces and Heterogeneous Catalysts* (D. A. King and D. P. Woodruff, eds.), p. 174, Elsevier, Amsterdam (1988).
126. L. J. Brillson, Chemical reaction and charge redistribution at metal–semiconductor interface, *J. Vac. Sci. Technol.* **15**(4), 1378 (1978).

127. A. M. Cowley and S. M. Sze, Surface states and barrier height of metal–semiconductor systems, *J. Appl. Phys.* **36**(10), 3212 (1965).
128. H. Krupp, Physical models of the static electrification of solids, *Static Electrification*, p. 1, Institute of Physics Conference Series No. 11 (1971).
129. C. Dervos and W. S. Truscott, Physical processes for contact charge transfer, *J. Electrostatics* **16,** 137 (1985).
130. C. Messmer and J. C. Billeto, The surface energy of Si, GaAs and GaP, *J. Appl. Phys.* **52**(7), 4623 (1981).
131. D. M. Mattox, Thin film metallization of oxides in microelectronics, *Thin Solid Films* **18,** 173 (1973).
132. C. Weaver, Adhesion to high energy surfaces, in: *Adhesion: Fundamentals and Practices*, p. 46, Ministry of Technology (U.K.), London (1969).
133. J. D. Venables, Review—Adhesion and durability of metal–polymer bonds, *J. Mater. Sci.* **19,** 2431 (1984).
134. D. A. Hardwick, J. S. Ahearn, and J. D. Venables, Environmental durability of aluminum adhesive joints protected with hydration inhibitor, *J. Mater. Sci.* **19,** 223 (1984).
135. K. H. Johnson and S. V. Pepper, Molecular-orbital model for metal–sapphire interfacial strength, *J. Appl. Phys* **53**(10), 6634 (1982).
136. Y. H. Kim, Y. S. Chang, N. J. Chou, and J. Kim, Adhesion of titanium thin film to oxide substrates, *J. Vac. Sci. Technol.* **A5**(5), 2890 (1987).
137. S. A. Varchenya, A. Simanovskis, and S. V. Stolyarova, Adhesion of thin metallic films to non-metallic substrates, *Thin Solid Films* **164,** 147 (1988).
138. D. H. Buckley, Ceramic microstructure and adhesion, *J. Vac. Sci. Technol.* **A3**(3), 762 (1985).
139. L. E. Murr, Interfacial energetics metal–metal, metal–ceramic, metal–semiconductor, and related solid–solid and liquid–solid systems, in: *Surfaces and Interfaces in Ceramic and Ceramic-metal Systems* (J. A. Pask and A. Evans, eds.), p. 107, Plenum Press, New York (1980).
140. W. Naka, M. Kubo, and I. Okamoto, Wettability of silicon nitride by aluminum, copper and silver, *J. Mater. Sci. Lett.* **6,** 965 (1987).
141. P. Nikolopoulos and D. Sotiropoulou, Wettability between zirconia ceramics and the liquid metals copper, nickel and cobalt, *J. Mater. Sci. Lett.* **6,** 1429 (1987).
142. M. Shimbo, M. Naka, and I. Okamoto, Wettability of silicon carbide by aluminum, copper and silver, *J. Mater. Sci. Lett.* **8,** 663 (1989).
143. J. G. Li, D. Chafain, L. Coudurier, and N. Eustathopoulos, Wettability of sapphire by Sn–Al alloy, *J. Mater. Sci. Lett.* **7,** 961 (1988).
144. J. T. Klomp and P. J. Vrugt, Interfaces between metals and ceramics, in: *Surfaces and Interfaces in Ceramic and Ceramic-metal System* (J. Pask and A. Evans, eds.), p. 97, Plenum Press, New York (1980).
145. M. Morita, K. Suganuma, and T. Okamoto, Effect of pre-heat treatment and testing atmosphere on the joint strength of $Si_3N/Al/Si_3N$, *J. Adhes. Sci. Technol.* **2**(1), 43 (1988).
146. Y. Iino and N. Taguchi, Interdiffusing metals layer technique of ceramic–metal bonding, *J. Mater. Sci. Lett.* **7,** 981 (1988).
147. S. Baumgartl, W. Warnecki, and R. Holm, A contribution to the investigation of enamel/metal interface, *Surf. Interface Anal.* **3**(4), 189 (1981).
148. Y. Iino, T. Takamori, M. Mohri, N. Saito, and M. Suzuki, Joining of ceramic to metal by use of an electric discharge machined surface. *J. Mater. Sci. Lett.* **8,** 493 (1989).
149. Y. Iino and G. Elssner, *Trans. Jpn. Soc. Mech. Eng. A* **53,** 33 (1987) (in Japanese).
150. A. J. Pedraza, M. J. Godbole, D. H. Lowndes, and J. R. Thompson, Jr., Enhanced metal–ceramic adhesion by sequential sputter deposition and pulsed laser melting of copper films on sapphire substrates, *J. Mater. Sci. Lett.* **24,** 115 (1989).
151. K. Miyoshi, Adhesion in ceramic and magnetic media, in: *Proceedings of 5th International Congress on Tribology* (K. Holmberg and I. Nieminen, eds.), Vol. 3, p. 228, Finnish Society for Tribology, Espoo, Finland (June, 1989).
152. J. G. Bednorz and K. A. Müller, Possible high T_c superconductivity in the Ba–La–Cu–O-System, *Z. Phys. B* **64,** 189 (1986).
153. C. W. Chu, P. H. Hor, R. L. Meng, L. Gao, Z. J. Huang, and Y. Q. Wang, Evidence for superconductivity above 40K in the La–Ba–Cu–O compound system *Phys. Rev. Lett.* **58,** 405 (1987).
154. M. K. Wu, J. R. Ashburn, C. J. Thorng, P. H. Hor, R. L. Meng, L. Gao, Z. J. Huang, Y. Q. Wang, and C.

W. Chu, Superconductivity at 93K in a new mixed-phase Y–Ba–Cu–O compound system at ambient pressure, *Phys. Rev. Lett.* **58,** 908 (1987).
155. W. Y. Lee, J. Salem, V. Lee, C. T. Rettner, G. Lim, R. Savoy, and V. Deline, *AIP Conf. Proc.* **165,** 95 (1988).
156. W. Y. Lee, J. Salem, C. T. Rettner, G. Gorman, R. Savoy, V. Deline, T. Huang, and D. W. Chung, Effects of deposition conditions on the superconducting properties of R.F. and D.C. magnetron sputter-deposited $YBa_2 Cu_3 O_{7-x}$ films, *Thin Solid Films* **166,** 181 (1988).
157. S. J. Shah, C. R. Fincher, M. W. Duch, D. A. Beames, K. M. Unruh, and C. P. Swann, Growth and properties of $Y Ba_2 Cu_3 O_{7-x}$ superconducting thin films, *Thin Solid Films* **166,** 171 (1988).
158. C-A. Chang, C. C. Tsuei, C-C. Chi, and T. R. McGuire, Thin-film YBaCuO superconductors formed by $Cu/BaO/Y_2O_3$ layer structure, *Appl. Phys. Lett.* **52**(1), 72 (1988).
159. C-A. Chang, Effect of silver and gold on the YBaCuO superconducting thin films with the use of $Ag/Cu/BaO/Y_2O_3$ and $Au/Cu/BaO/Y_2O_3$ structures, *Appl. Phys. Lett.* **52**(11), 924 (1988).
160. C. T. Cheung and E. Ruckenstein, Superconductor–substrate interactions of the Y–Ba–Cu oxide, *J. Mater. Res.* **4**(1), 1 (1989).
161. J. T. Klomp, Bonding of metals to ceramics and glasses, *Ceram. Bull.* **51**(9), 683 (1972).
162. P. Benjamin and C. Weaver, The adhesion of evaporated metal films on glass, *Proc. Roy. Soc. London, Ser. A* **261,** 516 (1961).
163. P. Hrma, The kinetics of the adhesion of glass to a solid, *Glass Technol.* **18**(4), 117 (1977).
164. J. A. Pask, Glass–metal 'interfaces' and bonding, UCRL-10611, Lawrence Radiation Laboratory, Berkeley, California (1963).
165. P. K. Bachman and R. Messier, Emerging technology of diamond thin films, *Chem. Eng. News* 24 (May 15, 1989).
166. S. V. Pepper, Effect of electronic structure of the diamond surface on the strength of the diamond–metal interface, *J. Vac. Sci. Technol.* **20**(3), 643 (1962).
167. S. V. Pepper, Transformation of the diamond (110) surface, *J. Vac. Sci. Technol.* **20**(2), 213 (1962).
168. K. L. Mittal (ed.), *Adhesive Joint,* Plenum Press, New York (1982).
169. C-A. Chang, Interface interactions relevant to package technology, *Thin Solid Films* **166,** 97 (1988).
170. K. L. Mittal, Adhesion aspects of metallization of organic polymer surfaces, *J. Vac. Sci. Technol.* **13**(1), 19 (1976).
171. W. A. Zisman, Surface energetics of wetting, spreading and kinetics, *J. Paint Technol.* **44**(564), 43 (1972).
172. L. H. Lee, Relationships between surface wettability and glass temperatures of high polymers, *J. Appl. Polym. Sci.* **12,** 719 (1968).
173. E. G. Shafrin, *Polymer Handbook,* 2nd ed. (J. Bandrupt and E. H. Immergut, eds.), Section III, p. 221, Gordon and Breach, New York (1975).
174. L. H. Lee, Adhesions of high polymers. IV. Relationships between surface wettability and bulk properties of high polymers, in: *Interactions of Liquids at Solid Substrates* (R. F. Gould, ed.), p. 106, Advances in Chemistry Series No. 87, American Chemical Society, Washington, D.C. (1968).
175. W. A. Dukes and A. J. Kinloch, Preparing low-energy surfaces for bonding, in: *Developments in Adhesives—1* (W. C. Wake, ed.), Applied Science Publ., London (1977).
176. D. Dwight and W. M. Riggs, Fluoropolymer surface studies, *J. Colloid Interface Sci.* **47,** 650 (1974).
177. E. H. Andrews and A. J. Kinloch, Mechanics of adhesive failure. I, *Proc. R. Soc. London, Ser. A* **332,** 385 (1973); II, *Proc. R. Soc. London, Ser. A* **332,** 401 (1973).
178. N. J. DeLollis and O. Montoya, Surface treatments for difficult-to-bond plastics, *Adhes. Age* **6**(1), 32 (1963).
179. M. Brenman and Ch. H. Lerchenthal, Increase of adhesive bond strength through the mechano-chemical creation of free radicals: I, *Polym. Eng. Sci.* **16**(11), 747 (1976).
180. Ch. H. Lerchenthal and M. Brenman, Increase of adhesive bond strength through the mechano-chemical creation of free radicals. Part II. Shear and peel test results, *Polym. Sci. Eng.* **16**(11), 760 (1976).
181. H. Schonhorn and F. W. Ryan, Adhesion of polytetrafluoroethylene, *J. Adhes.* **1,** 43 (1969).
182. R. F. Roberts, F. W. Ryan, H. Schonhorn, G. M. Sessler, and J. E. West, Increase of gold–Teflon FEP joint strength by removal of deposited aluminum prior to gold deposition for electrical applications, *J. Appl. Polym. Sci.* **20,** 255 (1976).

183. R. F. Roberts and H. Schonhorn, Surface modification of polymers by deposition of evaporated metals. I. Teflon® FEP, *Polym. Prepr.* **16,** 146 (1975).
184. S. L. Vogel and H. Schonhorn, Adhesion of evaporated metallic films onto polyethylene and poly(tetrafluoroethylene): importance of surface crosslinking, *J. Appl. Polym. Sci.* **23,** 495 (1979).
185. A. I. Vilenskii, E. E. Virlich, N. N. Stefanovich, and N. A. Krotova, Mechanism of adhesion interaction of a metallic coating with activated PTFE, *Sov. Plast.* (*Engl. Transl.*), No. 5, 68 (1973).
186. R. Michael and D. Stulik, Surface topography development on Teflon® under KeV Xe atom bombardment, *J. Vac. Sci. Technol.* **A4**(4), 1861 (1986).
187. Chin-an Chang, J. E. E. Baglin, A. G. Schrott, and K. C. Lin, Enhanced Cu–Teflon adhesion by presputtering prior to the Cu-deposition, *Appl. Phys. Lett.* **51**(2), 103 (1987).
188. Chin-an Chang, Enhanced Cu–Teflon adhesion by presputtering treatment: Effect of topography changes, *Appl. Phys. Lett.* **51**(6), 1236 (1987).
189. C. Haag and H. Suhr, Improved adhesion of Cu on pre-etched polytetrafluoroethylene by PECVD-deposited thin metallic layers, *Appl. Phys.* **A47,** 199 (1988).
190. Y. Yamada, D. R. Wheeler, and D. H. Buckley, Adhesion between polymers and evaporated gold and nickel films, NASA Technical Paper No. 2360 (August, 1984).
191. D. R. Wheeler and S. V. Pepper, Effect of x-ray flux on polytetrafluorethylene in x-ray photoelectron spectroscopy, *J. Vac. Sci. Technol.* **20**(2), 226 (1982).
192. B. T. Werner, T. Vreeland, Jr., M. H. Mendenhall, Y. Qui, and T. A. Tombrello, Enhanced adhesion from high energy ion irradiation, *Thin Solid Films* **104,** 163 (1983).
193. S. Jocobson, B. Jönsson, and B. Sundqvist, The use of fast heavy ions to improve thin film adhesion, *Thin Solid Films* **107,** 89 (1983).
194. D. H. Buckley and W. A. Brainard, The atomic nature of polymer–metal interactions in adhesion, friction, and wear, in: *Advances in Polymer Friction and Wear* (L. H. Lee, ed.), Vol. A., p. 315, Plenum Press, New York (1974).
195. V. A. Karagounis, Role of free energies of formation and bond dissociation energies in adhesion of metals in polymers (unpublished).
196. F. M. Fowkes, Donor–acceptor interactions at interfaces, in: *Recent Advances in Adhesion* (L. H. Lee, ed.), p. 39, Gordon and Breach, New York (1973).
197. F. M. Fowkes, Acid–base interactions in polymer adhesion, in: *Physicochemical Aspects of Polymer Surfaces* (K. L. Mittal, ed.), Vol. 1, Plenum Press, New York (1981).
198. D. E. Packham, K. Bright, and B. W. Malpass, Mechanical factors in the adhesion of polyethylene to aluminum, *J. Appl. Polym. Sci.* **18,** 3237 (1974).
199. B. W. Malpass, D. E. Packham, and K. Bright, A study of the adhesion of polyethylene to porous alumina films using the scanning electron microscope, *J. Appl. Polym. Sci.* **18,** 3249 (1974).
200. A. McGregor and L. E. Perrins, Some factors influencing the adhesion of electroplated copper to polypropylene, *Plast. Polym.* **38,** 192 (1970).
201. L. E. Perrins and K. Pettett, Mechanism for the adhesion of electroplated copper to polypropylene, *Plast. Polym.* **39,** 391 (1971).
202. M. M. Kadash and C. G. Seefried, Jr. Closer characterization of corona-treated PE surfaces, *Plast. Eng.*, 45 (Dec. 1985).
203. D. Briggs, D. M. Brewis, and M. B. Konieczko, X-ray photoelectron spectroscopy studies of polymer surfaces. Part III. Flame treatment of polyethylene, *J. Mater. Sci.* **14,** 1344 (1979).
204. D. Briggs and C. R. Kendall, Chemical basis of adhesion to electrical discharge treated polyethylene, *Polym. Commun.* **20,** 1053 (1979).
205. J. E. Carley and P. T. Kitze, Corona-discharge treatment of polymeric films, II: Chemical studies, *Polym. Eng. Sci.* **20**(5), 330 (1980).
206. L. J. Gerenser, L. F. Erman, M. G. Mason, and J. M. Pochan, E.s.c.a. studies of corona-discharge-treated polyethylene surfaces by use of gas-phase derivatization, *Polymer* **26,** 1162 (1985)
207. F. Garbassi, E. Occhiello, and F. Polato, Surface effect of flame treatments on polypropylene, *J. Mater. Sci.* **22,** 207 (1987).
208. A. F. Lewis and L. J. Forrestal, Chemical nature of polymer-to-metal adhesion, Special Technical Publication No. 360, p. 59, American Society for Testing and Materials (1963).
209. D. Briggs, XPS studies of polymer surface modifications and adhesion mechanisms, *J. Adhes.* **13,** 287 (1982).

210. C-G. Golander and B.-A. Sultan, Surface modification of polyethylene to improve its adhesion to aluminum, *J. Adhes. Sci. Technol.* **2**(2), 125 (1988).
211. J. Schultz, A. Carré, and C. Mazeau, Formation and rupture of grafted polyethylene/aluminum interfaces, *Int. J. Adhes. Adhes.* **4**(4), 163 (1984).
212. B. M. DeKoven and P. L. Hagans, XPS studies of metal/polymer interfaces—thin films of Al on polyacrylic acid and polyethylene, *Appl. Surf. Sci.* **27,** 199 (1986).
213. P. Bodö and J.-E. Sundgren, Adhesion of evaporated titanium to polyethylene: Effects of ion bombardment pretreatment, *J. Vac. Sci. Technol.* **A2**(4), 1498 (1984).
214. P. Bodö and J.-E. Sundgren, Adhesion of evaporated titanium films to ion-bombarded polyethylene, *J. Appl. Phys.* **60**(3), 1161 (1986).
215. P. Bodö and J.-E. Sundgren, The adhesion of evaporated metal films to polyethylene substrate, *Surf. Interface Sci.* **9,** 437 (1986).
216. L. J. Gerenser, An X-ray photo-emission spectroscopy study of chemical interactions at silver/plasma modified polyethylene interfaces: correlation with adhesion, *J. Vac. Sci. Technol.* **A6**(5), 2897 (1988).
217. D. Bekiarov, B. Pashmakov, E. Vateva, and C. Alexieva, A Study of the adhesion of vacuum-evaporated film to aluminum, *Thin Solid Films* **157,** 43 (1988).
218. P. Bodö and J.-E. Sundgren, Titanium deposition onto ion-bombarded and plasma-treated polydimethylsiloxane:surface modification, interface and adhesion, *Thin Solid Films* **136,** 147 (1986).
219. K. Heymann, W. Riedel, and G. Woldt, Electroplating of plastics in theory and practice, *Angew. Chem.* **82,** 412 (1970).
220. T. Tsukada and N. Hosokawa, Adhesion of copper films on ABS polymers deposited in an internal magnetron sputtering system, *J. Vac. Sci. Technol.* **16**(2), 348 (1979).
221. J. M. Burkstrand, Formation of a copper–oxygen–polymer complex on polystyrene, *Appl. Phys. Lett.* **33**(5), 387 (1978).
222. J. M. Burkstrand, Core-level spectra of chromium and nickel atoms on polystyrene, *J. Appl. Phys.* **50**(2), 1152 (1979).
223. J. M. Burkstrand, Formation of metal–oxygen–polymer complexes on polystyrene with nickel and chromium, *J. Vac. Sci. Technol.* **16**(4), 1072 (1979).
224. J. M. Burkstrand, Copper–polyvinyl alcohol interface: A study with XPS, *J. Vac. Sci. Technol.* **16**(2), 363 (1979).
225. J. M. Burkstrand, Substrate effects on the electronic structure of metal overlayers—an XPS study of polymer–metal interfaces, *Phys. Rev. B* **20**(12), 4853 (1979).
226. J. M. Burkstrand, Metal–polymer interfaces: Adhesion and X-ray photoemission studies, *J. Appl. Phys.* **52**(7), 4795 (1981).
227. J. M. Burkstrand, Summary abstract: Chemical interactions at polymer–metal interfaces and the correlation with adhesion, *J. Vac. Sci. Technol.* **20**(3), 440 (1982).
228. J. M. Burkstrand, 'Hot' atom interactions with polymer surfaces, *J. Vac. Sci. Technol.* **21,**(1), 70 (1982).
229. J. M. Burkstrand, Chemical interactions at polymer–metal interfaces: Studies with X-ray photoemission, in: *Physicochemical Aspects of Polymer Surfaces* (K. L. Mittal, ed.), Vol. 2, p. 1035, Plenum Press, New York (1983).
230. J. L. Grant, D. S. Dunn, and D. J. McClure, Argon and oxygen sputter etching of polystyrene, polypropylene and poly(ethylene terephthalate) thin films, *J. Vac. Sci. Technol.* **A6**(4), 2213 (1988).
231. B. Leclercq, M. Sotton, A. Baszkin, and L. Ter-Minassian-Saraga, Surface modification of corona treated poly(ethylene terephthalate) film: adsorption and wettability studies, *Polymer* **18,** 675 (1977).
232. D. Briggs, D. G. Rance, C. R. Kendall, and A. R. Blythe, Surface modification of poly(ethylene terephthalate) by electrical discharge treatment, *Polymer* **21,** 895 (1980).
233. J. M. Pochan, L. J. Gerenser, and J. F. Elman, An e.s.c.a. study of the gas-phase derivatization of poly(ethylene terephthalate) treated by dry-air and dry-nitrogen corona discharge, *Polymer* **27,** 1058 (1986).
234. J. H. Cross, M. W. LeMay, and D. J. McClure, Texturing of poly(ethylene terephthalate) film surfaces by sputter etching, *J. Vac. Sci. Technol.* **A3,** 495 (1985).
235. Y. DePuydt, P. Bertrand, and P. Lutgen, Study of the Al/PET interface in relation with adhesion, *Surf. Interface Anal.* **12,** 486 (1988).
236. C. Boiziau and G. Lecayon, Modeling of the chemistry mechanisms involved in grafting polymers on metals—relationship between structure and properties of the resulting product, *Int. J. Adhes. Adhes.* **6**(4)

207 (1986); Adhesion of polymers to metals: a review of the results obtained studying a model system, *Surf. Interface Anal.* **12,** 475 (1988).

237. H. Leidheiser, Jr., and P. D. Deck, Chemistry of the metal–polymer interfacial region, *Science* **241,** 1176 (1988).
238. A. Vértes, I. Czakó-Nagy, P. Deck, and H. Leidheiser, Jr., Emission Mössbauer study of the interface between a cobalt substrate and a polyimide coating, *J. Electrochem. Soc.* **134**(7), 1629 (1987).
239. N. J. Chou and C. H. Tang, Interfacial reaction during metallization of cured polyimide: An XPS study, *J. Vac. Sci. Technol.* **2**(2), 751 (1984).
240. N. J. Chou and C. H. Tang, A thermodynamic model for predicting formation of chemical bonds between metals and cured polyimides during metallization, in: *Surface and Colloid Science in Computer Technology* (K. L. Mittal, ed.), p. 287, Plenum Press, New York (1987).
241. S. G. Anderson, H. M. Meyer, III, and J. H. Weaver, Temperature-dependent x-ray photoemission studies of metalstable Co/polyimide interface formation, *J. Vac. Sci. Technol.* **A6**(4), 2205 (1988).
242. J. L. Jordan, P. N. Sanda, J. F. Morar, C. A. Kovac, F. J. Himpsel, and R. A. Pollak, Synchrotron-radiation excited carbons photoemission study of Cr/organic polymer interface, *J. Vac. Sci. Technol.* **A4**(3), 1046 (1986).
243. F. K. LeGoues, B. D. Silverman, and P. S. Ho, The microstructure of metal–polyimide interfaces, *J. Vac. Sci. Technol.* **A6**(4), 2200 (1988).
244. C. A. Kovac, J. L. Jordan-Sweet, M. J. Goldberg, J. G. Clabes, A. Viehbeck, and R. A. Pollak, Chemistry at interfaces: electropositive metals on polymer surfaces, *IBM J. Res. Dev.* **32**(5), 603 (1988).
245. P. S. Ho, B. D. Silverman, R. A. Haight, R. C. White, P. N. Sanda, and A. R. Rossi, Delocalized bonding at the metal–polymer interface, *IBM J. Res. Dev.* **32**(5), 658 (1988).
246. J. W. Bartha, P. O. Hahn, F. K. LeGoues, and P. S. Ho, Photoemission spectroscopy study of aluminum–polyimide interface, *J. Vac. Sci. Technol.* **A3**(3), 1390 (1985).
247. Lj. Atanasoska, S. G. Anderson, H. M. Meyer III, Z. Lin, and J. H. Weaver, Aluminum/polyimide interface formation: An X-ray photoelectron spectroscopy study of selective chemical bonding, *J. Vac. Sci. Technol.* **A5**(6), 3325 (1987).
248. J. G. Clabes, Correlation of interface chemistry and growth made of Ce on polyimide (pyromellitic dianhydride-oxydianiline), *J. Vac. Sci. Technol.* **A6**(5), 2887 (1988).
249. F. S. Ohuchi and S. C. Freilich, Metal–polyimide interface: a titanium reaction mechanism, *J. Vac. Sci. Technol.* **A4**(3), 1039 (1986).
250. P. Bodö and J–E. Sundgren, Ion bombardment and titanium film growth on polyimide, *J. Vac. Sci. Technol.* **A6**(4), 2396 (1988).
251. Lj. Atanasoska, H. M. Meyer, III, S. G. Anderson, and J. H. Weaver, Semiconductor/polyimide interface formation: An X-ray photoelectron spectroscopy study of germanium chemical bonding, *J. Vac. Sci. Technol.* **A6**(4), 2175 (1988).
252. J. Currie, P. Depelsenaire, R. Groleau, and E. Sacher, A nuclear scattering evaluation of organometallic bonding in the adhesion of metallization to polymers, *J. Colloid Interface Sci.* **97**(2), 410 (1984).
253. H. M. Meyer, III, S. G. Anderson, Lj. Atanasoska, and J. H. Weaver, X-ray photoemission investigations of clustering and electron emission, injection, and trapping at the gold/polyimide interface, *J. Vac. Sci. Technol.* **A6**(1), 30 (1988).
254. A. L. Ruoff, E. J. Kramer, and C.-Y. Li, Improvement of adhesion on polyimide by reactive ion-beam etching, *IBM J. Res. Dev.* **32**(5), 626 (1988).
255. W. J. van Ooij, R. H. Brinkhuis, and J. M. Mark, Adhesion of copper and chromium films to polyimide studied by static SIMS of the interface, *Surf. Interface Anal.* **12,** 505 (1988).
256. V. R. Gray, *Surface aspects of solid and adhesion,* Chem. Ind., 969 (June, 1965).
257. S. A. Chambers and K. K. Chakravorty, Oxidation at the polyimide/Cu interface, *J. Vac. Sci. Technol.* **A6**(5), 3008 (1988).
258. S. R. Cain and L. J. Matienzo, Acid–base properties of copper oxides and fluorides by means of band calculations: Relevance to metal–polymer adhesion, *J. Adhes. Sci. Technol.* **2**(5), 395 (1988).
259. K. W. Paik and A. L. Ruoff, Adhesion enhancement of polyimide on copper by Ar ion beam etching of copper, *J. Adhes. Sci. Technol.* **2**(4), 245 (1988).
260. M. C. Burrell, P. J. Codella, J. A. Fontana, J. J. Chera, and M. D. McDonnell, Interfacial reactions at copper surfaces coated with polyimide films prepared from poly(amide-acid) precursors, *J. Vac. Sci. Technol.* **A7**(1), 55 (1989).

261. W. R. Salaneck, S. Stafström, J. L. Brédas, S. Anderson, P. Bodö, S. P. Kowalczyk, and J. J. Ritsko, Phthalimide on copper: a model system to address certain site-specific interactions at the polyimide–copper interface, *J. Vac. Sci. Technol.* **A6**(6), 3134 (1988).
262. Y.-H. Kim, J. Kim, G. F. Walker, C. Feger, and S. P. Kowalczyk, Adhesion and interface investigation of polyimide on metals, *J. Adhes. Sci. Technol.* **2**(2), 95 (1988).
263. L. H. Lee, Adhesion of high polymers. II. Wettability of elastomers, *J. Polym. Sci. A-2* **5,** 1103 (1967).
264. W. J. van Ooij, Mechanisms of rubber-to-brass adhesion: Effect of rubber composition on the adhesion, *Rubber Chem. Technol.* **52,** 605 (1979).
265. W. J. van Ooij, Mechanism and theories of rubber adhesion to steel tire cores—An overview, *Rubber Chem. Technol.* **57,** 421 (1984).
266. J. R. Ball, Optimization of bonding formulation using a ternary alloy-coated steel cord in tire construction, *Elastomerics* **119,** 16 (1987).
267. H. Leidheiser, Jr., S. Music, and G. W. Simmons, Metal-organic coating interface as studied by emission Mössbauer spectroscopy, *Nature* **297,** 667 (1982).
268. J. F. Watts and J. E. Castle, The application of X-ray photoelectron spectroscopy to the study of polymer-to-metal adhesion. Part 1, Polybutadiene coated mold steel, *J. Mater. Sci.* **18,** 2987 (1983).
269. H. Bauser, W. Klöpffer, and H. Rabenhorst, On the charging mechanism of insulating solids, in: *Advances in Static Electricity,* Vol. 1, p. 2, Proceedings of the 1st International Conference on Static Electricity, Vienna (May, 1970).
270. W. T. Morris, Static electrification of polymers: a review, *Plast. Polym.,* 41 (Feb. 1970).
271. W. R. Harper, *Contact and Frictional Electrification,* Clarendon Press, Oxford (1967).
272. J. A. Medley, The electrostatic charging of some polymers by mercury, *Br. J. Appl. Phys.* **4,** Suppl. 2, S28 (1953).
273. B. V. Deryagin, N. A. Krotova, and V. P. Smilga, *Adhesion of Solids* (R. K. Johnson, transl.), Consultant Bureau, New York (1978).
274. J. Lowell, Tunnelling between metals and insulators and its role in contact electrification, *J. Phys. D.* **12,** 1541 (1979).
275. A. D. Roberts, Surface charge contribution in rubber adhesion and friction, *J. Phys. D* **10,** 1801 (1977).
276. W. Possart and A. Röder, Measurement of electrical potential distribution in a polymer near the contact to a metal by means of scanning electron microscopy, *Phys. Status Solidi A* **84,** 319 (1984).
277. W. Possart and I. Müller, The electrical double layer at the polymer–metal contact—A phenomenological model, *Phys. Status Solidi A* **106,** 525 (1988).
278. W. Possart, Experimental and theoretical description of the electrostatic component of adhesion at polymer/metal contacts, *Int. J. Adhes. Adhes.,* 79 (April, 1988).
279. S. Strella, unpublished data cited by J. M. O'Reilly and P. Erhardt, in: *Electrophotography,* p. 95, Second International Conference, Society of Photographic Science and Engineering, Washington, D.C., (1974).
280. J. Henniker, Triboelectricity of polymers, *Nature* **196,** 474 (1962).
281. C. F. Gallo and W. L. Lama, Classical electrostatic description of the work function and ionization energy of insulators, *IEEE Trans.* **1A-12,** 7 (1976).
282. D. Hays, The contact potential and charge exchange at a mercury–polymer surface, Inst. Phys. Conf. Series No. 48, p. 265, Institute of Physics (1979).
283. D. Hays, Contact electrification between metals and polymers: Effect of surface oxidation, in: *Proceedings of the International Conference on Modern Electrostatics* (Li Ruinian, ed.), p. 327, International Academic Publishers (1988).
284. L. H. Lee (ed.), *Adhesive Chemistry—Developments and Trends,* Plenum Press, New York (1985).
285. R. L. Patrick (ed.), *Treatise on Adhesion and Adhesives,* Vols. 1–6, Marcel Dekker, New York (Vol. 1, 1967; Vol. 2, 1969; Vol. 3, 1973; Vol. 4, 1976; Vol. 5, 1981; Vol. 6, 1988).
286. L. H. Lee (ed.), *Adhesives, Sealants, and Coatings for Space and Harsh Environments,* Plenum Press, New York (1988).
287. L. H. Lee (ed.), *Adhesion Science and Technology,* Plenum Press, New York, (1975).
288. L. H. Lee (ed.), *Adhesion and Adsorption of Polymers,* Plenum Press, New York (1980).
289. S. Wu, *Polymer Interface and Adhesion,* Marcel Dekker, New York and Basel (1982).
290. G. Schneberger (ed.), *Adhesives in Manufacturing,* Marcel Dekker, New York and Basel (1983).
291. I. Skeist (ed.), *Handbook of Adhesives,* 3rd ed. Van Nostrand Reinhold, New York (1989).
292. W. C. Wake (ed.), *Developments in Adhesives,* I, Applied Science Publ., London (1977).

293. D. M. Brewis and D. Briggs (eds.), *Industrial Adhesion Problems,* Wiley–Interscience, New York (1985).
294. C. V. Cagle, *Handbook of Adhesive Bonding,* McGraw-Hill, New York (1973).
295. A. H. Landrock, *Adhesive Technology Handbook,* Noyes Publications, Park Ridge, New Jersey (1985).
296. A. J. Kinloch (ed.), *Durability of Structural Adhesives,* Applied Science Publ., London and New York (1983).
297. L. H. Lee (ed.), *Adhesive Bonding,* Plenum Press, New York (1990).
298. L. H. Lee, Recent studies in adhesion mechanisms of polymers, Chapter 1 in Reference 297.
299. I. M. Krieger, Flow properties of adhesives, Chapter 2 in Reference 297.
300. R. J. Good and R. K. Gupta, The coupling of interfacial, rheological, and thermal control mechanisms in polymer adhesion, Chapter 3 in Reference 297.
301. R. K. Gupta, Extension rheometry of polymer melts, Chapter 4 in Reference 297.
302. S. G. Chu, Dynamic mechanical properties of pressure-sensitive adhesives, Chapter 5 in Reference 297.
303. G. D. Davis, Characterization of surfaces, Chapter 6 in Reference 297.
304. J. A. Filbey and J. P. Wightman, Surface characterization in polymer/metal adhesion, Chapter 7 in Reference 297.
305. H. M. Clearfield, D. K. McNamara, and G. D. Davis, Adherend surface preparation for structural adhesive bonding, Chapter 8 in Reference 297.
306. J. D. Minford, Durability evaluation of adhesive bonded structures, Chapter 9 in Reference 297.
307. R. B. Krieger, Jr., Testing structural adhesives for properties necessary for stress analysis, Chapter 10 in Reference 297.
308. J. A. Greenwood and K. L. Johnson, The mechanics of adhesion of viscoelastic solids, *Philos. Mag.* **43**(3), 697 (1981).
309. D. Maugis, Adherence and fracture mechanics, Chapter 11 in Reference 297.
310. E. H. Andrews, Fracture mechanics approach to adhesion, Chapter 12 in Reference 297.
311. J. N. Reddy and S. Roy, Finite-element analysis of adhesive joints, Chapter 13 in Reference 297.
312. J. T. Dickinson, Fracto-emission from adhesive failure, Chapter 14 in Reference 297.
313. J. L. Rose, Ultrasonic nondestructive evaluation technology for adhesive bond and composite material inspection, Chapter 15 in Reference 297.
314. J. R. Smith and T. V. Cianciolo, Impurity effects on adhesive energies, *Surf. Sci.* **210,** L229 (1989).
315. J. R. Smith, G. Bozzolo, A. Banerjea, and J. Ferrante, Avalanche in adhesion, *Phys. Rev. Lett.* **63**(12), 1269 (1989).
316. L. H. Lee, Relevance of the density-functional theory to surface interactions and solid adhesion, presented to the Symposium on Acid–Base Interactions: Relevance to Adhesion Science and Technology, held at Lehigh University, Bethlehem, Pa., and to be published in *J. Adhes. Sci. Technol.*
317. B. W. Petley, The values of the fundamental physical constants, *Sci. Prog. (Oxford)* **73,** 231 (1989).

2

Thermodynamics of Adhesion

W. Gutowski

1. MOLECULAR FORCES

1.1 General

This work outlines the mechanism of interfacial interactions with respect to different types of forces involved, discusses a transition between theoretical considerations and practical applications, e.g., in the area of adhesive bonds, and describes experimental techniques necessary for characterization of surfaces with respect to thermodynamic parameters that are significant for the bond design.

1.2. Types of Intermolecular and Interatomic Forces, Their Range and Magnitude

The literature on intermolecular and interatomic forces is extensive[1–11] and the author refers the reader to works of Dzyaloshinskii, Lifshitz, and Pitaevskii,[1] Israelachvili,[2] Good,[4] Kaelble,[5] Pauling,[7] and Fowkes.[9]

With respect to the type of interactions, the intermolecular and interatomic bonds are classified as illustrated in Table 1. The strength of these bonds and their range can be estimated using classical approaches of electrostatics and electrodynamics,[12,13] if the spatial structure and distribution of molecules and atoms are known. The relationship between the strength and range of the typical intermolecular and interatomic forces (for the single bonds) which are listed in Table 1 is shown in Figure 1.

Based on a known statistical spatial distribution of atoms and structure of molecules in analyzed substances, one can determine the following:

- types of interatomic and intermolecular interactions (bonds),
- potential energy of interaction for the above bonds,
- the number of interacting sites and therefore the number of attractive forces of given types.

W. Gutowski • Division of Building, Construction and Engineering, Commonwealth Scientific and Industrial Research Organization (CSIRO), Melbourne, Australia.

TABLE 1
Typical Interatomic and Intermolecular Forces

Type	Range (nm)	Energy (kJ/mol)
I. Interatomic bonds		
(a) Ionic	0.15–0.24	335–1050
hydrogen	0.26–0.30[a]	8.0–42[b]
(b) Covalent	0.15–0.24	63–920
(c) Metallic	0.26–0.30	110–350
II. Intermolecular bonds (indefinite range)		
(a) Dipole–dipole (excluding hydrogen bonds)		4.0–21
(b) London (disperson)		4.0–42
(c) Dipole–induced dipole		~2.0

[a] 0.186 nm for H_2O and 0.367 nm for crystalline hydrazine.
[b] 42 kJ/mol for HF_2^-.

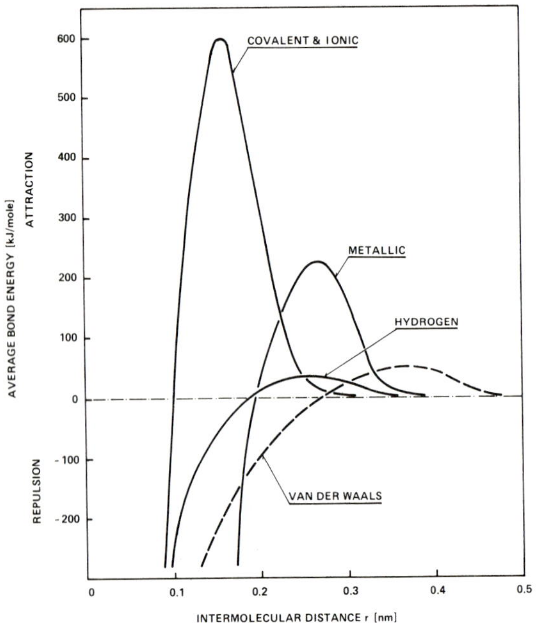

FIGURE 1. Strength and range of the typical intermolecular and interatomic forces.

It is frequently proposed that for a simple system consisting of two molecules separated by distance r, the potential energy of interaction U can be expressed by the following equation:

$$-U = \frac{A}{r^m} - \frac{B}{r^n} \tag{1}$$

where A and B are the attractive and repulsive constants, respectively.

The force of interaction between analyzed substances is determined as a derivative of the overall potential energy function, with respect to the separation distance, e.g.,

$$F = -\mathrm{d}U/\mathrm{d}r \tag{2}$$

where U is the total energy of interaction between bodies, molecules, or atoms under consideration.

1.3. Classical (Microscopic) Theory of Interatomic and Intermolecular Forces

1.3.1. Balance Between Attractive and Repulsive Forces

A general model of an interatomic bond can be illustrated based on an example of the ionic bond, formed by positively and negatively charged ions (understood as nuclei surrounded by a distribution of electrons). Depending upon the separation distance r, an attractive or repulsive force prevails in this system (see Figure 2). At greater distances, a Coulombic attractive force dominates, i.e.,

$$F_{\text{attr}} = q^+Q^-/r^2 \tag{3}$$

However, the contribution of F_{attr} decreases spontaneously with shorter distance of separation (where repulsive forces are predominant), up to the level where thermodynamic equilibrium is achieved. This relates to the minimum energy corresponding to the equilibrium distance r_0. Any further reduction of the separation distance (i.e., below r_0) requires external work to be done on the system, as some electrons must be promoted to the higher energy levels.

The amount of this external work is approximately proportional to the inverse power of the separation distance,

$$E_{\text{ext}} = R/r^n \tag{4}$$

and is associated with the repulsive force calculated from reference to Eqs. (2) and (4), i.e.,

$$F_{\text{rep}} = -nR/r^{n+1} \tag{5}$$

The overall force of interaction for the analyzed system is

$$F = F_{\text{attr}} + F_{\text{rep}} = +\frac{q^+\,Q^-}{r^2} - \frac{nR}{r^{n+1}} \tag{6}$$

This equation is illustrated by a continuous line in the upper part of Figure 2, while the corresponding energy function [see Eq. (2)] is shown in the lower portion of this figure.

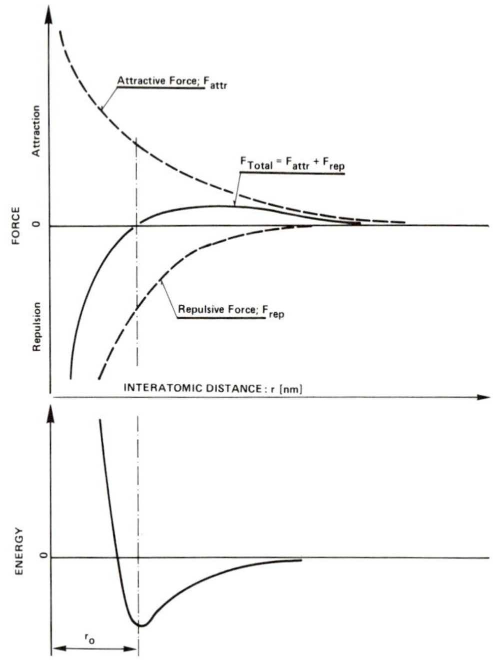

FIGURE 2. Attractive and repulsive components of forces of interaction between two molecules.

The total intermolecular pair potential is frequently represented by the Lennard-Jones[(4)] equation

$$U = -\frac{A}{2}\left(\frac{2}{r^6} - \frac{r_0^6}{r^{12}}\right) \tag{7}$$

where r_0 is the equilibrium distance corresponding to the minimum energy, i.e.,

$$U_o = A/2r_0^6 \tag{8}$$

1.3.2. Interatomic Bonds

1.3.2.1. Ionic Bonds. As explained in Section 1.3.1, an ionic bond is formed due to electrostatic forces between positive and negative ions. Such ions are formed by interaction of metallic and nonmetallic elements due to the fact that metallic elements tend to lose their outer electrons, after which they become positively charged i.e., cations.

The most stable configuration of any atom (except for H, He, Li, Mg, and Al) is achieved when its outer shell contains eight electrons. Accordingly, in a simplified way we can explain a process of creation of NaCl (see Figure 3) as

- donation of the electron from the outer shell of a Na atom to the outer shell of a Cl atom containing seven electrons,

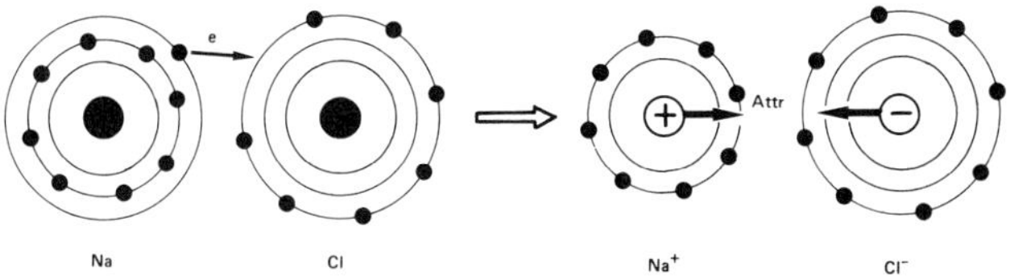

FIGURE 3. Process of formation of an ionic bond between Na and Cl.

- transformation of previously neutral atoms Na and Cl into positively and negatively charged ions, which attract each other with a Coulombic force of attraction relevant to Eq. (3).

The energy of a pure ionic bond (which is idealization) is

$$U = -q^{+}Q^{-}/r \tag{9}$$

while the force of interaction is given by Eq. (2). The Coulombic force between two charges provides the strongest of the physical forces—in many cases it is stronger than most chemical bonds.

It is seen from Eqs. (2) and (9) that the force of interaction in ionic bonds decays with the inverse square of the separation distance. This means that the overall ionic bond energy must account for a large number of atoms, not only the nearest neighbors. The ionic bond is nondirectional, which indicates that the bond strength is the same in any direction.

A pure ionic bond does not exist as there is always a covalent component in the bond character. A percent of the ionic character can be estimated using the relationship[5]

$$\text{Ionic character (\%)} = \frac{100\ \Delta E_{12}}{E_{12}} \tag{10}$$

where ΔE_{12} is an energy excess from the difference of electronegativity X[7] of elements 1 and 2, i.e.,

$$\Delta E_{12} = 23.06\ (X_1 - X_2)^2 \tag{11}$$

The energy E_{12} of a single bond between two atoms can be estimated using Pauling's formula[7]

$$E_{12} = \frac{1}{2}\,(E_{12} + E_{21}) + \Delta E_{12} \tag{12}$$

The above is equivalent to

$$E_{12} = \int_{r_0}^{\infty} (F_{\text{attr}} + F_{\text{rep}})\ \mathrm{d}r \tag{13}$$

Table 2 presents data on the ionic character of some simple bonds, as calculated by Kaelble[5] using Eq. (10).

TABLE 2
Comparison Between Experimental and Calculated Values for % Ionic Character in Selected Single Bonds of Diatomic Molecules[5] (Elements I and II)

I–II	$X_I - X_{II}$	Eq. (10)	Experimental
IBr	0.3	5	4
HI	0.4	5	5
ICl	0.5	11	8
HBr	0.7	15	12
HCl	0.9	19	18
HF	1.9	54	45
LiI	1.5	52	55
LiBr	1.8	67	60
LiCl	2.0	68	73
NaCl	2.1	73	75
KBr	2.0	76	78
KF	3.2	90	87
LiF	3.0	87	92

Ionic bonds occur rarely in macromolecular substances, except for some thermoplastics containing ionizable carboxyl groups (ionomers).

1.3.2.2. Covalent Bonds. Covalent bonds, which are predominant in polymers, are formed when unpaired electrons of outer shells are shared between atoms. Pure covalent bonds exist between like atoms, where there is no charge excess from either side. The six basic multivalent elements which interact through mainly covalent bonds are C, N, O, Si, P, and S. Also, some crystalline substances such as diamond, gray tin, germanium, and silicon are bound together with purely covalent bonds.[4,7]

Single or multiple covalent bonds are formed, depending upon the number of shared electrons. When a pair of electrons is shared by two nuclei, then single bonding takes place. This particular type of covalent bond provides a possibility of free rotation resulting in structures (e.g., polymers) with very good rheological properties.

Sharing of more electrons results in multiple covalent bonds. Increased overlapping of orbitals in these bonds leads to a decrease of bond length which increases the bond energy. Constraints from multiple bonds reduce the flexibility of molecules and substances.

A covalent bond is highly directional, and its strength and directionality can be deduced from the spatial distribution of wave function ψ^2. Binding energy of the covalent bond can be well represented by the minimum energy required to dissociate the bond, as expressed by Eq. (8). Table 3 presents some data (from Pauling[7]) on the length and energy of single and multiple bonds between selected elements.

With regard to adhesion between two phases, the covalent bond can be created across an interface providing one of the most significant components of the gross attractive force. For example, the hydroxy, amino, and carboxy groups in the substrate are involved in

Table 3
Length and Energy of Selected Single and Multiple Covalent Bonds[7]

Element and bond type	Bond length (nm)	Bond Energy (kJ/mol)
C—C	0.154	347.5
C═C	0.134	615.5
C≡C	0.120	812.3
N—N	0.148	159.1
N═N	0.124	418.7
N≡N	0.110	946.3
O—O	0.148	138.2
O═O	0.124	402.0
C—O	0.141	351.7
C═O		686.7–728.5
C—N	0.147	293.1
C═N		615.5
C≡N		866.7–891.8

formation of covalent bonds with relevant functional groups of the polymer, such as epoxy, nitrile, etc., across the interface.

1.3.2.3. Mutual Ionic–Covalent Bonds. It has been mentioned that a pure ionic bond does not exist while a pure covalent bond can be formed if homonuclear molecules are involved, where there is no charge excess from either side. Accordingly, partially ionic–partially covalent bonds are formed when two atoms of different electronegativity are attracted and their orbitals mutually overlap, resulting in sharing of some electrons. Table 2 shows some examples of mutual ionic–covalent bonds, with the emphasis on their ionic character (%) as estimated by Kaelble.[5]

1.3.3. Intermolecular (van der Waals) Bonds

Three types of intermolecular bonds which are considered in this group are formed due to the following interactions: (1) dipole–dipole, (2) dispersion (London), and (3) dipole–induced dipole. The common feature of these interactions is that the bond energy is proportional to the inverse sixth power of the separation distance.

1.3.3.1. Dipole–Dipole Interactions. Any molecule consisting of two dissimilar atoms of different electronegativity exhibits some degree of ionic character (see Section 1.3.2) and can be regarded as a dipole. Dipoles are characterized by a so-called dipole moment

$$\mu = ql \tag{14}$$

where q is the electronic charge and l is the distance between charges.

Examples of dipole moments for some molecules, molecular groups, and bonds are given in Table 4. The energy of interaction between two dipoles separated by distance r and at a fixed orientation corresponding to the minimum energy level (see Figure 4a) is given by[3,4]

$$U_{12} = -2\mu_1\mu_2/r_{12}^3 \tag{15}$$

The example in Figure 4a relates to dipolar molecules under conditions where interactions of other molecules restrict free rotations. However, in the general case of interactions between two molecules they can be free to rotate in an optional direction (see Figure 4b) if the chemical bonding with neighbor groups allows it. The overall energy of dipole–dipole interactions (taking into account attractive and repulsive forces as well as an average value of the kinetic energy due to thermal motion) is best described by the Keesom potential[3,11]

$$U_{D-D} = -2\mu_1^2\mu_2^2/3kTr^6 \tag{16}$$

In multimolecular interactions, however, the dipole–dipole energies are not pairwise additive, because the most stable configuration of dipoles is not very common in the lowest-energy structure of three or more dipoles.[4]

It can then be deduced that in the condensed phases, e.g., substrate and cured adhesive, conflicting dipole fields are created which significantly reduce the net force of dipole–dipole interactions.

1.3.3.2. Dipole–Induced Dipole Interactions. A dipole–induced dipole interaction exists between a permanent dipole (1) and a molecule (2) with zero permanent dipole

TABLE 4
Dipole Moments of Selected Molecules, Bonds, and Molecular Groups[2,53][a]

Compound	Dipole moment μ(Cm) $\times 10^{30}$	Compound	Dipole moment μ(Cm) $\times 10^{30}$
Molecules			
n–Hexane	0	Dichloromethane	6.171
Benzene	0	Chlorobenzene	5.270
Toluene	1.434	Nitromethane	11.675
o–Xylene	1.801	Butylonitrile	12.008
m–Xylene	1.067	*o*–Nitrotoluene	12.108
Bromoform	3.302	*m*–Nitrotoluene	14.176
Bonds			
C—C	0	$O—H^+$	1.51
C═C	0	$N—H^+$	1.31
C^+—O	0.74	C^+—N	0.22
C^+═O	2.5	C^+—Cl	1.6
Molecular Groups			
C^+—OH	1.65		
$C^+—NO_2$	3.45		

[a] 1 debye = 3.33564×10^{-30} Cm.

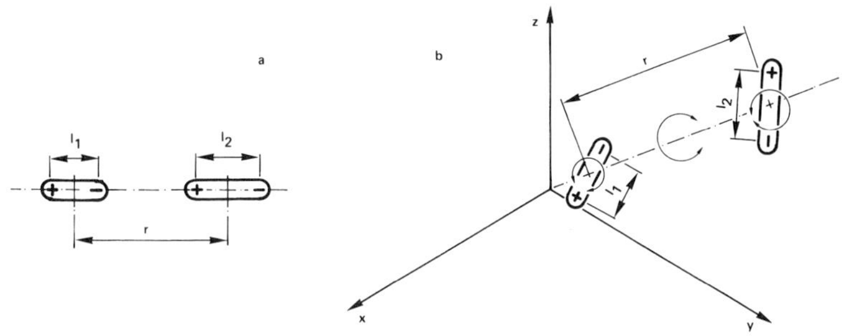

FIGURE 4. Interaction of two dipoles: (a) in configuration corresponding to the minimum energy of interaction; (b) general case, where molecules are free to rotate.

moment which becomes transformed into a secondary (induced) dipole under the influence of the electric field existing around the permanent dipole. The dipole moment μ_2 of this secondary dipole is

$$\mu_2 = \alpha_2 E_e \tag{17}$$

where E_e is the strength of the electric field and α is the polarizability.

The energy of interaction (i.e., U_D) between a permanent dipole (1) and a neutral (nonpolar) molecule (2) before polarization is expressed by Debye's formula[3,10]:

$$U_D = -\mu_1^2\alpha_2/r^6 \tag{18}$$

A newly polarized molecule starts to attract the permanent dipole (1). Accordingly, the overall energy of interaction between the primary and secondary dipoles is[3]

$$U_{D\text{-}ID} = -\frac{\alpha_1\mu_2^2 + \alpha_2\mu_1^2}{r^6} \tag{19}$$

The energy field arising from dipole–induced dipole interactions is not isotropic,[14] because the polarizability of chemical bonds is[15,16]

$$\alpha = \frac{3}{4\pi}\frac{V}{N}\left(\frac{n^2 - 1}{n^2 + 2}\right) \tag{20}$$

(where V is a molar volume, N is Avogadro's number, and n is the real part of the complex refractive number), which is different whether measured parallel or perpendicular to the bond. For instance, α_{parallel} for the C—C bond is 1.88×10^{-30} m^3 while $\alpha_{\text{perpendicular}}$ is 0.02×10^{-30} m^3, as given by Good.[4]

1.3.3.3. Dispersion-Type (London) Interactions. This type of interaction is created between otherwise neutral atoms due to the constant motion of electrons in the surrounding orbitals. This results in an instantaneous dipole of periodically varying magnitude. As described in Section 1.3.3.2, an instantaneous dipole induces a secondary dipole in neighboring atoms.

The overall energy of interaction due to the above mechanism depends upon the polarizability of the involved molecules (analogously to dipole–induced dipole interaction) and the energy of oscillation of the atom surrounded by electrons

$$U_{osc} = I \simeq h\nu_o \tag{21}$$

For two dissimilar atoms the total energy of interaction due to dispersive forces is given by[2,3]

$$U_L = -\frac{3}{2}\frac{\alpha_1 \alpha_2}{r^6}\frac{I_1 I_2}{I_1 + I_2} \tag{22}$$

or

$$U_L = -\frac{3}{2}\frac{\alpha_1 \alpha_2}{r^6}\frac{h\,\nu_1 \nu_2}{\nu_1 + \nu_2} \tag{23}$$

where $I \cong U_{osc}$ as given by Eq. (21) is valid for most substances except water.[8,4]

In general, the total contribution of London (dispersion) forces to cohesive and/or adhesive forces is proportional to the number of atomic nuclei in the main unit of the polymer and the number of electrons forming ions in these units. In metals, it is related to the number of electrons forming the ion core.[4]

It has been pointed out by Stachurski[17] that the strength of polymers (e.g., amorphous polyethylene) is dictated by the strength of the interchain van der Waals bonds, an observation which satisfies the well-known principle that the strength of a structure depends upon the strength of its weakest links. Fowkes[9] estimated that London forces are responsible for about 41% of the cohesive strength of mercury. In relation to polymers, Mark[18] estimated that London forces contribute about 75 to 100% of the total cohesion, except for very polar molecules (e.g., NO_2 or HCl) and when hydrogen bonding is taken into account (e.g., H_2O, NH_3, or HF_2^-).

1.3.3.4. Hydrogen Bonds. a. General. A hydrogen bond (H bond) exists between two electronegative atoms or groups, one of which (in the state that existed before formation of the H bond) has a hydrogen atom covalently bound to it. Schematically, this is described as follows:

$$\underset{\nearrow\ \text{covalent bond}}{\text{A—H}} \cdots \underset{\nwarrow\ \text{H bond}}{\text{B}} \rightleftarrows \text{A}^-\text{H}^+\text{—B} \tag{24}$$

The strength of the hydrogen bond depends upon the electronegativity of the attached atom. Accordingly, it can be ranked in the following order:

$$\text{F} \gg \text{O} > \text{N} > \text{Cl} \tag{25}$$

The energy of a hydrogen bond is 8 to 25 kJ/mol for bonds with oxygen, nitrogen, and chlorine. However, the H···F hydrogen bond is stronger and for HF_2^- its energy is 42 kJ/mol.

The energy of a hydrogen bond is higher than that of intermolecular bonds formed due to van der Waals interactions (dipole–dipole, dipole–induced dipole, and dispersive forces). It is attributed to the very small size of the hydrogen atom (r = 0.03 nm) and to the fact that strong Coulombic forces act at the smaller intermolecular distances relevant to the H bond.

According to theory of Lennard-Jones and Pople,[19] the H bond in water is bendable around the O—H···O axis. The total energy of this bond depends upon the externally incurred bending angle.

The bendability of the H bond is also confirmed by Coulson,[20,21] Slater,[22] and Sparnaay.[23]

Owing to the relatively small energy required for its formation, the hydrogen bond plays an important role in room-temperature reactions, e.g., adsorption of silanol groups from silane solutions on surface hydroxyl groups of oxidized metals and oxide-based ceramics, adsorption of polymers from solvents, flotation, etc.

b. Importance of the Hydrogen Bond in Interfacial Phenomena. It was mentioned in Section 1.2 that the total force of adhesion between adjacent phases depends upon the strength of single bonds associated with the type of intermolecular and interatomic interactions across the interface, the number of interaction sites, and the range of bonds.

Under ambient conditions, however, Bolger and Michaels[24] pointed out that the surface of metals is covered with a relatively thick layer of oxides (2 to 20 nm). It follows that the surfaces of most structurally important materials, such as metals and ceramics, hydrate under ambient conditions to form a surface with densely distributed active hydroxyl groups. Consequently, it can be seen by comparing a typical range of intermolecular forces with the thicknesses of surface oxide and hydroxide layers that interactions other than those due to hydrogen-bond forces are probably negligible in the adhesion of polymers with hydroxyl-rich surfaces. Although the oxidized surface can be dehydrated and dehydrogenated under high temperature, it can be seen that under ambient conditions the presence of surface hydroxyl groups is a feature of metallic and oxide-ceramic materials, as illustrated in Figure 5.

Based on the above it was first suggested by Bolger and Michaels[24] and Bolger[25] that hydrogen bonding plays a significant role in adhesion and that acid–base interactions can be used to predict the magnitude of hydrogen-bond energy between polymer and hydroxyl-rich surfaces. The details of the hydrogen bond will be discussed by Good *et al.* in Chapter 4.

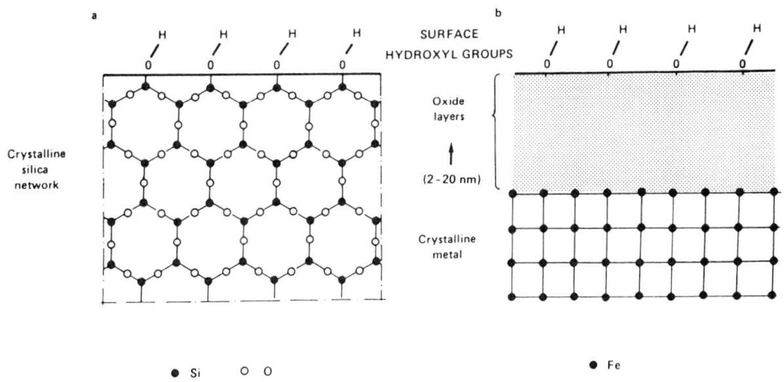

FIGURE 5. Schematic structure of a solid's surface with exposed hydroxyl groups: (a) silica; (b) iron (covered with a layer of an oxide).

1.4. Application of Quantum Field Theory for Estimation of Interaction Force and Energy Between Two Solids (Macroscopic Theory)

1.4.1. General

It was explained earlier that atoms and molecules can be regarded as dipoles which emanate an electromagnetic field around them. This field not only exists in the interior of the material but extends beyond its boundaries.

Therefore, as originally suggested by Lifshitz,[26] the interaction between two condensed bodies treated as a system of atoms acting as isotropic oscillators can be considered as taking place through the fluctuating electromagnetic fields. In Lifshitz theory derived on the basis of quantum electrodynamics, the atomic structure of interacting bodies is ignored. The overall force of interaction between bodies treated as a continuum is shown to be dependent upon the bulk properties of these bodies, such as complex dielectric permittivities.

1.4.2. Energy of the Particle and Condensed Body

For a particle of mass M not under the influence of any external field, the kinetic energy is given by[27]

$$E_k = \frac{Mv^2}{2} = \frac{P^2}{2M} \tag{26}$$

where v is the speed of the particle and P is the momentum.

If the particle moves freely in three-dimensional space, its kinetic energy is described by an unperturbed Hamiltonian H_0, i.e.,

$$H_0 = \frac{1}{2M}(\bar{P}_x^2 + \bar{P}_y^2 + \bar{P}_z^2) = -\frac{\hbar^2}{2M}\Delta \tag{27}$$

where

$$\Delta = \frac{\partial}{\partial x^2} + \frac{\partial}{\partial y^2} + \frac{\partial}{\partial z^2} \tag{28}$$

is the Laplacian operator and $\hbar$ is (Planck's constant)/2π, i.e., $\hbar = h/2\pi$.

In an external field, the particle energy consists of two components[27]: the kinetic energy H_0 and potential energy $U(x,y,z)$, i.e.,

$$H = -\frac{\hbar^2}{2M}\Delta + U(x,y,z) \tag{29}$$

By considering a condensed body as a system of charged particles interacting with an electromagnetic field, we can express the energy of the system as

$$H = H_o + H_{int} \tag{30}$$

where H_o is the Hamiltonian of free particles and photons describing the energy of an unperturbed system, and H_{int} is the interaction operator (Hamiltonian) representing the energy of interaction of particles and photons with the electromagnetic field. The value of the interaction operator H_{int} in Eq. (30) is given by

$$H_{int} = -\int A_\alpha(r) j_\alpha(r) d^3r \tag{31}$$

where $A_\alpha(r)$ is the field operator and $j_\alpha(r)$ the particle current operator.

Physically, quantity H_{int} expresses an increment in the energy of the condensed body due to long-wave fluctuations of the electromagnetic field. Substitution of Eq. (31) into Eq. (30) yields the energy of a body interacting with an electromagnetic field, i.e.,

$$H = H_o - \int A_\alpha(r) j_\alpha(r) d^3 r \tag{32}$$

1.4.3. General Formula for the Force of Interaction Between Two Condensed Bodies

The general formula for interaction forces and energies between two condensed bodies has been developed by Dzyaloshinskii, Lifshitz, and Pitaevskii,[(1)] and is applicable to van der Waals forces between two bodies 1 and 2 separated by a fluid 3 (see Figure 6).

As mentioned earlier, a condensed body can be treated as a system of atoms acting like harmonic oscillators of characteristic absorption frequency $\omega_j = f(i\xi)$. A fluctuating electromagnetic field emanated from body 1 interacts with body 2; this interaction refers not only to the surface of body 2 but also to the substance beyond it.

The response [a secondary (or induced) electromagnetic field] of body 2 depends upon its dielectric permittivity $\epsilon_2 = \epsilon_2(i\xi)$ as well as upon the permittivity of the fluid in the gap between the bodies, i.e., $\epsilon_3 = \epsilon_3(i\xi)$, where $i\xi$ is the imaginary frequency. This secondary field in turn is reflected across the gap and interacts with body 1 [characterized by $\epsilon_1 = \epsilon_1(i\xi)$] giving rise to additional terms in the expression for the forces of interaction.

In a thermodynamic sense, the effects described are related to the nonadditivity of the free energies of components of the system, as molecules (bodies) are effected not only through the field directly radiated from other molecules (bodies) but also through the reflected fields. In practical terms it means that the overall forces of interaction are greater than would be expected from pairwise additive (not enhanced) interactions.

The interaction of particles and photons (present in a condensed body) with the fluctuating electromagnetic field (expressed by a momentum flux) creates an additional

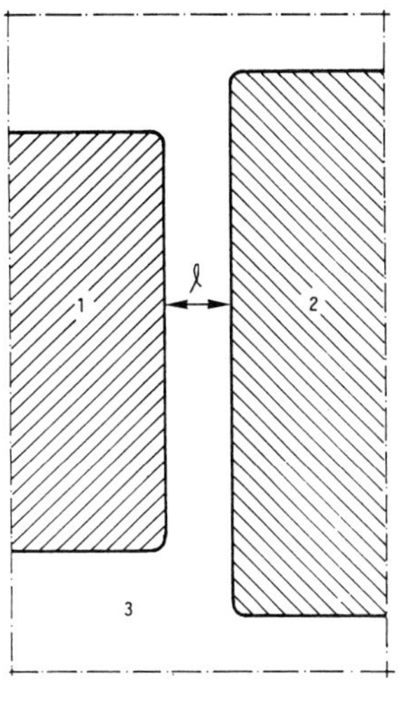

FIGURE 6. Schematic illustration of the system of interacting bodies analyzed by Dzialoshinskii, Lifshitz, and Pitaevskii.

pressure on this body. This pressure, described by an additional stress tensor, results in a change of free energy of the body considered. Consequently, it has been shown[1] that the additional stress σ, and therefore the force F acting on unit area (i.e., stress) between bodies 1 and 2 separated by a gap of width l occupied by medium 3, is

$$\sigma \equiv F(l) = \frac{kT}{\pi c^3} \sum_{n=0}^{\infty}{}' \epsilon_3^{3/2} \xi_n^3 \int_1^{\infty} p^2 \left(\frac{1}{\Delta} + \frac{1}{\bar{\Delta}} \right) dp \tag{33}$$

The parameters in Eq. (33) are defined as follows:

$$\Delta = \left[\left(\frac{s_1 + p}{s_1 - p} \right) \left(\frac{s_2 + p}{s_2 - p} \right) \exp \left(\frac{2p\xi_n}{c} l\epsilon_3^{1/2} \right) - 1 \right] \tag{34}$$

$$\bar{\Delta} = \left[\left(\frac{s_1 + p\epsilon_1/\epsilon_3}{s_1 - p\epsilon_1/\epsilon_3} \right) \left(\frac{s_2 + p\epsilon_2/\epsilon_3}{s_2 - p\epsilon_2/\epsilon_3} \right) \exp \left(\frac{2p\xi_n}{c} l\epsilon_3^{1/2} \right) - 1 \right] \tag{35}$$

$$\xi_n = \frac{2\pi nkT}{\hbar} \text{ is frequency} \qquad [\xi = \xi(i\omega)] \tag{36}$$

k is Boltzmann's constant while ϵ is a dielectric permittivity, being a function of rotational frequency ω, i.e., $\epsilon = \epsilon(\omega)$,

$$s_1 = \left(\frac{\epsilon_1}{\epsilon_3} - 1 + p^2 \right)^{1/2} \quad \text{and} \quad s_2 = \left(\frac{\epsilon_2}{\epsilon_3} - 1 + p^2 \right)^{1/2} \tag{37}$$

l is the gap width between bodies 1 and 2, and n is the quantum number of the relevant oscillation. A prime on the summation sign indicates that, for $n = 0$, the sum is $\frac{1}{2}$ the obtained value.

It is customary in the literature to use a symbol F for the force of interaction instead of σ [see, for instance, Eqs. (33), (46), (49), etc.], but it is necessary to remember that it is the force per unit area (i.e., stress).

1.4.4. Dielectric Permittivity of the Body

In order to calculate the force of interaction by Eq. (33), the complex dielectric permittivities $\epsilon(i\xi)$ of the media involved must be known.

The full spectrum, i.e., the dielectric permittivity ϵ for $\epsilon(i\xi) = 0$ to ∞, is expressed by the Kramers–Kronig equation[1] where ϵ' is the real part and ϵ'' the imaginary part (dissipation term) of the complex permittivity.

$$\epsilon(i\xi) = 1 + \frac{2}{\pi} \int_0^{\infty} \frac{\omega\epsilon''(\omega)}{\omega^2 + \xi^2} d\omega \tag{38}$$

or

$$\epsilon(\omega) = \epsilon(i\xi) = \epsilon'(\omega) + i\epsilon''(\omega) \tag{39}$$

When electromagnetic waves travel across the body, the loss of energy appears as a consequence of a reduction in amplitude. This damping component is represented by an absorption coefficient κ, which together with the refractive index n constitute the complex refractive index of value identical to the complex permittivity, i.e.,

$$\epsilon^{1/2} = n + i\kappa \tag{40}$$

From Eqs. (39) and (40) we have

$$\epsilon'(\omega) = n^2(\omega) - \kappa^2(\omega) \tag{41}$$

and

$$\epsilon''(\omega) = 2n(\omega)\kappa(\omega) \tag{42}$$

It follows from Eq. (42) that $\epsilon''(\omega)$ is real and positive, and so is the dielectric permittivity of the body for the imaginary frequency $i\xi$ expressed by Eq. (38). The value of $\epsilon(i\xi)$ as expressed by Eq. (38) decreases monotonically[1,28–32] with respect to ξ in the manner illustrated in Figure 7 (curve I).

Frequently, it suffices to use a simplified expression for dielectric permittivity[33] that includes frequencies from the far infrared to the far ultraviolet, i.e.,

$$\epsilon(i\xi) = 1 + \frac{\epsilon_0 - \epsilon_{1/2}}{1 + (\xi/\omega_1)} + \frac{\epsilon_{1/2} - n^2}{1 + (\xi/\omega_2)^2} + \frac{n^2 - 1}{1 + (\xi/\omega_3)^2} \tag{43}$$

where ϵ_0 is the static dielectric constant, $\epsilon_{1/2}$ is the dielectric constant at the frequency corresponding to the boundary of MW and IR, ω_1 is the major absorption frequency for the microwave region (MW), ω_2 the major absorption frequency for the infrared region (IR), and ω_3 the major absorption frequency for the ultraviolet region (UV), while n is the refractive index in the visible region.

Figure 7 (curve II) illustrates the dielectric permittivity spectrum of water, calculated by Eq. (43).[28]

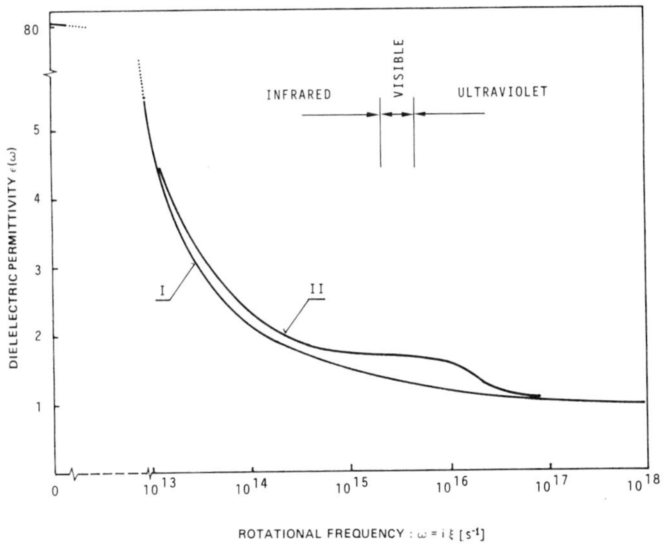

FIGURE 7. Relationship between dielectric permittivity of the body and frequency: (I) graphical representation of the Kramers–Kronig equation for water [see Eq. (38)]; (II) dielectric permittivity spectrum of water calculated by Eq. (43). (Data from Ref. 28.)

1.4.5. Practical Solutions of Equations for the Force of Interaction Between Two Condensed Bodies

The theory is applicable to any material despite its molecular nature, whether liquid or solid or whether these are polymers, ceramics, or metals.

As regards retardation effects, when interacting atoms (or bodies) are very close to each other then the time necessary for the magnetic field emanated from atom (body) 1 to reach atom (body) 2 and return is very short. Thus, interactions of the primary and secondary (induced) fields are in phase, which leads to the enhancement of the interaction forces.

This is known as a nonretarded interaction and the condition for it is given by

$$l \ll \lambdabar \tag{44}$$

where l is the distance between the two atoms (or bodies) and $\lambdabar$ is the wavelength. However, when the atoms are an appreciable distance apart, i.e., when

$$l \gg \lambdabar \tag{45}$$

then the electromagnetic field returned to atom (body) 1 finds that the direction of the instantaneous dipole is now out of phase, i.e., interaction is less favorable. Thus, the resultant force of interaction is smaller than that in the previous case. This is known as a "retardation effect."

It is necessary to stress that the retardation affects only dispersion forces while zero-frequency forces (dipole–dipole and dipole–induced dipole) are not retarded at all separations. Therefore, with an increase in separation distance the dispersion component of the total van der Waals forces becomes less significant.

1.4.5.1. Force of Interaction Over a Short Distance (Nonretarded Force). Dzyaloshinskii, Lifshitz, and Pitaevskii[1] have noted that the temperature usually has a negligible effect on the interaction force. Accordingly, we have from Eq. (33) that the force of interaction for this case is

$$F = \frac{\hbar}{2\pi^2 c^3} \int_0^\infty \int_1^\infty p^2 \xi^3 \epsilon_3^{3/2} \left(\frac{1}{\Delta} + \frac{1}{\bar{\Delta}} \right) \mathrm{d}p\mathrm{d}\xi \tag{46}$$

This equation is valid for separations $l \ll c\hbar/kT$; at room temperature this gives a distance of up to 10^{-4} cm.

It can be seen from Eqs. (34) and (35) describing parameters Δ and $\bar{\Delta}$ that those values of p for which

$$p\xi l/c \simeq 1 \tag{47}$$

are dominant in the integration of Eq. (46) with respect to dp.

From Eq. (47) we have

$$p \gg 1 \tag{48}$$

Introducing relation (48) in Eq. (37) yields $s_1 \simeq s_2 \simeq p$. It follows that $\Delta = \infty$ [see Eq. (34)], and subsequently Eq. (46) is simplified to

$$F = \frac{\hbar}{16\pi^2 l^3} \int_0^\infty \int_0^\infty x^2 \left[\left(\frac{\epsilon_1 + \epsilon_3}{\epsilon_1 - \epsilon_3} \right) \left(\frac{\epsilon_2 + \epsilon_3}{\epsilon_2 - \epsilon_3} \right) e^x - 1 \right]^{-1} dx d\xi \tag{49}$$

where

$$x = p(2l\xi\epsilon_3^{1/2}/c) \tag{50}$$

It should be noted that the expression in parentheses in Eq. (50) represents the ratio of the travel time of the signal of frequency ξ across the gap from body 1 to 2 and back [i.e., $2l/(c/\epsilon_3^{1/2})$] to the period of oscillation (i.e., $1/\xi$).

Equation (49) can be further simplified without significant loss of accuracy[1] to

$$F \simeq \frac{\hbar\bar{\omega}}{8\pi^2 l^3} \tag{51}$$

where

$$\bar{\omega} = \int_0^\infty \left[\left(\frac{\epsilon_1 - \epsilon_3}{\epsilon_1 + \epsilon_3} \right) \left(\frac{\epsilon_2 - \epsilon_3}{\epsilon_2 + \epsilon_3} \right) \right] d\xi \tag{52}$$

and

$$\xi_j = \epsilon_j(i\xi) \tag{53}$$

It is apparent from Eqs. (49) and (51) that the force of interaction between two semi-infinite flat bodies at a close distance decays with the inverse third power of the separation distance. It is the same as predicted by the classical van der Waals theory for the forces between two atoms of a rarefied body (i.e., a gas). The result similar to Eq. (51) has been obtained by Langbein[34] and Israelachvili.[2]

1.4.5.2. Forces of Interaction Over a Long Distance (Retarded Forces). The long-distance interactions correspond to a distance large compared with the wavelength, i.e., for $l \gg \lambda\!\!\!^{-}$. The distance, however, is such that the assumption $l \ll c\hbar/kT$ is still valid. It has been shown by Dzyaloshinskii, Lifshitz, and Pitaevskii[1] that under those circumstances (i.e., $l \gg \lambda\!\!\!^{-}$) the following applies:

- ϵ_1, ϵ_2, and ϵ_3 can be replaced by their values for $i\xi = 0$, e.g., by electrostatic constants ϵ_{10}, ϵ_{20}, and ϵ_{30}.
- Integration in Eq. (33) is carried out with respect to variables p and $x = 2pl\xi/c$.

Accordingly, the force of interaction is[1]

$$F = \frac{3\hbar c}{32\pi^2 l^4 \epsilon_{30}^{1/2}} \int_0^\infty \int_1^\infty \frac{x^3}{p^2} \left\{ \frac{1}{\Delta'} + \frac{1}{\bar{\Delta}'} \right\} dp dx \tag{54}$$

where

$$\Delta' = \left[\left(\frac{s_{10} + p}{s_{10} - p} \right) \left(\frac{s_{20} + p}{s_{20} - p} \right) e^x - 1 \right] \tag{55}$$

$$\bar{\Delta}' = \left[\left(\frac{s_{10} + p\epsilon_{10}/\epsilon_{30}}{s_{10} - p\epsilon_{10}/\epsilon_{30}} \right) \left(\frac{s_{20} + p\epsilon_{20}/\epsilon_{30}}{s_{20} - p\epsilon_{20}/\epsilon_{30}} \right) e^x - 1 \right] \tag{56}$$

$$s_{10} = \left(\frac{\epsilon_{10}}{\epsilon_{30}} - 1 + p^2 \right)^{1/2} \quad \text{and} \quad s_{20} = \left(\frac{\epsilon_{20}}{\epsilon_{30}} - 1 + p^2 \right)^{1/2} \tag{57}$$

Considerable accuracy is maintained if Eq. (54) is simplified to

$$F = \frac{3\hbar c}{16\pi^2 l^4 \epsilon_{30}^{1/2}} \int_1^\infty \frac{1}{p^2} \left[\left(\frac{s_{10} - p}{s_{10} + p} \right) \left(\frac{s_{20} - p}{s_{20} + p} \right) + \left(\frac{s_{10} - p\epsilon_{10}/\epsilon_{30}}{s_{10} + p\epsilon_{10}/\epsilon_{30}} \right) \left(\frac{s_{20} - p\epsilon_{20}/\epsilon_{30}}{s_{20} + p\epsilon_{20}/\epsilon_{30}} \right) \right] dp \tag{58}$$

where s_{10} and s_{20} are given by Eq. (57) [see also Eqs. (37)].

It is clear from Eqs. (54) and (58) that for long distances of separation (retardation effect) the decay of the interaction forces is faster than in the case of nonretarded interactions [see Eqs. (49) and (51)].

1.4.5.3. Influence of Temperature. When the influence of temperature on interaction force cannot be omitted, we have from Eq. (33) after replacing p by a new variable

$$x = p(2l\, \xi_n\, \epsilon_{30}^{1/2}/c) \tag{59}$$

that

$$F = \frac{kT}{16\pi l^3} \int_0^\infty x^2 \left[\left(\frac{\epsilon_{10} + \epsilon_{30}}{\epsilon_{10} - \epsilon_{30}} \right) \left(\frac{\epsilon_{20} + \epsilon_{30}}{\epsilon_{20} - \epsilon_{30}} \right) e^x - 1 \right]^{-1} dx \tag{60}$$

The expression in parentheses in Eq. (59) represents the ratio of the travel time of the signal of frequency ξ_n across the gap l from body 1 to 2 and back [i.e., $2l/(c/\epsilon_{30}^{1/2})$] to the period of oscillation (i.e., $1/\xi_n$).

1.4.5.4. Conditions for Attractive and Repulsive Force of Interaction. Analysis of Eqs. (49), (51), (58) and (60) leads to the following conclusions:

- When bodies 1 and 2 are identical ($\epsilon_1 = \epsilon_2$), the expression under the integral sign is always positive. It follows that two identical bodies in a vacuum always attract each other and the force of attraction decreases monotonically with increase in the gap.
- When the gap is filled with some fluid (i.e., when $\epsilon_3 \neq 1$), the force may be attractive or repulsive depending upon ϵ_3, i.e., when

$$\epsilon_1 < \epsilon_3 \quad \text{and} \quad \epsilon_2 > \epsilon_3 \tag{61}$$

  the interaction results in repulsive forces.

1.4.6. Interactions of Atoms in Rarefied Bodies

It has been shown by Dzyaloshinskii, Lifshitz, and Pitaevskii[1] that the equations shown earlier are also applicable for estimating interaction forces and energies between atoms, e.g., in rarefied bodies. They have shown that the results achieved using their equations were the same as those derived using a classical approach.

For small distances of separation (see Section 1.4.5.1) it is shown[1] that the force of interaction between individual atoms in a vacuum (i.e., when $\epsilon_3 = 1$) is

$$F = \frac{\hbar}{16\pi^3 l^3} \int_0^\infty \int_0^\infty \frac{\epsilon_1''(\omega_1)\epsilon_2''(\omega_2)}{\omega_1 + \omega_2}\, d\omega_1 d\omega_2 \tag{62}$$

which corresponds to the interaction energy given by[1]

$$U = -\frac{3\,\hbar}{8\pi^4 l^6\, N_1 N_2}\int_0^\infty\int_0^\infty \frac{\epsilon_1''(\omega_1)\epsilon_2''(\omega_2)}{\omega_1+\omega_2}\,d\omega_1 d\omega_2 \tag{63a}$$

where N is the number of atoms per unit volume of the body.

By taking into account that the imaginary part of the dielectric permittivity $\epsilon''(\omega)$ is related to the oscillator strength through[27]

$$\omega\epsilon''(\omega) = \frac{2\pi^2 e^2}{M} Nf(\omega) \tag{64}$$

where e is the electronic charge and M the mass of the atom, we have from Eq. (63a) that the energy of interaction is

$$U = -\frac{3\,\hbar e^4}{2M^2 l^6}\int_0^\infty\int_0^\infty \frac{f_1(\omega_1)f_2(\omega_2)}{\omega_1\omega_2(\omega_1+\omega_2)}\,d\omega_1 d\omega_2 \tag{63b}$$

Equation (63b) is identical to that obtained by London[35] using a classical approach. The same result has been obtained by Israelachvili[36] for the low-temperature energy of interaction of two atoms.

For long-range interactions (see Section 1.4.5.2), the force of attraction is[1]

$$F = \frac{\hbar c}{l^4}\cdot\frac{23}{640\pi^2}(\epsilon_{10}-1)(\epsilon_{20}-1) \tag{65}$$

Considering that

$$\epsilon_0 = 1 + 4\pi N\alpha \tag{66}$$

where α is the static polarizability of the atom, we have from Eq. (65) that the energy of interaction is

$$U = -\frac{23\hbar c}{4\pi l^7}\,\alpha_1\alpha_2 \tag{67}$$

It is the same result as that obtained by Casimir and Polder[37] for two atoms distant enough to consider retardation effects.

2. ACID–BASE (DONOR–ACCEPTOR) INTERACTIONS

2.1. Brønsted and Lewis Concepts of Acid–Base Interactions

The concept of the use of acid–base interactions for analyzing interfacial phenomena related to adhesion has recently received much attention in the literature, mainly due to the developments by Bolger and Michaels,[24] Gutmann,[38] and Fowkes and co-workers.[39–42] (Also, see Chapters 3 and 4 by Good *et al.*).

Historically, the first significant approach was that of Brønsted where an acid (proton donor) has the capacity to lose a proton while a base (proton acceptor) is capable of accepting it, i.e.,

$$A \rightleftarrows B^- + H^+ \tag{68}$$

where A is the Brønsted acid, B the conjugate base, and H^+ the proton.

In the contemporary Lewis concept, the acid is the molecule or ion being an electron-pair acceptor due to its incomplete electronic arrangements, associated with the formation of a covalent bond with another species. A Lewis base is a donor of an electron pair with the formation of covalent bond.

There is some difference in Brønsted and Lewis theories in regard to concepts of acid but not in regard to those of base, i.e.,

- in Brønsted theory acid contains protons,
- in Lewis theory acid does not necessarily contain protons.

2.2. Acidity and Basicity of Solid Surfaces

The acidity of a solid's surface corresponds to the ability of the surface to convert an adsorbed base into its conjugate acid. Accordingly, the basicity of the solid surface corresponds to its ability to convert an acid into its conjugate base.

The acidity and basicity of a solid surface is expressed by the isoelectric point of the surface (I_s), defined as the pH of the solution containing suspended particles of a given solid, at which the number of positive charges at the solid surface equals the number of negative charges—this corresponds to a zero value of the zeta potential. Typical values of the isoelectric point for selected oxides (corresponding to surfaces of typical substrates) are given in Table 5.

According to Pauling[(7)] the hydrogen bond has partially covalent–partially ionic character as demonstrated by Eq. (24). The energy of this increases when the ionic contribution to the bond increases, i.e., when the equilibrium of reaction (24) is moving to the right side. Based on the above, Bolger and Michaels[(24)] proposed that a qualitative assessment of the energy of the hydrogen bond can be carried out by assessing the equilibrium constants of acid–base interactions between surface active groups with the polymers (e.g., adhesives) categorized as organic bases and acids. Possible acid–base interactions were categorized[(24,25)] into two typical groups.

(1) A solid provides the basic groups that interact with an adhesive acting as organic acid. In this case the relative probability Δ_A of nonionic or ionic reaction associated with the formation of the H bond is

$$\Delta_A \equiv \log K_A = I_s - pk_a \tag{69}$$

where K_A is the equilibrium constant of the acidic reaction between the substrate and adhesive, I_s the isoelectric point of the substrate's surface, and k_a the acidic dissociation constant of the adhesive.

(2) A solid provides acidic groups which interact with an adhesive acting as organic base, hence

$$\Delta_B \equiv \log K_B = pk_b - I_s \tag{70}$$

where K_B is the equilibrium constant of the basic reaction between the substrate and adhesive, and k_b is the basic dissociation constant of the adhesive.

The theoretical[(24)] relationship between the energy of the hydrogen bond and the parameter Δ in acid–base interactions between the solid surface and organic bases and acids is illustrated in Figure 8a. Large positive values of Δ indicate strong ionic interac-

TABLE 5
Isoelectric Points (I_s) for Selected Oxides(24,32)

Oxide	Isoelectric point I_s
SiO_2	1.8–2.2
AlOOH	
Diaspore	5.0–7.5
Bayerite	5.4–9.3
Boehmite	6.5–8.8
α-Al_2O_3	6.6–9.2
γ-Al_2O_3	7.4–8.6
Amorphous Al_2O_3	7.5–8.0
Fe_2O_3	6.5–6.9
Fe_3O_4	6.7–8.6
TiO_2	6.0–6.3
MgO	12.0–12.7

tions, which are associated with high strength of created hydrogen bonds, while large negative values of Δ correspond to negligible ionic forces associated with very weak hydrogen bonds. If an organic compound interacts with the solid surface in a mutual mode, the dominant mode (acid or base) is given by a larger value of parameter Δ.

Table 6 contains some data(24) on values of pk_a and pk_b of selected organic acids and

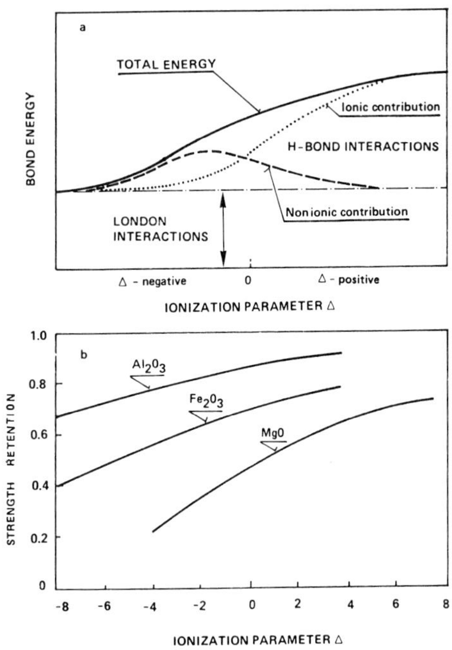

FIGURE 8. Relationship between the energy of the hydrogen bond and parameter Δ in acid–base interactions between a solid and organic base and acid; (a) theoretical prediction(24) and (b) experimental results for adhesive bonds(32) (the coordinate "strength retention" refers to the ratio wet strength/dry strength; wet strength was determined after 7 days immersion of joints in 25 °C water).

TABLE 6

Acid–Base Interactions of Selected Oxides (Acidic, SiO_2; Slightly Basic, Al_2O_3; Basic, MgO) with Organic Acids and Bases

				$\Delta_A \equiv \log K_A = I_s - pk_a$			$\Delta_B \equiv \log K_B = -I_s + pk_b$		
No.	Organic acid	pk_a	pk_b	SiO_2 ($I_s = 2$)	Al_2O_3 ($I_s = 8$)	MgO ($I_s = 12$)	SiO_2 ($I_s = 2$)	Al_2O_3 ($I_s = 8$)	MgO ($I_s = 12$)
1.	Amines	20	11	−18	−12	−8	+8	+2	−2
2.	Phenols	10	−7	−8	−2	+2	−9	−15	−19
3.	Carboxylic acids	4.5	−6	−2.5	+3.5	+7.5	−8	−14	−18
4.	Water	15	−1.7	−13.7	−7.7	−3.7	−3.7	−9.7	−13.7
5.	Dodecyl sulphonic acid	−1		3	9	13			
6.	Acetic acid	4.7	−6.1	−2.7	3.3	7.3	−8.1	−14.1	−18.1
7.	Acetone	20	−7.2	−18	−12	−8	−9.2	−15.2	−19.2
8.	Ethyl acetate	26		−24	−18	−14			
9.	Toluene	37		−35	−29	−25			
10.	Ammonium hydroxide		12.5				10.5	4.5	0.5
11.	Urea		1.0				−1.0	−7	−11

bases, including functional groups (such as phenolics, carboxylic acids, and primary amines) present in typical adhesives. It also shows values of I_s of typical substrates in adhesion, e.g., acidic SiO_2, slightly basic Al_2O_3, and basic MgO.

Based on these data, the values of parameters $\Delta_A \equiv \log K_A$ and $\Delta_B \equiv \log K_B$ were calculated and are applicable for the assessment of the expected behavior of adhesive bonds made with adhesives containing particular functional groups in the environment of interest, e.g., water.

Depending upon the character of the substrate's surface, water can interact in the following modes:

- acidic—in contact with basic surfaces, e.g., MgO;
- basic—in contact with acidic surfaces, e.g., SiO_2;
- mixed, basic and acidic—in contact with neutral surfaces, e.g., Fe_2O_3 and Al_2O_3.

Accordingly, it can be seen from Table 6 that the best durability of adhesive bonds under water attack can be expected when adhesives contain the following functional groups:

- for Fe or Al substrate—adhesives with carboxylic acids or amines,
- for silica-glass substrates—adhesives with amines,
- for Mg substrates—adhesives with carboxylic acids and phenols.

The above conclusions have been supported by experimental evidence published by Wu,[(32)] who gives data on strength retention of lap-shear bonds (see Figure 8b) made with Mg, Al, and Fe substrates bonded by polyethylenes containing functional groups cited above, i.e., amine, carboxy, and OH.

It is seen that there is a strong correlation between the value of parameter $\Delta \equiv \log K$ and strength retention, and it is exactly as predicted in Figure 8a.

2.3. Quantitative Assessment of Acid–Base Interactions

Significant progress in the area of the application of acid–base interactions to adhesion is attributed to Fowkes and co-workers[39,41,42] and Gutmann,[38] who developed methods allowing one to estimate the enthalpy of these interactions. The Fowkes[39] method is based on the earlier Bolger–Michaels[24] assumption that only dispersion and hydrogen-bond forces (the latter being a subset of acid–base interactions) are significant in adhesion between adjacent phases.

The thermodynamic work of adhesion is assumed by Fowkes[39] to be

$$W_A = W_A^d + W_A^{ab} \tag{71}$$

where W_A^{ab} is the contribution of acid–base interactions to W_A. Since a dispersive component of the total work of adhesion is

$$W_A^d = 2\,(\gamma_1^d\,\gamma_2^d)^{1/2} \tag{72}$$

and the total work of adhesion is as expressed by the Young–Dupré equation, namely

$$W_A = \gamma_2\,(1 + \cos\theta) + \pi_{e(1)} \tag{73}$$

the acid–base component [by reference to Eq. (71)] is

$$W_A^{ab} = \gamma_2\,(1 + \cos\theta) - 2\,(\gamma_1^d\,\gamma_2^d)^{1/2} + \pi_{e(1)} \tag{74}$$

All the parameters in Eq. (74) are easy to determine experimentally from wettability studies (see Section 4).

Experimental results published by Fowkes[39] confirm his theoretical predictions.

In many cases polymers (such as adhesives or protective coatings) are applied from solutions and therefore the whole spectrum of possible interactions (i.e., polymer–solid or solvent–solid) as well as interactions between polymer and solvent must be analyzed in order to predict the conditions favoring adsorption of the polymer (but not the solvent) onto the relevant receptive sites of the solid surface.

It has been proposed[39] that there is a similarity between expression (71) for thermodynamic work of adhesion and that for the enthalpy of interaction between adjacent phases (see Gardon[43] and Hansen[44]), i.e.,

$$\Delta H_{12} = \Delta H_{12}^d + \Delta H_{12}^{ab} \tag{75}$$

Taking into account this similarity as well as the fact that the enthalpy of acid–base interactions was shown to be independent of dipole–dipole interactions (see Table 7), it was subsequently hypothesized by Fowkes that the Drago[45,46] expression for enthalpy of acid–base interactions ΔH_{12}^{ab} is suitable for predicting the enthalpy of polar and hydrogen-bond interactions and, furthermore, conditions for the maximum adhesion.

According to Edwards[47] and Drago,[45,46] the free energy[47] or enthalpy of acid–base interactions can be predicted by the use of new constants expressing basic (B) and acidic (A) properties of the substrate and polymer. In Drago's approach, the enthalpy of acid–base interactions is given by the following expression:

$$-\Delta H_{12}^{ab} = C_A\,C_B + E_A\,E_B \tag{76}$$

TABLE 7

Comparison of Enthalpies of Acid–Base Interactions[39] (Acetone with Acids and *p*-Chlorophenol with Bases) with the Energy of Dipole–Dipole Interactions ($U = 2\mu_1^2\mu_2^2/3kTr^6$)

	Acetone (2)				*p*-Chlorophenol (2)		
Acid (1)	$2\mu_1^2\mu_2^2$ (Cm × $10^{-30})^4$	ΔH_{E+C} (kJ/mol)	ΔH_{Exp} (kJ/mol)	Base (1)	$2\mu_1^2\mu_2^2$ (Cm × $10^{-30})^4$	ΔH_{E+C} (kJ/mol)	ΔH_{Exp} (kJ/mol)
Iodine	0	13.81	13.81	Trimethylamine	421	39.78	40.19
Aluminium trimethyl	512	83.74	85.00	Butyl ether	1535	26.38	25.96
Chloroform	2130	15.07	15.07	Ethyl sulfide	2624	20.93	20.93
Water	7032	13.40	13.40	Ethyl acetate	3491	20.93	21.35
p-Chlorophenol	9136	22.61	22.61	Acetone	9136	21.35	21.35

where C_A, E_A are covalent (C) and electrostatic, i.e., ionic (E), constants for acids; and C_B, E_B are covalent (C) and electrostatic, i.e., ionic (E), constants for bases.

It has been proved by Drago[45] that Eq. (76) allows predictions of ΔH_{12}^{ab} with an accuracy ± 0.8 kJ/mol. Taking into account that the average value of hydrogen-bond enthalpy is 8.0 to 42 kJ/mol, this accuracy corresponds to only a 2 to 9% difference between experimental and theoretical results.

A hypothesis established by Fowkes[39] is that the enthalpy of acid–base interactions ΔH_{12}^{ab} equals that of adsorption ΔH_{12}^{ads}. For an organic base or acid adsorbed on a given solid, this can be calculated from the temperature dependence of the equilibrium constants $k(T)$ derived from the Arrhenius reaction-rate equation, assuming that over a small change of temperature, ΔH_{12}^{ads} is constant:

$$-\Delta H_{12}^{ab} = -\Delta H_{12}^{ads} = \frac{RT_1T_2}{T_2 - T_1}\,\ell n\left[\frac{k(T_2)}{k(T_1)}\right] \tag{77}$$

Adsorption equilibrium constants for experiments carried out at temperatures T_1 and T_2, e.g., $k(T_1)$ and $k(T_2)$ in Eq. (77), are determined by reference to the straight-line Langmuir adsorption plots (see Figure 9b):

$$\frac{C}{\Gamma} = \frac{1}{\Gamma_m k(T)} + \frac{C}{\Gamma_m} \tag{78}$$

where Γ is the surface concentration for adsorption, Γ_m is the maximum value for Γ, C is the volume concentration, and

$$k(T) = \frac{\Gamma}{(\Gamma_m - \Gamma)C} \tag{79}$$

Acidity and basicity of the solid is determined from adsorption studies, in which two or more organic acids and/or bases are used with known values of Drago's parameters: $C_{B(1)}$, $E_{B(1)}$ and $C_{B(2)}$, $E_{B(2)}$. Having determined the enthalpies of adsorption $\Delta H_{12(1)}^{ads}$ for the first organic acid/base and $\Delta H_{12(2)}^{ads}$ for the second organic acid/base [see Eqs. (77)–(79)], we can estimate values of parameters E_A and C_A related to the basic and acidic properties of the solid by solving two simultaneous Drago equations,

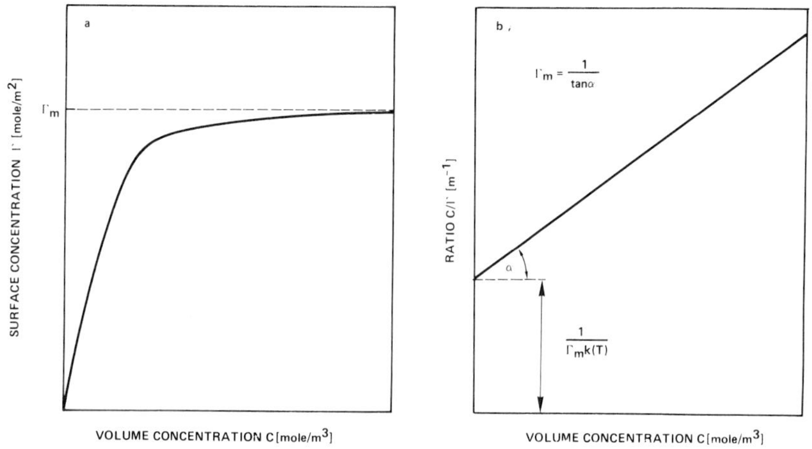

FIGURE 9. (a) Langmuir's adsorption isotherm. (b) Representation of Eq. (78) by a straight-line Langmuir plot with explanation of the sense of the parameters of Eq. (78).

$$E_A = -\frac{\Delta H^{ads}_{12(1)}}{E_{B(1)}} - C_A\left(\frac{C_{B(1)}}{E_{B(1)}}\right) \tag{80}$$

and

$$E_A = -\frac{\Delta H^{ads}_{12(2)}}{E_{B(2)}} - C_A\left(\frac{C_{B(2)}}{E_{B(2)}}\right) \tag{81}$$

It follows from Eqs. (80) and (81) that

$$C_A = \left(\frac{\Delta H^{ads}_{12(2)}}{E_{B(2)}} - \frac{\Delta H^{ads}_{12(1)}}{E_{B(1)}}\right) \Bigg/ \left(\frac{C_{B(1)}}{E_{B(1)}} - \frac{C_{B(2)}}{E_{B(2)}}\right) \tag{82}$$

The value of E_A is calculated by further substitution of Eq. (82).

A significant development of donor–acceptor concepts has been made by Gutmann.[38] This approach considers not only electronic changes (in charge distribution) associated with acid–base interactions, but also changes of bond geometry (bond length and bond angle) brought about by mutual interactions of the considered molecule with the other constituents of the system. These changes, indicated by bond lengths or spectroscopic properties, are correlated with thermodynamic and kinetic data. In this approach, the nature of molecular forces is therefore not important because changes in all types of bonds (e.g., covalent, ionic, metallic, hydrogen, and van der Waals) are governed by the same rules as defined by Gutmann.

The donor–acceptor properties of a substance are characterized by the donor number (DN) and acceptor number (AN), while an approximate measure of bond energy is given by[38]

$$\Delta H \simeq \frac{\text{DN} \times \text{AN}}{100} \tag{83}$$

Numerical values of DN and AN are determined experimentally. The donor number is determined from calorimetric measurements and is defined as the enthalpy of 1:1 molecular adduct formation[38] in the reaction between the donor (D) and $SbCl_5$ in 1,2 - dichloroethane (the reference solvent; DN $\equiv$ 0), i.e.,

$$DN \equiv -\Delta H_{D\cdot SbCl_5}; \qquad D + SbCl_5 \rightleftarrows D\cdot SbCl_5 \tag{84}$$

The enthalpy of adduct formation indicates approximately the energy of the covalent bond between the donor atom and the Sb atom in $SbCl_5$.

Also, the free energy is given by

$$\Delta G^{\circ} = -RT \ln K \tag{85}$$

The acceptor number (AN) is determined from spectroscopic data[38] and defined as a dimensionless number associated with the ^{31}P-NMR chemical shift δ_{corr} of P in triethylphosphine oxide (Et_3PO) in the solvent of interest (with hexane as a reference solvent characterized by AN = 0), which is related to the ^{31}P-NMR chemical shift of $Et_3PO\cdot SbCl_5$ adduct dissolved in 1,2 -dichloroethane ($\delta_{corr(Et_3PO\cdot SbCl_5)}$), i.e.,

$$AN \equiv \frac{\delta_{corr} \cdot 100}{\delta_{corr(Et_3PO\cdot SbCl_5)}} = \delta_{corr} 2.348 \tag{86}$$

Gutmann[38] pointed out that there is a correlation between the ^{31}P-NMR chemical shift and pK_a values. Values of DN and AN for selected chemicals are given in Table 8.

Use of inverse gas chromotography (IGC) for determining the enthalpy of acid–base interactions has been reported in recent work.[48–51] In this technique, the substrate is used as a stationary phase and its interactions with vapors of known solvents at "infinite

TABLE 8

Donor Numbers (DN) and Acceptor Numbers (AN) of Selected Chemicals[38]

Chemical	DN	AN
1,2-Dichloroethane	0.0	
Hexane		0.0
Benzene	0.1	8.2
Nitromethane	2.7	20.5
Nitrobenzene	4.4	14.8
Benzonitrile	11.9	15.5
Dioxane	14.8	10.8
Propandiol-(1,2)-carbonate	15.1	18.3
Acetone	17.0	12.5
Water	18.0	54.8
Tetrahydrofuran	20.0	8.0
Formamide	26.6	39.8
N-methyl-pyrrolidionone	27.3	13.3
N-dimethyl acetamide	27.8	13.6
Dimethylsulfoxide	29.8	19.3
Pyridine	33.1	14.2
Hexamethylphosphoric triamide	38.8	10.6

dilution" are recorded (temperature dependence of the specific retention volumes). It is assumed that over the small change of temperature, the value of ΔH^{ab} is constant. The enthalpy of acid–base interactions is given by[49]

$$-\Delta H^{ab} = (\Delta H_S^a - \Delta H_S^M)_{sample} - (\Delta H_S^a - \Delta H_S^M)_{ref} \quad (87)$$

where a, b, and M refer to acid, base, and model compound, respectively.

The enthalpies ΔH_S at infinite dilution are calculated[49] by

$$\Delta H_S = -R \frac{d\ln(V_g^o)}{d\ln(1/T)} \quad (88)$$

where V_g^o is the specific retention volume, i.e.,

$$V_g^o = (t_r F / w_L) J_2^3 \quad (89)$$

Here, t_r is the net retention time, F the corrected flow rate, w_L the mass of the substrate in the column, and J_2^3 the correction for the pressure drop.

2.4. Acid–Base Interactions in Adhesion

As mentioned earlier, polymers (such as adhesives, primers, or protective coatings) are frequently applied to surfaces of materials not as 100% solid- containing polymers but as solutions. Therefore, the adsorption of polymers in such a system is a very complex process which includes the following major problems:

- Mutual interactions between the constituents of the system, i.e., solvent–polymer, solvent–solid, and polymer–solid interactions.
- Conformation of the polymer molecule (or rather chain segments) on the surface.

Within the scope of this work some aspects of the first point are discussed below.

A solvent (as well as the polymer) must be regarded as an organic acid and base, which can compete for access to basic and acidic sites on the surface of a solid. Therefore, the best results in respect to a polymer's adsorption are achieved when a neutral solvent is used, i.e., when $\Delta H_{12}^{ab} = 0$ for acid–base interactions of the solvent with the solid.

The above has been investigated by Fowkes and co-workers.[39,41,42] Also, the results reported by Schreiber[48] and Ponce[50] contain interesting implications regarding some aspects of selective adsorption. Figure 10 illustrates the relationship[39] between the amount of adsorbed basic polymer, polymethylmethacrylate (PMMA), on the acidic surface of silica, from basic and acidic solvents listed in Table 9. The acidity of solvents was determined as their enthalpy of acid–base interactions (i.e., ΔH_{12}^{ab}) with ethyl acetate, while basicity was determined as ΔH_{12}^{ab} of solvents with t-butanol. As expected, the most effective was PMMA adsorption from a neutral solvent—carbon tetrachloride.

Figure 10 shows that apart from an optimum for CCl_4, the amount of adsorbed PMMA decreases significantly:

- with increase in solvent acidity, because acidic solvent does not want to "give away" the molecules of the basic polymer to the less basic solid, and
- with increase in solvent basicity, because both more basic solvent and less basic polymer compete for access to the acidic sites on silica.

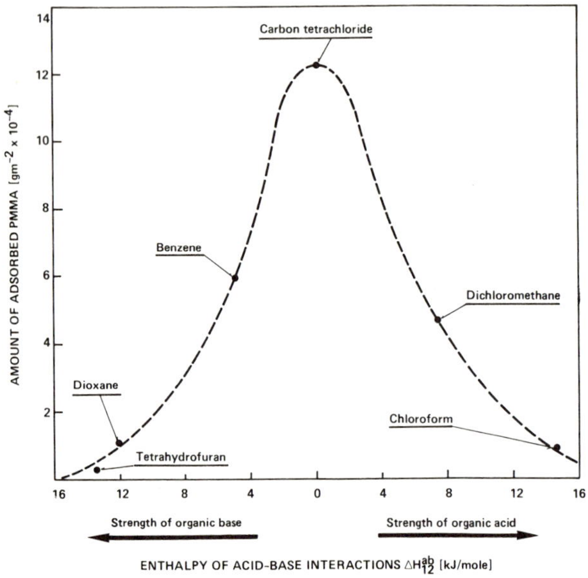

FIGURE 10. Relationship between the amount of PMMA adsorbed onto an acidic silica surface, depending upon the basicity or acidity of the organic solvent.[39]

The practical implications of the competition between basic polymer and basic solvent for access to the acidic SiOH sites on silica are illustrated in Figure 11.

With reference to the strength of adhesion between PMMA films and silica, Figure 11a explains that upon evaporation of the solvent less basic than PMMA (i.e., CCl_4), the polymer's basic groups are hydrogen-bonded to acidic sites of silica and good strength is expected. However, when PMMA film is cast from a solvent (i.e., THF) more basic than the polymer (see Figure 11b), the solvent molecules are more readily adsorbed to the surface SiOH groups. Thus, upon evaporation of solvent, the PMMA-basic groups are not bound to silica-OH groups as they were occupied by THF; resultant adhesion is therefore

TABLE 9
Acidity and Basicity of Solvents Used in Fowkes's Experiment Illustrated in Figure 10

Solvent	Acidity ΔH^{ab} vs. ethyl acetate (kJ/mol)	Basicity ΔH^{ab} vs. *t*-butanol (kJ/mol)
Tetrahydrofuran		13.6
p-Dioxane		12.2
Benzene		5.0
Carbon tetrachloride	0	0
Dichloromethane	7.4	
Chloroform	14.5	

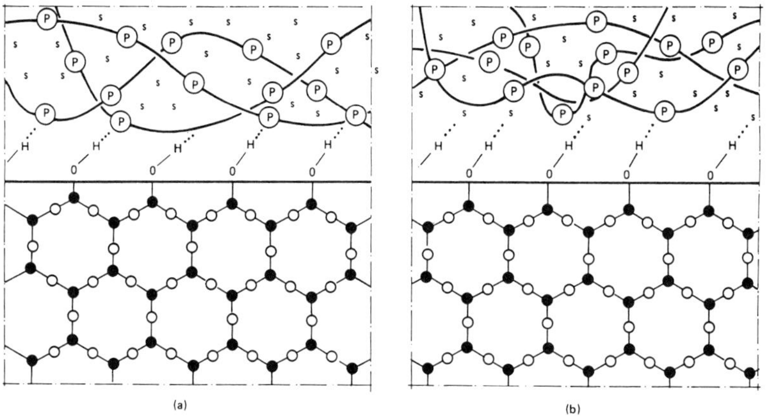

FIGURE 11. Competition between basic polymer (PMMA) and (a) slightly basic solvent or (b) more basic solvent, for access to the acidic SiOH sites on silica (P denotes basic groups of polymer, S denotes solvent molecules).

poor. It is exactly the same as the situation achieved experimentally by Fowkes and Maruchi[41] in their studies (see Figure 12) on the strength of PMMA–silica composites, where the polymer was adsorbed from THF and less basic CH_2Cl_2. A broken line in Figure 12 illustrates a hypothetical result, if neutral CCl_4 was used as a solvent for PMMA adsorbed on silica. Similar conclusions regarding acid–base interactions can be deduced from analysis of data reported elsewhere.[48,50]

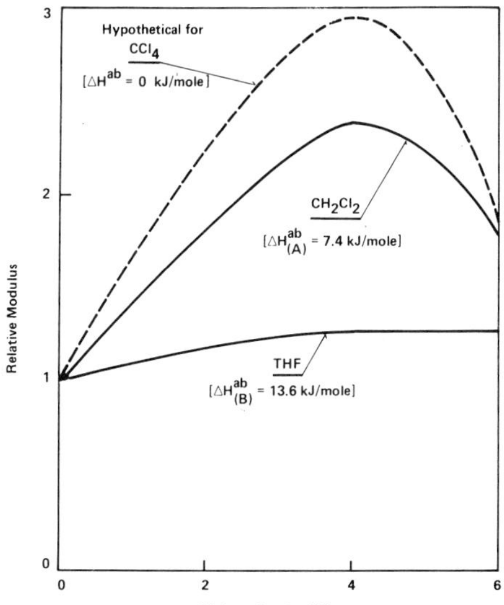

FIGURE 12. Influence of basicity of solvents on the strength of composites: PMMA (basic polymer)/silica (acidic substrate).[39]

3. THERMODYNAMICS OF SURFACES

3.1. Surface Energy, Thermodynamic Work of Adhesion, and Energy of Cohesion

In classical terms the thermodynamic work of cohesion is defined as work associated with the mechanical separation (in adiabatic conditions) of a primarily monolithic material and further adiabatic displacement of a separated part to infinity; this process is by definition reversible. If the stress associated with the separation is σ, then the energy of interaction is

$$-U = 2\gamma = \int_{r_0}^{\infty} \sigma dr \tag{90}$$

When the two phases are dissimilar, the work required for separation is known as the work of adhesion ($-U_{12} = W_A = \gamma_1 + \gamma_2 - \gamma_{12}$), while for separation of a primarily monolithic phase it is referred to as the work of cohesion.

According to the quantum electrodynamics approach (see Section 1.4) the energy of interaction of two condensed bodies depends upon their dielectric permittivities and the separation distance. For two dissimilar, semi-infinite bodies 1 and 2 acting through a layer of fluid 3 we have (as given by van Oss, Good, and Chaudhury[52] that

$$U_{132} = -W_{132} = -\frac{kT}{8\pi l^2} \sum_{n=0}^{\infty}{}' \left(\frac{\epsilon_1 - \epsilon_3}{\epsilon_1 + \epsilon_3}\right)\left(\frac{\epsilon_2 - \epsilon_3}{\epsilon_2 + \epsilon_3}\right) \tag{91}$$

For interactions over small distances we have, from Eqs. (51) and (2), that

$$U_{132} = -W_{132} = -\frac{\hbar\bar{\omega}}{16\pi^2 l^2} \tag{92}$$

where $\bar{\omega}$ is given by Eq. (52).

In applying the above to the energy of cohesion of the condensed bodies 1 and 2, we have $\epsilon_3 = 1$. It follows from Eq. (92) that

$$U_{11} = -\frac{\hbar\bar{\omega}_1}{16\pi^2 r_{0(1)}^2} \quad \text{and} \quad U_{22} = -\frac{\hbar\bar{\omega}_2}{16\pi^2 r_{0(2)}^2} \tag{93}$$

where $r_{0(1)}$ and $r_{0(2)}$ are the equilibrium separation distances for atoms or molecules in body 1 and 2, respectively. The values of $\bar{\omega}_1$ and $\bar{\omega}_2$ in Eq. (93) are given by

$$\bar{\omega}_1 = \int_0^{\infty} \left(\frac{\epsilon_1 - 1}{\epsilon_1 + 1}\right)^2 d\xi \quad \text{and} \quad \bar{\omega}_2 = \int_0^{\infty} \left(\frac{\epsilon_2 - 1}{\epsilon_2 + 1}\right)^2 d\xi \tag{94}$$

where $\epsilon_j = \epsilon_j(i\xi_j)$

The surface energies of the condensed bodies 1 and 2 are determined by reference to Eqs. (90), (93), and (94), i.e.,

$$\gamma_1 = -\frac{U_{11}}{2} = \frac{\hbar\bar{\omega}_1}{32\pi^2 r_{0(1)}^2} \quad \text{and} \quad \gamma_2 = -\frac{U_{22}}{2} = \frac{\hbar\bar{\omega}_2}{32\pi^2 r_{0(2)}^2} \tag{95}$$

It can be concluded from Eq. (91) that the attraction of two dissimilar bodies is always weaker than the attraction of two bodies made of the same substances.

3.2. Energy Equilibrium at the Interface

In 1805 Young expressed the conditions for equilibrium at a solid–liquid interface like that in Figure 13a as

$$\gamma_1 - \gamma_{12} = \gamma_2 \cos\theta_{21} \tag{96}$$

where γ_1 and γ_2 are surface energies of the solid (1) and liquid (2) in equilibrium with the saturated vapor of the liquid.

According to Dupré, the work of adhesion between two adjacent phases 1 and 2 is

$$W_A^{12} = \gamma_{1^o} + \gamma_2 - \gamma_{12} \tag{97}$$

where γ_{1^o} is a surface energy of phase 1 in vacuum. Combination of Eqs. (96) and (97) yields the well-known Young–Dupré equation

$$W_A^{12} = \gamma_2 (1 + \cos\theta_{21}) + (\gamma_{1^o} - \gamma_1) \tag{98}$$

where

$$\gamma_{1^o} - \gamma_1 = \pi_e \tag{99}$$

is the so-called equilibrium spreading pressure given by

$$\pi_e = RT\int_0^{p^o} \Gamma d(\ln p) \tag{100}$$

Here, Γ is the number of moles of vapor adsorbed on a unit area of substrate, p^o the saturated vapor pressure, and p the vapor pressure.

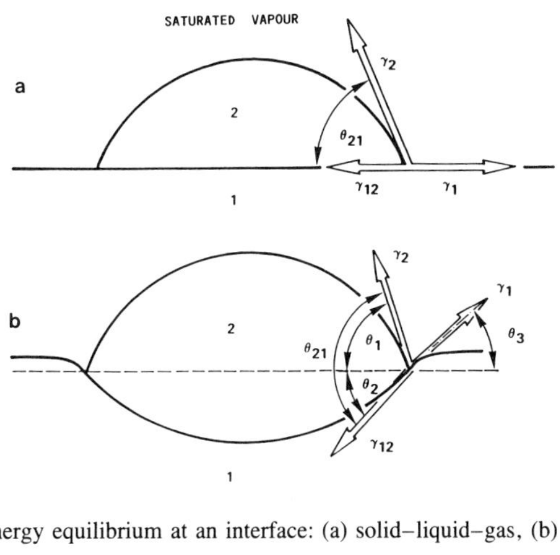

FIGURE 13. Energy equilibrium at an interface: (a) solid–liquid–gas, (b) liquid–liquid–gas.

It has been shown by Fowkes[39] that $\pi_e = 0$ for high-energy liquids on low-energy solids. For this reason (and for simplicity of further analysis) it can be assumed that, as a first approximation, the Young–Dupré equation is well represented in the following form:

$$W_A^{12} = \gamma_2 (1 + \cos\theta_{21}) \tag{101}$$

3.3. *Interfacial Energy*

3.3.1. Basic Equations

Interfacial energy γ_{12} can be calculated from rearranged Eqs. (96) and (101), i.e.,

$$\gamma_{12} = \gamma_1 - \gamma_2 \cos\theta_{21} \tag{102}$$

while the Dupré equation (97) yields it in the form

$$\gamma_{12} = \gamma_{1^\circ} + \gamma_2 - W_A^{12} \tag{103}$$

It is apparent from Eqs. (102) and (103) that, apart from surface energies of adjacent phases, either the equilibrium contact angle θ_{21} or the thermodynamic work of adhesion W_A^{12} must be determined in order to estimate the value of interfacial energy. Both of those parameters (e.g., θ_{21} and W_A^{12} are interrelated through Eqs. (98) or (101) and, since θ_{21} is easy to determine experimentally, it is possible to estimate the value of W_A^{12} and therefore the value of the corresponding interfacial energy γ_{12}.

It follows from Eq. (102) that interfacial energy can be expressed in the following dimensionless form:

$$\frac{\gamma_{12}}{\gamma_2} = \frac{\gamma_1}{\gamma_2} - \cos\theta_{21} \tag{104}$$

Equations (101)–(104) explain that $\cos\theta_{21}$ or W_A^{12} appears in explicit or implicit form in any equation for interfacial energy. Therefore, in order to solve equations for interfacial energy it is necessary to express $\cos\theta_{21}$ or W_A^{12} in terms of thermodynamic parameters describing surface properties of interacting phases, for instance their surface energies γ_1 and γ_2.

3.3.2. Theoretical Aspects of Interfacial Energy

The literature on experimental and theoretical aspects of interfacial energy is extensive.[9,32,53,54] The most comprehensive theory to date is that developed by Good and Girifalco,[53] and extended further by Good and Elbing.[54] In this theory, a thermodynamic function Φ_{12} is introduced and defined as the ratio of the free energy of adhesion ΔG_{12}^a between adjacent phases to the geometric mean of the energy of cohesion ΔG_1^c and ΔG_2^c of these phases, i.e.,

$$\Phi_{12} \equiv \frac{\Delta G_{12}^a}{(\Delta G_1^c \Delta G_2^c)^{1/2}} \equiv \frac{\gamma_1 + \gamma_2 - \gamma_{12}}{2(\gamma_1\gamma_2)^{1/2}} \tag{105a}$$

It follows from this equation that the interfacial energy (assuming $\pi_e = 0$) can be estimated by

$$\gamma_{12} = \gamma_1 + \gamma_2 - 2\Phi_{12}(\gamma_1\gamma_2)^{1/2} \tag{106}$$

It can be seen from Eq. (106) that the value of the interaction parameter must be known in order to determine interfacial energy γ_{12}.

According to Good and Elbing,(54) the interaction parameter can be estimated using the following theoretical equation:

$$\Phi_{12} = \frac{\frac{3}{4}\alpha_1\alpha_2 \frac{2I_1\delta_1 I_2\delta_2}{I_1\delta_1 + I_2\delta_2} + \frac{1}{2}(\alpha_1\mu_2^2 + \alpha_2\mu_1^2)\beta + \frac{2\mu_1^2\mu_2^2\beta^2}{3kT}}{\left[\left(\frac{3}{4}\alpha_1^2 I_1\delta_1 + \alpha_1\mu_1^2\beta + \frac{2\mu_1^4\beta^2}{3kT}\right)\left(\frac{3}{4}\alpha_2^2 I_2\delta_2 + \alpha_2\mu_2^2\beta + \frac{2\mu_2^4\beta^2}{3kT}\right)\right]^{1/2}} \tag{107}$$

Although relatively accurate, Eq. (107) in its original form has rather limited applicability for solving practical problems in adhesion science and technology. The reasons are obvious from the form of the equation. An accurate value of the interaction parameter Φ_{12} can be estimated alternatively using the quantum electrodynamics approach described in Section 1.4.

From Eq. (105a) we have that the interaction parameter Φ_{12} can be expressed alternatively as

$$\Phi_{12} = \frac{W_A^{132}}{2(\gamma_1\gamma_2)^{1/2}} = \frac{-U_{132}}{2(\gamma_1\gamma_2)^{1/2}} \tag{105b}$$

If Eqs. (91) and (95) are substituted into Eq. (105b) and we assume that (1) $\epsilon_3 = 1$ for interactions in vacuum and (2) the velocity of light is approximately the same in substances 1, 2, and 3, then the interaction parameter for nonmetallic materials is given by

$$\Phi_{12} = \frac{r_{0(1)}\, r_{0(2)}}{l^2} \left[\sum_{n=0}^{\infty}{}' \left(\frac{\epsilon_1 - 1}{\epsilon_1 + 1}\right)\left(\frac{\epsilon_2 - 1}{\epsilon_2 + 1}\right)\right] \left[\sum_{n=0}^{\infty}{}' \left(\frac{\epsilon_1 - 1}{\epsilon_1 + 1}\right)^2 \sum_{n=0}^{\infty}{}' \left(\frac{\epsilon_2 - 1}{\epsilon_2 + 1}\right)^2\right]^{-1/2} \tag{108}$$

which can be expressed as

$$\Phi_{12} = \Phi_{12}^{s}\, \Phi_{12}^{\epsilon} \tag{109}$$

where superscripts s and ϵ refer to the spatial structure and dielectric properties of substances 1 and 2, respectively. By assuming that the value of l can be quantified(53) by

$$l \simeq (r_{0(1)} r_{0(2)})^{1/2} \tag{110}$$

we have from Eqs. (108)–(110) that $\Phi_{12}^{s} \simeq 1.0$ and subsequently

$$\Phi_{12} \simeq \Phi_{12}^{\epsilon} = \left[\sum_{n=0}^{\infty}{}' \left(\frac{\epsilon_1 - 1}{\epsilon_1 + 1}\right)\left(\frac{\epsilon_2 - 1}{\epsilon_2 + 1}\right)\right] \left[\sum_{n=0}^{\infty}{}' \left(\frac{\epsilon_1 - 1}{\epsilon_1 + 1}\right)^2 \sum_{n=0}^{\infty}{}' \left(\frac{\epsilon_2 - 1}{\epsilon_2 + 1}\right)^2\right]^{-1/2} \tag{111}$$

as given by van Oss *et al.*(52)

It is apparent from Eqs. (108)–(111) that when Eq. (110) holds, the numerical value of the interaction parameter Φ_{12} is dependent upon the dielectric properties of the materials involved only.

It was shown[52] that for van der Waals interactions the value of Φ_{12} varies between 0.98 and 1, and thus in most cases it can be assumed to be 1.

An approximate method for estimating interaction parameter Φ_{12} was derived by Kaelble[5] following Fowkes,[40] who assumed that the dipole–induced dipole component of van der Waals interactions is negligible in comparison with the sum of all interactions. It is shown, based on the above, that the approximate value of Φ_{12} is given by the geometric-mean equation

$$\Phi_{12} \simeq (d_1 \times d_2)^{1/2} + (p_1 \times p_2)^{1/2} \tag{112}$$

where

$$d = \gamma^d/\gamma \tag{113}$$

$$p = \gamma^p/\gamma \tag{114}$$

and

$$p + d = 1 \tag{115}$$

In the alternative method derived by Wu[32] the total work of adhesion is taken into account and both W_A and γ_1 (as well as γ_2) are subdivided into their dispersive and nondispersive components. The resultant formula for Φ_{12} is given by the following harmonic-mean equation:

$$\Phi_{12} = 2\left[\frac{d_1\, d_2}{\left(\frac{\gamma_1}{\gamma_2}\right)^{1/2} d_1 + \left(\frac{\gamma_2}{\gamma_1}\right)^{1/2} d_2} + \frac{p_1\, p_2}{\left(\frac{\gamma_1}{\gamma_2}\right)^{1/2} p_1 + \left(\frac{\gamma_2}{\gamma_1}\right)^{1/2} p_2}\right] \tag{116}$$

According to Wu,[32] Eq. (116) leads to the best correlation of experimental and theoretical values of Φ_{12} as justified by experimental and theoretical values of interfacial energies γ_{12}. It is necessary to observe, however, that there is no reference to the correlation of the experimental and theoretical values of Φ_{12} for systems in which one of the components forms hydrogen bonds.

In their work, Good and Elbing[54] provided a solution of an expression for interfacial energy in the form of Eq. (106) and in the following dimensionless form:

$$\frac{\gamma_{12}}{\gamma_1} = 1 + \frac{\gamma_2}{\gamma_1} - 2\Phi_{12}\left(\frac{\gamma_2}{\gamma_1}\right)^{1/2} \tag{117}$$

where the interaction parameter Φ_{12} appears in a general form, independent of the procedure of estimation.

Figure 14 illustrates the plot of Eq. (106) for a solid of surface energy $\gamma_1 = 24$ mJ/m^2 and a series of Φ_{12} = constant, and the solution of Eq. (117) in the form presented by Good and Elbing in their work.[54]

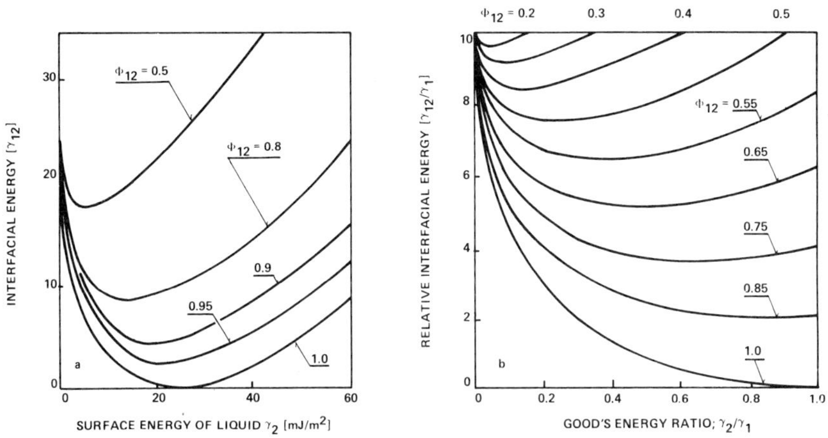

FIGURE 14. Plots of Eqs. (106) and (117) as given by Good and Elbing[54]: (a) Eq. (106) for an example of a solid of surface energy $\gamma_1 = 24$ mJ/m^2; (b) dimensionless Eq. (117).

The general formula for interfacial energy is given alternatively by Eqs. (103) or (104). From Good's definition of the interaction parameter Φ_{12} [see Eq. (105a)] the interfacial energy can be expressed by the following dimensionless equation:

$$\frac{\gamma_{12}}{\gamma_2} = \frac{\gamma_1}{\gamma_2} + 1 - 2\Phi_{12}\left(\frac{\gamma_1}{\gamma_2}\right)^{1/2} \tag{118}$$

where

$$2\Phi_{12}(\gamma_1/\gamma_2)^{1/2} - 1 = \cos\theta_{21} \tag{119}$$

Comparison of Eq. (118) with the general formula for γ_{12} [see Eq. (104)] shows that the original Good's Eq. (106) implicitly involves $\cos\theta_{21}$. This fact has significant implications for the solution of equations for interfacial energy [including Eqs. (104), (106), (117), and (118)], as

$$\cos\theta_{21}^{\max} = 1 \qquad \text{for } \gamma_2 \leq \gamma_{c(1)} \tag{120}$$

where γ_c is the critical surface tension for wetting.

With regard to Good's Eq. (106), it follows from our condition (120) and Eq. (104) or (119) that

$$\gamma_{12} = \begin{cases} \gamma_1 + \gamma_2 - 2\Phi_{12}(\gamma_1\gamma_2)^{1/2} & \text{for } \gamma_2 \geq \gamma_{c(1)} \quad (121) \\ \gamma_1 - \gamma_2 & \text{for } \gamma_2 \leq \gamma_{c(1)} \quad (122) \end{cases}$$

In the case of interactions between the nonpolar liquid and nonpolar solid we have $\Phi_{12} = 1.0$ and $\gamma_{c(1)} = \gamma_1^d = \gamma_1$, from which it follows that for the purely dispersive interactions (i.e., between nonpolar materials), the following applies:

$$\gamma_{12} = \begin{cases} \gamma_1 + \gamma_2 - 2(\gamma_1\gamma_2)^{1/2} & \text{for } \gamma_2 \geq \gamma_1 \quad (123) \\ \gamma_1 - \gamma_2 & \text{for } \gamma_2 \leq \gamma_1 \quad (124) \end{cases}$$

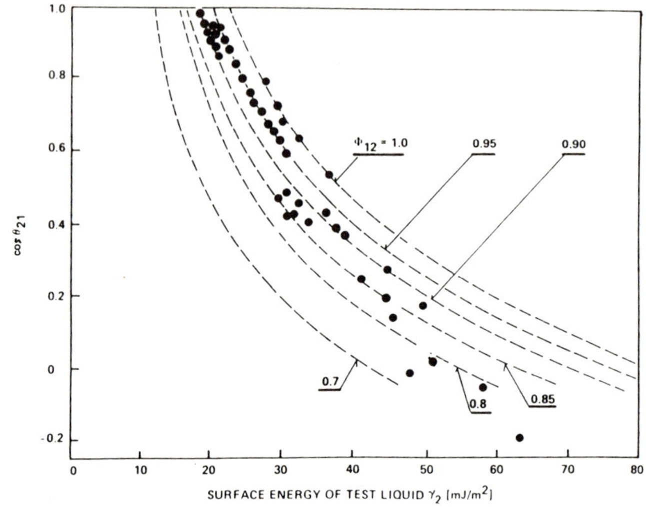

FIGURE 15. Wettability data for PTFE with reference to Φ_{12}.

Conditions governing the minimum interfacial energy for systems consisting of the solid of surface energy γ_1 = constant and a series of liquids with variable surface energy γ_2 are found from the first derivative of Eq. (121) with respect to γ_2, i.e.,

$$\frac{d\gamma_{12}}{d\gamma_2} = 1 - \Phi_{12}\left(\frac{\gamma_1}{\gamma_2}\right)^{1/2} - 2(\gamma_1\gamma_2)^{1/2}\frac{d\Phi_{12}}{d\gamma_2} \tag{125}$$

It is obvious that the term $d\Phi_{12}/d\gamma_2$ appears in the derivative of Eq. (106). Hence only in the case where Φ_{12} = constant is the term $d\Phi_{12}/d\gamma_2$ zero and minimum interfacial energy occurs:

$$\gamma_{12}^{\min} \text{ at } \left(\frac{\gamma_1}{\gamma_2}\right) = 1/\Phi_{12}^2 \tag{126}$$

Good and Elbing[54] have shown that the relationship between $\cos\theta_{21}$ and the surface properties of phases 1 and 2 is as follows:

$$\cos\theta_{21} = 2\Phi_{12}\left(\frac{\gamma_1}{\gamma_2}\right)^{1/2} - 1 - \frac{\pi_{e(1)}}{\gamma_2} \tag{119}$$

and for $\pi_{e(1)} = 0$ it has the form illustrated in Figure 15.

This relationship has been confirmed experimentally by Kaelble and Uy[56] for a series of test liquids producing constant values of Φ_{12} [when calculated by Eq. (112)] with PTFE, i.e., when the following is applied simultaneously within a given series[57]:

$$\gamma_2^{(1)} \neq \gamma_2^{(2)} \neq \cdots \neq \gamma_2^{(n)} \tag{127}$$

$$d_2^{(1)} = d_2^{(2)} = \cdots = d_2^{(n)} \tag{128}$$

From Eqs. (120) and (119) we derive that $\cos\theta_{21}^{\max} = 1$ is achieved[58] when the ratio of surface energies γ_1/γ_2 obtains its special value $(\gamma_1/\gamma_2)_s$, i.e.,

$$(\gamma_1/\gamma_2)_s = 1/\Phi_{12}^2 \tag{129}$$

Further introduction of the term (129) into Good's Eq. (117) leads to the following general expressions for interfacial energies:

$$\frac{\gamma_{12}}{\gamma_1} = \begin{cases} 1 + \dfrac{\gamma_2}{\gamma_1} - 2\Phi_{12}\left(\dfrac{\gamma_2}{\gamma_1}\right)^{1/2} & \text{for } \gamma_2 \geq \gamma_c \quad (130) \\[2ex] 1 - \dfrac{\gamma_2}{\gamma_1} & \text{for } \gamma_2 \leq \gamma_c \quad (131) \end{cases}$$

The minimum interfacial energy, as determined from Eq. (130), is achieved at the surface energy ratio $(\gamma_1/\gamma_2)_{\min}$ given by

$$(\gamma_1/\gamma_2)_{\min} = 1/\Phi_{12}^2 = (\gamma_1/\gamma_2)_s \tag{132}$$

i.e., where the contact angle θ_{21} is zero [see Eq. (129)].

The above approach [see Eqs. (118), (130), and (131)] can now be employed to present (see Figure 16) the amended solutions of the Good–Elbing equations (106) and (117) for interfacial energy (as originally given by Good and Elbing in the form illustrated in Figure 14).

4. EXPERIMENTAL TECHNIQUES

It is clear from Sections 1–3 that a knowledge of thermodynamic properties of solids and liquids is a necessary condition for any analysis of interfacial phenomena. The following techniques exist for determining values of surface energies at the liquid–liquid or liquid–gas interface.[32,59,60]

- Techniques independent of contact angle, such as the pendant drop or sessile drop and/or bubble methods.

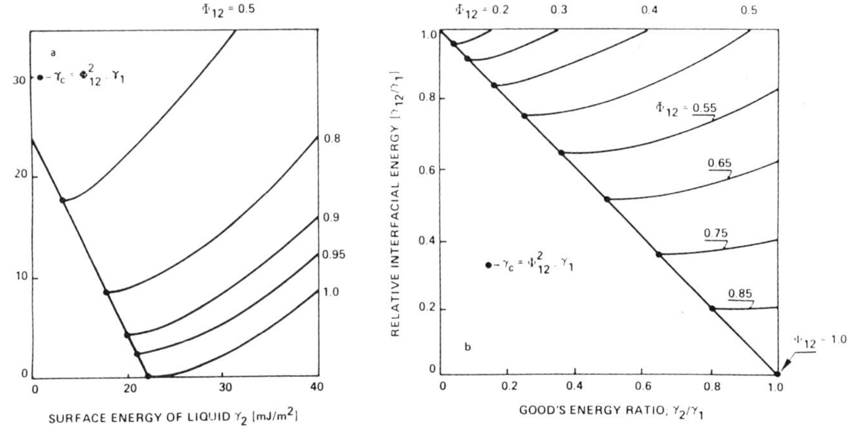

FIGURE 16. Amended plots of the Good–Elbing equations (106) and (117) previously given in Figure 14: (a) solution of Eq. (106); (b) solution of Eq. (117).

- Techniques dependent on contact angle, associated with the use of the Young equation (97) or the Young–Dupré (101) equation.
- Techniques based on acid–base interaction concepts.

4.1. *Interfacial Energy at the Liquid–Liquid and Liquid–Gas Interface by Techniques Independent of Contact Angle*

4.1.1. Pendant Drop Method

It has been found by Bashforth and Adams[61] and further analyzed by Adamson[59] that interfacial energy at the liquid–liquid or liquid–gas interface can be determined from the geometrical parameters of a nonspherical but axially symmetric pendant drop at its geometrical equilibrium state (illustrated in Figure 17a) using the following relationship:

$$\gamma_{12} = g\Delta\rho(d_e)^2/H \tag{133}$$

where

$$\Delta\rho = \rho_1 - \rho_2 \tag{134}$$

$$1/H = f(S = d_s/d_e) \tag{135}$$

and g is the acceleration of gravity.

Tabulated data on $1/H = f(S)$ are available.[59] Specific densities of phase 1 and phase 2 can be determined with accuracy using a mercury dilatometer.[62] The author refers the reader elsewhere[63–67] for extensive examples on applications of this technique.

4.1.2. Sessile Drop (or Bubble) Method

The principles of the method are similar to the pendant drop method, as illustrated in Figures 17b and 17c. Interfacial energy is given by

$$\gamma_{12} = \frac{g\Delta\rho r^2}{BF^2} = \frac{g\Delta\rho h^2}{BG^2} \tag{136}$$

where B, F, and G are functions of the ratio r/h. Values of parameters B, F, and G are tabulated elsewhere.[68–70]

4.2. *Techniques Based on Measurement of the Equilibrium Contact Angle*

4.2.1. Equilibrium Contact Angle and Contact Angle Hysteresis

On the smooth, homogeneous and rigid surface of solid 1 the contact angle exhibited by liquid 2 is the equilibrium contact angle θ_{21}; it obeys the Young equation given by expression (96). The equilibrium, however, is not achieved instantaneously[71]—it takes some time for the contact angle to become effectively constant. Analysis of the kinetics of the process reveals that when the liquid front is advancing, the contact angle after an advance is θ_A while after retreat of the liquid front it is θ_R. Usually, values of θ_A and θ_R are different and the difference

$$H = \theta_A - \theta_R \tag{137}$$

is commonly known as the hysteresis of the contact angle.

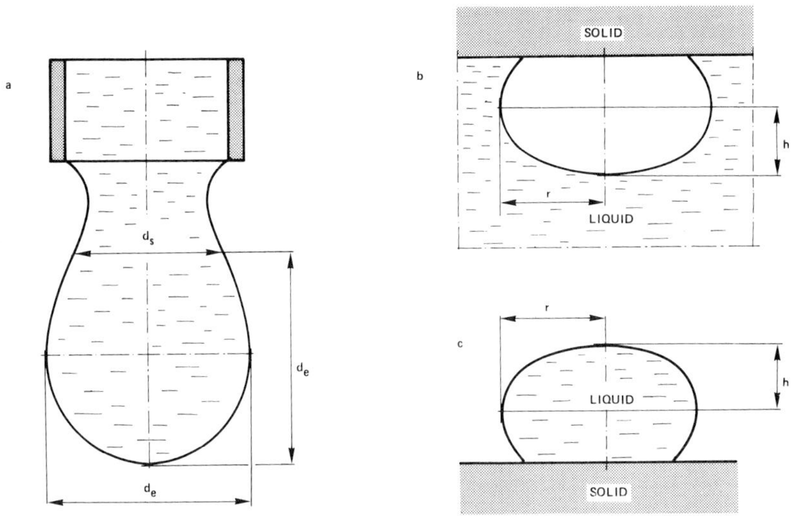

FIGURE 17. Illustration of the geometric parameters necessary for determining interfacial energy from: (a) the pendant drop method, (b) the pendant bubble method, and (c) the sessile bubble method.

The major reasons for this phenomenon are the surface roughness and heterogeneity. The influence of roughness on the value of the contact angle has been quantified by Wenzel,[72] who has found that

$$r = \frac{\cos\theta_W}{\cos\theta} = \frac{\text{true surface area}}{\text{apparent planar area}} \tag{138}$$

where θ_W is Wenzel's angle (contact angle on a rough surface), θ is the contact angle on an ideally smooth surface, and r is the surface roughness. The effect of roughness on contact angle hysteresis has been studied in detail by Johnson and Dettre.[73]

The equilibrium contact angle on heterogeneous surfaces has been proposed by Cassie[74] to be

$$\cos\theta_C = f_1 \cos\theta_{e1} + f_2 \cos\theta_{e2} \tag{139}$$

where f_i is the fraction of area with intrinsic equilibrium contact angle θ_{ei} and $f_1 + f_2 = 1$.

The Cassie contact angle may be approximated by[75]

$$\cos\theta \simeq \frac{1}{2}(\cos\theta_R + \cos\theta_A) \tag{140}$$

which is frequently used for expressing the approximate value of the equilibrium contact angle.

There are numerous methods for determining the contact angle; they are discussed thoroughly by Neumann and Good.[60] For convenience and accuracy we describe below the three most popular methods, which are illustrated in Figure 18, i.e., the tangent, Wilhelmy, and capillary-rise methods. They differ from each other in regard to substrate geometry and basic principle. Although the contact angle θ_{21} is not a direct measure of the

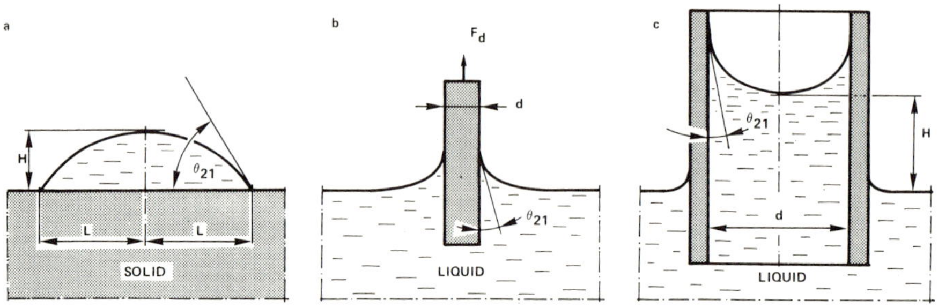

FIGURE 18. Illustration of the principles governing three methods for determining contact angle θ_{21}: (a) sessile drop method, (b) Wilhelmy method, (c) capillary-rise method.

surface energy of a solid or liquid alone, it provides a very useful tool for estimating these parameters.

4.2.1.1. Significance of the Accuracy of Determining the Contact Angle. It is obvious from Sections 1–3 that the contact angle appears in basic equations for adhesion as $\cos\theta_{21}$. In order to achieve scientific results it is necessary to determine the contact angle with an accuracy not less than $\pm$ 0.3 to 1.0°, depending upon the magnitude of the contact angle.

The significance of this requirement is apparent from Figure 19, which illustrates the influence of the three levels of accuracy in determining θ_{21} (e.g., $\pm$ 0.5°, $\pm$ 1°, and $\pm$ 2°) on the value of $\cos\theta_{21}$.

The error in estimating $\cos\theta$ is determined by

$$\Delta\cos\theta = \frac{\cos\theta_{\text{True}} - \cos\theta_{\text{Measured}}}{\cos\theta_{\text{True}}} \ (\%) \tag{141}$$

An example [see Figure 19 and Eq. (141)] refers to the true value of the contact angle $\theta_{\text{True}} = 60°$ (i.e., $\cos\theta = 0.5$), and relevant points in Figure 19 correspond to: (1) $\theta_{\text{True}} =$

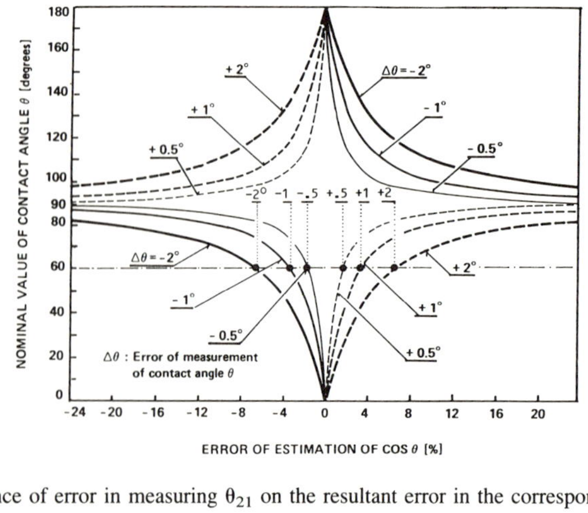

FIGURE 19. Influence of error in measuring θ_{21} on the resultant error in the corresponding value of $\cos\theta_{21}$.

60°; (2) $\theta_M = 60^\circ \pm 0.5^\circ$, i.e., 59.5° and 60.5°; (3) $\theta_M = 60^\circ \pm 1^\circ$, i.e., 59° and 61°, and (4) $\theta_M = 60^\circ \pm 2^\circ$, i.e., 58° and 62°.

It is seen from Figure 19 that for contact angles $\theta = 0$–40° and $\theta = 140^\circ$–180°, a $\pm 1^\circ$ accuracy of a technique for determining θ may be regarded as sufficient. However, for the region 70°–110° there is a need for a technique with much greater accuracy: $\pm 0.2^\circ$ to $\pm 0.3^\circ$ if the values of the resultant $\cos\theta$ are to be taken as significantly close to $\cos\theta_{\text{True}}$.

4.2.1.2. Direct Measurement of Contact Angle from the Drop Profile. In this method, a drop of the test liquid is placed on top of the substrate or a gas bubble (or a drop of an immiscible liquid) is attached to the surface of the substrate immersed in a test liquid (Figure 18a). The value of the contact angle can be determined by the following alternative methods:

- Direct measurement employing a goniometer or image analysis techniques; there is no need for the drops or bubbles to be spherical in this case.
- Indirect method based on the geometrical parameters of the drop or bubble, which must be spherical in this case, i.e.,

$$\theta_{21} = 2\tan^{-1}(H/L) \tag{142}$$

The accuracy of the tangent method can be increased by photographing the drop or bubble, from which a contact angle can be read with a protractor or calculated using Eq. (142). The reproducibility of results (in the case of direct readings of θ_{21} on the goniometer) is dependent upon the size of the drop; it is proposed that a constant drop volume of 2–4 μl be used as an optimum.

A drop of liquid placed on a substrate undergoes dynamic changes, described by the equation[71]

$$\cos\theta_\infty = \cos\theta_{21}[1 - a \exp(-ct)] \tag{143}$$

where a and c are experimental constants and t is time.

Accordingly, a suitable length of time is required for a droplet to achieve equilibrium and must be determined experimentally.

4.2.1.3. Wilhelmy Method for a Fiber. The principles of the Wilhelmy method are illustrated in Figure 18b, where a cylindrical fiber of diameter d is partially immersed in a test liquid of surface energy γ_2. The downward force exerted on the fiber by the liquid is

$$F_d = \pi\gamma_2 \, d \cos\theta_{21} \tag{144}$$

This force is balanced by buoyancy F_b,

$$F_b = V\rho g \tag{145}$$

where V is the volume of the displaced liquid and ρ the density of the liquid.

The Wilhelmy method provides the possibility of a very accurate estimation of θ_{21} by using an electronic microbalance in a setup illustrated in Figure 20. The technique is based on the fact that θ_{21} = constant for very slow motion of the liquid–solid interface. Thus, if the mass recorded by the microbalance during immersion or emersion of the substrate is M, then the contact angle from Eq. (144) is given by

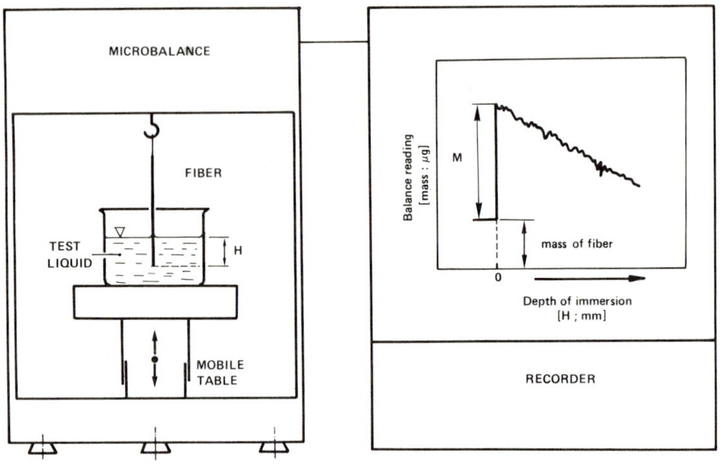

FIGURE 20. Typical setup for the Wilhelmy balance method.

$$\theta_{21} = \cos^{-1}(Mg/\pi d\gamma_2) \tag{146}$$

The Wilhelmy balance method can also be used to determine the surface energy of a liquid, assuming that θ_{21} in Eq. (144) is known.

4.2.1.4. Capillary-Rise Method. By reference to Figure 18c, equilibrium between capillarity and the weight of the liquid yields

$$\rho g H = 4\gamma_2 \cos\theta_{21}/d \tag{147}$$

It follows that for a liquid of known surface energy γ_2, the contact angle is

$$\theta_{21} = \cos^{-1}(\rho g d H/4\gamma_2) \tag{148}$$

4.2.2. Estimation of Surface Energies of Solids

Equations describing adhesion phenomena, such as Young or Young–Dupré equations, involve the surface energy of the substrate and of a cured adhesive. This necessitates a need to estimate γ_1 and γ_2. There are several concepts regarding estimation of surface energies of solids that do not yet provide an unequivocal solution to the problem.

The commonly accepted procedure is based on the additivity approach(76) according to which the total thermodynamic work of adhesion between two substances in contact comprises contributions arising from various types of interactions, i.e.,

$$W_A = W_A^d + W_A^{n\text{-}d} \tag{149}$$

where W_A^d is the dispersion term of the work of adhesion W_A, while $W_A^{n\text{-}d}$ refers to all nondispersion interactions, e.g., dipole–dipole, dipole–induced dipole, and acid–base, including hydrogen-bond interactions.

It was also suggested by Fowkes that the only significant interactions across an interface are those which are common to both phases in contact. Since the work of cohesion is expressed by a formula similar to Eq. (149), we have, considering Eq. (90), that the surface energy of a material can be expressed accordingly by

$$\gamma_{\text{total}} = \gamma^{\text{d}} + \gamma^{\text{n-d}} \tag{150}$$

Fowkes[77] has shown that the dispersion component of W_A is well predicted by

$$W_A^{\text{d}} = 2(\gamma_1^{\text{d}}\ \gamma_2^{\text{d}})^{1/2} \tag{151}$$

where subscripts 1 and 2 refer to materials 1 and 2 in contact.

Determining the dispersion-force component of the total surface energy of a solid, namely γ_1^{d}, is straightforward, since the following relationship applies for the solid–liquid system interacting through dispersion forces only:

$$\cos\theta_{21} = -1 + 2(\gamma_1^{\text{d}})^{1/2}[(\gamma_1^{\text{d}})^{1/2}/\gamma_2] \tag{152}$$

A plot of $\cos\theta_{21}$ vs. $(\gamma_1^{\text{d}})^{1/2}/\gamma_2$ gives a straight line with origin at $\cos\theta_{21} = -1$ and slope $2(\gamma_1^{\text{d}})^{1/2}$ (see Fowkes[78] for particulars). Thus, contact-angle measurements with one test liquid are sufficient to determine γ_1^{d}.

Successful use of Eq. (151) has led many researchers[79–83] to suggest that the nondispersion interactions can also be quantified by the geometric-mean expression, i.e.,

$$W_A^{\text{n-d}} = 2(\gamma_1^{\text{n-d}}\ \gamma_2^{\text{n-d}})^{1/2} \tag{153}$$

Ultimately, it followed from Eqs. (149) and (151), and the proposed relationship (153), that the total work of adhesion can be given by

$$W_A = 2(\gamma_1^{\text{d}}\ \gamma_2^{\text{d}})^{1/2} + (\gamma_1^{\text{n-d}}\ \gamma_2^{\text{n-d}})^{1/2} \tag{154}$$

Fowkes commented[84] that Eq. (154) is erroneous since, for instance, the hydrogen-bond formers, such as ethers, esters, or aromatic hydrocarbons, cannot themselves form hydrogen bonds and thus γ^{h} is zero for these materials, despite the fact that they have large W_A^{h} values with other compounds. A similar argument has been used by Panzer.[85]

According to Fowkes's suggestions,[39] the nondispersion component of W_A can be related to acid–base interactions and, for $\pi_{e(1)} = 0$, it is given by

$$W_A^{\text{ab}} = W_A - 2(\gamma_1^{\text{d}}\ \gamma_2^{\text{d}})^{1/2} = \gamma_2\ (1 + \cos\theta_{21}) - 2(\gamma_1^{\text{d}}\ \gamma_2^{\text{d}})^{1/2} \tag{74}$$

where superscript ab refers to acid–base interactions.

It was alternatively proposed by Fowkes that W_A^{ab} can be quantified by

$$W_A^{\text{ab}} = -f\Delta H^{\text{ab}} n \tag{155}$$

where f is the correction factor (close to unity) converting enthalpy values into free-energy values, $-\Delta H^{\text{ab}}$ is the enthalpy of acid–base interactions, and n is the number of acid–base pairs at the interface per unit area. By combining Eqs. (74) and (155) and considering Drago's proposal[45,46] on enthalpy of acid–base interactions, i.e.,

$$-\Delta H^{\text{ab}} = C_A\ C_B + E_n\ E_B \tag{156}$$

Fowkes obtained the following expression for the thermodynamic work of adhesion:

$$W_A = 2(\gamma_1^{\text{d}}\ \gamma_2^{\text{d}})^{1/2} - (C_A\ C_B + E_A\ E_B)\,fn \tag{157}$$

where C and E are constants related to covalent and electrostatic interactions, respectively, while subscripts A and B refer to the acidic and basic properties.

Although accurate according to Fowkes's experiments,[39,84] the approach given by Eq. (157) has currently limited applicability due to the lack of sufficient data on C and E for potential substrates and adhesives. Also, there is still a question to be answered on how to represent $-\Delta H^{ab}$, and thus W_A^{ab}, as a product of nondispersion components of γ_1 and γ_2, i.e.,

$$W_A^{ab} = (C_A C_B + E_A E_B) fn = [f(\gamma_1^{n\text{-}d}, \gamma_2^{n\text{-}d}) = (?)] \tag{158}$$

Due to the above-mentioned shortcomings of the approach expressed by Eqs. (157) and (158), the geometric-mean approach discussed by Owens and Wendt,[79] Schultz *et al.*,[82] and Kaelble,[83] given by Eq. (154), is frequently employed to estimate a solid's surface energy, despite criticisms outlined in the summary of this section.

The technique is based on the use of test liquids of known total surface energies and their dispersive (d) and nondispersive components [where the simpler notation p replaces n-d used in Eqs. (150) to (158)], i.e.,

$$\gamma_2 = \gamma_2^d + \gamma_2^p \tag{159}$$

which are used for determining equilibrium contact angles by static or dynamic techniques.

Subsequently, the simultaneous equations

$$\begin{cases} 0.5\gamma_2^{(1)} (1 + \cos\theta^{(1)}) = (\gamma_1^d \gamma_2^{(1)d})^{1/2} + (\gamma_1^p \gamma_2^{(1)p})^{1/2} & (160) \\ 0.5\gamma_2^{(2)} (1 + \cos\theta^{(2)}) = (\gamma_1^d \gamma_2^{(2)d})^{1/2} + (\gamma_1^p \gamma_2^{(2)p})^{1/2} & (161) \end{cases}$$

are solved for each pair of test liquids 1 and 2, which produce contact angles $\theta^{(1)}$ and $\theta^{(2)}$ respectively during contact with the solid of unknown surface energy, i.e.,

$$\gamma_1 = \gamma_1^d + \gamma_1^p \tag{162}$$

The solution of Eqs. (160) and (161) yields values of γ_1^d and γ_1^p. The total surface energy is estimated from Eq. (162). Table 10 provides data on typical test liquids, as given by Fowkes.[77,78] Although it can be argued that the dispersive and nondispersive components of the total surface energy calculated via Eqs. (160) and (161) involve some errors due to Kaelble's assumptions, it can be shown that the *total* surface energy obtained in this manner is very close to values computed by other techniques.

As mentioned earlier, the geometric-mean approach has been criticized by Fowkes[39,84] and Panzer.[85] At the same time, however, it has also achieved theoretical and experimental verification.[32,56,78–81,86–88] It can be shown that the numerical values obtained by the use of geometric-mean equations are in reasonable agreement with those obtained from molten polymer studies[32] or by use of Wu's harmonic-mean approach.[32]

In the light of the available theoretical and experimental evidence, the geometric-mean procedure described in this work can be helpful in the analysis of adhesion phenomena. However, the ultimate concept regarding estimation of $\gamma^{n\text{-}d}$ and the total surface energy of a solid material is yet to be unequivocally established.

TABLE 10
Total Surface Energy and Its Dispersive and Nondispersive Components for Selected Test Liquids (t = 20°C) as Determined by Fowkes[77]

No.	Test liquid	γ_2 (mJ/m²)	γ_2^d (mJ/m²)	γ_2^p (mJ/m²)
1	Water	72.8	21.8	51.0
2	Glycerol	63.4	37.0	26.4
3	Formamide	59.2	39.5	18.7
4	Methylene iodide	50.8	48.5	2.3
5	Trichlorobiphenyl	45.3	44.0	1.3
6	α-Bromonaphthalene	44.6	44.6	0
7	Tricresyl phosphate	40.9	39.2	1.7
8	n-Hexadecane	27.6	27.6	0

4.3. Techniques Based on Acid–Base Interaction Concepts

A procedure for determining the dispersive component of a solid's surface energy, namely γ_1^d, has been developed by Schultz *et al.*[51] The technique is based on the fact that the thermodynamic work of adhesion, W_A^{12}, is related to the free enthalpy of adsorption (or desorption) at infinite dilution in inverse gas chromatography (IGC) through the following approximated equation:

$$\Delta G_o \simeq N\, a\, W_A^{12} \tag{163}$$

where N is Avogadro's number and a is the surface area of the probe molecule.

Thermodynamic considerations related to IGC yield

$$\Delta G_D^o = -\,\Delta G_A^o = RT\,\ln\left(\frac{V_N\, P_o}{Sg\,\pi_o}\right) \tag{164}$$

where V_N is the net retention volume, π_o is the two-dimensional spreading pressure of the adsorbed film to a reference gas-phase state defined by partial pressure P_o, S is the specific surface area, and g is the weight of the material tested.

For the chosen reference states[51] Eq. (164) can be simplified to

$$\Delta G^\circ = RT\,\ln V_N + K \tag{165}$$

where K is a constant depending on a reference state.

For the dispersive interactions we have[77]

$$W_A^d = 2(\gamma_1^d\,\gamma_2^d)^{1/2} \tag{151}$$

so Eqs. (163) and (165) yield

$$RT\,\ln V_N = 2N(\gamma_1^d)^{1/2}\,(\gamma_2^d)^{1/2}\,a + C \tag{166}$$

It follows that for tests carried out with a series of n-alkanes, the dispersive component of a solid's surface energy γ_1^d can be calculated from the slope of an experimental graph plotted in the coordinates $RT \ln V_N$ vs. $a\,(\gamma_2^d)^{1/2}$, where γ_2^d is the surface tension of n-alkanes used in the tests. A similar technique for determining γ_1^d has been described by Dorris and Gray.(89)

NOMENCLATURE

α	Polarizability
β	ρ_o/r_o
c	Speed of light
C	Volume concentration (mol/m^3)
C_A, C_B	Drago's covalent constant for acid (A) and base (B)
d_1, d_2	Dispersive fraction of total surface energy of phase 1 or 2
δ	Correction term multiplying I (ionization energy) in the Pitzer–Donath theory of dispersion forces
E_A, E_B	Drago's electrostatic (ionic) constant for acid (A) and base (B)
E_{11}, E_{12}, E_{22}	Energy of a single bond between like and unlike molecules
ΔE_{12}	Energy excess from the difference of electronegativity
ϵ	Dielectric permittivity
ϵ_o	Dielectric permittivity in vacuum
F	Force of interaction
F_{attr}	Attractive force
F_{rep}	Repulsive force
Φ_{12}	Interaction parameter
γ_1, γ_2	Total surface energy of phase 1 or 2
γ_c	Critical surface tension for wetting
γ_{12}	Interfacial energy
γ^d	Dispersive component of total surface energy
γ^p or $\gamma^{n\text{-}d}$	nondispersive component of total surface energy
(γ_{12}/γ_2), (γ_{12}/γ_1)	relative interfacial energy
Γ	Surface concentration for adsorption [mol/m^2]
Γ_m	Maximum value of Γ (total number of adsorption sites for the solute)
h	Planck's constant
$\hbar = h/2\pi$	(Planck's constant)/2π
ΔH_{12}	Enthalpy of interaction between phases 1 and 2
ΔH_{12}^{ab}	Acid–base component of ΔH_{12}
ΔH_{12}^{d}	Dispersive component of ΔH_{12}
I	Ionization energy
I_s	Isoelectric point of the surface
k	Boltzmann's constant
K_A, K_B	Equilibrium constant of acid–base interactions with organic acid (A) and organic base (B)
k_a, k_b	Acidic (a) and basic (b) dissociation constant
$k(T)$	Adsorption constant at temperature T
κ	Absorption coefficient
ξ	Frequency

l	Separation distance between semi-infinite bodies
λ	Wavelength
M	Mass
μ	Dipole moment
n	Refractive index of the substance
N	Avogadro's number
ν_0	Natural frequency
P	Momentum of particle or photon
ω	Rotational frequency
p_1,p_2	Nondispersive fraction of the total surface energy of phase 1 or 2
π_e	Equilibrium spreading pressure
ψ	Wave function
q,Q	Charge of an ion
r	Separation distance between atoms
r_0	Equilibrium separation distance
ρ	Density
ρ_o	Equilibrium intercore distance
σ	Stress
T	Temperature
θ_{21}	Equilibrium contact angle corresponding to an interaction of phase 2 with the solid or liquid phase 1
U	Potential energy of interaction
W_A	Thermodynamic work of adhesion
W_A^d	Dispersive component of W_A
W_A^{ab}	Acid–base interaction component of W_A
$[W_A/\gamma_2]$, $[W_A/\gamma_1]$	Relative work of adhesion

REFERENCES

1. I. E. Dzyaloshinskii, E. M. Lifshitz, and L. P. Pitaevskii, *Adv. Phys.* **10,** 165 (1961).
2. J. N. Israelachvili, *Intermolecular and Surface Forces,* Academic Press, London (1985).
3. F. London, *The General Theory of Molecular Forces, Trans. Faraday Soc.* **33,** 8 (1937).
4. R. J. Good, Intermolecular and interatomic forces, in: *Treatise on Adhesion and Adhesive,* Vol. 1: *Theory,* Marcel Dekker Inc., New York (1967).
5. D. H. Kaelble, *Physical Chemistry of Adhesion,* Wiley–Interscience, New York (1971).
6. J. O. Hirschfelder, C. F. Curtis, and R. B. Bird, *Molecular Theory of Gases and Liquids,* Wiley–Interscience, New York (1954).
7. L. Pauling, *The Nature of the Chemical Bond and the Structure of Molecules and Crystals,* 3rd edn., Cornell University Press, Ithaca, New York (1960).
8. H. C. Hamaker, *Physica* **4,** 1058 (1937).
9. F. M. Fowkes, Attractive forces at interfaces, *Ind. Eng. Chem.* **56,** 40–52 (1964).
10. P. Debye, Die Van der Waalsschen Kohäsionkräfte, *Phys. Z.* **21,** 178–187 (1920).
11. W. H. Keesom, Die Van der Waalsschen Kohäsionkräfte, *Phys. Z.* **22,** 129–141 (1921).
12. H. Hellmann, *Einfürung in die Quantentechemie,* Franz Deuticke, Leipzig (1937).
13. R. P. Feynman, *Phys. Rev.* **56,** 340 (1939).
14. K. G. Denbigh, *Trans. Faraday Soc.* **36,** 939 (1940).
15. C. P. Smyth, *Dielectric Behaviour and Structure,* McGraw-Hill, New York (1955).

16. A. R. Von Hippel, *Dielectrics and Waves,* Wiley–Interscience, New York (1954).
17. Z. H. Stachurski, *J. Mater. Sci.* **21,** 3231 (1986).
18. H. F. Mark, Cohesive and adhesive strength of polymers, Part 1, *Adhes. Age* (July 1979), pp. 35–40; Part 2, *Adhes. Age* (Sept. 1979), pp. 45–50.
19. J. Lennard-Jones, and J. A. Pople *Proc. R. Soc. London, Ser. A* **205,** 155 (1951).
20. C. A. Coulson, *Valence,* 2nd edn., Oxford University Press, London (1961).
21. C. A. Coulson, *Research* **10,** 149 (1957).
22. J. C. Slater, *Electronic Structure of Molecules,* Vol. 1, McGraw-Hill, New York (1963).
23. M. J. Sparnaay, *J. Colloid Interface Sci.* **22,** 23 (1966).
24. J. C. Bolger and A. S. Michaels, Molecular structure and electrostatic interaction at polymer–solid interface, in: *Interface Conversion for Polymer Coatings* (P. Weiss and G. Dale Cheever, eds.), Elsevier, New York (1968).
25. J. C. Bolger, Acid–base interactions between oxide surfaces and polar organic compounds, in: *Adhesion Aspects of Polymeric Coatings* (K. L. Mittal, ed.), Plenum Press, New York and London (1983).
26. E. M. Lifshitz, *J. Exp. Theor. Phys.* **29,** 94 (1955).
27. L. D. Landau and E. M. Lifshitz, *Quantum Mechnics,* Pergamon Press, London–Paris (1959).
28. V. A. Parsegian, Long range van der Waals forces, in: *Physical Chemistry: Enriching Topics From Colloid and Surface Science* (H. van Olphen and K. J. Mysels, eds.), IUPAC, Theorex (1975).
29. R. M. Pashley, *J. Colloid Interface Sci.* **62,** 334 (1977).
30. S. Nir, S. Adams, and R. Rein, *J. Colloid Interface Sci.* **49,** 196 (1974).
31. W. A. Donners, J. B. Rijnbout, and A. Vrij, *J. Colloid Interface Sci.* **60,** 540 (1977).
32. S. Wu, *Polymer Interface and Adhesion,* Marcel Dekker Inc., New York and Basel (1982).
33. B. W. Ninham and V. A. Parsegian, *J. Chem. Phys.* **52,** 4578 (1970).
34. D. Langbein, *J. Adhes.* **1,** 237 (1969).
35. F. London, *Phys. Z.* **60,** 491 (1930).
36. J. N. Israelachvili, *Proc. R. Soc. London, Ser. A* **331,** 39 (1972).
37. H. B. C. Casimir and D. Polder, *Phys. Rev.* **73,** 360 (1948).
38. V. Gutman, *Donor–Acceptor Approach to Molecular Interaction,* Plenum Press, New York and London (1978).
39. F. M. Fowkes, Acid–base interactions in polymer adhesion, in: *Physico-Chemical Aspects of Polymer Surfaces,* **2** (K. L. Mittal, ed.), Plenum Press, New York and London (1983).
40. F. M. Fowkes, Attractive forces at solid liquid interface, in: *Wetting,* SCI Monograph No. 25, London (1967).
41. F. M. Fowkes and S. Maruchi, *Org. Coat. Plast. Chem. Prep.* **37,** 605 (1977).
42. F. M. Fowkes and M. A. Mostafa, *Ind. Eng. Chem., Prod. Res. Dev.* **17,** 3 (1978).
43. J. L. Gardon, *J. Paint Technol.* **38,** 43, (1966).
44. C. M. Hansen, *J. Paint Technol.* **39,** 104 (1967).
45. R. S. Drago, G. C. Vogel, and T. E. Needham, *J. Am. Chem. Soc.* **93,** 6014 (1970).
46. R. S. Drago, L. B. Parr, and C. S. Chamberlain, *J. Am. Chem. Soc.* **99,** 3203 (1977).
47. J. O. Edwards, *J. Am. Chem. Soc.* **76,** 1540 (1954).
48. H. P. Schreiber, C. Richard, and M. R. Werthemeir, Acid–base considerations of surface interactions in polymer system control by microwave plasma treatment, in: *Physico-Chemical Aspects of Polymer Surfaces,* **2** (K. L. Mittal, ed.), Plenum Press, New York and London (1983).
49. P. Koening, T. C. Ward, R. D. Allen, and J. E. McGrath, *Polymer Prepr.* **26**(1), 189 (1985).
50. S. Ponce, R. Bazinet, and H. P. Schreiber, Component interaction and properties of pigmented epoxy systems, presented at the 1987 International Meeting of the Adhesion Society, Williamsburg, VA (1987).
51. J. Schultz, L. Lavielle, and C. Martin, The role of interface in carbon fibre–epoxy composites, *J. Adhes.* **23,** 45 (1987).
52. C. J. van Oss, R. J. Good, and M. K. Chaudhury, *J. Colloid Interface Sci.* **111,** 378 (1986).
53. R. J. Good and L. A. Girifalco, *J. Phys. Chem.* **64,** 561 (1960).
54. R. J. Good and E. Elbing, Generalization of the theory for estimation of interfacial energies, in: *Chemistry and Physics of Interfaces—II,* pp. 74–96, American Chemical Society Publications, Washington, DC (1971).
55. R. J. Good and C. J. Hope, *J. Chem. Phys.* **53,** 540 (1970).
56. D. H. Kaelble and K. C. Uy, *J. Adhes.* **2,** 51 (1970).

57. W. Gutowski, Physico-chemical criteria for maximum adhesion, Part I. Theoretical concepts and experimental evidence, *J. Adhes.* **19,** 29 (1985).
58. W. Gutowski, Physico-chemical criteria for maximum adhesion, Part II. A new comprehensive analysis, *J. Adhes.* **19,** 51 (1985).
59. A. W. Adamson, *Physical Chemistry of Interfaces,* Wiley–Interscience, New York (1967).
60. W. Neumann and R. J. Good, Techniques for measuring contact angles, in: *Surface and Colloid Science,* Vol. II (R. J. Good and R. R. Stromberg, eds.), Plenum Press, New York (1979).
61. F. Bashforth and J. C. Adams, *An Attempt to Test the Theory of Capillary Action,* Cambridge University Press and Deighton, Bell and Co. (1892).
62. N. Bekkedehl, *J. Res. Natl. Bur. Stand.* **42,** 145 (1949).
63. S. Wu, *J. Macromol. Sci., Rev. Macromol. Chem.* **31,** 153 (1969).
64. R. J. Roe, *J. Phys. Chem.* **72,** 2013 (1968).
65. R. J. Roe, V. L. Bacchetta, and P. M. Wong, *J. Macromol. Sci., Rev. Macromol. Chem.* **71,** 4190 (1967).
66. C. A. Smolders, PhD Thesis, Rijksuniversiteit, Utrecht (1961).
67. C. A. Smolders and J. T. G. Overbeck, *Recl. Trav. Chim. Pays-Bas* **80,** 695 (1961).
68. D. N. Staicopolus, *J. Colloid Sci.* **17,** 439 (1962).
69. D. N. Staicopolus, *J. Colloid Sci.* **18,** 793 (1963).
70. D. N. Staicopolus, *J. Colloid Sci.* **23,** 453 (1967).
71. B. W. Cherry and S. Muddaris, Wetting, kinetics and strength of adhesive joints, *J. Adhes.* **2,** 42 (1970).
72. R. N. Wenzel, *Ind. Eng. Chem.* **28,** 988 (1936).
73. R. E. Johnson Jr. and R. H. Dettre, in: *Contact Angle, Wettability and Adhesion* (F. M. Fowkes, ed.), *Adv. Chem. Ser.* **43,** American Chemical Society (1964).
74. A. B. D. Cassie, *Discuss. Faraday Soc.* **3,** 11 (1948).
75. R. E. Johnson Jr. and R. H. Dettre, *J. Adhes.* **2,** 3 (1970).
76. F. M. Fowkes, *J. Phys. Chem.* **67,** 2538 (1963).
77. F. M. Fowkes, Donor–acceptor interactions at interfaces, *J. Adhes.* **4,** 155–159 (1972).
78. F. M. Fowkes, in: *Treatise on Adhesion and Cohesion* (R. L. Patrick, ed.), Vol. 1, Chapter 9, Marcel Dekker Inc., New York (1967).
79. D. K. Owens and R. C. Wendt, Estimation of surface free energy of polymers, *J. Appl. Polym. Sci.* **13,** 1741–1747 (1969).
80. D. K. Owens, Some thermodynamic aspects of polymer adhesion, *J. Appl. Polym. Sci.* **14,** 1725–1730 (1970).
81. R. J. Good, Surface free energies of solids and liquids, thermodynamics, molecular forces, and structure, *J. Colloid Interface Sci.* **59,** 398–419 (1977).
82. J. Schultz, K. Tsutsumi, and J. D. Donnet, Surface properties of high energy solids. II: Determination of the nondispersive component of the surface free energy of mica and its energy of adhesion to polar liquids, *J. Colloid Interface Sci.* **59,** 277–282 (1977).
83. D. H. Kaelble, Dispersion—Polar surface tension properties of organic solids, *J. Adhes.* **2,** 66–81 (1970).
84. F. M. Fowkes, Role of acid–base interfacial bonding in adhesion, *J. Adhes. Sci. Technol.* **1,** 7–27 (1987).
85. J. Panzer, Components of solid surface free energy from wetting measurements, *J. Colloid Interface Sci.* **44,** 142–161 (1973).
86. A. F. M. Barton, *J. Adhes.* **14,** 33 (1982).
87. A. Baszkin and L. Ter-Minassian-Saraga, *Polymer* **19,** 1083 (1978).
88. S. Wu, Polar and nonpolar interactions, *J. Adhes.* **5,** 39–55 (1973).
89. G. M. Dorris and D. G. Gray, *J. Colloid Interface Sci.* **77,** 353 (1980).

3

Theory of Adhesive Forces Across Interfaces
1. The Lifshitz–van der Waals Component of Interaction and Adhesion

Robert J. Good and Manoj K. Chaudhury

1. INTRODUCTION

It is widely recognized that hydrogen bonding, or (more generally) acid–base interaction, across an interface, can make a major contribution to joint strength. It has recently been found[1,2] that, under certain conditions, hydrogen bonding can lead to a negative interfacial free energy between condensed phases.

This chapter, and the chapter that follows, will review recent work on this subject. Before we can go into the relationships that have been found and that pertain directly to hydrogen bonding, it will be necessary to discuss in some detail the components of interfacial energy that are *not* due to hydrogen bonding.

Certain concepts that have not appeared in the adhesion literature will be presented in this chapter. One of the most important new items is the demonstration that the assumption of pairwise additivity of forces, between dipolar molecules in *condensed* phases, is not an acceptable approximation. In Section 4 of this chapter, we will develop the consequences of this fact.

The basis for our treatment is the Lifshitz theory[3] of interaction between macroscopic bodies, which was made accessible by Parsegian and Ninham[4–6] a few years ago. See also the extensive paper by Krupp.[7] A secondary source is the preceding chapter, by Gutowski.

2. THEORY OF VAN DER WAALS INTERACTIONS ACROSS INTERFACES

2.1. The Pairwise-Addition Approximation

In order to explain the limitations of the pairwise-addition approximation, we must briefly review the standard theory of intermolecular force—what we will call the London–

Robert J. Good • Department of Chemical Engineering, State University of New York at Buffalo, Buffalo, New York 14260. *Manoj K. Chaudhury* • Dow Corning Corporation, Midland, Michigan 46868.

Debye–Keesom (LDK) theory[6–10] of the force across an interface. We will then review the Lifshitz theory of this force. It will be pointed out that the LDK theory amounts to what might be called the "dilute gas" approximation.

A general way of expressing the LDK theory for *bimolecular* interactions is as follows. An electric field exists at any point in the neighborhood of any molecule, due to the local nonuniformity of charge distribution in the molecule. If the molecule has a permanent dipole moment, the field will fluctuate as the molecule rotates or oscillates.

If a dipolar molecule is at the neighboring point, it will interact with the field, according to the laws of electrostatics. For two freely rotating dipoles, the attraction depends on their mutual orientation at any instant; and, averaged over time, the free energy is given by the Keesom–Israelachvili (K–I) equation[11]

$$U_{ab}(\text{K–I}) = \frac{-\mu_a^2\mu_b^2}{3kTr_{ab}^6} \tag{1a}$$

where μ is the dipole moment and r the distance. The Keesom equation for the coefficient, A_K, of r^{-6} is

$$A_K = \frac{-2\mu_a^2\mu_b^2}{3kT} \tag{1b}$$

The K–I equation was adopted by Good, in 1976, for calculations of surface energies.[12] The development below, leading to Eq. (17), shows that this choice was valid in the dilute gas approximation. But the further development, which we give in Eqs. (21)–(27), and the results obtained in calculations using Eq. (27), show that both Eqs. (1) and (17) are now superseded.

Any molecule, whether or not it has a permanent dipole moment, possesses a polarizable electron cloud. The polarizability α is, in general, a function of the frequency of the polarizing field: $\alpha = \alpha(\omega)$. There is (in classical terms) a "natural frequency" ν_0 for the polarizable electron cloud, and oscillations of electric displacement at this frequency are always present. London proposed the approximation $h\nu_o \cong I$, where I is the ionization energy. The field due to oscillations in one molecule induce synchronous oscillations of the charge cloud in a neighbor molecule, which are in the correct phase for maximum attraction. For two unlike molecules, the London[8] expression for the energy of that attraction is

$$U_{ab}(\text{London}) = -\frac{3}{4}\,\alpha_a\alpha_b\,\frac{2I_aI_b}{(I_a + I_b)}\cdot\frac{1}{r_{ab}^6} \tag{2a}$$

For like molecules, i.e., when a and b are identical, Eq. (2a) becomes

$$U(\text{London}) = -\,3\alpha^2I/2r^6 \tag{2b}$$

There is a cross-term, between the London and Keesom energies, the theory of which was given by Debye[9] and by van der Avoird and Hoflich[13]

$$U_{ab}(\text{Debye}) = -\,\frac{\alpha_a\mu_b^2 + \alpha_b\mu_a^2}{2r_{ab}^6} \tag{3}$$

This term is always small compared to whichever is larger of U(London) and U(Keesom), and it is commonly neglected. The three terms, i.e., the London, Keesom, and Debye energies, contribute in a simple, additive fashion to the interaction energy between two molecules a and b. See below.

The r^{-6} dependence of energy in Eqs. (3)–(5) should be noted. Hamaker[14] employed the r^{-6} bimolecular potential function to derive a theory for the energy of interaction across an interface. He integrated (summed) the force that acts between molecules, *as independent pairs,* for interactions across a phase boundary (Figure 1). Mathematically, this integration converts the r^{-6} dependence that appears in Eqs. (1)–(3) into the l^{-2} dependence in Eqs. (4) and (6) below. The bimolecular approximation is valid (as will be shown below) only when the two phases are both low-density gases.

Hamaker's theory for the attractive energy between parallel bodies of phases 1 and 2, as a function of the distance l (cf. Figure 2), led to the equation

$$\Delta G_{12}(l) = -A_{12}/12\pi l^2 \tag{4}$$

where A is known as the Hamaker coefficient. Equation (4) holds for systems in which l is appreciably greater than the equilibrium distance l_0. When the intermolecular force that corresponds to the repulsion term in the Lennard-Jones potential

$$U = -\frac{a}{r^6} + \frac{b}{r^{12}} \tag{5}$$

are included, it has been shown[15] that at the equilibrium separation, i.e., when $l = l_0$,

$$\Delta G_{12}^{\mathrm{a}} = -A_{12}/16\pi l_0^2 \tag{6}$$

When the separability of the energy into the Keesom, Debye, and London components[16] was recognized, it became possible to write the Hamaker coefficient as the sum of three terms,[17–19]

$$\mathrm{A}_{12} = A_{12}(\text{London}) + A_{12}(\text{Debye}) + A_{12}(\text{Keesom–Israelachvili}) \tag{7}$$

$$\Delta G_{12}^{\mathrm{a}} = -\frac{1}{16\pi l_0^2}\left[\frac{3\alpha_\mathrm{a}\alpha_\mathrm{b}}{2}\,\frac{I_\mathrm{a}I_\mathrm{b}}{(I_\mathrm{a}+I_\mathrm{b})} + \frac{\alpha_\mathrm{a}\mu_\mathrm{b}^2+\alpha_\mathrm{b}\mu_\mathrm{a}^2}{2} + \frac{\mu_\mathrm{a}^2\mu_\mathrm{b}^2}{3kT}\right] \tag{8}$$

The physical reasoning behind Eq. (8) led Fowkes[20,21] to hypothesize that the surface free energy of a liquid or solid could be subdivided into components due to the London

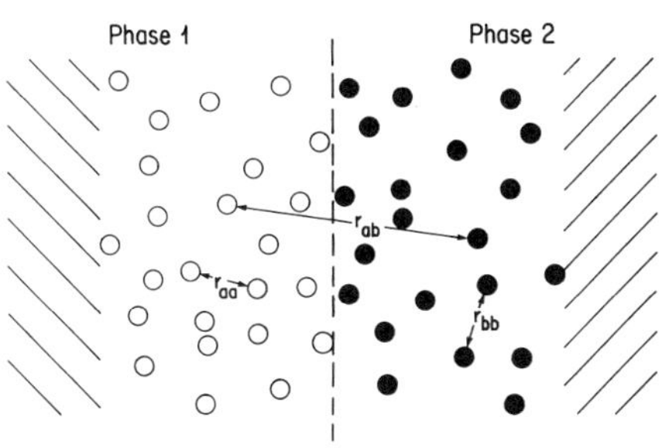

FIGURE 1. Diagram for the "dilute gas" approximation: two gaseous half spaces, in contact. The intermolecular distances for nearest-neighbor pairs of molecules are all considerably larger than the diameters of the molecules a and b. The structure shown here is, of course, unstable because of interdiffusion. It is useful, here, for illustration of one stage in the development of the theory.

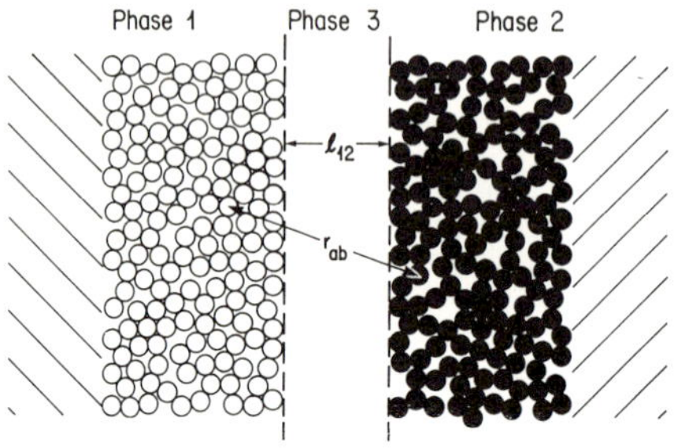

FIGURE 2. Two bodies consisting of condensed matter (phases 1 and 2) separated by a third phase. Phase 3 may be a vacuum (for the purposes of the theory) or it may contain matter, whether solid, liquid, or gas. When phase 3 is absent, $l_{012} \cong r_{0ab}$, the equilibrium distance between molecules of species a and b, as nearest neighbors in the condensed state.

dispersion force γ^d, the Debye induction force γ^i, and the Keesom dipole–dipole force γ^p. He also included a component due to hydrogen bonding, γ^h. For phase i,

$$\gamma_i = \gamma_i^d + \gamma_i^i + \gamma_i^p + \gamma_i^h \tag{9}$$

2.2. The Lifshitz Theory: General Considerations

Lifshitz[(3)] developed a treatment that was related to that of London, which took into account all displacements of charge, of whatever frequency, in a macroscopic body. This included the rotations of permanent dipoles. Unlike the bimolecular approximation, the Lifshitz theory is applicable to condensed phases. Let us consider a phase with a planar surface, and effectively infinite thickness and lateral extension: a "half-space." See Figure 2, which applies to pairs of real, condensed phases.

The electrical displacements within each phase include the oscillations of molecular electron clouds that were treated by London, and the dipolar rotations of Keesom. But they also include displacements that exist over regions of space, in each phase, that are larger than atomic or molecular dimensions. An extreme example is the conduction electrons of a metal; displacement of these electrons may occur over the entire volume of the metal. The so-called "image force" of an ion or dipole, adsorbed on a metal, is a case in which electrons may be drawn from, or repelled to, regions that are well removed from the zone closest to the adsorbed molecule.

The electronic fluctuations that give rise to the London energy, for molecular solids or liquids, are considered to be almost completely uncorrelated, beyond the two-body level. So the isotropic, bimolecular approximation is a good one. This approximation is usable even though the polarizability of bonds and of groups in molecules is not, in fact, symmetrical. For aromatic compounds, it is sometimes worth the trouble to include explicitly, in intermolecular potential functions, the fact that attraction is stronger when two aromatic rings are parallel than when they are edge-on to each other. But for the greatest fraction of substances, α is taken as an average, scalar property, i.e., of equal value in all directions around a center of attraction.

The energy of attraction, for two bodies which are of larger than atomic dimensions, is then the consequence of the mutual attraction of the (oscillating) electric displacements within the two bodies. These oscillations may be synchronous (as when a displacement in one body induces a displacement in the other, and then attracts it) or they may include thermal randomness, as with the Keesom bimolecular energy. [The factor $1/kT$ should be noted, in Eq. (1).]

The (essentially) scalar nature of the molecular polarizability of a molecule is in contrast to the vectorial character of the dipole moment. The energy of attraction for two dipoles, at any instant, depends strongly on orientation (Figure 3a). The mutual potential energy of two dipoles is given by[22]

$$U_{ab}(\text{dipole–dipole}) = -(\mu_a\mu_b/r^3)[2\cos\theta_a\cos\theta_b - \sin\theta_a\sin\theta_b\cos(\phi_a - \phi_b)] \quad (10)$$

The lowest-energy orientation for two dipoles is head-to-tail, i.e., $\theta_a = \theta_b = 0$. But for a cluster of three or more dipoles, the lowest-energy configuration is more complex (Figure 3b). If two of the molecules are in the lowest-energy mutual orientation, then no more than one of them can be in the lowest-energy orientation with respect to the third. In a crystal, the magnitude of the dipole–dipole energy can (in principle) be calculated if the crystal structure is known; and it is very considerably lower than that given by the pairwise approximation.

In a liquid, too, the attractive energy is very much lower than that given by the bimolecular approximation. If the dipole moments are appreciable, then the energy [see Eq. (10)] when summed over all the important neighbors, in all possible orientations, is sufficiently larger than kT that rotation is highly restricted; and the thermal effects, which are accounted for by the $1/kT$ term in Eq. (3), do not randomize the orientations very effectively. (Even when the dipole moments are small, their vectorial nature must be taken into account when integrating over all the molecules.) This electrostatic effect is supplemented by effects due to molecular geometry, e.g., that the effective center of a group dipole is seldom at the molecular center.[18] Also, the important question of the distance of nearest approach of dipoles (considered as point centers of force) has been raised by Fowkes.[26] All of these effects join together to reduce the magnitude of the dipole–dipole energy across a phase boundary, well below the value estimated by the bimolecular approximation.

Thus, for nonpolar molecules, the interaction of n nearest-neighbor (NN) molecules (whether or not of the same species, and whether or not interaction across an interface is included) is given quite well by the London approximation with the assumption of pairwise additivity. If we sum only over the nearest neighbors, then

$$U(\text{NN}) \cong zU_{ab}(\text{London})/2 \quad (11)$$

where z is the number of nearest neighbors.

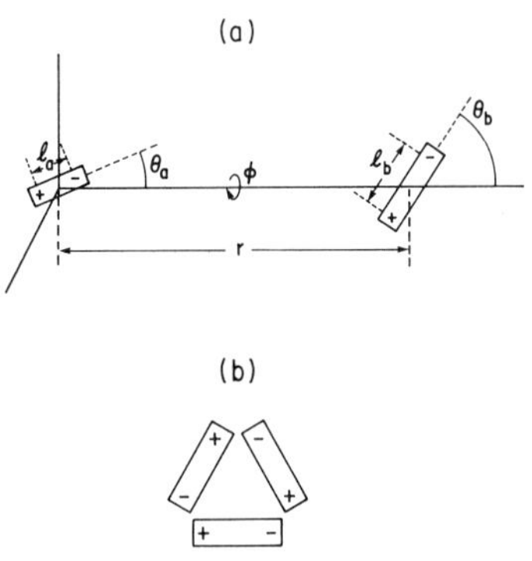

FIGURE 3. (a) Interaction of two dipoles, with center-to-center distance, r; l_a and l_b are the distances between centers of charge, within molecules a and b (from Reference 23, p. 22), (b) Interaction of three dipoles in their lowest-energy configuration. See Reference 23, p. 27, and also References 24 and 25.

But the dipole energy for polar molecules in a condensed system is greatly reduced on account of the effect shown in Fig. 3b, below the value estimated using the pairwise-additivity approximation. Using again the nearest-neighbor approximation,

$$U(\text{condensed phase}) < zU_{ab}(\text{dipole–dipole})/2 \tag{12}$$

The Lifshitz theory, which will be discussed next, bears out this conclusion.

3. INTERACTIONS BETWEEN CONDENSED PHASES

We will first analyze the problem using McLachlan's theory of molecular interactions.[27] The McLachlan expression is appropriate for the dilute gas case; and, for such systems, it gives the same result as the pairwise-addition approximation, which was used by Good and Elbing.[18] We will then develop the Lifshitz theory for interactions across an interface. It will be shown that, for dilute gases, the theory yields the same result as that given by McLachlan; this fact has not been explicitly written out before. The McLachlan equation for the free energy of interaction of two molecules in free space is

$$g(r) = -\frac{6kT}{r^6}\sum_{0}^{\infty}{}' \alpha_1(i\omega_n)\alpha_2(i\omega_n) \tag{13}$$

where α_1 and α_2 are the dielectric polarizabilities (per molecule) of the two phases. For organic and inorganic compounds, α_1 and α_2 are real numbers and are functions of frequency ω_n, where n is the quantum number of the relevant oscillation; see below. They are expressed as functions of $i\omega_n$, a notation that indicates the use of the complex frequency axis. The quantity, ω_n, is related to n by

$$\omega_n \equiv n\left(\frac{4\pi^2 kT}{h}\right) \tag{14}$$

$$= 2.43 \times 10^{14} n \text{ rad/sec at room temperature}$$

where k is Boltzmann's constant, T is temperature, and h is Planck's constant. The prime on the summation symbol in Eqs. (13) and (14) specifies that the $n = 0$ term is given half weight in the sum.

Using this expression, the Hamaker coefficient for interaction between two phases is found to be

$$A_{12} = 6\pi^2 N_1 N_2 kT \sum_{n=0}^{\infty}{}' \alpha_1(i\omega_n)\alpha_2(i\omega_n) \tag{15}$$

where N_1 and N_2 are the densities (in molecules per unit volume) of phases 1 and 2.

For gaseous substances, $\alpha(i\omega_n)$ can be expressed as a sum of two terms, one arising from the dipolar response and the other from the electronic response:

$$\alpha'(i\omega_n) = \frac{\mu^2}{3kT\left(1 + \dfrac{\omega_n}{\omega_{\text{rot}}}\right)} + \frac{\alpha_e^{(0)}}{1 + \left(\dfrac{\omega_n}{\omega_e}\right)^2} \tag{16}$$

where the prime in the term α' denotes limitation to the gaseous state, and $\alpha_e(0)$ is static polarizability; ω_e is the angular frequency that corresponds to the classical ν_o, the fundamental frequency of the electronic cloud of a molecule. (We recall the term $h\nu_o$, which London approximated by I in his "dispersion force" theory.[8]) The quantity ω_{rot} is the angular frequency of rotation of a molecule in a low-pressure gas.*

Equation (16) is not rigorously usable for liquids or solids. The experimental term corresponding to the first term on the right in Eq. (16) can, however, be measured by dielectric or spectroscopic methods. The absorption occurs over a much broader range of frequencies than that found for gas molecules. Moreover, the presence of the μ^2 term in Eq. (16) is based on the interaction of the dipole being only with the external field. It ignores the interaction and shielding due to the other dipoles in a condensed phase, which was referred to above. So the maximum intensity of the spectroscopic absorption is not given correctly.

Returning to the dilute gas approximation, we may make a first step in applying Eq. (14), by setting $n = 0$. Equation (14) then reduces to the zero-frequency component of the Hamaker coefficient:

$$A_{12}(n = 0) = \pi^2 N_1 N_2 \left\{ \frac{\mu_1^2 \mu_2^2}{3kT} + [\mu_1^2 \alpha_{e2}(0) + \mu_2^2 \alpha_{e1}(0)] \right\} \tag{17}$$

It is clear that the first term inside the brackets in Eq. (17) gives the Keesom component of force; compare Eq. (1). The second term is the Debye component; compare Eq. (3).

A second step in applying Eq. (14) is to replace the summation, for $n > 0$, by an integral:

$$A_{12} = A_{12}(n = 0) + 6\pi^2 N_1 N_2 kT \int_0^\infty \alpha_1'(i\omega)\alpha_2'(i\omega)d\omega \tag{18}$$

where $\alpha'(i\omega_n)$ corresponds to $\alpha(i\omega_n)$ in Eq. (16). This yields the higher-frequency component of the Hamaker coefficient, which corresponds to the London component of force:

$$A_{12}(n > 0) = \pi^2 N_1 N_2 \left[\frac{3h}{2} \frac{\omega_{e1}\omega_{e2}}{\omega_{e1} + \omega_{e2}} \alpha_{e1}(0)\alpha_{e2}(0) \right] \tag{19}$$

If we associate the right side of Eq. (17) with the dipolar component of surface tension of a liquid, γ^p, and the right side of Eq. (19) with the London component, γ^d, then the ratio of these two components is simply

$$\frac{\gamma^p}{\gamma^d} = \frac{\dfrac{\mu_1^2 \mu_2^2}{3kT} + [\mu_1^2 \alpha_{e2}(0) + \mu_2^2 \alpha_{e1}(0)]}{\dfrac{3h}{2} \dfrac{\omega_{e1}\omega_{e2}}{(\omega_{e1} + \omega_{e2})} \alpha_{e1}(0)\alpha_{e2}(0)} \tag{20}$$

However, Eq. (20) is valid only for interactions of rarefied media; see Figure 1. For

*Actually, a freely rotating molecule has, in general, three angular frequencies corresponding to the three moments of inertia I_x, I_y, and I_z. If, say, E_x is a specified energy of rotation about the x-axis, the classical relation is $\omega_x = \sqrt{2E_x/I_x}$; and the quantized energy is $E_x = BJ(J+1)$, $J = 0,1,2,\cdots$ and $B = h/2I_x$. We need not go into the theory of rotations in a low-pressure gas, because it is not directly relevant to liquids and solids. For example, the rotational energy levels in a liquid or solid are usually quite unsharp. See, e.g., P. W. Atkins, *Physical Chemistry*, p. 554, W. H. Freeman and Co., San Francisco (1978).

condensed phases, as indicated above, the interaction energy of an arbitrary pair of dipoles is seriously reduced. A particular dipole in phase 1 is not free to take up any orientation relative to that of any particular dipole in phase 2.

For liquids, the general *ab initio* calculation of this reduction in interaction energy, below that in the rarefied media approximation, is far too complicated to carry out; and the calculation would be strongly dependent on the physical details of the molecules involved. Therefore, no general theoretical estimate of the reduction in dipolar interaction energy can be made at this time.

Fortunately, the Lifshitz theory of the interaction energy provides a way out of the impasse that we have just described. Strictly, this application of the Lifshitz theory must be regarded as an approximation, because it is rigorous only when the separation of the two interacting phases is large. The dielectric susceptibilities of the two phases, as measured at wavelengths that are large compared to molecular dimensions, may not be valid at short distances. But the strategy employed in Eq. (27) below should minimize the seriousness of this approximation.

The interaction of dipoles in a condensed medium with an electromagnetic field is the same, whether the field is due to molecules in an adjoining phase, or to an applied, external field such as is employed in measuring dielectric susceptibilities. In other words, measurements of the frequency-dependent dielectric constants of the phases, over the whole electromagnetic spectrum, will yield the needed information.

For liquids or solids, the Lifshitz theory shows that Eq. (14) must be replaced with a more general expression, in terms of the measured (macroscopic) dielectric susceptibilities of the two phases, ϵ_1 and ϵ_2. Then

$$A_{12} = \frac{3}{2} kT \sum_{n=0}^{\infty}{}' \sum_{j=1}^{\infty} \left\{ \left[\frac{\epsilon_1(i\omega_n) - 1}{\epsilon_1(i\omega_n) + 1} \right] \left[\frac{\epsilon_2(i\omega_n) - 1}{\epsilon_2(i\omega_n) + 1} \right] \right\}^j (1/j^3) \tag{21}$$

For numerical computations, it is generally acceptable to retain only the first term in the second summation, i.e., the term for which j = 1. (However, Hough and White[28] recommend carrying out the sum for terms in $j > 1$.) Then

$$A_{12} = \frac{3}{2} kT \sum_{n=0}^{\infty}{}' \left[\frac{\epsilon_1(i\omega_n) - 1}{\epsilon_1(i\omega_n) + 1} \right] \left[\frac{\epsilon_2(i\omega_n) - 1}{\epsilon_2(i\omega_n) + 1} \right] \tag{22}$$

The relation between Eqs. (15) and (22) is clear: The quantities, α, in Eq. (15) are replaced by the functions in parentheses, in Eq. (22), multiplied by $kT/4\pi N_1 N_2$.

If both phases are gases, then ϵ_1 and ϵ_2 can be set equal to unity in the denominator of Eq. (22); and the Lorenz–Lorentz equation can be used to express $\epsilon(i\omega_n)$ as a function of the polarizability $\alpha(i\omega_n)$:

$$\epsilon(i\omega_n) = 1 + 4\pi N\alpha'(i\omega_n) \tag{23}$$

where (as indicated above) α' is the polarizability of the *gaseous* molecules. Then, by employing Eq. (23), Eq. (22) reduces to

$$A_{12} = 6\pi^2 N_1 N_2 kT \sum_{n=0}^{\infty}{}' \alpha_1'(i\omega_n)\alpha_2'(i\omega_n) \tag{24}$$

which has the same form as Eq. (15) with α replaced by α'.

If the phases are liquids or solids, we must use Eq. (22). By analogy to Eq. (17), we may expect the zero-frequency term in Eq. (22) to give the magnitude of the Keesom and Debye interactions, together with the dipolar interactions for fixed orientation, that correspond to Eq. (10). The ratio of the zero-frequency term to the sum of the higher-frequency terms will again be the ratio of γ^p to γ^d. For a single component, Eq. (22) becomes

$$A_{11} = \frac{3}{2} kT \left\{ \frac{1}{2} \left[\frac{\epsilon(0) - 1}{\epsilon(0) + 1} \right]^2 + \sum_{n=1}^{\infty} \left[\frac{\epsilon(i\omega_n) - 1}{\epsilon(i\omega_n) + 1} \right]^2 \right\} \tag{25}$$

where we have separated out the zero-frequency term from the summation. The ratio γ^p/γ^d for a single component is then given by

$$\frac{\gamma^p}{\gamma^d} = \frac{1}{2} \left[\frac{\epsilon(0) - 1}{\epsilon(0) + 1} \right]^2 \Big/ \sum_{n=1}^{\infty} \left[\frac{\epsilon(i\omega_n) - 1}{\epsilon(i\omega_n) + 1} \right]^2 \tag{26}$$

Equation (26) was evaluated numerically for water, using the methods of Parsegian and Ninham[(14–16)] and of Krupp.[(7)] The result was the value 0.07, which is significantly lower than the value 2.43 obtained by the pairwise-addition approximation.[(18)] Then, the true value of γ_p for water must be about 1.4 mJ/m^2, instead of the commonly accepted value 51 mJ/m^2. The balance of the term, 51 − 1.4 mJ/m^2, is due to hydrogen bonding, which will be discussed in chapter 4. *In comparison to γ^d and to γ^h, the dipolar components γ^p and γ^i [see Eq. (9)] are second-order effects.*

The interaction parameter, Φ_{12},[(15)] can now be computed by means of the expression

$$\Phi = \frac{\sum_{n=0}^{\infty}{}' \left[\frac{\epsilon_1(i\omega_n) - 1}{\epsilon_1(i\omega_n) + 1} \right] \left[\frac{\epsilon_2(i\omega_n) - 1}{\epsilon_2(i\omega_n) + 1} \right]}{\left\{ \sum_{n=0}^{\infty}{}' \left[\frac{\epsilon_1(i\omega_n) - 1}{\epsilon_1(i\omega_n) + 1} \right]^2 \sum_{n=0}^{\infty}{}' \left[\frac{\epsilon_2(i\omega_n) - 1}{\epsilon_2(i\omega_n) + 1} \right]^2 \right\}^{1/2}} \tag{27}$$

The errors in the calculation that lead to the numerator in Eq. (27) can probably be approximated by a multiplying factor. The same is true of the denominator; and the two multiplying factors will tend to cancel out.

Sine the hydrogen-bond component of surface free energy is not included in the theory that leads to Eq. (27), the values of Φ_{12} computed by this equation will only be directly usable in estimating interfacial free energy when neither of the phases has an appreciable capacity for hydrogen bonding.

The values of Φ_{12} for a few pairs of liquids, as computed via Eq. (27), are given in Table 1. Details of these calculations will be given in a separate communication. It may be seen that the value of Φ_{12} is very close to unity in all cases. Krupp[(7)] has carried out similar calculations for pairs of metals, and found $\Phi \cong 1$, which is in agreement with the experimental result of Good and Givens[(29)] for mercury vs. gallium.

4. APPLICATION TO INTERFACIAL TENSION AND FREE ENERGY

The conclusions that we have just reached enable us to modify the existing theory of the components of interfacial tension and free energy. First, we note that the interaction

TABLE 1
Values of the Interaction Parameter

Material combinations	Φ_{12}
Pentane–water	0.98
Hexane–water	0.98
Heptane–water	0.98
Octane–water	0.97
Decane–water	0.97
Dodecane–water	0.98
Hexadecane–water	0.98
Teflon FEP–water	0.98
Teflon FEP–dodecane	0.99
Teflon FEP–benzene	0.999
Teflon FEP–glycerol	0.99

parameter Φ^d for the London dispersion force component alone should practically never be *exactly* equal to unity. The presence of the $2I_aI_b/(I_a + I_b)$ factor in Eq. (2a), and of I in Eq. (2b), lead to the general expression for Φ_{12}^d given by

$$\Phi_{12}^d = \frac{\Delta G_{12}^d}{\sqrt{\Delta G_1^d \, \Delta G_2^d}} = \frac{2(I_aI_b)^{1/2}}{(I_a + I_b)} \tag{28}$$

This is the ratio of a geometric mean to an arithmetic mean of two terms; and if the two terms are not exactly equal, this ratio is always less than 1. Values of I have been tabulated elsewhere[18] for water and a number of organic compounds. Using $I_w = 12.67$ eV and $I_{organic} = 9$ to 11 eV, we obtain $\Phi_{12}^d = 0.985$ to 0.998. Thus, the commonly used expression for the London dispersion force component of the free energy of adhesion

$$\Delta G_{12}^a = -2\sqrt{\gamma_1^d\gamma_2^d} \tag{29}$$

is uncertain to less than 2%. This small correction is commonly taken to be a second-order effect, and the approximation

$$\Phi_{12}^d \cong 1 \tag{30}$$

is generally employed.

The Lifshitz theory, which we have discussed above, supersedes the London theory, and we have shown in Table 1 that Φ_{12} is within 3% of unity for a number of two-phase systems. So it is reasonable to lump the second-order terms γ^p and γ^i in Eq. (11) into a single term, along with γ^d:

$$\gamma^d + \gamma^p + \gamma^i \equiv \gamma^{LW} \tag{31}$$

The superscript LW denotes "Lifshitz–van der Waals." In terms of LW components of surface free energy, we may write

$$\Phi_{12}^{LW} = \frac{\gamma_1^{LW} + \gamma_2^{LW} - \gamma_{12}^{LW}}{2\sqrt{\gamma_1^{LW}\gamma_2^{LW}}} \tag{32}$$

and express the results of the calculations by Eq. (27), in Table 1, using the approximation

$$\Phi_{12}^{LW} \cong 1 \tag{33}$$

In terms of the superscript LW (instead of d) Eq. (29) becomes

$$\Delta G^{aLW} = -2\sqrt{\gamma_1^{LW}\gamma_2^{LW}} \tag{34}$$

The LW component of the free energy of adhesion is

$$\Delta G^{LW} = -\gamma_1^{LW} - \gamma_2^{LW} + \gamma_{12}^{LW} \tag{35}$$

The solution of Eqs. (34) and (35) for γ_{12}^{LW} yields

$$\gamma_{12}^{LW} \cong \gamma_1^{LW} + \gamma_2^{LW} - 2\sqrt{\gamma_1^{LW}\gamma_2^{LW}} \tag{36a}$$

$$\cong [(\gamma_1^{LW})^{1/2} - (\gamma_2^{LW})^{1/2}]^2 \tag{36b}$$

We can now take an important step in advancing the understanding of polar interactions across interfaces. We observe that for binary systems such as water/hydrocarbon or glycerol/Teflon, hydrogen bonding is present in one component, and that the *complete* interaction parameter Φ_{12} is given by

$$\Phi_{12} = \frac{\gamma_1 + \gamma_2 - \gamma_{12}}{2\sqrt{\gamma_1\gamma_2}} \tag{37}$$

For these systems, the experimental value of Φ is appreciably less than unity. So we will attribute the difference between the total surface tension, γ, and the LW component, γ^{LW}, to hydrogen bonding.

For greater generality, we will use the superscript AB for "acid–base," since hydrogen bonding can be regarded as an example of (Lewis) acid–base interaction. Then

$$\gamma = \gamma^{LW} + \gamma^{AB} \tag{38}$$

This expression is, formally, the same as that proposed by Fowkes.[26] In chapter 4, we will examine the consequences of Eq. (38). We will also introduce a new combining rule, which takes into account the complementary nature of the contributions of the acidic component and the basic component to the hydrogen bond.

APPENDIX. A NOTE ON ELECTROSTATIC AND ELECTROMAGNETIC UNITS IN THE THEORY OF ADHESION

The Editor of this volume has pointed out that in a recent, highly authoritative book, *Intermolecular and Surface Forces* (Academic Press, New York, 1985), J. N. Israelachvili employs the factor $\frac{1}{4}\pi\epsilon_o$ throughout the derivations that he gives. In Chapter 3 we do not employ this term, and we are in conformity with most writers on intermolecular forces in surface chemistry and adhesion in omitting the term. At the Editor's request, we have furnished the following, brief treatment of the origin and relevance of the $\frac{1}{4}\pi\epsilon_o$ factor.

The present discussion is adapted from a section called "A note on dipoles" in a booklet entitled *Materials Science Experiments and Notes on Selected Topics in Materials Science and Corrosion* by the present author (published by the State University of New York at Buffalo for use in an undergraduate course). The topic of dipole–dipole energy is quite appropriate, as it provides a frame for explaining the factor $\frac{1}{4}\pi\epsilon_o$, because the dipolar component of interaction energy has been discussed in many theoretical treatments of adhesion. See, for example, our Eq. (10) in Chapter 3, which may be compared with Eqs. (A2) and (A4) below.

When Peter Debye[(30)] did his pioneering work on dipole moments in chemistry and physics, he used the cgs system of units, and expressed charge in electrostatic units or esu. (The esu of charge is the statcoulomb.) The majority of workers in the field have done the same ever since. Dipole moments are still, today, commonly described and tabulated in units that are based on the esu of charge. The debye is defined as the dipole moment (ql) possessed by a system of two opposite charges (q) each being one esu of charge separated by a distance l of one angstrom. Thus, 1 debye $= 1.0 \times 10^{-18}$ esu-cm. The electronic charge is 4.18×10^{-10} esu, so a dipole consisting of one electronic charge (negative) separated 1×10^{-8} cm from a positive charge (of the same size) would have a dipole moment of 4.18 debyes.

Now, in electrostatic units, Coulomb's law for two equal charges of opposite sign is

$$U = -q^2/\kappa r \tag{A1}$$

where κ is the dielectric constant (permittivity) of the medium and q is the magnitude of the charge. Typical values of κ at 20 °C are given in Table A1. For two dipoles μ_a and μ_b in head-to-tail orientation, the energy is

$$U = -2\mu_a\mu_b/\kappa r^3 \tag{A2}$$

Thus if $\mu_a = \mu_b = 1.0$ debye and $r = 10$ Å, then in a vacuum

$$U = \frac{-2 \times (1 \times 10^{-18})^2}{1 \times (10 \times 10^{-8})^3} = -2.000 \times 10^{-15} \text{ erg}$$

In SI units, charge is expressed in coulombs and 1 esu of charge $\equiv 3 \times 10^9$ coulombs. The unit of distance is the meter. The debye is then 3.336×10^{-30} coulomb-meters.

Coulomb's law in SI is

$$U = -q^2/4\pi\epsilon_o\kappa r \tag{A3}$$

where ϵ_o is the permittivity of free space (more loosely, the dielectric constant of a vacuum) and κ is the ordinary dielectric constant, i.e., as listed in Table A1. The value of ϵ_o is 8.854×10^{-12} farads per meter. Israelachvili, on p. 23 of his book cited at the beginning of this Appendix, uses the same form as Eq. (A3) for Coulomb's law.

The reason why the SI version of Coulomb's law contains $\frac{1}{4}\pi\epsilon_o$ lies in a choice made by physicists for the development of Maxwell's equations of electrodynamics. (The solution of Maxwell's set of four differential equations provided the prediction that electromagnetic waves would exist and would have velocity c, the velocity of light.) When

TABLE A1
Typical Values of the Dielectric Constant

Medium	κ
Vacuum	1.000
Hexane	1.874
Benzene	2.283
Water	78.0
Ethyl acetate	6.4

Coulomb's law is used in the SI form (along with the corresponding SI equation for magnetic interactions) the result is a simplification of Maxwell's equations: The term $\frac{1}{4}\pi\epsilon_o$ no longer appears explicitly.

The energy of a system of two dipoles in head-to-tail orientation is then

$$U = -2\mu_a\mu_b/4\pi\varepsilon_o\kappa r^3 \tag{A4}$$

For the same distance and dipole moments used above

$$\begin{aligned} U &= \frac{-2 \times (3.336 \times 10^{-30})^2}{4\pi \times 8.854 \times 10^{-12} \times (10 \times 10^{-1)})^3} \\ &= -2.000 \times 10^{-22} \text{ joule} \\ &= -2.000 \times 10^{-15} \text{ erg} \end{aligned}$$

It is noteworthy that in the Good–Girifalco methodology, it is irrelevant whether electrostatic or SI units are used in order to conduct absolute calculations of the free energies of adhesion and cohesion. The calculated *macroscopic* energy of adhesion or of cohesion must be the same, whichever set of units is used. Moreover, when the ratio of these free energies is set up, e.g., to form the term $\Delta G^a_{12}/\sqrt{\Delta G^c_1 \Delta G^c_2}$, the coefficient $\frac{1}{4}\pi\epsilon_o$, which is present because of the units used, cancels out.

To answer the question raised in the opening paragraph, the factor $\frac{1}{4}\pi\epsilon_o$ in Israelachvili's book arose because the author employed SI units. If he had used cgs units and charges in esu, this factor would have been replaced by unity.

W. K. H. Panofsky and M. Phillips, in Appendix I of their book *Classical Electricity and Magnetism* (Addison Wesley Publ. Co., Reading, MA, 1956) present a slightly more advanced discussion of electrostatic and electromagnetic units.

NOMENCLATURE

A	Hamaker coefficient
B	$h/2I_x$
E	Rotational energy
G	Free energy
I_a, I_b	Ionization energy of molecule a or b
I_x, I_y, I_z	Moments of inertia

J	Quantum number of a rotator
N_1, N_2	Number of molecules of species 1 or 2 per unit volume
U	Total energy
U(NN)	Total energy of interaction of nearest-neighbor molecules
T	Absolute temperature (Kelvin)
$g(r)$	Free energy of interaction of two molecules
h	Planck's constant
i	$\sqrt{-1}$
j	Summation index, Eq. (21)
k	Boltzmann's constant
l	Distance between boundaries of phases 1 and 2
n	Quantum number
r	Intermolecular distance
z	Number of nearest neighbors
Φ	Interaction parameter, Eq. (27)
α	Polarizability
γ	Surface tension or surface free energy
θ	Angle between dipole axis and line connecting molecular centers
ϕ	Angle between dipole axis and a plane containing the line connecting molecular centers
ν	Frequency
ω	Angular frequency
Superscripts	*used with* γ
d	London dispersion force
i	Debye induction force
p	Dipole–dipole force
h	Hydrogen bonding
LW	London–van der Waals component
AB	Acid–base component

REFERENCES

1. C. J. van Oss, M. K. Chaudhury, and R. J. Good, *Adv. Colloid Interface Sci.* **28,** 35–64 (1987).
2. C. J. van Oss, R. J. Good, and M. K. Chaudhury, *J. Protein Chem.* **5,** 385 (1986).
3. E. M. Lifshitz, *Sov. Phys. JETP* **2,** 73 (1956).
4. V. A. Parsegian and B. W. Ninham, *Nature* **224,** 1197 (1969).
5. B. W. Ninham and V. A. Parsegian, *Biophys. J.* **10,** 646 (1970).
6. V. A. Parsegian and B. W. Ninham, *J. Colloid Interface Sci.* **32,** 332 (1971).
7. H. Krupp, *Adv. Colloid Interface Sci.* **1,** 111–239 (1967).
8. F. London, *Trans. Faraday Soc.* **33,** 8 (1937).
9. P. Debye, *Phys. Z.* **21,** 178 (1920); **22,** 302 (1921).
10. W. H. Keesom, *Phys. Z.* **22,** 126 (1921); **23,** 225 (1922).
11. J. Israelachvili, *Q. Rev. Biophys.* **6,** 341 (1974).
12. R. J. Good, *J. Colloid Interface Sci.* **59,** 398–417 (1977).
13. A. van der Avoird and F. Hoflich, *J. Chem. Phys.* **47,** 1563 (1967).
14. H. C. Hamaker, *Physica* **4,** 1058 (1937).
15. L. A. Girifalco and R. J. Good, *J. Phys. Chem.* **61,** 904–909 (1957).

16. R. H. Fowler and E. A. Guggenheim, *Statistical Thermodynamics,* Oxford University Press (1939).
17. R. J. Good, in: *Contact Angle, Wettability and Adhesion,* Advances in Chemistry no. 43, Am. Chem. Soc., Washington, D.C. (1964).
18. R. J. Good and E. Elbing, *Ind. Eng. Chem.* **62,** 54–78 (1970).
19. P. E. Berghausen, R. J. Good, G. Kraus, B. Podolsky, and W. Soller, WADC Report no. 55-44 (1955).
20. F. M. Fowkes, *J. Phys. Chem.* **66,** 382 (1962).
21. F. M. Fowkes, *Ind. Eng. Chem.* **56**(12)**,** 40 (1964).
22. J. O. Hirschfelder, C. F. Curtis, and R. B. Bird, *Molecular Theory of Gases and Liquids,* Wiley–Interscience, New York (1954).
23. R. J. Good, in: *Treatise on Adhesion and Adhesives* (R. L. Patrick, ed.), pp. 9–68, Marcel Dekker, New York (1967).
24. R. J. Good and M. C. Phillips, in: *Proc. Conf. on Wetting, Bristol, England,* Soc. Chem. Ind. (London) Monograph no. 25, pp. 44–48 (Sept. 1966).
25. M. C. Phillips, R. J. Good, D. A. Cadenhead, and H. F. King, *J. Colloid Interface Sci.* **37,** 437 (1971).
26. F. M. Fowkes and M. A. Mostafa, *Ind. Eng. Chem. Prod. Dev.* **17,** 3 (1978).
27. A. D. McLachlan, *Proc. R. Soc. London, Ser. A* **271,** 387–401 (1963); **274,** 80–90 (1963); *Mol. Phys.* **6,** 423–427 (1963); *Discuss. Faraday Soc.* **40,** 239–245 (1965).
28. D. B. Hough and L. R. White, *Adv. Colloid Interface Sci.* **14,** 3 (1980).
29. R. J. Good and W. G. Givens, in: *Contact Angle, Wettability and Adhesion,* Advances in Chemistry no. 43, Am. Chem. Soc., Washington, D.C. (1964).
30. P. Debye, *Polar Molecules,* Chemical Catalog Co., New York (1929).

4

Theory of Adhesive Forces Across Interfaces
2. Interfacial Hydrogen Bonds as Acid–Base Phenomena and as Factors Enhancing Adhesion

Robert J. Good, Manoj K. Chaudhury, and Carel J. van Oss

1. INTRODUCTION

The theory of the apolar components of interfacial forces was examined in the previous chapter of this volume.[1] It has been possible to develop that theory of apolar components at this time owing to the existence of quantitative, mathematically formulated theories of forces between molecules (e.g., the London theory) together with the Lifshitz electromagnetic theory of the interaction of macroscopic bodies. (See the previous chapter for references.)

The concept of acid–base interactions is many decades older than the London dispersion force theory. Jensen[2,3] has recently provided an excellent review of acid–base concepts. Hydrogen bonds[4] constitute a subclass of acid–base interactions. It was informally known, within a few years after hydrogen bonding was recognized, that hydrogen bonding was of profound significance with regard to interfaces. But a quantitative, general treatment of hydrogen bonds *at interfaces* has not become available until very recently.

A theory that fills the need was recently proposed.[5,6] We will point out below some characteristics of hydrogen bonding that set it apart from the apolar interfacial forces. The new methodology that has been developed leads, among other things, to the establishment of the conditions in molecular structure under which the interfacial free energy between condensed phases can be negative. It will be shown, in Section 7 of this chapter, that the existence of negative interfacial tension has important consequences in regard to practical adhesion, which have not been recognized heretofore.

Robert J. Good • Department of Chemical Engineering, State University of New York at Buffalo, Buffalo, NY 14260. *Manoj K. Chaudhury* • Dow Corning Corporation, Midland, Michigan 46868. *Carel J. van Oss* • Department of Chemical Engineering and Department of Microbiology, State University of New York at Buffalo, Buffalo, New York 14214.

Fowkes[7–9] first suggested that surface tension could be resolved into components due to the dispersion force, the induction and dipole–dipole forces, and hydrogen bonding. We have reviewed this subject in the previous chapter. Fowkes *et al.*[10,11] also pioneered the explicit application of Lewis acid–base theory to interfaces between condensed phases.

We have shown in the previous chapter[1] that the equations

$$\gamma_i = \gamma_i^{\mathrm{LW}} + \gamma_i^{\mathrm{AB}} \tag{1}$$

and

$$\gamma_{ij}^{\mathrm{LW}} = (\sqrt{\gamma_i^{\mathrm{LW}}} - \sqrt{\gamma_j^{\mathrm{LW}}})^2 \tag{2}$$

can be used in describing the surface and interfacial behavior of condensed phases. Here, γ_i is surface free energy and γ_{ij} is interfacial free energy. The superscript LW designates "Lifshitz–van der Waals", and AB designates acid–base interaction. As a formalism, γ^{LW} corresponds to the "dispersion component of surface tension," of Fowkes.[7–9] (The previous chapter elaborated on the physical meaning of γ^{LW}, in terms of a new, quantitative assessment of the contribution of dipole–dipole and dipole–induced dipole interactions to the energy of cohesion of a condensed phase, and to the energy of adhesion across an interface.)

There is an intrinsic difference between the two terms, γ_i^{LW} and γ_i^{AB}. We will show that, in general, no expression *in the form of Eq. (2)* can be written as a combining rule for acid–base interactions. The LW interactions possess a physical and mathematical symmetry which acid–base interactions, or hydrogen-bond interactions, lack.

Formally in Eq. (2), the designations i and j can be interchanged. The LW interaction of two unlike bodies, i.e., the free energy of adhesion $\Delta G_{ij}^{\mathrm{aLW}}$, can be constructed from the respective free energies of cohesion, $\Delta G_{ii}^{\mathrm{cLW}}$ and $\Delta G_{jj}^{\mathrm{cLW}}$, introduced independently of each other.

By contrast, this type of symmetry does not exist in acid–base interactions. Instead, a complementarity of function controls behavior. In an isolated, pure phase of a substance such as chloroform, the proton-donating (or electron-accepting) capacity of the compound does not contribute anything to the cohesion of the liquid, because of the absence of a proton acceptor. But for such a substance, in a two-phase system with another substance that can play the complementary role in hydrogen bonding, there is a major contribution due to proton donation, in the energy of adhesion between the phases.

Systems of this kind cannot be accommodated by the formalism of Eq. (2). The complementary roles of acid and base cannot be represented by such a symmetrical relationship. In general, one molecular contributor to a hydrogen bond must be a proton donor, and the other a proton acceptor. A new formalism will be required to accommodate this fact.

2. *MONOPOLAR AND BIPOLAR SUBSTANCES, AND HYDROGEN BONDS* per se

An isolated molecule (or group) may have the potentiality to be exclusively a proton acceptor in a hydrogen bond, as is the case with a carbonyl group. Or it may be exclusively a proton donor, as with chloroform. We refer the reader to the discussion of nonideality of chloroform–acetone solutions, given by Hildebrand and Scott[12], the adduct structure may written $Cl_3C\text{–}H \cdots\cdots O{=}C(CH_3)_2$.

We define the term *monopolar* to mean having either exclusively proton-donor or exclusively proton-acceptor character.[15] More generally, we will use the term, monopolar, to denote molecules (or groups) that have exclusively Lewis acid or Lewis base[2,3] character. (The Lewis acid–base terminology has the advantage that extension to non-hydrogen-bond interactions is facilitated.) The acidic or basic character of monopolar molecules or groups can be manifested only in the presence of molecules or groups of the opposite type.

Probably more common than the cases of monopolar character, in an isolated molecule, are the cases where both acidic (e.g., proton-donor) and basic (e.g., proton-acceptor) character are present, as with water. We will call such molecules *bipolar*. Again, the generalization to Lewis acids and Lewis bases can be made directly.

A few years ago Drago *et al.*[13] represented the energy of acid–base interactions by an algebraic product formulation. He separated out two components of the interaction; and for each substance, he postulated a parameter for electrostatic interaction as well as a parameter for covalent interaction. Fowkes and Mostafa[10,11] have used Drago's formalism to describe interfacial energies. In modern molecular orbital theory,[14] the results of which were prefigured by Small,[15] hydrogen-bond energy is represented as the product of two difference terms: the difference between values of a single parameter that characterizes the acid functionality for each of the two components, and the corresponding difference for a single parameter for the basic functionality. In his extensive review of the Lewis acid–base concept, Jensen[2] points out that the separation of a Lewis acid–base interaction into electrostatic and covalent components is not necessary when an appropriate choice of atomic orbitals is made for the construction of molecular orbitals. We will follow References 2 and 14, 15 in this regard, in the next few paragraphs; this will lead to a simpler and much more tractable formulation than that based directly on Reference 11, and one that has the further advantage of having a theoretical base in molecular orbital theory.

Lewis acid–base interactions[16,17] require that one chemical component be an electron acceptor (acid) and the other an electron donor (base). The electron donor furnishes an unshared pair of electrons, while the electron acceptor furnishes an orbital system that can accommodate an additional pair of electrons. See Jensen's book[2] for details as to the kinds of donor and acceptor orbitals that can take part in an acid–base interaction.

We note in passing that the Brønsted–Lowry acid–base theory[18,19] may be regarded as a special case of the Lewis acid–base theory. A Brønsted base (proton acceptor) is a Lewis base, and a Brønsted acid is a Lewis acid in which the electron acceptor is a proton.

It is well known that hydrogen bonding constitutes a subclass of Lewis acid–base bonding. This subclass is distinct within the general class, because of the presence of a proton *in* the bond structure. (That proton was, in the isolated molecule that entered into hydrogen-bond formation, covalently bound to an electronegative atom or group.) Hence, a *two-center* treatment based on atomic orbitals, for a hydrogen bond, has serious deficiencies. In principle, a three-center molecular orbital treatment should be possible. But an interpretation of such a treatment in terms of covalent bonding concepts that are broadly familiar, such as σ or π bonds, or an identifiable electrostatic component, may be so complex as to be unprofitable. (An early term for "hydrogen bond" was "hydrogen bridge.") Perhaps it may be best to conclude that hydrogen bonds are *sui generis,* i.e., in a class by themselves. This seems to be the attitude taken in the very recent, important paper of Nelson *et al.*[(20)]

For the present, we will consider it permissible to use whichever of the three terminologies (hydrogen bonds, Lewis acids/bases, Brønsted acids/bases) is appropriate for the chemical system that is being discussed.

The necessary conditions for hydrogen bonding are given by Pimentel and McClellan[(4)]: (a) An electronegative atom or group, with a hydrogen atom covalently bound to it, and (b) an atom or group, usually also electronegative, with an unshared pair of electrons, or a π-electron system as in aromatic compounds. Both (a) and (b) functionalities may be present in the same molecule, as is the case with water or glycerol.

These two conditions, while necessary, may not amount to a sufficient condition for hydrogen bonding. This is shown by the very recent results of Nelson *et al.*,[(20)] who found spectroscopic and other evidence that the hydrogens of ammonia do not act as electron acceptors in hydrogen bonding, although the unshared pair electrons on the nitrogen constitute a strong electron-donor site. Until the time of this report, it had been widely thought that ammonia was analogous to water, in having both kinds of functionality. The fact that the boiling point of NH_3 is -33 °*C* (which is 133 °*C* below that of water) appears to have been held less important, in regard to this analogy, than the contrast with methane, which boils at -161.5 °*C*.

We may use the term *Lewis neutralization* to describe interaction of a proton-bearing Lewis acid with a Lewis base, in which the proton is not fully transferred to the base. The acidic and basic components become associated with each other, strongly enough that the term "compound formation" can often be used.[(12)]

A question can be raised, as to whether it is possible to distinguish, in terms of macroscopic phenomena such as adhesion and surface tension, between nonprotonic electron-donor/acceptor interactions and hydrogen bonds. At present, it seems that this is not possible. That distinction can, however, be made with the aid of a different class of measurements, e.g., spectroscopy. Or it may be made with the aid of molecular orbital theory, in the future.

In view of the (present) operational difficulty of distinguishing between protonic and nonprotonic acid–base interactions, we will simply refer to the phenomenon as Lewis acid–Lewis base interaction.

Another question that arises is whether the surface manifestations of hydrogen bonding must correspond exactly to the bulk manifestations. For example, let us consider a solution of a species that contains a hydroxyl group in a liquid polyether. From spectroscopic shifts in absorption frequencies and departures from Raoult's law,[(12)] it may be inferred that hydrogen bonding exists. Now, in interactions across an interface, it might be expected that whenever hydroxyl groups are present in one phase (e.g., an aqueous solution) and ether groups in the other, hydrogen bonds will form. But this is not necessarily so. The protons of the hydroxyl groups may interact with the unshared pair of electrons of other hydroxyl groups in the same phase, strongly enough that they are not available for Lewis neutralization across the interface. Or, on the other hand, the hydroxyls may interact more strongly with surface atoms or groups of another phase, so that the hydroxyl compound is adsorbed at the interface.

We have introduced the term "monopolar" to designate a substance that has a structure such that it can act as either a Lewis acid or a Lewis base, but not both. (More precisely, if it has an appreciable capability of acting as one type in Lewis acid–base interaction, then its capability of acting as the other type is negligible.) A bipolar substance, such as water, has the capability of acting both as a Lewis acid and as a Lewis base. As we have just noted, a polarity function which one would expect would be operational may not be manifested in surface behavior. So a compound that one would expect, structurally, to be bipolar may be effectively monopolar.[5] An apolar compound has neither kind of capability, under any circumstances.

3. CLASSIFICATION OF BINARY SYSTEMS

The complimentarity of function of acids and bases leads us to propose, now, a system of classification of chemical substances and groups. The operational property γ_i^{AB} in Eq. (1) needs to be broken down into two terms.

As a heuristic introduction, we may compare the surface tension data of $CHC1_3$ and $CC1_4$ and their interfacial tensions vs. water (see Table 1).

The free energy of adhesion in Table 1 is given by the equation

$$-\Delta G_{ij}^{a} = \gamma_i + \gamma_j - \gamma_{ij} \tag{3}$$

The surface tensions of $CC1_4$ and $CHC1_3$ differ by a negligible amount, 0.18 mN/m; but the interfacial tensions vs. water differ by 13.4 mJ/m^2. The difference between the free energies of adhesion, 13.6 mJ/m^2, is due to hydrogen bonding, and corresponds to a bond energy of about 3 kJ/mol of interfacial bonds.

We have already mentioned the fact that the form of Eq. (2) is not *in general* applicable to acid–base interactions. If it were applicable, we could write without restriction

$$\gamma_{ij}^{AB} = (\sqrt{\gamma_i^{AB}} - \sqrt{\gamma_j^{AB}})^2 \tag{a}$$

for the acid–base interaction, and consequently,

$$\gamma_{ij} = (\sqrt{\gamma_i^{LW}} - \sqrt{\gamma_j^{LW}})^2 + (\sqrt{\gamma_i^{AB}} - \sqrt{\gamma_j^{AB}})^2 \tag{b}$$

for the Lifshitz–van der Waals and acid–base interactions, combined.

We may test Eqs. (a) and (b) with the chloroform–water and chloroform–Teflon systems. The contact angle of chloroform on Teflon FEP is 50°, and the surface free energy

TABLE 1

Substance	Surface tension (mN/m)	Interfacial tension vs. water (mN/m)	Free energy of adhesion to water (mJ/m^2)
$CHCl_3$	27.13	31.6	−68.3
CCl_4	26.95	45.0	−54.7

of Teflon FEP is $\gamma_s^{LW} = 18$ mJ/m^2.[21] Using the theory of Good and Girifalco[22–24] and Fowkes[7–9,25,26] for the contact angle of a liquid on an apolar solid, we may write

$$\gamma_l^{LW} = 4\gamma_s^{LW}/(1+\cos\theta)^2 \tag{4a}$$

or

$$\gamma_s^{LW} = \gamma_l^{LW}(1+\cos\theta)^2/4 \tag{4b}$$

From these data, the calculated value of γ^{LW} for chloroform is 26.7 mJ/m^2. Subtracting this from the observed surface tension (27.13 mN/m) we obtain $\gamma_{CHCl_3}^{AB} = 0.4$ mJ/m^2, which is indistinguishable from zero within the precision of the measurements employed in deriving it. The calculated value of γ_{12} by Eq. (a) for $CHC1_3$ vs. water would be $(\sqrt{26.7} - \sqrt{21})^2 + (\sqrt{0.4} - \sqrt{51.8})^2$, or 42.7. (If $\gamma_{CHCl_3}^{AB}$ is taken to be zero, the computed value of γ_{12} is found to be 51.4 mJ/m^2.) These calculations of γ_{12} disagree with the observed value, 31.6 mJ/m^2. The reason for the disagreement is the fact that (as indicated above) hydrogen bonding makes a negligible contribution to the surface tension of $CHC1_3$ (i.e., $\gamma_l = \gamma_l^{LW}$) and none to the free energy of adhesion of $CHC1_3$ to Teflon. But it makes a major contribution to the interfacial tension with water. Hence, Eqs. (a) and (b) are not valid for this system. There are many other pairs of systems, besides $CHC1_3/H_2O$ vs. $CC1_4/H_2O$, that can be used to demonstrate this point about Eqs. (a) and (b).

We can now make the following generalizations about the liquid/vapor surface tensions of *pure* substances that have any Lewis acidic or Lewis basic character.

(1) For a monopolar Lewis acid or base in its pure state, only an acidic component, or only a basic component, is present; so the AB component of the surface free energy is zero. The total surface tension of a pure, monopolar Lewis acid or Lewis base is determined solely by its LW interactions:

$$\begin{aligned} \gamma_i^{AB} &= 0 \\ \gamma_i &= \gamma_i^{LW}, \qquad \text{if substance } i \text{ is monopolar} \end{aligned} \tag{5}$$

(2) For a bipolar substance such as water, which can act both as a Lewis acid and as a Lewis base, $\gamma_i^{AB} > 0$:

$$\gamma_i = \gamma_i^{LW} + \gamma_i^{AB}, \qquad \text{substance } i \text{ bipolar} \tag{6}$$

Next, we can classify binary systems consisting of components 1 and 2. We arrange these classes in order of increasing generality (see Table 2).

In all seven kinds of systems, the LW component of interfacial tension will be given by Eq. (2). We will see, below, that the AB component of γ_{ij} is never given by Eq. (a).

In systems of types I to IV, the free energy of adhesion is due solely to LW interactions; and so

$$\gamma_{ij}(\text{Types I to IV}) = \gamma_{ij}^{LW} \tag{7}$$

$$\gamma_{ij}^{AB}(\text{Types I to IV}) = 0 \tag{8}$$

$$\Delta G_{ij}^{a}(\text{Types I to IV}) = -2\sqrt{\gamma_i^{LW}\gamma_j^{LW}} \tag{9}$$

Two-phase systems of types V, VI, and VII usually do not have stable interfaces

TABLE 2

Classification of Systems Consisting of Components 1 and 2, With Respect to the Polarity of the Substances

I. Both 1 and 2 apolar.
II. One component apolar and the other monopolar in either sense, i.e., either acidic or basic.
III. One component apolar, the other bipolar.
IV. Both 1 and 2 monopolar in the same sense.
V. Both 1 and 2 monopolar, but in the opposite sense, i.e., one acidic and the other basic, as with chloroform/acetone. See below regarding solubility questions, particularly for Type V systems.
VI. Either 1 or 2 monopolar (in either sense) and the other bipolar.
VII. Both 1 and 2 bipolar.

when both phases are liquids. (The $CHCl_3$/water system is one of the exceptions.) The interfacial tension may be negative (see below) and then the interface will have no mechanical stability. An interface between two miscible liquids may have a transient existence such that an interfacial tension can be measured. An example of this has been reported by Smith *et al.*[27] If the interfacial tension is positive, the interface will decay by interdiffusion. Solid–solid interfaces of types V to VII can be stable because of the bulk mechanical strength of the solids. Solid–liquid interfaces can have a transient existence if the solid is soluble in the liquid, and a permanent character if solubility of the solid is low. They may also exist if the solid surface layer is a strongly adsorbed monolayer on a stable solid. If the substrate is a polymer that has been swollen (to a gel) by the liquid, a stable interface may exist.[28] It is possible to determine the interfacial energy between two miscible liquids, by forming a gel of one of the liquids with a polymer and measuring the contact angle of the other liquid on the gel surface. (This measurement must be made rapidly, before diffusion out of the gel changes the concentration of liquid in it.) Varying the initial concentration of liquid in the gel, and extrapolating to zero concentration of the polymer, makes it possible to estimate the interfacial tension between the two pure, unmixed liquids.

For a system of type V, since AB interactions make no contribution to the surface free energy of either pure substance,

$$\gamma_i^{\mathrm{AB}} = \gamma_j^{\mathrm{AB}} = 0 \qquad \text{(Type V)} \tag{10}$$

We can estimate that AB component of the free energy of adhesion, i.e., $(\Delta G_{ij}^{\mathrm{a}})^{\mathrm{AB}}$, if we know $\epsilon_{ij}^{\mathrm{AB}}$, the hydrogen-bond energy (defined to be a positive quantity) between i and j. If there are N_{s} acid–base bonds per unit area across the interface, then

$$(\Delta G_{ij}^{\mathrm{a}})^{\mathrm{AB}} = -N_{\mathrm{s}}\,\epsilon_{ij}^{\mathrm{AB}} \qquad \text{(Type V)} \tag{11}$$

(This relation is approximate; it neglects the entropy of adhesion.) We can then write, for the total free energy of adhesion,

$$\Delta G_{ij}^{a} = -2\sqrt{\gamma_i^{\mathrm{LW}}\gamma_j^{\mathrm{LW}}} - N_{\mathrm{s}}\epsilon_{ij}^{\mathrm{AB}} \qquad \text{(Type V)} \tag{12}$$

$$\gamma_{ij} = \left(\sqrt{\gamma_i^{\mathrm{LW}}} - \sqrt{\gamma_j^{\mathrm{LW}}}\right)^2 - N_{\mathrm{s}}\epsilon_{ij}^{\mathrm{AB}} \qquad \text{(Type V)} \tag{13}$$

$$\gamma_{ij}^{\mathrm{AB}} = -N_{\mathrm{s}}\epsilon_{ij}^{\mathrm{AB}} \qquad \text{(Type V)} \tag{14}$$

The squared term on the right in Eq. (13) is generally small, and the last term may be large, so γ_{ij} may be negative in type V systems.

In the last two types of systems, γ^{AB}(bipolar) $\neq 0$, so that

$$\Delta G^{a}_{ij} = -2\sqrt{\gamma_i^{LW}\gamma_j^{LW}} - N_s\epsilon_{ij}^{AB} \tag{15a}$$

$$\gamma_{ij}^{AB} = -N_s\epsilon_{ij}^{AB} + \gamma_j^{AB} \qquad \text{(Type VI)} \tag{15b}$$

where j is the bipolar component. Also,

$$\Delta G^{a}_{ij} = -2\sqrt{\gamma_i^{LW}\gamma_j^{LW}} - N_s\epsilon_{ij}^{AB} \tag{16a}$$

$$\gamma_{ij}^{AB} = -N_s\epsilon_{ij}^{AB} + \gamma_i^{AB} + \gamma_j^{AB} \qquad \text{(Type VII)} \tag{16b}$$

Again, the interfacial tension is likely to be negative, if an interface exists at all.

4. SURFACE PARAMETERS $\gamma^{\oplus}$ AND $\gamma^{\ominus}$ THAT GIVE DIRECT CHARACTERIZATION OF ACIDIC AND BASIC BEHAVIOR

We now introduce a new set of symbols, $\gamma^{\oplus}$ and $\gamma^{\ominus}$, which will distinguish the acidic and basic properties, and which can be used to reveal some important properties of the interfaces (see Table 3). The polarities of the symbols, $\oplus$ and $\ominus$, are of course somewhat arbitrary. They are based on the charge of the donor functionality; thus, it is reasonable to assign a positive character to a proton donor and a negative character to an electron donor. The symbols $\oplus$ and $\ominus$ cannot be employed as superscripts on a symbol that refers to an operationally measurable property of an interface between condensed phases, such as γ_{ij}.*

We postulate a combining rule that is related to the simple product rule of Small[13] and Kollman[14] for short-range interfacial forces. The acid–base component of the free energy of adhesion is given by

$$(\Delta G^{a}_{ij})^{AB} = -2\sqrt{\gamma_i^{\oplus}\gamma_j^{\ominus}} - 2\sqrt{\gamma_i^{\ominus}\gamma_j^{\oplus}} \qquad (i \text{ and } j \text{ bipolar}) \tag{17}$$

The two terms on the right-hand side of this equation, rather than a single term as in Eq. (a) above, are required because of the complementarity of the $\gamma^{\oplus}$ and $\gamma^{\ominus}$ functions. The right-hand side of Eq. (17) is obviously not a geometric mean and $\gamma^{\oplus}$ and $\gamma^{\ominus}$ appear only in products of terms of opposite polarity. If one of the components is monopolar and the other bipolar, or monopolar in the opposite sense, then eq. (17) reduces to

$$(\Delta G^{a}_{ij})^{AB} = -2\sqrt{\gamma_i^{\oplus}\gamma_j^{\ominus}}$$

$$(i \text{ and } j \text{ monopolar in the opposite sense, or one of } i \text{ and } j \text{ bipolar}) \tag{18}$$

The Dupré equation for $(\Delta G^{a})^{AB}$ can be put in the form

$$\gamma_{ij}^{AB} = (\Delta G^{a}_{ij})^{AB} + \gamma_i^{AB} + \gamma_j^{AB} \qquad (\text{substances } i \text{ and } j \text{ bipolar}) \tag{19}$$

$$= (\Delta G^{a}_{ij})^{AB} + \gamma_i^{AB} \qquad (\text{substance } j \text{ monopolar and } i \text{ bipolar}) \tag{20}$$

*These parameters bear a relationship to the Gutman scales[29] of "donor numbers" and "acceptor numbers." It does not at present seem possible to make this relationship quantitative, however.

TABLE 3
Symbols to Identify Acidic and Basic Properties

	Acid/base definition	Monopolar surface parameter
(1)	{Lewis acid = electron acceptor; Brønsted acid = proton donor}	$\gamma^{\oplus}$
(2)	{Lewis base = electron donor; Brønsted base = proton acceptor}	$\gamma^{\ominus}$

We also have, for a single substance i,

$$\gamma_i^{AB} = 2\sqrt{\gamma_i^{\oplus}\gamma_i^{\ominus}} \qquad \text{(substance } i \text{ bipolar)} \tag{21a}$$

$$\gamma_i^{AB} = 0 \qquad \text{(substance } i \text{ monopolar or apolar)} \tag{21b}$$

On combining Eqs. (16), (18), and (20) for bipolar substances, we obtain

$$\gamma_{ij}^{AB} = 2(\sqrt{\gamma_i^{\oplus}\gamma_i^{\ominus}} + \sqrt{\gamma_j^{\oplus}\gamma_j^{\ominus}} - \sqrt{\gamma_i^{\oplus}\gamma_j^{\ominus}} - \sqrt{\gamma_i^{\ominus}\gamma_j^{\oplus}})$$

$$= 2(\sqrt{\gamma_i^{\oplus}} - \sqrt{\gamma_j^{\oplus}})(\sqrt{\gamma_i^{\ominus}} - \sqrt{\gamma_j^{\ominus}}) \tag{22}$$

This equation stands in contrast to Eq. (a) above.

It is clear from Eq. (23) that if $\gamma_i^{\oplus} > \gamma_j^{\oplus}$ and $\gamma_i^{\ominus} < \gamma_j^{\ominus}$, or if $\gamma_i^{\oplus} < \gamma_j^{\oplus}$ and $\gamma_i^{\ominus} > \gamma_j^{\ominus}$, then γ_{ij}^{AB} will be negative.

For a ternary system consisting of an initiali/j interface into which a multimolecular layer of substance k is interposed, displacing j from the surface of i, the acid–base component of the free energy of displacement may be expressed in the form

$$\Delta G_{ikj}^{AB} = 2[\sqrt{\gamma_k^{\oplus}}(\sqrt{\gamma_i^{\ominus}} + \sqrt{\gamma_j^{\ominus}} - \sqrt{\gamma_k^{\ominus}}) +$$

$$\sqrt{\gamma_k^{\ominus}}(\sqrt{\gamma_i^{\oplus}} + \sqrt{\gamma_j^{\oplus}} - \sqrt{\gamma_k^{\oplus}}) - \sqrt{\gamma_i^{\oplus}\gamma_j^{\ominus}} - \sqrt{\gamma_i^{\ominus}\gamma_j^{\oplus}}] \tag{23}$$

The LW component of the free energy change is

$$\Delta G_{ikj}^{LW} = 2(\gamma_k^{LW} - \sqrt{\gamma_i^{LW}\gamma_k^{LW}} - \sqrt{\gamma_j^{LW}\gamma_k^{LW}} + \sqrt{\gamma_i^{LW}\gamma_j^{LW}}) \tag{24a}$$

$$= 2(\sqrt{\gamma_j^{LW}} - \sqrt{\gamma_i^{LW}})\,(\sqrt{\gamma_k^{LW}} - \sqrt{\gamma_i^{LW}}) \tag{24b}$$

For liquid l on solid s, contact angles follow the Young–Dupré equation:

$$\gamma_l\,(1 + \cos\theta l_s) = -\Delta G_{ls}^{a} \tag{25}$$

The total free energy of adhesion is given by

$$\Delta G_{ls}^{a} = (\Delta G_{ls}^{a})^{LW} + (\Delta G_{ls}^{a})^{AB} \tag{26}$$

Now

$$(\Delta G_{ls}^{a})^{LW} = -2\sqrt{\gamma_l^{LW}\gamma_s^{LW}} \tag{27}$$

and $(\Delta G_{ls}^{a})^{AB}$ is given by an expression in the form of Eq. (17) for a bipolar liquid on a bipolar solid. We can combine Eqs. (17) and (25)–(27) to yield

$$\gamma_l(1 + \cos\theta_{ls}) = 2\sqrt{\gamma_l^{LW}\gamma_s^{LW}} + 2\sqrt{\gamma_l^{\oplus}\gamma_s^{\ominus}} + 2\sqrt{\gamma_l^{\ominus}\gamma_s^{\oplus}} \tag{28}$$

If either the liquid or the solid is monopolar, then ΔG_{ls}^{aAB} is given by a relation in the form of Eq. (18); and one or the other of the last two terms in Eq. (28) is zero.

5. APPLICATION OF THE LEWIS ACID–BASE PARAMETERS

It should be possible to apply these equations if a set of γ^{LW}, $\gamma^{\oplus}$, and $\gamma^{\ominus}$ parameters can be established. The γ^{LW} terms, for use in Eq. (28), can be determined in a straightforward manner,[7–9] which we now review.

By definition, apolar solids such as Teflon and apolar liquids such as alkanes have surface free energies that are purely apolar:

$$\gamma_{i(\mathrm{apolar})} \equiv \gamma_i^{\mathrm{LW}} \tag{29}$$

Fowkes has pointed out[7–9] that polar forces, such as those due to hydrogen bonds, do not contribute to the interfacial free energy of a polar liquid vs. an apolar liquid or solid. Equation (2) is valid regardless of whether phase i or phase j has the capability of forming hydrogen bonds. If both phases are liquid, we may measure the interfacial tension, and use Eq. (2), rearranged in the form

$$\gamma_i^{\mathrm{LW}} = \gamma_{ij}^{\mathrm{LW}} - \gamma_j^{\mathrm{LW}} + 2\sqrt{\gamma_i^{\mathrm{LW}}\gamma_j^{\mathrm{LW}}} \tag{30}$$

The apolar component of the surface free energy of a polar solid can be determined by measuring the contact angle of an apolar liquid, and using the equation

$$\gamma_{\mathrm{s}}^{\mathrm{LW}} = \frac{\gamma_{l(\mathrm{apolar\ liquid})}(1 + \cos\theta)^2}{4} \tag{31}$$

In employing Eqs. (4) and (31), we ignore (for the present) the equilibrium spreading pressure π_{e}.[22,30] If π_{e} is appreciable, separate measures must be employed to determine it; and we must use the form

$$\gamma_{\mathrm{s}}^{\mathrm{LW}} = \frac{[\gamma_l(1 + \cos\theta) + \pi_{\mathrm{e}}]^2}{4\gamma_l}, \qquad l \text{ apolar} \tag{32}$$

An important, practical restriction on the use of Eqs. (28) and (31) is the fact that the liquid must have a surface tension larger than the critical surface tension of the solid.[31] The $\gamma_{\mathrm{s}}^{\mathrm{LW}}$ values for a large number of important polymers are in the neighborhood of 40 $\mathrm{mJ/m^2}$. For such solids, either of two liquids which have high surface tensions but very small propensities for AB interactions may be used; these are diiodomethane ($\gamma = 50.8$ $\mathrm{mJ/m^2}$) and α-bromonaphthalene ($\gamma = 44.4$ $\mathrm{mJ/m^2}$).

For a pair of polar liquids, 1 and 2, γ_l^{LW} is determined for liquids l_1 and l_2 by means of the contact angle on an apolar solid such as Teflon, and the use of Eq. (4).

The contact angles of the two polar liquids on solid s can now be measured: $\theta_{l_1\mathrm{s}}$ and $\theta_{l_2\mathrm{s}}$. Since at this point we have the values of γ_{l_1}, γ_{l_2}, $\theta_{l_1\mathrm{s}}$, $\theta_{l_2\mathrm{s}}$, $\gamma_{l_1}^{\mathrm{LW}}$, $\gamma_{l_2}^{\mathrm{LW}}$, and $\gamma_{\mathrm{s}}^{\mathrm{LW}}$, we may rearrange Eq. (28) into the form with the "unknowns" $\gamma^{\oplus}$ and $\gamma^{\ominus}$ on the right:

$$\gamma_{l_1}(1 + \cos\theta_{l_1\mathrm{s}}) - 2\sqrt{\gamma_{l_1}^{\mathrm{LW}}\gamma_{\mathrm{s}}^{\mathrm{LW}}} = 2\sqrt{\gamma_{l_1}^{\oplus}\gamma_{\mathrm{s}}^{\ominus}} + 2\sqrt{\gamma_{l_1}^{\ominus}\gamma_{\mathrm{s}}^{\oplus}} \tag{33a}$$

$$\gamma_{l_2}(1 + \cos\theta_{l_2\mathrm{s}}) - 2\sqrt{\gamma_{l_2}^{\mathrm{LW}}\gamma_{\mathrm{s}}^{\mathrm{LW}}} = 2\sqrt{\gamma_{l_2}^{\oplus}\gamma_{\mathrm{s}}^{\ominus}} + 2\sqrt{\gamma_{l_2}^{\ominus}\gamma_{\mathrm{s}}^{\oplus}} \tag{33b}$$

To solve for $\gamma_s^\oplus$ and $\gamma_s^\ominus$ we need four more relations. Two of these can be obtained with the aid of other contact angle measurements, as will be explained below: these are the ratio $\gamma_{l_1}^\oplus/\gamma_{l_2}^\oplus$, which we may call β, and the corresponding term, $\gamma_{l_1}^\ominus/\gamma_{l_2}^\ominus$.

There does not seem to be any method available at this time to obtain the needed fifth and sixth independent equations. Therefore, we will resort to the selection of a bipolar reference liquid that has a high surface tension, for which an arbitrary ratio, $\alpha = \gamma^\oplus/\gamma^\ominus$, is chosen. This methodology is analogous to the tactics used in thermodynamics, in which an arbitrary zero of energy is selected. (The pH scale is an example.) The use of an arbitrary value of α, reasonably close to unity, does not affect the determination of operational quantities, as will be demonstrated below.

We will use the convention that, for water as the reference liquid, $\alpha = 1$, $\gamma_W^\oplus = \gamma_W^\ominus$. This, together with Eq. (21) for a bipolar liquid, yields

$$\gamma_W^{AB} = \gamma_W - 21.8 = 51.0 \text{ mJ/m}^2$$
$$\gamma_W^\oplus = \gamma_W^\ominus = 25.5 \text{ mJ/m}^2$$

These constitute the last two equations, in $\gamma_W^\oplus$ and $\gamma_W^\ominus$, that are needed.

5.1. $\gamma^\oplus$ *and* $\gamma^\ominus$ *Values of Selected Liquids*

We may explain the determination of the $\gamma_{l_1}^\oplus/\gamma_{l_1}^\oplus$ and $\gamma_{l_1}^\oplus/\gamma_{l_2}^\ominus$ ratios by an illustration with the liquid pair, water and glycerol. The AB component of the free energy of adhesion of a bipolar liquid to a monopolar solid, such as PMMA, may be obtained by use of Eqs. (25)–(27). Thus

$$(\Delta G_{ls}^a)^{AB} = -\gamma_l(1 + \cos\theta_{ls}) + 2\sqrt{\gamma_l^{LW}\gamma_s^{LW}} \tag{34}$$
$$= -2\sqrt{\gamma_l^\oplus\gamma_s^\ominus}, \quad \gamma_s^\oplus = 0 \quad \text{(solid a monopolar base)} \tag{35a}$$
$$= -2\sqrt{\gamma_l^\ominus\gamma_s^\oplus}, \quad \gamma_s^\ominus = 0 \quad \text{(solid a monopolar acid)} \tag{35b}$$

It turns out experimentally (see below) that we can use Eq. (35a) with several solids.

Then, for water (W) and glycerol (GL),

$$\frac{\gamma_W^\oplus}{\gamma_{GL}^\oplus} = \left[\frac{\gamma_W(1 + \cos\theta_{W/s}) - 2\sqrt{\gamma_W^{LW}\gamma_s^{LW}}}{\gamma_{GL}(1 + \cos\theta_{GL/s}) - 2\sqrt{\gamma_{GL}^{LW}\gamma_s^{LW}}}\right]^2 \equiv \beta \tag{36a}$$

$$\gamma_{GL}^\oplus = \gamma_W^\oplus/\beta \tag{36b}$$

Combining Eqs. (21) and (36), and using the convention that $\alpha = 1$, we obtain

$$\gamma_{GL}^\ominus = (\gamma_{GL}^{AB})^2\beta/4\gamma_W^\oplus \tag{37}$$

We may illustrate the use of these equations with the water/PMMA and glycerol/PMMA systems. On one sample of PMMA, the contact angle of α-bromonaphthalene was found to be 26°. Using Eq. (31) (i.e., assuming α-bromonaphthalene to be apolar) and $\gamma_{\alpha BN} = 44.4$, the value of $\gamma_{PMMA}^{LW} = 40$ mJ/m^2 was obtained. The contact angle of glycerol was 69.0°, and for water, 74.5°. Since γ^{AB} of water is 21.8 mJ/m^2 and γ^{AB} of glycerol is 34 mJ/m^2, Eq. (36a) yields $\beta = 6.35$.

The average value of β for glycerol, obtained from measurements of this type on 12 solids, was 6.505. For formamide, $\beta = 11.2$; for DMSO, $\beta = 36.4$. Details of these measurements are given elsewhere.[32] With this value of β, and Eq. (37) together with the convention that $\alpha = 1$, $\gamma^{\oplus} = 25.5$ mJ/m², we have obtained the data in Table 4.

In References 5 and 33, we have employed the symbol, δ, for the relative polarities of a pair of liquids. Thus,

$$\delta^{\oplus}_{ij} = \sqrt{\gamma_i^{\oplus}/\gamma_j^{\oplus}} \tag{38a}$$

$$\delta^{\ominus}_{ij} = \sqrt{\gamma_i^{\ominus}/\gamma_j^{\ominus}} \tag{38b}$$

The first of these equations is equivalent to Eq. (36b), with $(\delta^{\oplus}_{i\mathrm{W}})^2 = 1/\beta$. The use of the dimensionless parameters, $\delta^{\oplus}$ and $\delta^{\ominus}$, is mathematically equivalent to the use of the convention, $\gamma^{\oplus}_{\mathrm{W}} = \gamma^{\ominus}_{\mathrm{W}}$. Some values of the δ parameters are given in Ref. 33.

5.2. Application to Selected Solid Surfaces

As indicated above, Eq. (35a) can be used, and hence also Eq. (36), for monopolar solids that are Lewis bases. An independent basis for choosing solids that are monopolar bases is that contact-angle measurements with a *monopolar liquid* that is a Lewis base yield the same apparent γ_s^{LW} value as does an *apolar liquid*. Thus, the value of γ_s^* is calculated using

$$\gamma_s^* = \gamma_{l\,(monopolar)}^{\mathrm{TOT}}\,(1 + \cos\theta)^2/4 \tag{39}$$

This is compared with the value obtained using Eq. (31). If $\gamma_s^* = \gamma_s^{\mathrm{LW}}$, then the solid is either apolar or has the same polarity as the monopolar liquid. If $\gamma_s^* > \gamma_s^{\mathrm{LW}}$, then the solid is either monopolar with the opposite polarity, or else it is bipolar. Dimethyl sulfoxide (DMSO) is a suitable test liquid, although its small $\gamma^{\oplus}$ (as shown in Table 4) may be a cause for uncertainty. On the basis of its structure, it should be a preponderantly Lewis base monopole. Using this test liquid, we have found that the following polymers appear to be $\gamma^{\ominus}$ monopoles: PMMA, cellulose acetate, polyoxyethylene, sucrose, zein, gelatin,

TABLE 4
Surface Tension Components of Some Liquids: $\gamma^{\oplus}$ and $\gamma^{\ominus}$Values Based on $\gamma^{\oplus}_{\mathrm{W}} \equiv \gamma^{\ominus}_{\mathrm{W}}$ Reference Convention

	Surface tension components (mJ/m²)					
	γ_l	γ_l^{LW}	γ_l^{AB}	$\gamma_l^{\oplus}$	$\gamma_l^{\ominus}$	Reference
H_2O (W)	72.8	21.8	51	25.5	25.5	
Glycerol (GL)	64	34	30	3.92	57.4	32
Formamide (FO)	58	39	19	2.28	39.6	32
Dimethyl sulfoxide (DMSO)	44	36	8[a]	0.5	32	33
α-Bromonaphthalene (ABN)	44.4	43.5				
Diiodomethane (DIM)	50.8	50.8				

[a]Values of $\gamma^{\oplus}$ for DMSO range from 0.07 to 0.7 mJ/m². Fowkes (Colloids and Surfaces **43,** 367–387, 1990) estimates γ^{LW} for DMSO as 34.9 mJ/m² and γ^{AB} as 11.6 mJ/m². This value of γ^{AB} is slightly larger than the value in Table 4, which is 8.0 mJ/m².

TABLE 5
Surface Parameters for Some Polymers

Solid	γ^{LW} (mJ/m^2)	$\gamma^{\oplus}$ (mJ/m^2)	$\gamma^{\ominus}$ (mJ/m^2)	Reference
	(a) Based on advancing angles			
Poly(methylmethacrylate), Cast film	39 to 43	(0)	9.5 to 22.4	*a, b*
Poly(vinylchloride)	43	0.04	3.5	*b*
Poly(oxyethylene): PEG 6000	45	(0)	66	*a, b*
Cellulose acetate	35	0.3	22.7	*c*
Cellulose nitrate	45	0	16	*c*
Agarose	41	0.1	24	*c*
Gelatin	38	0	19	*c*
Human serum albumin (dry)	41	0.15	18	*d*
Polystyrene	42	0	1.1	*a*
	(b) Based on advancing and receding angles			
Polyethylene (commercial film)*e*				
Based on advancing angles	33	0	0.1	*a*
Based on receding angles	42	2.1	30 to 50	

[a] Recent measurements in this laboratory.
[b] References 32 and 35.
[c] References 5 and 33.
[d] Reference 32.
[e] These results should not be taken as necessarily representative of highly purified polyethylene.

RNA, and human serum albumin and many other proteins. Bipolar solids were: glucose, cellulose, and DNA. Several solids that were $\gamma^{\ominus}$ monopoles when fully dried exhibited $\gamma^{\oplus}$ polarity as well when hydrated.[(34)]

These conclusions, as to whether a solid was $\gamma^{\ominus}$ monopolar, $\gamma^{\oplus}$ monopolar, or bipolar, were supported by the values of the surface parameters determined using Eqs. (33a) and (33b). Liquid 1 was water and liquid 2 was glycerol. Table 5 shows some data for a variety of polymers.

The equations that relate the new parameters $\gamma^{\oplus}$ and $\gamma^{\ominus}$ to phenomena such as interfacial tension and adhesion, together with values of the parameters such as those reported in Table 5, lead to some important conclusions.

5.3. *Discussion of $\gamma^{\oplus}$ and $\gamma^{\ominus}$ Values in Table 5*

Probably the most striking characteristic of Table 5(a) is the fact that none of these solids has a large $\gamma^{\oplus}$; five of the solids are monopolar Lewis bases, by the criterion that we have proposed, and for three others, $\gamma^{\oplus}$ is very small. See also the list given earlier, of polarities as determined using DMSO.

In terms of properties that could reasonably be expected on the basis of structure, PMMA and PEG 6000 are "well behaved," i.e., monopolar basic. Aside from the range of the $\gamma^{\ominus}$ values of PMMA (which, no doubt, is due to variations in sample preparation) it is to be expected that ester groups would be Lewis bases, and $\gamma^{\oplus}$ should be zero. This is, indeed, what was observed.

For polyoxyethylene (PEG 6000) the polar moiety is the ether group. (The ratio of hydroxyls to ether groups in this polymer is less than 0.01.) Hence, dominant Lewis base behavior, with a large $\gamma^{\ominus}$, is unsurprising. It is to be expected that the few hydroxyls in the polymer should be buried below the surface, forming hydrogen bonds (by Lewis neutralization) to ether groups, which are present in a large excess.

It is surprising to find that certain carbohydrates such as sucrose and agarose, and also cellulose acetate and nitrate, are effectively $\gamma^{\ominus}$ monopoles. On the basis of structure, these should be bipolar; the protons of the hydroxyl groups should be electron acceptors. As suggested above, in certain carbohydrate structures there may be no hydroxyls pointed "outward" to the adjacent phase. In these structures, all the hydroxyls of the surface carbohydrate rings are Lewis-neutralized by hydrogen bonding to the Lewis base oxygen atoms of other hydroxyls, etc. When a stronger Lewis base is present in the other phase, or in a solvent, it might be expected that hydroxyls so bonded to oxygen atoms could be taken away by the external Lewis base, with the hydroxyls turning from "inward-to-the-bulk phase" to "outward" orientation.

The possibilities that we have just raised indicate that a dependence of the values of the surface parameters on crystallinity (as opposed to amorphous behavior, at least, at the surface) and on temperature may be found. (The temperature dependence arises because higher temperature would lead to a greater randomization, as between "inward" and "outward" orientation. It is clear that a complete theory must include entropy-of-adhesion terms.) Also, a dependence of $\gamma^{\oplus}$ and $\gamma^{\ominus}$ values on the intrinsic base or acid strength of the individual polar groups in the test liquids might be expected, since a weakly basic liquid would be less effective in causing rearrangement of hydroxyls than would a strongly basic liquid.

Poly(vinyl chloride) is another case in which there is a surprising lack of evidence for effective $\gamma^{\oplus}$ behavior. The value $\gamma^{\oplus} = 0.04$ mJ/m^2 is scarcely distinguishable from zero. True, a hydrogen on the same carbon atom as a single chlorine should be less acidic than the H in $CHCl_3$. However, the smallness of the $\gamma^{\oplus}$ value may be due to Lewis neutralization, too. The $\gamma^{\ominus}$ value of this polymer, though small, is probably significant, and is probably due to chlorine atoms acting as electron donors.

Gelatin and serum albumin are also, effectively, $\gamma^{\ominus}$ monopoles. This fact may well be due to the orientation of NH groups toward Lewis base groups in the bulk, as suggested above, in regard to OH groups in certain carbohydrates. It is unlikely that the absence of $\gamma^{\oplus}$ character has the same cause as the lack of Lewis acid behavior of NH_3[(20)] mentioned above. Amides have not yet been subjected to the same kind of microwave spectroscopic study as ammonia, so we cannot be sure that behavior analogous to ammonia will be present or absent. The existence of amide hydrogen bonds is commonly inferred from the mechanical strength of proteins and Nylon polymers. Moreover, we can draw on the difference in the boiling points of acetamide (222 °C) vs. propylamine (48.7 °C) and acetone (56.6 °C) to conclude that the NH group of an amide behaves differently, in hydrogen bonding, from an NH in an amine.

Polystyrene has an appreciable $\gamma^{\ominus}$ character. This is no doubt due to the π electrons of the aromatic ring. Good and Elbing,[(36)] 20 years ago, reported evidence for hydrogen bonding between water and the π electrons of benzene. The energy of this bond was estimated to be 2.4 kcal/mole.

Some results for commercial polyethylene are shown in Table 5(b) to illustrate the difference in behavior that arises when receding angles, as opposed to advancing angles, are measured.[37] No purification was carried out in preparing the polyethylene films for this measurement. The three parameters γ^{LW}, $\gamma^{\oplus}$, and $\gamma^{\ominus}$, based on the receding contact angles, are notably different from those based on advancing angles. It has been suggested[37] that the receding angle corresponds to the interaction of liquid with a higher-energy component of the solid surface, and that the advancing angle corresponds to interaction with the lower-energy component. Table 5(b) indicates that if this two-component surface model is valid, then the higher-energy component of this surface differs qualitatively from that with lower energy, both in its apolar (LW) interactions and in its acid–base behavior.

6. AN ANOMALY

There is a mathematical possibility that arises from Eqs. (33a) and (33b) which can lead to some puzzlement when application of this theory is attempted. The situation in which this problem arises is real—in the sense that we have found, that it arose out of real measurements of contact angles.

In a small number of cases it was observed that, in the solution to Eqs. (33a) and (33b), a negative value was obtained for $\sqrt{\gamma_s^{\oplus}}$. (The absolute values of $\sqrt{\gamma_s^{\oplus}}$ were below 1.5 $mJ^{1/2}/m$ in every case found so far.) We rarely observed negative values of $\sqrt{\gamma_s^{\ominus}}$ using real data.

There can be no physical meaning to the positive sign associated with a value of $\gamma_s^{\oplus}$ obtained by squaring a negative value of $\sqrt{\gamma_s^{\oplus}}$. Four possible explanations can be suggested. The most obvious is experimental error in measurement of contact angles on the solid polymer. One source of error arises in the measurement of the contact angle of water on highly hydrophilic solids, such as polyoxyethylene, dextran, and some hydrophilic proteins. If there is more than a few seconds (e.g., 10 seconds) delay after stopping adding water to the drop, some of the liquid may diffuse into these polymers, and swell them by an uncontrolled amount. This would change the liquid–solid interfacial tension, γ_{sl}, by an uncontrolled amount, and render equations, such as (2) and (22), and hence Eq. (28), inapplicable.

A second possibility is film pressure, π_e, which is neglected in the form of Young's equation that we have used. A third is bulk interaction of the liquid on the solid, e.g., swelling or tending to dissolve the solid. A fourth is a systematic matter: that the selected value of $\gamma_W^{\oplus}/\gamma_{GL}^{\oplus}$, obtained from contact angle data, may need to be revised.

The fifth explanation is that the result is "real," and there is no anomaly. By Eq. (21a), γ_s^{AB} would be negative (see below).

7. APPLICATIONS

7.1. Negative Interfacial Tensions: An Illustration

We may use the equations developed above, and data from Tables 4 and 5, in order to estimate the interfacial free energy for a solid/liquid pair in which hydrogen bonds form across an interface. By combining Eqs. (1), (2), and (22), we obtain for solid s and liquid ℓ

$$\gamma_{sl} = (\sqrt{\gamma_s^{LW}} - \sqrt{\gamma_l^{LW}})^2 + 2(\sqrt{\gamma_s^{\oplus}} - \sqrt{\gamma_l^{\oplus}})(\sqrt{\gamma_s^{\ominus}} - \sqrt{\gamma_l^{\ominus}}) \tag{40}$$

For water, we may use $\gamma_l^{LW} = 21.8 \text{ mJ/m}^2$ and $\gamma_l^{\oplus} = \gamma_l^{\ominus} = 25.5 \text{ mJ/m}^2$, in combination with the data in Table 1 for PEG 6000. The result is, $\gamma_{sl} = -26.2 \text{ mJ/m}^2$ for the water/PEG interface.

As already indicated, the operational γ_{sl} obtained from a calculation of this type is independent of the reference value assumed for $\gamma_W^{\oplus}/\gamma_W^{\ominus}$. This invariance could be checked by assuming another value for this ratio and calculating new values of $\gamma^{\oplus}$ and $\gamma^{\ominus}$ to replace those in Tables 4 and 5. It was found, assuming $\gamma_W^{\oplus}/\gamma_W^{\ominus} = 2$, that the calculated value of γ_{sl} for water on PEG 6000 was -26.3 mJ/m^2, which is in close agreement with the previous calculation. It is shown elsewhere[6] that this invariance is general and exact.

In the previous section, we have noted that negative values of $\gamma_s^{\oplus}$ could lead to negative values of γ_s^{AB}. The solid surface would still be stable if $\gamma_s^{LW} > \gamma_s^{AB}$; so that the existence of a negative square root of $\gamma_s^{\oplus}$ does not necessarily mean a negative γ_s.

7.2. Application to Adhesion and Adhesive Bonding

The concepts presented above have important applications to the direct problem of adhesion. It has long been known (since the 1920s) that the formation of hydrogen bonds across interfaces is conducive to the development of strength in adhering systems. Our new treatment provides some fresh, quantitative conclusions about interfacial hydrogen bonds.

We have already noted that Eq. (23) delineates the conditions under which the interfacial tension between phases i and j will be negative. Now, there are two ways in which negative interfacial tension favors adhesive joint strength.

The first way is by making the free energy of adhesion, ΔG_{ij}^{a}, more negative. The direct importance of a large, negative free energy of adhesion is intuitively obvious, so we will not review that here. Good and Gupta[38] have analyzed the micromechanics of detachment of a polymer from a solid, and introduced a new dimensionless number which, we believe, clarifies the true role of the free energy of adhesion in controlling the process of separating a thermoplastic polymer from a solid.

The second way is via a mode of action which has not been discussed before—probably because the very possibility of a negative interfacial free energy has not been recognized in the past.

An adhesive or coating for a solid is ordinarily applied as a liquid. Zisman[39] has pointed out that pores and surface cavities in a solid surface must be filled in, if an interface that is free from voids and flaws is to be achieved. Under tension, voids act as stress concentrators; and the detachment of the adhesive or coating is initiated at voids and flaws. This initiation occurs at a much smaller stress level than that required in the absence of voids. Zisman noted the need for the adhesive or coating, in its fluid form, to have a zero contact angle, if the formation of interfacial voids is to be prevented.

Zisman did not go beyond specifying zero contact angle as being an optimum property that leads to freedom from voids. It will be recalled that Young's equation,

$$\cos\theta = \frac{\gamma_{sv} - \gamma_{sl}}{\gamma_{lv}}, \qquad -1 \leqslant \cos\theta \leqslant 1 \tag{41}$$

when employed with a *single* liquid on a solid, is not capable of detecting the conditions in which γ_{sl} is negative; and this fact constituted a limitation on the Zisman methodology.

However, as shown above, when three properly chosen liquids are employed, it is possible to determine whether γ_{sl} will be negative or positive.

If the liquid/solid interfacial tension is negative, there will be a larger driving force for spreading into cavities and pores in the solid surface than if γ_{sl} is positive. The penetration of pores by a liquid that forms a zero contact angle is described by the well-known capillary rise equation, or the Washburn equation.[(40)] If the liquid/solid interfacial tension is negative, the rate of penetration will be faster (and, if $|\gamma_{sl}|$ is large, then very much faster) than the rates predicted by these equations; and penetration will be more complete.

Hence, large negative values of γ_{sl} will be particularly conducive to the formation of void-free adhesive/solid or coating/solid interfaces, and to the development of high, practical strength in adhering systems.

7.3. The Effective Range of Acid–Base Interactions

An important matter which we have not yet touched upon is the effective range of the forces due to hydrogen bonds and other acid–base interactions.

First, it is clear that for interactions of Type V, pairs of substances that are monopolar in the opposite sense, the *strongest* attractive force is limited to interaction of a molecule with one nearest neighbor. But monopolar acidic and basic groups can be solvated by bipolar molecules, and when this happens the effective range of action of the acidic or basic group may be increased. For example, in a polyether, an oxygen (which is basic) can hydrogen bond to a hydroxyl of water, the hydrogen of which is acidic. The second hydrogen of the water molecule may be in a hydrogen bond to another water molecule, and so on. Rabinovich and Derjaguin[(41)] have very recently reported observing an excess attractive force, out to a distance of over 100 nm. A molecule that was purely acidic in character could, of course, exhibit a corresponding kind of behavior.

Next, we consider interactions of Type IV, two substances that are monopolar in the same sense. In a pure substance, like-monopole interactions make no contribution to the cohesion, nor do they contribute to the adhesion between two unlike substances in a binary system. However, a strong and (often) effectively long-range repulsion may develop in the presence of a third component. If the third component is monopolar in the opposite sense, it will solvate a purely monopolar substance, and tend to dissolve it. Even if one or both of the components is not dissolved in a third component, the surface layers of both substances will become solvated, and the two materials will spontaneously separate [see Eq. (23)].

If the third component is bipolar, the solvation may develop into an oriented multilayer, thereby making the effective repulsion a long-range force. Estimates indicate a range that may be from the order of 2 nm to as large as over 1 μm.[(41,42)] Very recent papers [(43,44)] by the present authors have contributed further to this subject.

7.4. Other Developments and Applications

We give here a brief list of other applications of the new concepts that have been discussed above.

- Elucidation of the mechanism of so-called "hydrophobic bonding."[(35,45–47)]

- Chromatography[48] including "Southern blotting" of DNA and "western blotting" of proteins.[33] The bipolar character of DNA and the Lewis base ($\gamma^{\ominus}$) monopolar behavior of RNA, and of all the proteins examined so far, have important consequences in the "blotting" separation of these substances.
- Polymer solubility,[48] protein solubility,[49] coacervation,[50] and phase separation in binary polymer systems brought about by the introduction of a solvent in which both polymers are (individually) soluble.[45]
- Antigen–antibody interactions.[6]
- Cell–polymer and cell–cell interactions.[51–53]
- The molecular mechanism of cell membrane fusion.[54]
- The osmotic pressure of water-soluble polymers.[55]
- The surface behavior of coal and graphite.[56]

8. SUMMARY

We have reviewed and enlarged upon the recently proposed concepts of apolar, Lewis acid, and Lewis base components of surface free energy. Hydrogen bonding is a chemical manifestation of acid–base interaction that is of immediate significance to interfacial phenomena and adhesion. The electron donor (base) and proton donor (acid) make complementary, highly asymmetrical contributions to the interfacial energy.

The recognition of these facts has allowed us to examine several fundamental problems that had previously been intractable. Early attempts to treat surface polarity of nonionic systems had been limited by the use of a single molecular parameter, the dipole moment. In more recent studies, Fowkes[10,11] had pointed to a higher level of complexity by introducing acidic and basic character. We have introduced a way of looking at polar interfacial interactions, with independent scales of acidity and basicity.

We have written out explicitly the equations by which contact-angle data can be processed to yield values of the surface parameters, $\gamma^{\oplus}$ and $\gamma^{\ominus}$. We have carried out the appropriate calculations in a selection of cases, and reported some values of the three parameters γ^{LW}, $\gamma^{\oplus}$, and $\gamma^{\ominus}$.

We have established the conditions under which the interfacial free energy γ_{ij} of the interface between condensed phases may be negative. This has allowed us to go beyond the simple employment of Young's equation and treat quantitatively the condition $|(\gamma_{sv}-\gamma_{sl})/\gamma_{lv}| > 1$. It is now seen that the use of three, appropriately chosen liquids on a solid leads to a set of objectively determinable properties, for example, a negative value of γ_{sl}.

A novel consequence of the existence of negative values of γ_{sl} in adhesive bonding and coating technology has been pointed out. The long-range manifestations (both attractive and repulsive) of surface hydrogen bonding have been explained. Other recent applications of the new concepts have been listed briefly.

NOMENCLATURE

N_s	Number of acid–base bonds per unit area
G	Gibbs free energy

ΔG^{a}_{ij}	Free energy of adhesion of i to j
ΔG^{c}_{ii}	Free energy of cohesion of i
α	Conventional reference ratio of $\gamma^{\oplus}$ to $\gamma^{\ominus}$ for water
β	$\gamma^{\oplus}_{l_1}/\gamma^{\oplus}_{l_2}$ for the pair of liquids, 1 and 2, based on contact angles on solid s
γ_i	Surface tension or surface free energy of i
γ_{ij}	Interfacial tension or interfacial free energy of the i/j interface
$\gamma^{\oplus}_i$	Acidic (proton-donor or electron-acceptor) component, or surface parameter, for substance i
$\gamma^{\ominus}_i$	Basic (electron-donor) component, or surface parameter, for substance i
θ	Contact angle
Additional superscripts	
LW	Lifshitz–van der Waals component
AB	Acid–base component
Additional subscripts	
l	Liquid
s	Solid
so	Solid in contact with a vacuum or (only) its own vapor
sv	Solid in equilibrium with the vapor of the liquid ℓ

REFERENCES

1. R. J. Good and M. K. Chaudhury, Chapter 3, this Volume.
2. W. B. Jensen, *The Lewis Acid–Base Concept,* Wiley–Interscience, New York (1980).
3. W. B. Jensen, The relevance of Lewis acid–base chemistry to surface interactions, in: *Surface and Colloid Science in Computer Technology* (K. L. Mittal, ed.), Plenum Press, New York (1987).
4. G. C. Pimentel and A. L. McClellan, *The Hydrogen Bond,* Freeman & Co., San Francisco (1960).
5. C. J. van Oss, M. K. Chaudhury, and R. J. Good, *Adv. Colloid Interface Sci.* **28,** 35 (1985).
6. M. K. Chaudhury, C. J. van Oss, and R. J. Good, *Chem. Rev.* **88,** 927 (1988).
7. F. M. Fowkes, *J. Phys. Chem.* **66,** 682 (1962).
8. F. M. Fowkes, *Adv. Chem. Ser.* **43,** 99 (1964).
9. F. M. Fowkes, *J. Phys. Chem.* **72,** 3700 (1968).
10. F. M. Fowkes and M. A. Mostafa, *Ind. Eng. Chem., Prod. Res. Dev.* **17,** 3 (1978).
11. F. M. Fowkes, in: *Physicochemical Aspects of Polymer Surfaces* (K. L. Mittal, ed.), *Vol. 2,* p. 583, Plenum Press, New York (1983).
12. J. H. Hildebrand and R. L. Scott, *Solubility of Nonelectrolytes,* 3rd edn., Reinhold, New York (1950).
13. R. S. Drago, G. C. Vogel, and T. E. Needham, *J. Am. Chem. Soc.* **93,** 6014 (1971).
14. P. Kollman, J. McKelvey, A. Johannson, and S. Rothenberg, *J. Am. Chem. Soc.,* **5,** 955 (1975); P. Kollman, **99,** 4875–4984 (1977); P. Hobza and R. Zahradnik, *Weak Intermolecular Interactions in Chemistry and Biology,* Elsevier, New York (1980).
15. P. A. Small, *J. Appl. Chem.* **3,** 71 (1953).
16. G. N. Lewis, *Valence and the Structure of Atoms and Molecules,* The Chemical Catalog Co., New York (1923).
17. G. N. Lewis, *J. Franklin Inst.* **226,** 293 (1938).
18. J. N. Brønsted, *Recl. Trav. Chim. Pays-Bas* **42,** 718 (1923).
19. T. Lowry, *Chem. Ind.* (*London*) **42,** 274 (1923).
20. D. D. Nelson, G. T. Fraser, and W. Klemperer, *Science* **238,** 1670 (1987).
21. M. K. Chaudhury, Thesis, State University of New York at Buffalo (1984).
22. R. J. Good and L. A. Girifalco, *J. Phys. Chem.* **64,** 561 (1960).
23. L. A. Girifalco and R. J. Good, *J. Phys. Chem.* **61,** 904 (1957).

24. R. J. Good, and E. Elbing, *Ind. Eng. Chem.* **62**(3), 54 March (1970).
25. F. M. Fowkes *Ind. Eng. Chem.* **56**(12), 40 (1964).
26. F. M. Fowkes, *J. Adhes.* **4,** 155 (1972).
27. P. G. Smith, T. M. van de Ven, and S. G. Mason, *J. Colloid Interface Sci.* **80,** 302 (1981).
28. C. J. van Oss, M. J. Roberts, R. J. Good, and M. K. Chaudhury, *Colloids Surf.* **23,** 369 (1987).
29. V. Gutman, *The Donor–Acceptor Approach to Molecular Interactions,* Plenum Press, New York (1978).
30. R. J. Good, *J. Colloid Interface Sci.* **52,** 308 (1975).
31. H. W. Fox and W. A. Zisman, *J. Colloid Interface Sci.* **5,** 514 (1950).
32. C. J. van Oss, R. J. Good, and H. J. Busscher, *J. Dispersion Sci. Technol.*, **111,** 75–81 (1990).
33. C. J. van Oss, L. Ju, M. K. Chaudhury, and R. J. Good, *J. Colloid Interface Sci.* **128,** 313 (1989).
34. C. J. van Oss, R. J. Good, and M. K. Chaudhury, *J. Chromatogr.* **391,** 53 (1987).
35. C. J. van Oss, L. L. Moore, R. J. Good, and M. K. Chaudhury, *J. Protein Chem.* **4,** 425 (1985).
36. R. J. Good and E. Elbing, *Ind. Eng. Chem.* **62,** 54–78 (1970).
37. R. J. Good, in: *Colloid and Surface Science,* Vol. 11 (R. J. Good and R. R. Stromberg, eds.), Plenum Press, New York, 1979).
38. R. J. Good and R. K. Gupta, in: *Fundamentals of Adhesive Bonding* (L. H. Lee, ed.), Plenum Press, New York (1990).
39. W. A. Zisman, in: *Contact Angle, Wettability and Adhesion*, Advances in Chemistry no. 43, Am. Chem. Soc., Washington, D.C. (1964).
40. E. D. Washburn, *Phys. Rev.* **17,** 374 (1921).
41. Y. I. Rabinovich and B. V. Derjaguin, *Colloids Surf.* **30,** 243 (1988).
42. H. K. Christenson, *J. Dispersion Sci. Technol.* **9,** 1171 (1988).
43. C. J. van Oss and R. J. Good, *J. Protein Chem.* **7,** 179 (1988).
44. C. J. van Oss, R. J. Good, and M. K. Chaudhury, Third Chemical Congress of North America and 195th National ACS meeting, Toronto, Ontario (June 1988).
45. C. J. van Oss, R. J. Good, and M. K. Chaudhury, *J. Colloid Interface Sci.* **111,** 378 (1986).
46. C. J. van Oss, R. J. Good, and M. K. Chaudhury, *Sep. Sci. Technol.* **22,** 1 (1987).
47. C. J. van Oss and R. J. Good, *J. Dispersion Sci. Technol.*, **9,** 355 (1988).
48. C. J. van Oss, R. J. Good, and M. K. Chaudhury, *J. Chromatogr.* **376,** 111 (1986).
49. C. J. van Oss, R. J. Good, and M. K. Chaudhury, *J Protein Chem.* **5,** 385 (1986).
50. C. J. van Oss, M. K. Chaudhury, and R. J. Good, *Separ. Sci. Technol.* **24,** 15 (1988).
51. C. J. van Oss and W. T. Coakley, *Cell Biophys.* **13,** 141 (1988).
52. C. J. van Oss, *Cell Biophys.* **13,** 141 (1988).
53. C. J. van Oss, K. Arnold, and W. T. Coakley, *Cell Biophys.* **17,** 1 (1990).
54. C. J. van Oss, M. K. Chaudhury, and R. J. Good, in: *Molecular Mechanisms of Membrane Fusion* (S. Ohki, D. Doyle, T. D. Flanagan, S. W. Hui, and E. Mayhew, eds.), p. 113, Plenum Press, New York (1988).
55. C. J. van Oss, K. Arnold, R. J. Good, K. Gawrisch, and S. Ohki, *J. Macromol. Sci.–Chem.* **A27,** 563 (1990).
56. R. J. Good, N. R. Srivatsa, M. Islam, H. T. L. Huang and C. J. van Oss, *J. Adhes. Sci. & Tech.* (in press).

5

The Dynamics of Wetting

P. G. de Gennes

1. INTRODUCTION

The wetting properties of liquids against a solid surface are important for many adhesion purposes. A vast amount of knowledge on the statics of wetting has been collected by Zisman and his co-workers[1] mainly for the case of partial wetting, where contact angles can be measured, and related to interfacial energies. Situations of complete wetting are equally important, but have suffered from a certain lack of experimental information since there is no macroscopic angle to be measured.

In the present discussion, we shall examine mainly liquids which wet completely the solid surface of interest. We shall also assume that the liquid is nonvolatile, in order to avoid any adsorption from the vapor onto the solid. Our aim is to understand the spreading of a droplet on the solid in the absence of any vapor effects; we call this *dry spreading*.

A rather detailed analysis of dry spreading is given elsewhere.[2] Here we present only the general ideas and results, purposely avoiding most of the mathematics. The essential physical parameter controlling the spreading process is the spreading coefficient S, introduced long ago by Cooper and Nuttal[3] and given by

$$S = \gamma_{SO} - \gamma_{SL} - \gamma \tag{1}$$

where γ_{SO} and γ_{SL} are the solid/air and solid/liquid interfacial energies while γ is the surface tension of the liquid. We are interested here in cases of complete wetting, which correspond to positive S, and we shall try to understand first the final state of a spreading droplet (Section 2). It has been commonly stated that the final state is a monomolecular film, but we shall see that this need not be true, especially at small S, where the final state should be a relatively thick "pancake." In Section 3 some major features of the dynamics of dry spreading will be reviewed. Finally, certain special cases (polymers, Langmuir–Blodgett layers) are presented in Section 4. (See also Chapter 2 about the thermodynamics of adhesion, by Gutowski.)

P. G. de Gennes • Laboratoire de Physique de la Matière Condensée, Collège de France, F-75231 Paris Cedex 05, France.

2. THE FINAL STATE IN COMPLETE WETTING

An essential observation, going back to the early work of the Russian school,[4] is that thin liquid films are very sensitive to long-range forces. This can be described in terms of a long-range energy (per cm^2) $P(e)$, such that $P(e \to \infty) = 0$. The energy $P(e)$ depends both on the liquid and the solid, and on the film thickness e. Equivalently, one may describe the long-range forces through a "disjoining pressure":

$$\Pi(e) = -\frac{dP}{de} \tag{2}$$

For the simplest example (a pure van der Waals fluid)

$$P(e) = A/6\pi e^3 \tag{3}$$

where A is a Hamaker constant.[5] For more complex fluids, such as water, the exact structure of $P(e)$ is not fully understood at present but there exist some reliable data for certain solid substrates.[6]

We now consider a macroscopic droplet which has spread out and covered an area A of solid. The free energy $\mathcal{F}$ has two antagonist terms:

$$\mathcal{F} = -S\mathsf{A} + \mathsf{A}P(e) \tag{4}$$

Quantity $\mathcal{F}$ must be minimized at constant volume $\Omega = e\mathsf{A}$. The result[6] is an implicit equation for e:

$$S = e\Pi(e) + P(e) \tag{5}$$

The resulting $e(S)$ is a *decreasing* function of S. For instance, in the van der Waals case [Eq. (3)] we obtain

$$e = a\left(\frac{3\gamma}{2S}\right)^{1/2} \tag{6}$$

where a is a molecular size defined by

$$a^2 = A/6\pi\gamma \tag{7}$$

If $S/\gamma \sim 1$, Eq. (6) gives a thickness comparable to a, i.e., a monomolecular layer. But if $S/\gamma \ll 1$, we do expect a relatively thick "pancake" ($e \sim 100$ Å). This has many implications:

(a) Equation (5) or (6) gives us an interpretation of the Cooper–Nuttall rule. If, for instance, we want to spread an insecticide on leaves (the problem originally considered by Cooper and Nuttall), choosing a large S will give a thin film, and thus a much better coverage.

(b) If the final thickness e is measured (by ellipsometry or by some equivalent technique), Eqs. (5) and (6) give us an information on the spreading coefficient. In the future, these measurements (at positive S) may become just as relevant as the measurements of contact angles (at negative S).

We conclude this section with a technical comment concerning the edge of the pancake. The shape of the edge can be calculated simply by including capillary energies, van der Waals forces, and the S parameter.[2,7] The main parameter of interest is the

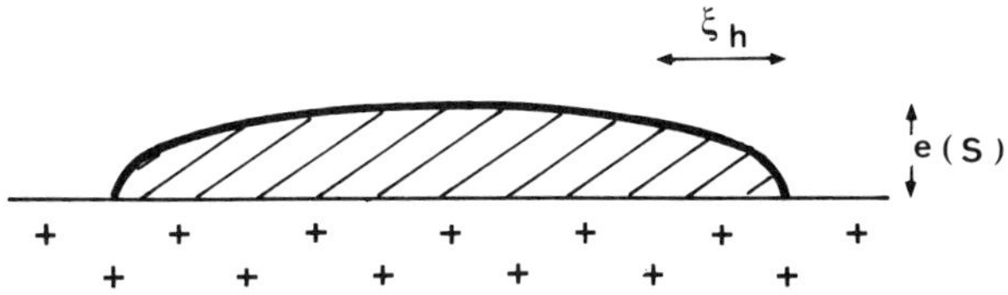

FIGURE 1. The final "pancake" profile in dry spreading. The thickness $e(S)$ is constant except in an edge region of width ξ_h. The thickness e is small; hydrostatic effects are negligible.

healing length ξ_h (Figure 1), which describes the horizontal size of the edge region and which turns out to be

$$\xi_h \sim e^2/a \tag{8}$$

for van der Waals forces. The edge has a line energy $\mathcal{L}$ which is of the order (omitting logarithmic factors) satisfying

$$\mathcal{L} \sim S\xi \sim \gamma a \tag{9}$$

where a is again a molecular size. These considerations are important for a discussion of dry spreading on a random surface (i.e., when S is slightly modulated from point to point); more details can be found elsewhere.[8,9]

3. THE DYNAMICS OF DRY SPREADING

The basic situation is illustrated in Figure 2. The advancing fluid displays a certain dynamic contact angle θ, which depends on the contact line velocity U. A number of experiments[10,2] indicate that for various liquids and various solids, the relation between U and θ is universal,

$$U = (\text{constant})\, \frac{\gamma}{\eta}\, \theta^3 \qquad (\theta \ll 1) \tag{10}$$

where η is the liquid viscosity. It is at first very surprising that the (U,θ) relation does not involve the spreading coefficient S. For instance, if we think in terms of irreversible

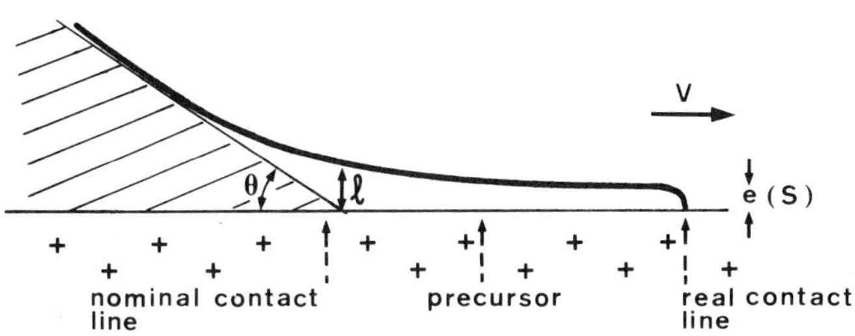

FIGURE 2. Detailed form of the advancing profile in dry spreading of a completely wetting fluid. The crossover thickness l is of order a/θ.

thermodynamics, the dissipation $T\mathring{\Sigma}$ (per unit length of line) is the product of a flux (U) by a force F, which is the "noncompensated Young's force": The three interfacial tensions, when projected on the allowed direction of motion, do not add up to zero but to a finite force,

$$F = \gamma_{SO} - \gamma_{SL} - \gamma \cos \theta \tag{11}$$

$$\cong S + \tfrac{1}{2}\gamma\theta^2 \qquad (\theta \ll 1) \tag{12}$$

In typical conditions ($\theta = 10^{-2}$, $S/\gamma = 10^{-1}$) the angular correction $\frac{1}{2}\gamma\theta^2$ amounts only to a very small fraction (10^{-3}) of the main force S. It is thus very puzzling that S does not appear in Eq. (10).

The explanation lies in the existence of a *precursor film* ahead of the macroscopic droplet. This film was detected long ago by Hardy and Harkins, and its properties are analyzed in a review by Marmur.[(10)] On the theoretical side, it is clear that these films depend on long-range forces. A detailed analysis has been carried out by two groups: Teletzke, Davis, and Scriven[(11)] mostly considered wet spreading and made much use of computer calculations; Hervet and the present author [(12,2)] concentrated on the dry spreading of simple van der Waals fluids (see Figure 2).

The main conclusion is that the large free energy S described in Eq. (12) is entirely burned off in the precursor film, while the remaining energy ($\frac{1}{2}\gamma\theta^2$) is spent by macroscopic viscous flow. The flows in an advancing wedge were discussed long ago by Huh and Scriven[(13)]: for small θ, the corresponding dissipation can be derived very simply using what is called in mechanics a "lubrication approximation" (i.e., taking the velocities parallel to the surface). The result is

$$T\mathring{\Sigma}_{\text{macro}} = kU^2/\theta \tag{13}$$

where k is nearly constant (a logarithmic function of θ). Equating $T\mathring{\Sigma}_{\text{macro}}$ to the work done by $\frac{1}{2}\gamma\theta^2$ immediately yields the experimental law.[10)]

This has a number of practical consequences:

- We see that the dynamics is finally independent of S (for $S > 0$). Thus the Cooper–Nuttall law cannot be explained by dynamic effects and depends entirely on properties of the final equilibrium state [see (5)].
- There are many pitfalls. For instance, it is completely incorrect to equate the larger dissipation term SU to the macroscopic losses $T\mathring{\Sigma}_{\text{macro}}$. Mistakes of this type have occurred in the literature.

The above discussion was restricted to horizontal spreading, with negligible gravity effects. The case of vertical spreading is also of some interest.[(14)] For instance, if a vertical plate is pushed down at *low* velocities U inside a vessel of wetting liquid, the fluid still builds up a stationary wetting film along the plate. In the lower part of this film, there is a quasi-static equilibrium between disjoining pressure and hydrostatic pressure, giving a thickness $e \sim h^{-1/3}$ (where h is the altitude) for van der Waals fluids. However, in the upper part of the film the thickness $e(h)$ decreases more rapidly, very much as in the horizontal case ($e \sim h^{-1}$). These considerations may be of some interest in soil physics.

4. SPECIAL SYSTEMS

4.1. Polymer Melts

These have very special spreading laws, as pointed out very early by Frisch and co-workers.[15] (See also References 16 and 17.) This can be understood from a remark made in 1979,[18] namely, that polymers flowing near a smooth surface tend to have a very anomalous slippage.

More precisely, let us assume that the polymer is *not chemically bound* to the solid, and that the polymer is *fluid* down to the last layer near the solid (as opposed to some liquids which order near a solid surface). Then slippage along the surface is preferable to shear in the bulk melt, because shearing an entangled system is difficult.

This process should occur even for slow flows. The first experimental observations of slippage were performed in rather strong flows (in a transparent extruder) by Galt and Maxwell.[19] More recently, rheometric studies by Keller and co-workers[20] (on molten polystyrene in a plate/plate geometry) did show some slippage in a much slower flow regime. The spreading of polymer droplets is an ideal example of slow flows. The consequences of this strong slippage for spreading have been worked out[21]: the predicted transient profiles for a droplet should show a characteristic "foot," which is indeed observed specifically with polymer melts. The thickness b of the foot is in the range 1–50 microns.

4.2. Langmuir–Blodgett Deposition

Monolayers can be transferred from a Langmuir trough to a vertical plate, which is moved upward or downward. We shall consider here the "ascending Y mode" where the polar heads from the fresh film stick to the polar heads of a preceding layer already deposited on the plate. The details of the deposition process are only partially understood. Some of the best quantitative data are due to Petrov *et al.*[22] using fluorescence quenching to monitor the contact between the two layers. Correct deposition occurs only when the vertical velocity of the plate (U) is below a certain maximum U_m. The value of U_m indicates the possibility (or impossibility) of practical deposition in a reasonable time interval.

Recently, we have constructed a relation between U_m and the thermodynamic parameters (interfacial energies) characterizing a given system.[23] Here we are dealing with a situation of *partial* wetting, with a certain equilibrium contact angle θ_e and a negative value of S. There is a dynamic contact angle θ; the driving force is still given by Eq. (12) and becomes

$$F = -\tfrac{1}{2}\gamma(\theta_e^2 - \theta^2) \qquad (\theta_e, \theta \ll 1)$$

while the velocity of the line with respect to the solids is $-U$. There is no precursor in situations of partial wetting; thus all the dissipation is due to macroscopic flow and described by Eq. (13). Equating this dissipation to $-FU$ gives

$$U \cong \frac{\gamma}{\eta}\,\theta\,(\theta_e^2 - \theta^2) \qquad (14)$$

The velocity U is very small if we are close to equilibrium ($\theta \rightarrow \theta_e$), but it is also very small if $\theta \rightarrow 0$ because the dissipation in a thin wedge is enormous. Between these two limits, there is a maximum

$$U_m = (\text{const}) \frac{\gamma}{\eta} \theta_e^3 \tag{15}$$

which yields the maximum pulling velocity. Since θ_e is related to the interfacial energies by Young's equation, Eq. (15) provides a relation between U_m and thermodynamic parameters. Detailed experimental tests of Eq. (15) are not yet available.

5. CONCLUSIONS

(a) In situations of dry spreading the final state is a pancake, which becomes thinner when the spreading coefficient S increases. This provides a fundamental explanation for the rule of Cooper and Nuttall, stating that liquids with large S spread better.

(b) The dynamics of dry spreading is remarkably insensitive to the value of S; all the free energy described by S is "burned" by viscous flow in the precursor film.

(c) The same pattern of interpretation may be extended to the motion of contact lines in situations of *partial wetting*. We discussed here the deposition of Langmuir–Blodgett layers, but there are many other applications, in particular to *drying* processes, where a solid is covered by a metastable film of liquid ($S < 0$) and a dry spot nucleates at one point of the surface.[24]

(d) Another domain of application is the problem of *fracture in a liquid environment*. In certain regimes of cavitation, there is a vapor region near the tip of the fracture and a contact line on each lip. The losses near the contact line have been analyzed recently using the same approach[25] and may sometimes be relevant.

ACKNOWLEDGMENTS

The work described here originated from collaboration between many participants (J.-F. Joanny, H. Hervet, and F. Brochard). It has also benefited from discussions with various experimentalists (A. M. Cazabat, M. Cohen Stuart, B. Legait, L. Léger, L. Ter Minassian, and F. Rondelez).

NOMENCLATURE

A	Hamaker constant
A	An area of solid
a	A molecular size, defined in detail by Eq. (7)
$P(e)$	Long-range energy of a liquid film of thickness e
S	Spreading coefficient
T	Temperature
$\mathcal{L}$	Line energy per unit length (in cases of complete wetting)

U_m	Maximum pulling velocity permissible in a Langmuir–Blodgett process
v	Velocity of contact line
γ	Surface tension of liquid
γ_{SL}	Interfacial tension between solid and liquid
γ_{SO}	Interfacial tension between solid and air
η	Viscosity
θ	Dynamic contact angle
θ_e	Equilibrium contact angle
ξ_h	Healing length
Π	Disjoining pressure
$\mathring{\Sigma}$	Entropy source (per unit length of contact line)

REFERENCES

1. W. A. Zisman, *Adv. Chem. Ser.* **43,** 1 (1964).
2. P. G. de Gennes, *Rev. Mod. Phys.* **57,** 827 (1985).
3. W. Cooper and W. Nuttall, *J. Agric. Sci.* **7,** 219 (1915).
4. B. V. Derjagin, *J. Colloid (USSR)* **17,** 191 (1955).
5. J. N. Israelashvili, *Intermolecular and Surface Forces,* Academic Press, New York (1985).
6. R. M. Pashley, *J. Colloid Interface Sci.* **78,** 246 (1980).
7. J.-F. Joanny and P. G. de Gennes, *C.R. Acad. Sci. Paris, Ser. B* **299,** 279 (1984).
8. P. G. de Gennes, *C.R. Acad. Sci. Paris,* **B300,** 129 (1985).
9. M. Robbins and J.-F. Joanny, *Europhys. Lett.* **3,** 729 (1987).
10. A. Marmur, *Adv. Colloid Interface Sci.* **19,** 75 (1983).
11. G. Teletzke, T. Davis, and L. Scriven, *J. Colloid Interface Sci.* (to appear).
12. H. Hervet and P. G. de Gennes, *C.R. Acad. Sci. Paris, Ser. B* **299,** 499 (1984).
13. C. Huh and L. Scriven, *J. Colloid Interface Sci.* **35,** 85 (1971).
14. P. G. de Gennes, *C.R. Acad. Sci. Paris Ser. B* **298,** 439 (1984); J.-F. Joanny and P. G. de Gennes, *J. Phys. (Paris)* **47,** 121 (1986).
15. H. Ghiradella, W. Radigan, and H. Frisch, *J. Colloid Interface Sci.* **51,** 522 (1978); this paper contains a list of earlier references by the same group.
16. G. Sawicki, in: *Wetting, Spreading and Adhesion* (J. Padday, ed.), p. 361, Academic Press, New York (1978).
17. V. Arslamov, T. Ivanova, and V. Ojar, *Dokl. Phys. Chem.* **198,** 502 (1971).
18. P. G. de Gennes, *C.R. Acad. Sci. Paris, Ser. B* **288,** 219 (1979).
19. J. Galt and B. Maxwell, *Mod. Plast.* Dec. (1964).
20. R. Burton, M. Folkes, K. Narh, and A. Keller, *J. Mater. Sci.* **18,** 315 (1983).
21. F. Brochard and P. G. de Gennes, *J. Phys. Lett. (Paris)* **45,** L-597 (1984).
22. J. Petrov, H. Kuhn, and D. Möbius, *J. Colloid Interface Sci.* **73,** 66 (1980).
23. P. G. de Gennes, *Colloid Polym. Sci.* **264,** 463 (1986).
24. For a related analysis see P. G. de Gennes, *C.R. Acad. Sci. Paris, Ser. B* **303,** 1275 (1986).
25. P. G. de Gennes, *C.R. Acad. Sci. Paris,* **B304,** 547 (1987).

6

Kinetics of Polymer–Polymer Interdiffusion

F. Brochard-Wyart

1. INTRODUCTION

Polymer–polymer interdiffusion has important applications in engineering, such as welding of plastics and of layered structures, preparing bulk thermoplastics from small resin particles via coalescence, crack healing, and creating structures in polymer mixtures.

We shall distinguish three types of polymer–polymer welding:

- The two polymers are *identical*. For instance, "Gaz de France" produces several thousand miles of polyethylene tubes which are connected by autoadhesion under heating.
- The two polymers are chemically identical, but the "A" chains are much larger than the "B" chains ($N_A > N_B$, where N is the number of monomers per chain).
- The two polymers are *compatible* (i.e., they possess negative enthalpy of mixing). This means that they strongly attract each other.

(See also Chapter 7 by Wool on welding.)

The main parameter controlling the kinetics of polymer–polymer interdiffusion is the *mutual diffusion coefficient* D_M.[1,2] On the other hand, if one monitors the motion of *one* labeled molecule, what is seen is the *self-diffusion coefficient* D_S.[3] We shall now define precisely D_M, D_S, and the relation between D_S and D_M if the two interdiffusive polymers are almost identical (labeled/unlabeled).

1.1. Interdiffusion Coefficient

1.1.1. Mutual Diffusion versus Self-Diffusion

1.1.1.1. Mutual Diffusion. We consider a binary mixture A/B. The composition can be characterized either by the concentration c in component A (for the case of a polymer A–polymer B mixture c will be the number of A monomers per unit volume) or by the

F. Brochard-Wyart • Physique de la Matière Condensée, Collège de France, F-75231 Paris Cedex 05, France and Structure et Réactivité aux Interfaces, Université Pierre et Marie Curie, F-75231 Paris Cedex 05, France.

volume fraction $\phi = ca^3$, where a^3 is the volume per monomer. If c (or ϕ) is not uniform, a net monomer flux is generated. The transfer of A monomers per unit area, $\mathbf{J}_c = c\mathrm{v}_A$, is proportional to the concentration gradient,

$$\mathbf{J}_c = -D_M(\phi)\,\nabla c \tag{1}$$

where $\mathbf{v}_A$ is the velocity of A,

$$\mathbf{J}_c = c \times \text{velocity},\ [\mathbf{J}_c] = [\mathrm{L}^{-2}\mathrm{t}^{-1}],\ \text{and}\ [D_M] = [\mathrm{L}^2\mathrm{t}^{-1}]\ (\mathrm{m}^2\mathrm{s}^{-1}).$$

One can also use the volume transfer $\mathbf{J}_A$ of A, $\mathbf{J}_A = \phi\mathrm{v}_A$, called $\mathbf{J}$, where

$$\mathbf{J} = -D_M\,(\phi)\nabla\phi \tag{2}$$

In the following, we shall assume no overall change of volume of the two liquids upon mixing. The currents $\mathbf{J}_A$ and $\mathbf{J}_B$ must satisfy the condition of constant density. The volume transfer of B per unit time across unit area is therefore

$$\mathbf{J}_B = -\mathbf{J}_A = -\mathbf{J} \tag{3}$$

i.e.,

$$(1 - \phi)\mathrm{v}_B = -\phi\mathrm{v}_A$$

More general situations with changes of volume are treated elsewhere.(18)

1.1.1.2. Self-Diffusion Coefficient D_S. Let us consider one test chain, suitably labeled to be recognized inside the melt. After time t this chain will have moved under Brownian motion through distance x of mean square given by

$$\langle x^2 \rangle = 2\,D_S t \tag{4}$$

where x is the displacement along one of three cartesian coordinates. In A/B mixtures, we have two self-diffusion coefficients D_{S_A} and D_{S_B} associated with the Brownian motion P_A and P_B.

For a binary mixture, we can measure three *distinct* diffusion coefficients D_{S_A}, D_{S_B}, and D_M. However, as $\phi \rightarrow 0$, the dilute A chains are entirely surrounded by B monomers and one can show easily that $D_M = D_{S_A}$. For the opposite case, $\phi \rightarrow 1$, $D_M = D_{S_B}$.

We now examine the case of a binary blend of two identical polymers, but one is labeled (e.g., H-PS and D-PS). If we assume that monomer A/monomer A and monomer A/monomer A* friction coefficients and interactions are identical, then one can check easily that $D_{S_A} = D_{S_{A^*}} = D_M$. (However, we do have evidence(50) for phase separation of H-PS and D-PS: at very high molecular weights, the assumption of chemical identity between A and A* must be checked carefully.)

Molecular theories of D_S and D_M are discussed in Sections 2 and 3 in the framework of the reptation model. We list below recent experimental techniques used to measure D_S and D_M.

1.1.2. Experimental Techniques (D_S, D_M)

The kinetics of diffusion are extremely slow in entangled polymer melts. Diffusivities range from 10^{-17} to 10^{-10} m^2/sec. According to Eq. (4), one requires either very long times or short spatial intervals x. To achieve experiments in a time of the order of one hour, diffusion must be examined at very small scales. Powerful techniques have been developed recently to measure diffusion on the micron scale.

1.1.2.1. Infrared Spectroscopy[4] (D_S, D_M). The labeled species was deuterated polyethylene (D-PE) diffusing in a protonated polyethylene melt (H-PE). The local *D* fraction was measured by infrared spectroscopy on thin slices of the nearly glassy polymer. The spatial resolution was about 100 μm and the time involved of the order of months.

1.1.2.2. Electron-Induced X-Ray Fluorescence[5] (D_M). This technique can be combined with scanning electron microscopy and energy dispersive X-ray analysis. It is valid only for unsymmetrical junctions and heavy atoms on one species.

1.1.2.3. Radioactive Tracers[6] (D_S, D_M). A thin layer of radioactive polymer is applied to the top surface of a thick film of unlabeled polymer. As the tagged polymer diffuses into the thick inactive substance, the intensity of radioactivity emanating from the surface decreases.

1.1.2.4. NMR Field-Gradient Technique[7] (D_S). In this method one does not need chemical labeling. In a field gradient, the Larmor frequency of a proton (attached to the molecule) depends on its location $X(t)$. The resulting signals (in an adequate echo sequence) can be measured directly as $\langle X^2(t) \rangle$. However, this technique is restricted to times t less than the nuclear relaxation times (milliseconds). Thus it probes motions on a more local scale.

1.1.2.5. Forced Rayleigh Scattering (FRS)[8,9] (D_S). The labeled chain carries a photochromic group. The sample is illuminated with a very bright optical fringe pattern which creates a spatially periodic distribution of photoexcited molecules (Figure 1). The net result is to have an absorption grating inside the sample. One then cuts off the illumination; the grating fades out by diffusion of the labeled species, and this fading out is monitored by a probing laser. Here the diffusion length is the spatial period of the grating (between about 1 and 100 μm) and the decay times are rather short (minutes). In practice, the optics is well controlled only if the photochromic groups are *dilute*.

1.1.2.6. Photobleaching (FRAPP)[10] (D_S). This technique is very similar to FRS, but here the molecules are labeled with fluorescent probes. Fluorescence is destroyed in the light fringes. One monitors the fluorescence redistribution after bleaching.

1.1.2.7. Donor–Acceptor Fluorescence Method[11] (D_S, D_M). If the two interdiffusing polymers are labeled respectively with donor D and acceptor A fluorophores, which exhibit nonradiative energy transfer, the emission spectrum will depend on whether the two polymeric species are in contact or segregated. The reaction

$$D^* + A \rightarrow D + A^*$$

(for example, D = carbazole and A = anthracene) is possible only if the distance between D and A is less than 50 Å. The interdiffusion is monitored by the increase in the

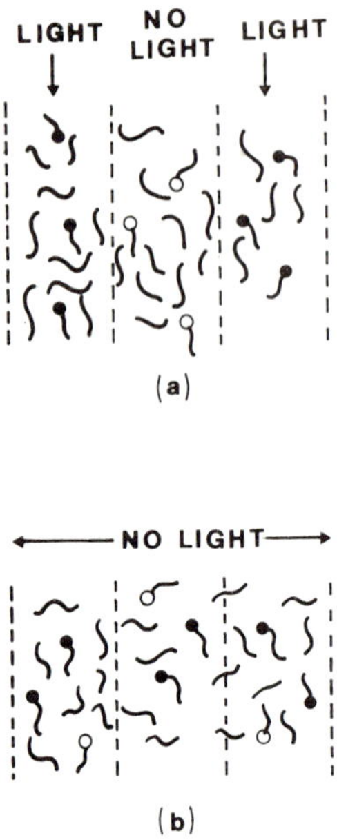

FIGURE 1A. Principle of the forced Rayleigh effect. A polymer solution (or melt) contains a few chains labeled with a photochromic dye (circle). (a) The solution receives a strong light intensity from a flash source; in the illuminated regions the dyes become dark, but in the nonilluminated regions the dyes are still transparent. (b) One then cuts off the flash light; the successive dark/transparent slabs fade out by diffusion. This fading out is monitored through the Bragg reflection of a low-power laser on the periodic structure.

nonradiative energy transfer S, where S is proportional to the number of donor–acceptor pairs ($S \sim D^{1/2}t^{1/2}$). This technique has high resolution ($X \sim 50$ Å) and sensitivity, but is still at an experimental stage.[11]

1.1.2.8. Forward Recoil Spectrometry (FRES)[12,13] *(D_S, D_M).* This technique has high resolution ($X \sim 800$ Å) and sensitivity. The principle of the experiment is shown in Figure 2. The sample is irradiated by a beam of α particles. The He^{++} ions collide with protons, or deuterons which are ejected. Since ^2H receive more energy than ^1H, ^2H and ^1H are well separated in energy. The energy losses of the scattered species depend upon the length of polymer which they meet before exit. Thus one can plot the fractions of ^2H and ^1H as a function of depth.

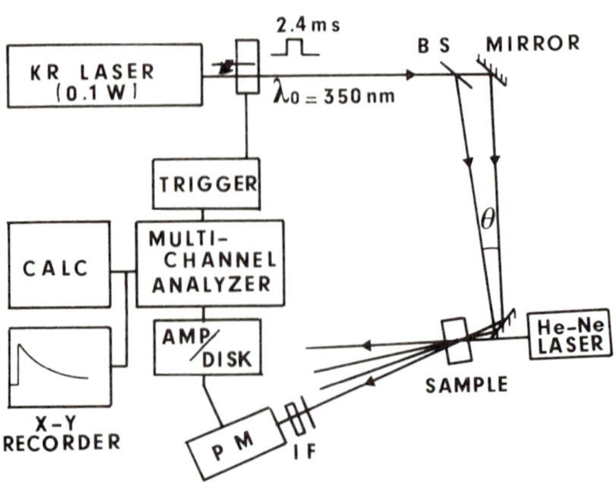

FIGURE 1B. Block diagram of the experimental setup. The exciting laser is a Coherent Inc. 500-K krypton laser delivering 100 mW at 350 nm. The fringe spacing can be adjusted from 5 to 100 μm by varying the angle θ between the two interfering beams. The mechanical chopper gives light pulses of 2.4 msec duration with a repetition rate adjustable from 200 msec to 200 sec. The diffracted intensity is collected with a cooled Philips "56 T.V.P." photomultiplier. (Reprinted with permission from Reference 8.)

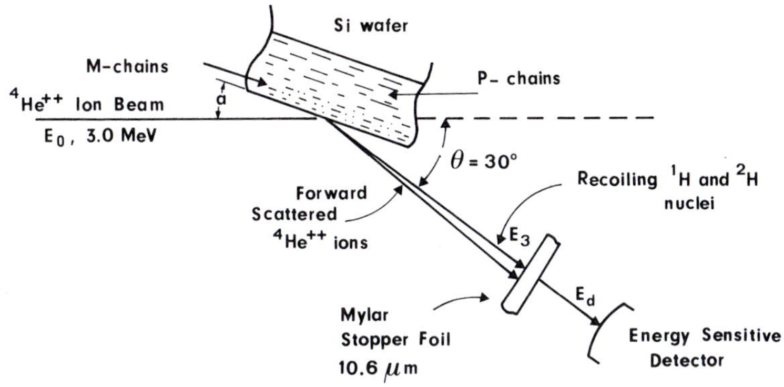

FIGURE 2. Experimental setup of forward recoil spectrometry experiment. (Reprinted with permission from Reference 12.)

1.1.2.9. Rutherford Back-Scattering.(14,15) Gold particles of diameter 50 Å are deposited at the interface. The motion of these particles is monitored by the attenuation of the energy of He^{++} ions reflected by the gold markers. The gold particles respond differently to the A and B fluxes. Hence they move, but the interpretation of their motion is quite delicate.

1.1.2.10. Photon Correlation Spectroscopy.(16) Measurements of D_M by inelastic scattering of laser light (of *wavelength* λ) have been performed recently on polymer blends of short chains. The light is scattered by concentration fluctuations of the wave vector **q**,

$$\phi(x,t) = \phi + \delta\phi_{\mathbf{q}}(t) \cos \mathbf{q}x \tag{5}$$

which relax exponentially,

$$\delta\phi_{\mathbf{q}}(t) = \delta\phi_{\mathbf{q}}(0) \exp(-D_M \mathbf{q}^2 t) \tag{6}$$

For a scattering angle θ, the scattering wave vector **q** is given by $\mathbf{q} = (4\pi/\lambda) \sin \frac{1}{2}\theta$. One then monitors the diffusion on length $x = 2\pi/\mathbf{q}$ of the order of the optical wavelength λ.

1.1.2.11. Small-Angle Neutron Scattering (SANS).(17) One can achieve alternating layers of several micron thickness of protonated and partially deuterated chains. The coherent SANS intensity $I_c(\mathbf{q})$ is given by

$$I_c(\mathbf{q}) = K \langle\phi(1 - \phi)\rangle P(\mathbf{q}) N \tag{7}$$

where **q** is the scattering wave vector, $P(\mathbf{q})$ the single-chain scattering function, while ϕ and $(1 - \phi)$ are the fractions of deuterated and protonated polymers. Contrast develops as the layers interdiffuse (see Figure 3). The function $F(t) = \langle\phi(1 - \phi)\rangle$ increases from 0 to $\frac{1}{4}$ when the mixture becomes uniform; $F(t)$ is simply related to the diffusion coefficient by solving the diffusion equation. By this technique, a rather small diffusion coefficient ($D \sim 10^{-17}$ m²/sec) can be measured in short diffusion times.

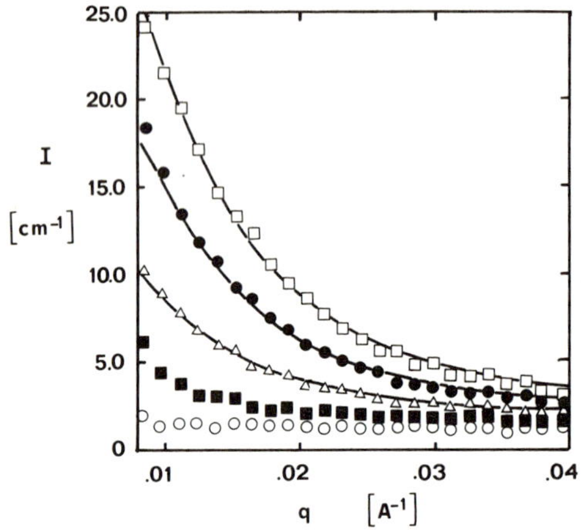

FIGURE 3. A representative set of total scattering curves for a layered film of H-PB and D-PB of molecular weight $M = 73{,}500$ taken at different diffusion times of 0 (■), 5 (△), and 20 (●) min at 125 °C. Also shown are scattering results for the compositionally homogeneous mixture, $t = \infty$ (□), and for the pure H-PB blend (○). (Reprinted with permission from Reference 17.)

This list is not exhaustive. Progress in interdiffusion measurements has been made. However, in most cases, the chains have to be labeled under some delicate chemical procedure. The labeling may also, in principle, modify the diffusion coefficient, but in most cases the fraction of labeled monomers is very small and the effect should not be too distinct; however, the effect can be checked by varying the labeled fraction.

1.2. Interdiffusion Profiles

We study the kinetics of interdiffusion of two slabs of polymer, P_A and P_B. The governing equations for an incompressible melt consist of Fick's law[2,3]

$$\mathbf{J} = \mathbf{J}_A = -\mathbf{J}_B = -D_M \nabla \phi \tag{8}$$

where ϕ is the volume fraction of A monomers, $\mathbf{J}_A$ the volume transfer of species A, and $\mathbf{J}_B$ of B. The mass conservation law leads to

$$\frac{\partial \phi}{\partial t} + \text{div } \mathbf{J} = 0 \tag{9}$$

If at time $t = 0$ we place a specimen of composition ϕ_1 in contact with a specimen of composition ϕ_2, then the concentration profile $\phi(x)$ is given by the diffusion equation resulting from Eqs. (8) and (9), namely

$$\frac{\partial \phi}{\partial t} = \frac{\partial}{\partial x} D \frac{\partial(\phi)}{\partial x} \tag{10}$$

with boundary conditions

$$\begin{aligned} \phi &= \phi_1, \qquad x \to -\infty \\ \phi &= \phi_2, \qquad x \to +\infty \end{aligned} \tag{11}$$

(For pure melts in contact, $\phi_1 = 1$ and $\phi_2 = 0$.)

1.2.1. Constant D Independent of Concentration

If D is constant, then the profile is classical,[18]

$$\frac{\phi(x,t) - \phi_1}{\phi_2 - \phi_1} = \frac{1}{2}\left(1 + \operatorname{erf} \frac{x}{2\sqrt{Dt}}\right) \tag{12}$$

where $\operatorname{erf} y = (2/\sqrt{\pi}) \int_0^y \exp(-u^2)\, du$.

The profile $\phi(x)$ measured at different times and plotted as a function of $x/\sqrt{t}$ should fit on one master curve presented in Figure 4. From this curve, one can derive the interdiffusion coefficient D.

1.2.2. Concentration-Dependent Diffusion

If D is strongly dependent upon ϕ, one must solve Eq. (10) numerically. As $\phi \to 0$, one can often write $D(\phi) \sim \phi^k$.

Let $k > 0$, $D(\phi) \to 0$. This case is called *hypodiffusive*.[19] The polymer current vanishes for $\phi \to 0$. It leads to sharp profiles, with a finite zone of mixing. The case $D = D_0\, \phi(1 - \phi)$, expected for the mutual diffusion coefficient of compatible pairs[20] (Section 3), leads to a rather surprising profile (Figure 5a,b) and $\phi(x)$ is just a straight line.

Let $k < 0$, $D(\phi) \to \infty$. This case is called *hyperdiffusive*. The flux of polymer increases and the profile becomes very flat and extended. This case arises for a junction of two polymers, when $N_A \gg N_B$.[21] An example is shown in Figure 5c. Thus most cases where D is composition-dependent lead to unusual profiles. Experimentally, the situation is delicate (as explained in Figure 15) because the real profiles are overshadowed by polydispersity effects.

In practice, one often assumes Eq. (12) to be valid and extracts a mean diffusion coefficient D^*. This procedure may lead to poor estimates if the profiles are rather asymmetrical (cf. Figure 5) and overshadowed by polydispersity.

The best way to measure $D(\phi)$ is to study a junction between two blocks of slightly different concentrations,

$$\begin{aligned} \phi_1 &= \phi \\ \phi_2 &= \phi + \Delta\phi, \qquad \Delta\phi \ll 1 \end{aligned} \tag{13}$$

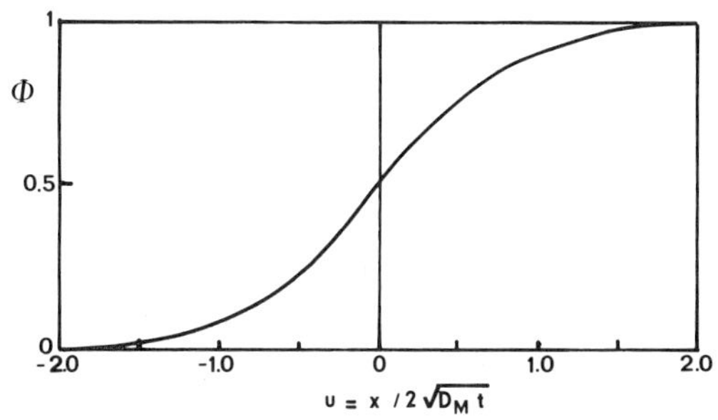

FIGURE 4. Classical Fickian concentration profile: D_M = Constant.

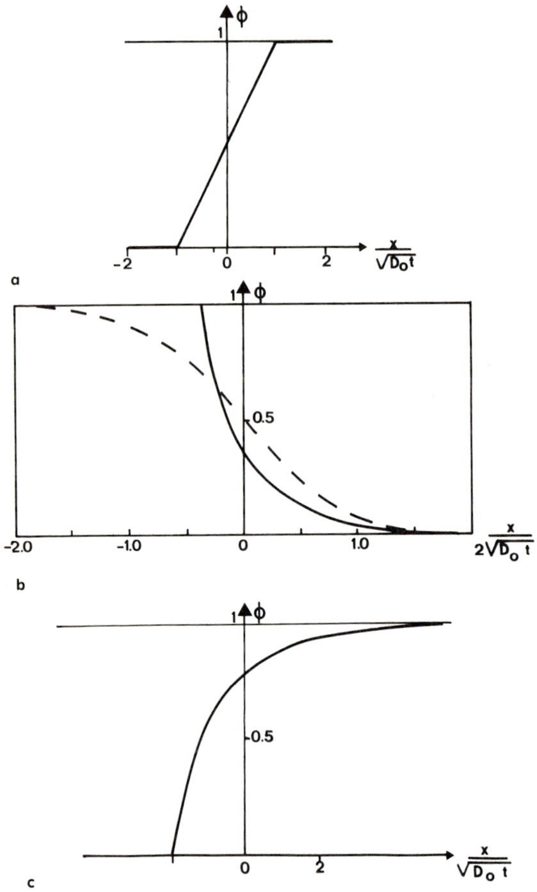

FIGURE 5. (a) Profile for pure melts of compatible polymers in contact. If $D_M = D_0\,\phi(1-\phi)$, the profile is a straight line. The zone of mixing is finite. (Reprinted with permission from Reference 20.) (b) Concentration profile of lower-molecular-weight chains (A) diffusing into higher-molecular-weight chains (B) computed numerically for two molecular-weight ratios of A to B (---) $N_A/N_B = 1$ ($D_M = D_0$); the profile is classical (———) for $N_A/N_B = 0.01$ [$D_M \sim D_0(1-\phi)^2$]. (Reprinted with permission from Reference 14.) (c) The concentration profile of the lower-molecular-weight chain B diffusing into higher-molecular-weight chains A computed numerically assuming $D_M = D_0\,\phi/(1-\phi)$. (Reprinted with permission from Reference 21.)

Quantity D is constant in the interval (ϕ, $\phi + \Delta\phi$) and the profile is given rigorously by Eq. (12). This has been carried out on one system by Klein and co-workers for compatible pairs.[22] When $D(\phi)$ is known experimentally, one can calculate numerically the profiles $\phi(x,t)$ when pure A and pure B are set in contact. The inverse procedure, $\phi(x,t) \rightarrow D(\phi)$, is much more difficult.

One must be very careful when extracting diffusion coefficients from concentration profiles. Very often what is measured is an apparent diffusion coefficient obtained by forcing the classical profile on the data. This may be misleading in the cases with hypo- or hyperdiffusion.

2. DYNAMICS OF PURE POLYMER MELTS—AUTOADHESION

Our first aim is to give a qualitative picture of the molecular motions for flexible chains in a melt.[23]

2.1. Polymer Melts: Statics and Dynamics

Let us consider a melt of long flexible and linear polymer chains of degree of polymerization N. Neutron scattering experiments[24] probing a few labeled chains (deuterated) in a melt of identical (protonated) chains have shown that the chains are ideal and Gaussian, as first predicted by Flory.[25] The size of one chain is $R_0 = aN^{1/2}$, where a is the statistical segment length. In the volume occupied by one chain, there are $\nu = (R_0^3/Na^3) \sim \sqrt{N}$ other chains; the chains are strongly interpenetrated. The dynamics of melts depend strongly on the length of the chains.

2.1.1. Small Chains ($N < N_e$)

If $N < N_e$ ($N_e \sim 200$), the chains are too small to form entanglements[26]; the melt behaves as a regular viscous fluid. Molecular motions can be simply described because there are no hydrodynamic interactions between monomers (complete screening), and there are no topological constraints. The chains have what is called Rouse behavior.[23]

(1) Under the action of a force $\mathbf{F}$, a chain moves with velocity $v = (s_0/N)\mathbf{F} = s_{ch}\mathbf{F}$, where $s_0 \sim (\eta_0 a)^{-1}$ is a monomer mobility. The chain mobility $s_{ch} = s_0/N$ is inversely proportional to the number of monomers N (the friction is proportional to N). The diffusion coefficient is related to the chain mobility by Einstein's law

$$D_S = \frac{s_0}{N} kT \sim N^{-1} \tag{14}$$

This N^{-1} law has been confirmed experimentally in a few cases.[35]

(2) The *relaxation time* of a chain (the time required to change all its conformation) is called the Rouse relaxation time T_R, which is equal to the time required for the chain to diffuse through a length equal to its own size R_0,

$$D_S T_R = R_0^2, \qquad \text{i.e., } T_R \simeq \tau_0 N^2 \tag{15}$$

(3) The *viscosity* of the melt is given by

$$\eta = \frac{kT}{Na^3} T_R = \eta_0 N \tag{16}$$

2.1.2. Entangled Polymer Chains ($N > N_e$)

A polymer melt of long chains has special mechanical properties (Figure 6). At very low frequencies, it flows like a viscous fluid; at higher frequencies, it bounces like rubber. The characteristic time T_{rep} separating these two domains has been widely studied.[27,28] It increases rapidly with the number of monomer per chains,

$$T_{rep} = \tau_0 N^a \tag{17}$$

where $a \sim 3$ to 3.4. For $N = 10^4$, $N^a \sim 10^{13}$ and $\tau_0 \sim 10^{-10}$ sec; one expects $T_{rep} \sim 10^3$ sec.

What is the origin of this viscoelastic behavior? The chains are knotted together (Figure 7). Upon being subjected to shear they tend to undo certain knots, but it takes time T_{rep}. For times $t < T_{rep}$, the knots act as crosslinks in a rubber and the melt behaves as an elastic network. The elastic modulus is $E \sim \nu kT$, where $\nu = 1/N_e a^3$ is the entanglement

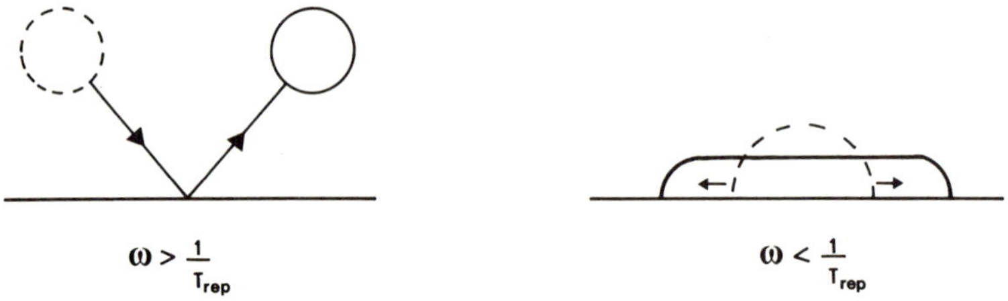

FIGURE 6. Viscoelastic behavior of a fluid polymer melt: a polydimethylsiloxane ball can bounce like rubber, but when set on a table for some minutes it flows.

density. On the other hand, for $t > T_{rep}$, the original knots fade out and the melt flows with a viscosity η related to E and T by a Maxwell law

$$\eta = E\,T_{rep} \sim N^{a} \tag{18}$$

These features are usually described in terms of the reptation concept, which we summarize below.

2.2. Reptation(29–31) and Self-Diffusion of Entangled Polymers

In a melt, the chains can move by Brownian motion, but they cannot intersect each other. A chain is confined in a tube (Figure 8). The diameter of the tube is $d \sim N_e^{1/2}a$, the distance between entanglements. The tube contour length is $L = (N/N_e)\,d$. The mobility of one chain (the "test chain") in its curvilinear motion along its fixed tube is $\mu_t = \mu_1/N$. Hence the curvilinear diffusion coefficient is given by

$$D_t = \frac{\mu_1}{N}\,kT = \frac{D_0}{N} \tag{19}$$

The mean-square displacement $S^2(t)$ due to Brownian motion of the chain in its tube has the standard form at time $t < T_{rep}$:

$$S^2(t) = 2D_t t \tag{20}$$

The time T_{rep} for disengagement of the chain is such that $S(T_{rep}) = L$, i.e.,

$$T_{rep} = \frac{L^2}{2D_t} = \frac{N^3a^2}{2N_e\mu_1 kT} = \frac{N^3\tau_1}{N_e} \tag{21}$$

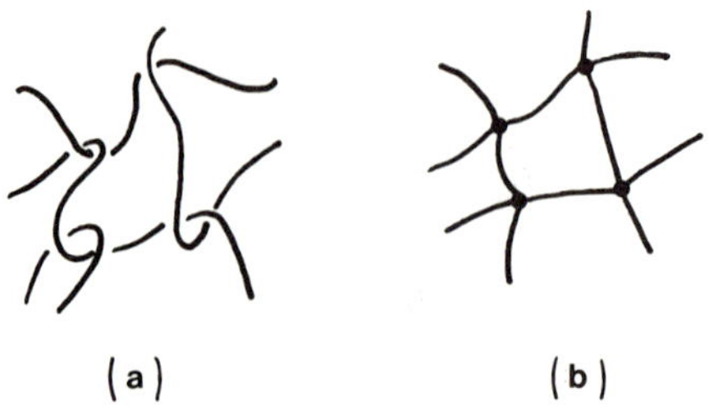

FIGURE 7. Idealized representation of an entangled melt (a) and of the equivalent network (b).

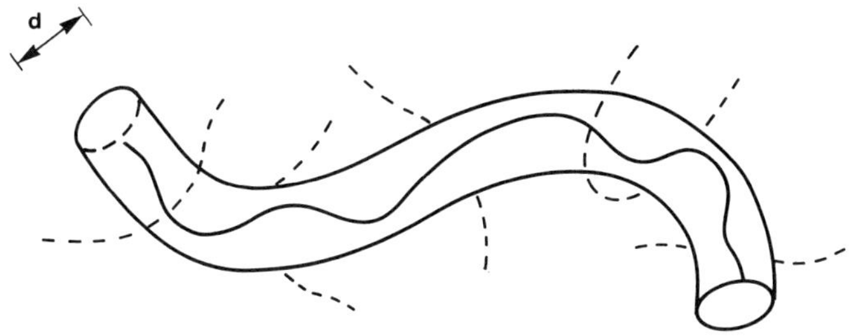

FIGURE 8. A polymer chain (full line) surrounded by other chains (dotted lines) is confined in a certain effective tube of diameter $d \sim 50$ Å [$d = N_e^{1/2}a$; L = total tube length = $(N/N_e)\, d$].

Over a time T_{rep}, the center of mass has moved by an end-to-end distance R_0, so that the diffusion coefficient is predicted to behave like

$$D_S = \frac{R_0^2}{T_{rep}} = \frac{N_e}{N^2} D_1 \sim N^{-2} \tag{22}$$

In entangled melts, the diffusion coefficient is therefore strongly reduced and D_s is very small (typically 10^{-16} m^2/sec.). The experimental data measured by techniques listed in Section 1 are in good agreement with the result in Eq. (22) as shown elsewhere.(3,4,9,17) The crossover to the nonentangled regime [Eq. (14)] is also confirmed experimentally.

This subsection is concluded with a few remarks on tube renewal.(32–38) The constraints defining a tube are themselves mobile. Thus a chain can sometimes escape from its original tube by the fluctuations of the tube. This effect become important if the chains are not very long ($N \sim N_e$) or if one long chain (N) is moving in a matrix of much smaller chains (P). We shall restrict the present discussion to situations where tube renewal can be neglected.

2.3. *Interdiffusion of Labeled P_A^*/Unlabeled P_A Polymers*

The junction between identical species can be monitored only if one of the polymers is labeled P_A^*, as, for instance, with the couple H-PS/D-PS. If the interactions are not modified by labeling, the interdiffusion constant is $D_S = D_1/N^x$, with $x = 1$ for small chains and $x = 2$ for entangled chains. The mutual diffusion coefficient is D_S and independent of composition. The interdiffusion profiles are then classical.

We do not have experiments on the symmetric case ($N_A = N_B$), but on junctions between pure A and *dilute* blends of B in A varying N_B. As shown in Section 3.2 below, $D_M = D_{S_B}$. We show in Figure 9 the pioneering experiment of Klein and Briscoe(4) on the D-PE/H-PE junction. The diffusion is observed in the long-time limit because the space scale of observation is of the order of one millimeter. The profiles shown in Figure 10 are classical; the interdiffusion coefficient is constant. By varying the molecular weight, the variation $D_S \sim N^{-2}$ is well verified (Figure 11).

The simple diffusion profile in Eq. (12) holds when the time scale is larger than the reptation time, but breaks down if $t \ll T_{rep}$. The small time limit is important in polymer

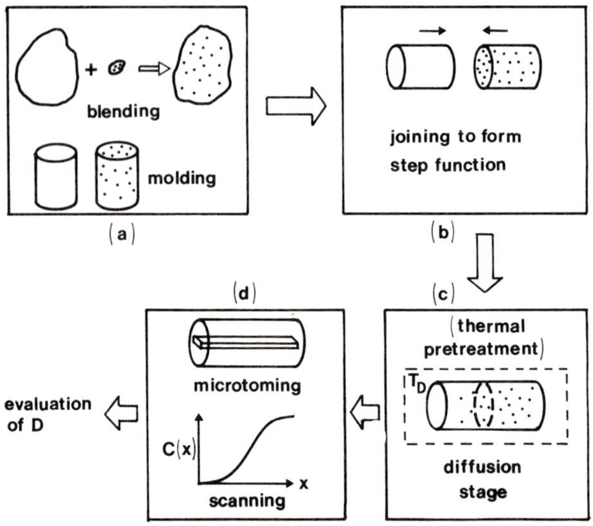

FIGURE 9. Schematic flowchart showing the stages involved in the experimental procedure. (a) Pure polymer is blended with the labeled diffusant to form a uniform dispersion, which is then molded into pellets. (b) Pellets of pure polymer and of the polymer/diffusant dispersion are joined as shown to form a coherent step function. (c) The diffusion stage, carried out in an inert atmosphere at temperature T_D, may be preceded by thermal pretreatment (e.g., annealing) of the polymer composite. (d) After a time t_D the diffusion is stopped by quenching the sample, and slices removed normal to the original interface. The slices are scanned in the IR microdensitometer to yield the broadened concentration profile, from which D is evaluated. See text for further details. (Reprinted with permission from Reference 4.)

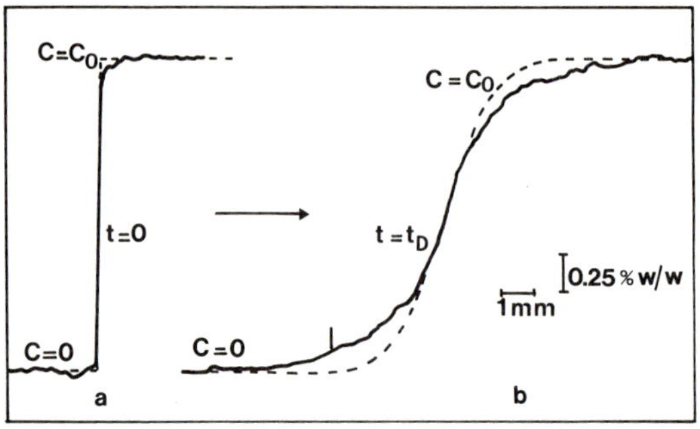

FIGURE 10. Typical concentration profiles of deuterated polyethylene (D-PE) in the protonated polyethylene matrix (H-PE). (a) Immediately subsequent to creation of a step function. $C_0 = 2\%$ w/w of fraction D-PE3 in H-PE. The broken line represents an ideal step function, and the continuous line is the profile as scanned in the infrared microdensitometer. The scanning frequency is that of the C–D stretching mode at 2.170 cm^{-1}. (b) The concentration profile of (a) following a diffusion run. Diffusion temperature $T_D = 176.0 \pm 0.3$ °C; $t_D = 1.114 \times 10^6$ sec; D, as evaluated from the slope at the position of the original interface, is $3.7 \pm 0.6 \times 10^{-9}$ cm^2/sec. The broken curve is the theoretical Fickian profile for a single diffusing species with diffusion coefficient D; the deviations from the experimental profile are due to the polydispersity of the actual diffusant. (Reprinted with permission from Reference 4.)

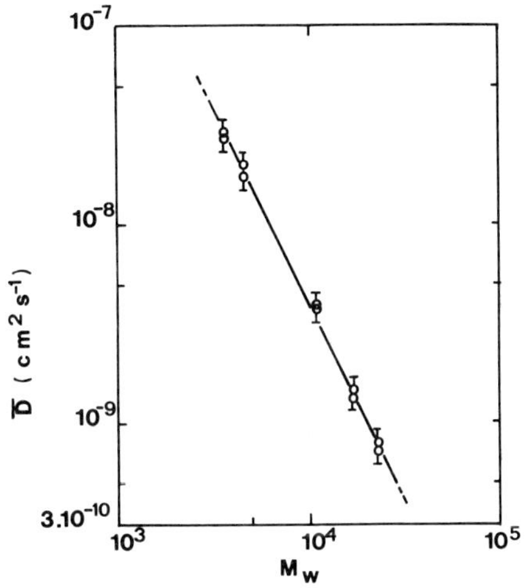

FIGURE 11. The variation of $\bar{D}$ with M_w for deuterated polyethylene diffusing in a protonated polyethylene matrix, at a diffusion temperature of 176.0 ± 0.3 °C. Each point represents a separate experiment, and is the mean of 10 separate profiles. The lengths of experimental runs (t_D) vary by a factor of about 3 within each pair of experiments for a given D-PE fraction. The least-squares best fit to the data is the relation $\bar{D} = 0.2\,M_W^{-2.0\pm0.1}$ and is shown. (Reprinted with permission from Reference 4.)

welding: after a time T_{rep} all the mechanical properties of a homogeneous plastic have been restored, since all entanglements are set up.

2.4. Autoadhesion: Polymer–Polymer Welding

The principle of the welding process is shown in Figure 12; two polymer blocks are brought in close contact, and maintained slightly above the glass temperature for a certain time t. Interdiffusion takes place, and the junction develops mechanical strength.[39–41] In the Lausanne experiments[39,40] this is monitored by fracturing the junction and measuring the fracture energy G_{Ic} as a function of the healing time t. In their experiments $G_{Ic}(t)$ saturates after one reptation time T_{rep} and increases at earlier times according to

$$G_{IC} \cong G_{IC/max}\left(\frac{t}{\tau}\right)^{1/2} \qquad (t < T_{rep}) \tag{23}$$

2.4.1. Description of the Interdigitated Region $e(t)$

An early microscopic picture of healing was based on macroscopic diffusion[42] (Figure 13). The real situation is quite different because we are now concerned with times

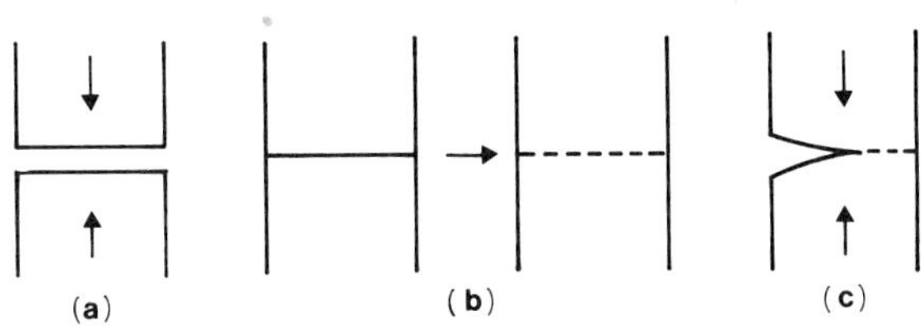

FIGURE 12. The Lausanne experiments on polymer–polymer welding. (a) Two polymer blocks are carefully brought into close contact. (b) During a long time t the two blocks remain in contact, at a fixed temperature slightly *above* the glass temperature; the junction progressively heals. (c) The state of healing is monitored by fracturing the junction.

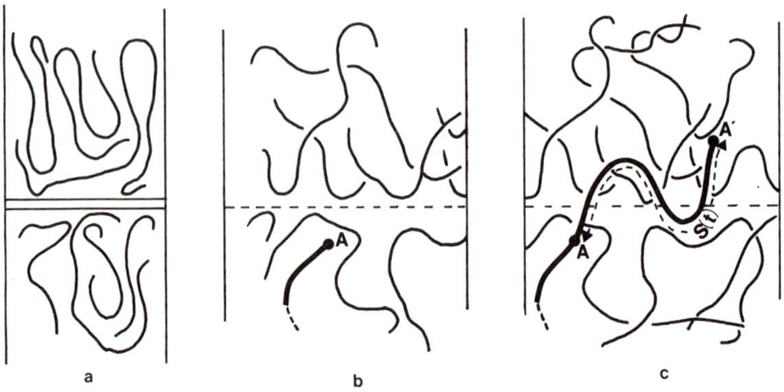

FIGURE 13. A microscopic picture of polymer–polymer welding. (a) The two blocks are put into contact. (b) Just after contact no chain crosses the boundary. (c) After some time t a chain may have reptated by a curvilinear length $s(t)$ and established some bridging.

$t < T_{\text{rep}}$. At this level, macroscopic diffusion does not hold. However, the process can be analyzed in the framework of the reptation model.[43–46]

2.4.1.1. Thickness e(t) *of the Interdigitated Region.* We choose a chain end located at point A (just below the contact surface) (Figure 12) which reaches a point A′ above this surface. The curvilinear length (AA′) = $l(t)$ is given by

$$\langle l^2(t) \rangle = 2D_t t \tag{24}$$

The actual distance AA′ = $\sqrt{la}$ gives the thickness $e(t)$, i.e.,

$$e(t) = R_0 \left(\frac{t}{T_{\text{rep}}} \right)^{1/4} \tag{25}$$

This law is very different from that for the macroscopic diffusion.

2.4.1.2. Concentration Profiles. The profiles were described elsewhere[43] by scaling laws and calculated in more detail.[45] Figure 14a presents the profile for the chain ends coming from below $\phi_t(z)$, which is regular and continuous; $\phi_{t_o} = 2/N$ is the density

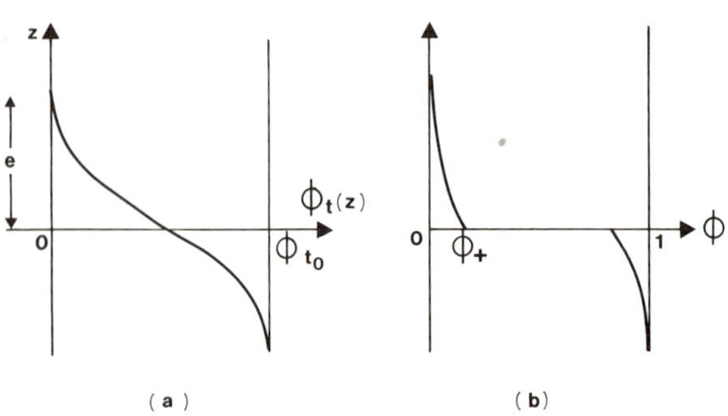

FIGURE 14. (a) Concentration profile for chain ends $\phi_t(z)$ coming from the low part. Note that $\phi_t(z)$ is continuous at the contact surface ($z = 0$). (b) Total concentration profile for monomers coming from the lower block $\phi(z)$. Here there is a discontinuity at $z = 0$. (Reprinted with permission from Reference 44.)

of chain ends before healing. On the other hand, the profile for a monomer coming from the lower block is discontinuous at $z = 0$ (Figure 14b). The value $\phi_+(z = 0 + \epsilon)$ is given by

$$\phi_+ \sim \frac{l}{a}\,\phi_t\,(z = e) \sim \frac{l}{a}\frac{2}{N} \sim \frac{l}{Na} \ll \frac{1}{2} \tag{26}$$

This shows that the concentration profile cannot be described by a diffusion equation: the monomer cannot diffuse freely because of topological constraints.

2.4.1.3. The Number of Bridges p(t) is directly related to ϕ_+ by

$$p(t) = \frac{\phi_+}{a^2} \sim \frac{l}{Na} \sim \left(\frac{t}{T_{rep}}\right)^{1/2} \tag{27}$$

2.4.1.4. The Number of Interconnecting Chains ν(t) is given by

$$\nu = 1/a^3 \int \phi_t(z)dz \sim e\,\phi_{t0} \sim e/Na^3 \tag{28}$$

2.4.2. Relation to Mechanical Measurements

There is no solid theory here, but several attempts have been made to relate G_{IC} to this microscopic description.

(1) In the simplest approach we assume $G_{IC} \sim p(t)$.[43,45] This leads to $G_{IC} \sim (t/T_{rep})^{1/2}$, in agreement with experimental data. However, the assumption that each bridge contributes a constant energy to G_{IC} is very bold. At early times these bridges connect one short piece with one long piece of chain, and G_{IC} might be sensitive to this feature. Then G_{IC} at fixed t predicted by $G_{IC} \sim p$ is not satisfactory.

(2) $G_{IC} \sim \nu l$[40,45] *where ν is the total number of connecting chains. This would lead to* $G \sim t^{3/4}$.

(3) $G_{IC} \sim \tau_c l$.[44] One assumes that fracture occurs when the stress τ has reached a critical value τ_c, leading to plastic flow of chains in their tube. The force f acting on one chain is such that $\nu f = \tau_c$. This force acts to extract a length l of chain. The total work to extract one chain is fl and G is νfl, hence

$$G = \nu fl = \tau_c l \sim l \tag{29}$$

This again gives $G_{IC} \sim t^{1/2}$, in agreement with the Kausch experiments. The derivation is somewhat more fundamental, because it incorporates some of the viscoelastic behavior which is known to be important in fracture. But the whole scheme works only when the strength at the virgin state G_0 is proportional to the molecular weight M [in Eq. (29), l becomes the tube length]. This is not true at very high M; in such cases a different analysis should be constructed: the overall size of the healing region becomes smaller than the coil size, and the healing time is modified.

(4) $G_{IC} \sim \tau_c R_0$.[61] Very recently, de Gennes improved his analysis of volume and energy balance in the suction process. This leads to a reduction of G by a factor of $N^{-1/2}$.

2.4.3. Discussion

- An ideal welding process takes place in one relaxation time T_{rep}. However, some evolution of $G_{IC}(t)$ at times much larger than T_{rep} has been reported.[41]
- The detailed nature of the fracture energy G_{IC} is not yet completely understood.

An interesting complication is the effect of surface orientation on the local reptation. More generally, we need some experimental characterization of the microscopic interdigitated region. This may be not altogether impossible by using very long chains and high-resolution techniques.

3. DYNAMICS OF P_A/P_B MIXTURES: HETEROJUNCTIONS

We consider now the interdiffusion of two compatible polymers (P_A/P_B). The formation of diffuse interfaces is important in welding, coextrusion, morphology of polymer blends, and so on. There are two cases of interest:

1. A and B chemically different,
2. A and B chemically identical, but $N_A \gg N_B$.

For polymer mixtures of *two* kinds of polymer, the mechanism of interdiffusion is very different from passive self-diffusion: the enthalpy of mixing gives a new driving force which accelerates the mixing. We discuss this driving force first.

3.1. The Flory-Huggins Free Energy of Mixing

In the liquid state, the free energy F of an A/B mixture is described reasonably by the Flory–Huggins[47] theory. Per site, F is given by

$$\frac{F}{kT} = \frac{\phi}{N_A} \ln \phi + \frac{1-\phi}{N_B} \ln(1-\phi) + \chi\phi(1-\phi) \tag{30}$$

where ϕ is the volume fraction of A while N_A and N_B are the degrees of polymerization of A and B. From Reference 1 one can derive the critical value of χ for segregation of A and B. In the symmetrical case ($N_A = N_B$), $\chi_c = 2/N$. For most A–B polymer pairs, χ is positive and larger than χ_c, and segregation occurs. The few mixed cases[48] are (a) almost chemically identical species (e.g., deuterated–nondeuterated pairs) ($\chi = 0$) and (b) some A–B pairs with specific interactions giving $\chi < 0$ (corresponding to a negative enthalpy of mixing).

The exchange chemical potential $\mu = \mu_A - \mu_B$ deduced from Eq. (30) is given by

$$\mu = \frac{\partial F}{\partial \phi} = kT\left[\frac{1}{N_A} \ln\phi - \frac{1}{N_B} \ln(1-\phi) + \chi(1-2\phi)\right] \tag{31}$$

and the driving force is

$$-\nabla\mu = \frac{\partial\mu}{\partial\phi}(-\nabla\phi)$$

where

$$\frac{1}{kT}\frac{\partial\mu}{\partial\phi} = \left[\frac{1}{N_A\phi} + \frac{1}{N_B(1-\phi)} + 2|\chi|\right] \tag{32}$$

For case (a) (zero enthalpy of mixing) the force is purely "entropic" and governed by the smallest species. For case (b) the force is controlled entirely by the enthalpic term $|\chi|$.

3.2. Microscopic Theory of Mutual Diffusion

One can construct[60] a general formula for the mutual diffusion and self-diffusion coefficients in terms of three microscopic frictional coefficients $z_{\mathring{A}A}$, $z_{\mathring{A}B}$, and $z_{\mathring{B}B}$ for monomers (A/A), (A/B), and (B/B), respectively ($z_{ij} \sim \eta_o a$, where η_o is a microscopic viscosity).

The entropic source can be written as

$$T\mathring{S} = -\mathbf{J}\nabla\mu = \frac{J^2}{D_M}\frac{\partial\mu}{\partial\phi} \tag{33}$$

We shall now calculate $T\mathring{S}$ in terms of the microscopic dissipation for short and long chains following Brochard.[58]

3.2.1. Interdiffusion of Small Unentangled Chains

Small unentangled chains behave as Rouse chains, where the frictional forces per A and B monomers can be written as

$$\begin{aligned} F_{v_A} &= z_{\mathring{A}B}\,(1-\phi)\,(v_A - v_B) \\ F_{v_B} &= z_{\mathring{A}B}\,\phi\,(v_B - v_A) \end{aligned} \tag{34}$$

The entropy source is $T\mathring{S} = F_{v_A} J_A + F_{v_B} J_B$, i.e.,

$$T\mathring{S} = z_{\mathring{A}B}\frac{J^2}{\phi(1-\phi)} \tag{35}$$

Equations (33)–(35) then lead to

$$D_M = \frac{1}{z_{\mathring{A}B}}\,\phi(1-\phi)\,\frac{\partial\mu}{\partial\phi} \tag{36}$$

The mutual diffusion of "Rouse" chains involves *one* frictional coefficient $z_{\mathring{A}B}$. On the other hand, in self-diffusion, one must include (A/A) or (B/B) frictional forces. As one A (or B) chain moves through the mixture, the frictional force is

$$F_{v_i} = N_i\tilde{z}_i^{\circ}\,v_i, \qquad i = A,B \tag{37}$$

with

$$\begin{aligned} \tilde{z}_{\mathring{A}} &= z_{\mathring{A}A}\phi + z_{\mathring{A}B}(1-\phi) \\ \tilde{z}_{\mathring{B}} &= z_{\mathring{A}B}\phi + z_{\mathring{B}B}(1-\phi) \end{aligned} \tag{38}$$

By using Einstein's relation between mobility and diffusion coefficient, one finds

$$D_{s_i} = \frac{kT}{N_i\tilde{z}_i^{\circ}}, \qquad i = A,B \tag{39}$$

For unentangled chains, D_M and $D_{s_{A,B}}$ are not related in the general case ($z_{\mathring{A}A} \neq z_{\mathring{A}B} \neq z_{\mathring{B}B}$).

3.2.2. Interdiffusion of Entangled Chains

When the two chain lengths are comparable, it is plausible to assume that both chains move by reptation. This assumption is satisfied if the longer partner A is not too long ($N_A < N_B^3/N_E$), to avoid tube renewal.[61]

We now calculate the entropy source for entangled chains within the framework of the reptation model. The chains are trapped in a matrix of tubes formed by other chains. We shall see that the chains move in a mobile matrix, while the tubes move with velocity v_T. Here v_T is deduced from the balance *between frictional forces* due to the chains and those acting on the matrix.

For A and B chains moving in their respective tubes, the entropy source is

$$T\mathring{S} = \tilde{z}_A^\circ \phi W_A^2 + \tilde{z}_B^\circ (1 - \phi) W_B^2 \tag{40}$$

where $\tilde{z}_i^\circ$ are defined by Eq. (38), while W_A and W_B are the curvilinear velocities of A and B trapped chains relative to their tubes. If v_T is the velocity of the tubes and v_A and v_B the velocities of the centers of mass of A and B, then the curvilinear velocities are given by

$$W_i^2 = \frac{N_i}{N_E} (v_i - v_T)^2, \qquad i = A,B \tag{41}$$

The entropy source can be written as

$$T\mathring{S} = F_{v_A} (v_A - v_T) + F_{v_B} (v_B - v_T) \tag{42}$$

by setting $F_{v_i} = \tilde{z}_i^\circ \phi_i N_i / N_E \ (v_i - v_T)$, $i = A,B$, in which case F_{v_i} then becomes the frictional forces per unit volume acting on the matrix. If the sum of these forces is assumed to be zero, an expression for v_T is obtained in the form

$$v_T = J \frac{\tilde{z}_A^\circ N_A - \tilde{z}_B^\circ N_B}{\tilde{z}_A N_A^\circ \phi + \tilde{z}_B N_B^\circ (1 - \phi)} \tag{43}$$

This relation for v_T shows that in the symmetrical case ($\tilde{z}_A^\circ N_A = \tilde{z}_B^\circ N_B$), the matrix is immobile. On the other hand, if $\tilde{z}_A^\circ N_A \gg \tilde{z}_B^\circ N_B$, the matrix moves at the velocity of the longer chains A. This effect corresponds to the "bulk flow" introduced by Crank.[18] In the Ruhterford back-scattering experiment (see Section 1.1.2.9), the motion of the gold particles may reflect that of the matrix.

Equations (33), (40), (41), and (43) yield general expression for D_M in the case of entangled melts:

$$D_M = N_E \phi (1 - \phi) \left(\frac{\phi}{N_B \tilde{z}_B^\circ} + \frac{1 - \phi}{N_A \tilde{z}_A^\circ} \right) \frac{\partial \mu}{\partial \phi} \tag{44}$$

This equation shows that D_M is controlled by the faster species.

We can also calculate through the reptation model the A and B self-diffusion coefficients. The mobility s_{T_A} of the A chain in its tube is $s_{T_A} = (N_A \tilde{z}_A^\circ)^{-1}$. The diffusion coefficient along the tube is $D_T^A = kTs_{T_A}$. Hence the macroscopic diffusion coefficient is

$$D_{s_A} = N_E \frac{kT}{N_A^2 \tilde{z}_A^\circ} \tag{45}$$

For B chains, one finds

$$D_{s_B} = N_E \frac{kT}{N_B^2 \bar{z}_B^\circ} \tag{46}$$

By combining Eqs. (44)–(46), one obtains a relation linking D_M, D_{s_A}, and D_{s_B} in the form

$$D_M = [\phi N_B D_{s_B} + (1 - \phi) N_A D_{s_A}] \phi(1 - \phi) \frac{1}{kT} \frac{\partial \mu}{\partial \phi} \tag{47}$$

This relation was first postulated by Crank.[18]

3.2.2.1. Remark 1. One finds a relation between D_M and $D_{s_{A,B}}$ for entangled chains because the frictional forces involved in both processes are the same for entangled polymer chains. On the other hand, this relation is not correct for nonentangled or semientangled polymers, for which the notion of a matrix disappears; the friction involved in the processes of interdiffusion and self-diffusion are then different.

3.2.2.2. Remark 2. In a previous work,[20] it was assumed implicitly that $v_T = 0$, i.e., pure reptation of A and B in a fixed matrix. This leads to

$$D_M^B = [(1 - \phi)(N_A D_{s_A})^{-1} + \phi(N_B D_{s_B})^{-1}]^{-1} \phi(1 - \phi) \frac{1}{kT} \frac{\partial \mu}{\partial \phi} \tag{48}$$

Equation (48) shows that D_M^B is dominated by the mobility of the slower species if the matrix is immobile. If the faster chains move more rapidly, the slower chains must also move to take their place and then the slower species control the kinetics of interdiffusion.

3.2.2.3. Remark 3. Expression (47) for D_M was first proposed by Crank.[18] Kramer *et al.*[15] then proposed a justification for Eq. (47) based on a model of vacancies having a constant chemical potential with varying density of the polymer melt corrected by an overall motion term. This derivation, valid for diffusion in alloys, was unacceptable for polymer melts as pointed out by de Gennes[54] and Binder[57] independently, but gives a good fit with their experimental data. Sillescu[60] has proposed a different derivation limited to the case $\chi = 0$.

3.3. *Interdiffusion of Small Chains* ($N_A, N_B < N_e$)

3.3.1. Mutual Diffusion Coefficient

For nonentangled polymer blends, we have simple Rouse behavior and mutual friction is controlled by $z_{AB}^\circ \cong \eta_0 a$, where η_0 is a microscopic viscosity and a a monomer size. It is noteworthy that η_0 can depend upon composition if the glass temperatures of A and B are different.[50] From Eqs. (32) and (39), D_M is predicted to be

$$D_M = D_0\, \phi(1 - \phi) \left[\frac{1}{N_A \phi} + \frac{1}{N_B(1 - \phi)} + 2|\chi| \right] \tag{49}$$

where $D_0 = kT/\eta_0 a$ is a monomer diffusion coefficient (about 10^{-9} m^2/sec). Here, we see in Eq. (49) that two very different situations can occur:

3.3.1.1. Compatible Pairs ($\chi < 0$). The entropic term is negligible as soon as $|\chi|\phi > 1/N$:

$$D_M = D_0\, \phi(1 - \phi) \tag{50}$$

The value of D_M is *large* and independent of molecular weight because the driving force and friction force are linear in N.(49) D_M is much larger than the self-diffusion coefficients ($D_M/D_{S_{A,B}} \sim N$). Recent experiments by Fytas *et al.*(62) seem to fit these predictions. Contrary to common belief, there are no simple relations between D_M and D_{S_A}, D_{S_B}.

3.3.1.2. Chemically Identical Pairs ($\chi = 0$). Here we have a very simple, additive structure,

$$D_M = D_{S_A}\,(1 - \phi) + D_{S_B}\phi \tag{51}$$

where $D_{S_A} = D_0/N_A$ and $D_{S_B} = D_0/N_B$ are the self-diffusion coefficients of A and B, respectively. If $N_A \gg N_B$, the mutual diffusion coefficient is equal to the self-diffusion coefficient of the small, *fast* species because the driving force is controlled by the small chains.

3.3.2. Concentration Profiles

Quantity D_M is composition-dependent and the profiles for two pure A and B blocks in contact at time $t = 0$ are not given by Eq. (12). Let us consider first the case $\chi \neq 0$, $N_A = N_B$, and assume that z_{AB} is independent of ϕ. This model leads to concentration profiles $\phi_t(x)$ which are very unusual.(20)

If we return to the diffusion equation

$$\frac{\partial \phi}{\partial t} = \frac{\partial}{\partial x}\left[D_0\, \phi(1 - \phi)\, \frac{\partial \phi}{\partial x}\right] \tag{52}$$

and set $u = x/\sqrt{D_0 t}$, then Eq. (11) becomes

$$-\frac{1}{2}\, u\, \frac{d\phi}{du} = \frac{d}{du}\, \phi(1 - \phi)\, \frac{d\phi}{du} \tag{53}$$

with boundary conditions $\phi_{-\infty} = 0$ and $\phi_{+\infty} = 1$. The solution of this latter equation is a sequence of straight lines:

$$\begin{aligned} \phi &= (u + 1)/2, && |u| < 1 \\ \phi &= 0, && u < -1 \\ \phi &= 1, && u > 1 \end{aligned} \tag{54}$$

The mixed layer is finite (no "tails" of diffusion) and has sharp edges (Figure 5a).

For $\chi = 0$, $N_A > N_B$, and $D_M \cong D_{S_B}\phi$, we expect a sharp profile (finite slope) on the side of the small chain where D_M becomes small ($D_M = D_{S_A}$) and the usual tail of diffusion is on the other side.

3.4. Interdiffusion of Entangled Polymers $(N_A, N_B > N_e)$

3.4.1. Compatible Pairs ($\chi \neq 0$) with Comparable Lengths

For compatible polymers with comparable lengths N, Eqs. (32) and (44) lead to

$$D_M = D_0 \frac{N_E}{N} |\chi| \phi(1 - \phi) \tag{55}$$

where D_0 is a monomer diffusion coefficient. For isotropic frictional coefficients ($z_{ij} \sim \eta_0 a$), $D_0 = kT/\eta_0 a$. For anisotropic coefficients $z_{\mathring{A}B} \gg z_{\mathring{A}A}, z_{\mathring{B}B}$ (in the case where strong A–B polar interactions assure A–B compatibility: the "sticking effect"), $D_0 = kT/z_{\mathring{A}B}$.

Quantity D_M is proportional to $\phi(1 - \phi)$, and concentration profiles for two blocks A and B in contact at time $t = 0$ are very unusual and given by Eq. (54). This relatively fast mutual diffusion [Eq. (55)] was first observed by Gilmore *et al.*(51) on PVC/PCL mixtures and interpreted by de Gennes.(52) By analogy with the reptation model for D_S, they were looking for a N^{-2} dependence but observed a N^{-1} law to their great surprise (Figure 15). This fast mutual diffusion D_M (ϕ) has been confirmed in recent experiments by Kramer and co-workers(53) and Klein and co-workers(22) (contact between blends of similar composition ϕ, $\phi + \Delta\phi$). A difficult problem with these mutual diffusion experiments on compatible pairs is that the glass transition temperature T_g of the mixture can depend strongly on its composition; this means that the monomer–monomer friction coefficient also becomes strongly dependent upon composition. When experiments are carried out at constant temperature T for a range of compositions, what Klein and Kramer have done is to correct the results with $(T - T_g)$ constant. This seems reasonable, but it is by no means certain that it is absolutely the correct procedure. Klein finds a deviation from the $\phi(1 - \phi)$ law, but a symmetrical curve. An experiment(55) based on the drift of gold particles

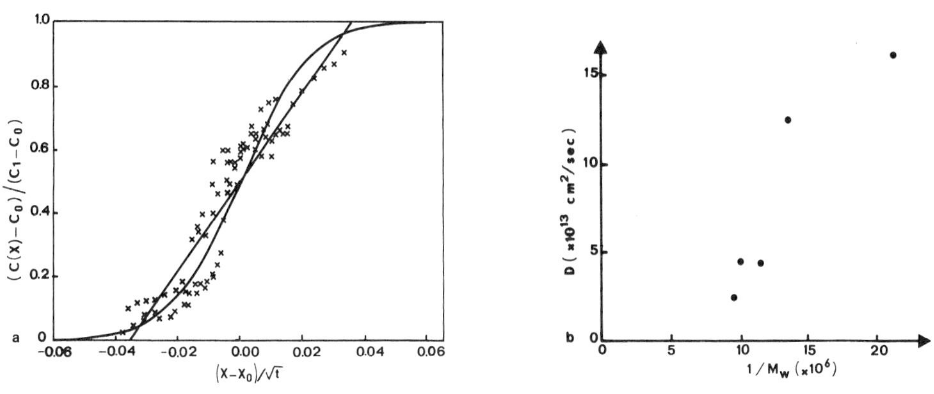

FIGURE 15. (a) Master curve for mutual diffusion in a blend of poly(vinyl chloride) (PVC) and poly(ϵ-caprolactone) (PCL) at 70 °C. The weight-average molecular weights are, respectively, 46,500 ($M_w/M_n = 2.5$) and 37,200 ($M_w/M_n = 1.9$). x is measured in μm and time in sec. We have presented the straight line corresponding to $D = D_0\phi(1 - \phi)$ and the classical curve for $D = D_0$, independent of ϕ. The scattering of experimental data is due to the high polydispersity of the sample. (b) D vs. $(1/M_w)$ for PVC, M_w(PCL) $= 3.7 \times 10^4$. (Reprinted with permission from Reference 51.)

gives a completely different result ($D_M \sim N^{-2}$) but, as already explained, the interpretation of this drift is not clear.

3.4.2. Interdiffusion of Long and Short Chains: Kramer's Experiments

The interdiffusion profiles for very asymmetrical junctions ($N_A \gg N_B$) has been studied indirectly via the motion of gold particles [12,13] and by radioactive tracers. From these experiments, one observes a rapid interdiffusion controlled by the fast (short) chains and independent of the slow (long) chains. In Cambridge, Jordan *et al.*[59] used infrared microdensitometry to study the mutual diffusion between two butadienes for different ratios N_A/N_B and found also that the mobility is controlled by the *faster* species.

All these data seem to fit the theoretical result deduced from Eqs. (32)–(44) with $\chi = 0$, $z^\circ_{ij} = z^\circ$:

$$D_M = N_E \frac{\phi^2(1-\phi)^2}{z^\circ} \left[\frac{1}{N_A\phi} + \frac{1}{N_B(1-\phi)} \right]^2 \tag{56}$$

The strong dependence of $D_M(\phi)$ leads to very asymmetrical concentration profiles at the junction between A and B blocks (Figure 5).

Expression (56) holds for long times ($t > T_{rep}$). At times $t < T_{rep}$, we have interpreted[56] this fast interdiffusion as a swelling of the network of the long chains by the more labile small chains.

3.5. P_A/P_B Welding (Short-Time Behavior)

In the short-time scale relevant for P_A/P_B welding, the case of homopolymers ot degree of polymerization N_A and N_B ($N_A > N_B$) is probably not different from the autoadhesion discussed in Section 2.4.

The case of compatible pairs, on the other hand, must be very different because now a driving force favors the mixing. This case has been discussed by de Gennes[43,44]: a chain A is sucked into the upper block because it prefers to be in contact with B rather than with A chains. If $l(t)$ is the length of A monomers of one chain immersed in the B region, the rate equation of $l(t)$ is not ruled by diffusion in a tube but by

$$\frac{dl}{dt} = s_{ch} kT \frac{|\chi|}{a} \tag{57}$$

where s_{ch} is the chain mobility in its tube ($s_{ch}\, kT = D_0 a/L$). Then $l(t) \sim t$. If we assume $G_{IC} \simeq \tau_c l(t)$, we expect a variation linear in time. However, the experiments[40] of heteroadhesion do not agree with this theoretical prediction ($G_{IC} \sim t^{1/2}$).

4. CONCLUSIONS

1. Experiments measuring the interdiffusion of polymer A (P_A)/polymer B (P_B) blends have been performed for three cases:

(a) P_A/P_A^* (unlabeled/labeled polymer). This case is relatively well understood. The mutual diffusion coefficient is equal to the self-diffusion coefficient. The theoretical predictions, namely $D_M \sim 1/N$ for small chains and $D_M \sim 1/N^2$ for large chains, appear to fit the data.

(b) P_A/P_B (compatible pairs). In this case, a large enthalpic driving force favors the mixing. This leads to large mutual diffusion coefficients, strongly dependent upon composition, and to unusual concentration profile. Experiments are in good agreement with theoretical predictions for both unentangled and entangled blends.

(c) P_A/P_B are *chemically identical* but the A chains are much larger than the B chains. We now have direct measurements of the full concentration profile(59) and indirect measurements by marker particles,(12,13) which show fast kinetics of interdiffusion controlled by smaller species.

This can be understood in terms of a reptation of A and B chains through a mobile matrix. This notion was initiated by Crank and has been developed by Kramer. In the present text, we derive the matrix velocity from a force-balance equation which avoids certain *ad hoc* assumptions of the previous treatments.

2. In all these experiments, one must distinguish between the long-time limit $t > \tau_e$, where τ_e is the largest relaxation time ($\tau_e = T_{\text{Rouse}}$ for short chains and $\tau_e = T_{\text{rep}}$ for long chains) and the short-time limit, where elastic stresses play a role. The condition $t > \tau_e$ can also be translated into a condition on space scale:

$$D_M \tau_e = l^2_M \tag{58}$$

which defines a "magic length." For $x > l_M$ we have a pure viscous regime. For $x < l_M$ one expects viscoelastic behavior(56) and discontinuities in the concentration profiles l_M can be large; for A/B compatible pairs, $l_M \simeq Na$ is much larger than a coil size (several microns for $N = 10^4$). For homopolymers with $N_A \gg N_B$, $l_M = (R_0)_A N_A/N_B$. When examining diffusion at very small scales x with powerful techniques, x is often smaller than l_M and viscoelastic properties must be taken into account.(62) Some of Kramer's experiments on long/short chain junctions are in this regime, where long chains behave like a gel.

3. The mutual diffusion coefficients depend upon composition in most cases. The diffusion profiles for pure blends in contact are then not given by error functions. The D^*_M values determined experimentally by assuming classical profiles, can be wrong. A better way is to measure $D_M(\phi)$ with junctions of similar compositions for which D_M is nearly constant.

4. Polydispersity effects alter the concentration profiles. Small species diffuse faster and give rise to diffusion tails, which may seem to contradict the sharp profiles expected theoretically.

5. The glass transition temperature can depend strongly upon comopsition, especially in the case of compatible polymers. It leads to a second cause for a variation in $D_M(\phi)$.

ACKNOWLEDGMENTS

The author thanks Matthew Tirrell and Pierre-Gilles de Gennes for their close scrutiny of this manuscript.

NOMENCLATURE

A,B	Species
c	Monomer concentration
D_M	Mutual diffusion coefficient
D_0	Monomer diffusion coefficient
D_{S_A}, D_{S_B}	Self-diffusion coefficients
E	Elastic modulus
G_{IC}	Fracture energy
$\mathbf{J}_B$	Volume transfer of B
$\mathbf{J} = \mathbf{J}_A = \phi \mathrm{v}_A$	Current J of A = volume transfer of A
N_A, N_B	Degree of polymerization of A,B
N_e	Threshold of entanglement
P_A, P_B	Polymer A, Polymer B
s_{ch}	Chain mobility
s_0	Monomer mobility
T_R	Rouse relaxation time
T_{rep}	Reptation relaxation time
v_A, v_B	Velocity of A, B
v_T	Tube (or matrix) velocity
W_A, W_B	Curvilinear velocity of chains trapped in tube
$z_{\mathring{A}A}$	Monoer A/monomer A (frictional coefficients)
$z_{\mathring{A}B}$	Monomer A/monomer B (frictional coefficients)
$z_{\mathring{B}B}$	Monomer B/monomer B (frictional coefficients)
η	Bulk viscosity
η_0	Local viscosity
$\phi = ca^3$	Volume fraction of A
χ	Flory interaction parameter

REFERENCES

1. A. R. Altenberger and M. Tirrell, *J. Polym. Sci., Polym Phys. Ed.* **22,** 909–910 (1984).
2. H. Sillescu, *Makromol. Chem. Rap. Commun.* **5,** 519–523 (1984).
3. M. Tirrell, *Rubber Chem. Technol.* **57,** 534–556 (1985).
4. J. Klein and B. Briscoe, *Proc. R. Soc. London, Ser. A* **365,** 53 (1979); J. Klein and B. Briscoe, *Nature* **271,** 143–145 (1978).
5. F. P. Price, P. T. Gilmore, E. L. Thomas, and R. L. Laurence, *J. Polym. Sci., Polym. Symp.* **63,** 33–34 (1978).
6. F. Bueche, W. M. Cashin, and P. Debye, *J. Chem. Phys.* **20,** 1956 (1952); F. Bueche, *J. Chem. Phys.* **48,** 1410 (1968); Y. Kumagai, H. Watanabe, K. Miyasaka, and H. Hata, *J. Chem. Eng. Jpn.* **12,** 1–8 (1979).
7. P. Callaghan and D. Pinder, *Macromolecules* **13,** 1085 (1980); P. Callaghan and D. Pinder, *Macromolecules* **14,** 1334 (1981); Evon Meerwall, *Adv. Polym. Sci.* **54,** 1 (1984).
8. L. Léger, H. Hervet, and F. Rondelez, *Phys. Rev. Lett.* **42,** 1681 (1979).
9. R. Bachus and R. Kimmich, *Polymer* **24,** 964–970 (1983); B. A. Smith, S. J. Mumby, E. T. Samulski, and L. P. Yu, *Macromolecules* **19,** 470 (1986).
10. L. M. Smith, B. A. Smith, and H. M. MacConnell, *Biochemistry* **18,** 2256 (1979).

11. C. Ayrault, F. Rondaux, J. F. Tassin, and C. Monnerie, private communication; H. Morawetz, *Polym. Eng. Sci.* **23,** 689–692 (1983); T. Shiah and H. Morawetz, *Macromolecules* **17,** 792–794 (1984); C. Ayrault, F. Rondaux, J. F. Tassin, and C. Monnerie, private communication.
12. P. J. Mills, P. F. Green, C. J. Palmström, J. W. Mayer, and E. J. Kramer, *Appl. Phys. Lett.* **45,** 9–16 (1984); P. J. Mills, P. F. Green, C. J. Palmström, J. W. Mayer, and E. J. Kramer, *Phys. Rev. Lett.* **53,** 2145–2148 (1984).
13. P. F. Green and E. J. Kramer, *Macromolecules* **19,** 1108–1116 (1986).
14. P. F. Green, C. J. Palmström, J. W. Mayer, and E. J. Kramer, *Macromolecules* **18,** 501–507 (1985).
15. E. J. Kramer, P. Green, and C. J. Palmström, *Polymer* **25,** 473 (1984).
16. U. Murschall, E. W. Fisher, Ch. Merktmactzky, and G. Fytas, *J. Polym. Sci., Polym. Lett. Ed.* (submitted).
17. C. R. Bartels, B. Crist, and W. W. Graessley, *Macromolecues* **17,** 2702 (1984).
18. J. Crank, *The Mathematics of Diffusion,* Clarendon Press, Oxford (1975).
19. This notation was introduced by P. G. de Gennes for the wetting of porous media, in: D. Adler, H. Fritzsche, and S. R. Ovshinsky (eds.), *Physics of Disordered Materials,* p. 227, Plenum Press, New York (1985).
20. F. Brochard, J. Jouffroy, and P. Levinson, *Macromolecules* **16,** 1638 (1983).
21. F. Brochard, J. Jouffroy, and P. Levinson, *J. Phys. (Paris), Lett.* **44,** L-455–L-460 (1983).
22. R. A. L. Jones, J. Klein, and A. M. Donald, *Nature* **321,** 161 (1986).
23. P. G. de Gennes, *Scaling Concept in Polymer Physics,* Cornell University Press, Ithaca, New York (1979).
24. J. P. Cotton, G. Jannink, J. des Cloizeaux, and P. G. de Gennes, *Macromolecules* **7,** 863 (1974).
25. P. Flory, *J. Chem. Phys.* **17,** 303 (1979).
26. S. M. Aharoni, *Macromolecules* **16,** 1722–1725 (1983).
27. J. D. Ferry, *Viscoelastic Properties of Polymers,* Wiley, New York (1970).
28. W. Graessley, *Adv. Polym. Sci.* **16,** 1–179 (1974).
29. P. G. de Gennes, *J. Chem. Phys.* **55,** 572 (1971).
30. S. F. Edwards, *Proc. Phys. Soc.* **92,** 9 (1967).
31. M. Doi and S. F. Edwards, *J. Chem. Soc., Faraday Trans.* **74,** 1789, 1802, 1818 (1978); *The Theory of Polymer Dynamics,* Clarendon Press, Oxford (1986).
32. J. Klein, *Polym. Prepr., Am. Chem. Soc., Div. Polym. Chem.* **22,** 105 (1979).
33. W. W. Graessley, *Adv. Polym. Sci.* **47,** 67 (1982).
34. M. Daoud and P. G. de Gennes, *J. Polym. Sci., Polym. Phys. Ed.* **17,** 1971 (1979).
35. P. F. Green and E. J. Kramer, *Macromolecules* **19,** 1108–1114 (1986).
36. M. Antonietti, J. Contandin, and H. Sillescu, *Macromolecules* **19,** 793–798 (1986).
37. R. A. Smith, E. Samulski, L. P. Yu, and M. Winnik, *Phys. Rev. Lett.* **52,** 45 (1984).
38. P. F. Green, P. J. Mills, C. Palmström, J. Mayer, and E. J. Kramer, *Phys. Rev. Lett.* **52,** 2145–2148 (1984).
39. K. Jud, H. H. Kausch, and J. G. Williams, *J. Mater. Sci.* **16,** 204 (1981).
40. K. Jud and H. H. Kausch, *Polym. Bull.* **1,** 697–707 (1979); Proceeding of the 5th International Conference on Fracture, Cannes (1986).
41. C. M. Roland and G. C. Böhm, *Macromolecules* **18,** 1310–1314 (1985).
42. S. S. Voyutskii, *Autoadhesion and Adhesion of High Polymers,* Wiley–Interscience, New York (1963).
43. P. G. de Gennes, *C.R. Acad. Sci. Paris* **291,** 219–221 (1980); **292,** 1505 (1981); **308** (II), 13 (1989).
44. J. M. Georges (ed.), *Microscopic Aspects of Adhesion and Lubrication,* pp. 355–367, Elsevier, Amsterdam (1982).
45. S. Prager and M. Tirrell, *J. Chem. Phys.* **75,** 5194–5201 (1981); D. Adolf, S. Prager, and M. Tirrell, *Springer Lecture Notes in Applied Math.* **1063,** 37–45 (1984); D. Adolf, S. Prager, and M. Tirrell, *J. Polym. Sci., Polym. Phys. Ed.* **23,** 413–427 (1985).
46. Y. H. Kim and R. P. Wool, *Macromolecules* **16,** 1115–1120 (1983).
47. P. J. Flory, *Principles of Polymer Chemistry,* Cornell University Press, Ithaca, New York (1953).
48. D. Klempner and K. Frisch (eds.), *Polymer Alloys,* Plenum Press, New York (1977); S. Krause, *J. Macromol. Sci., Rev. Macromol. Chem.* **C7,** 251 (1972); O. Olabishi, *Polymer–Polymer Miscibility,* Academic Press, New York (1979).
49. P. G. de Gennes, *Macromolecules* **9,** 587–594 (1976); P. G. de Gennes and F. Brochard, *Macromolecules* **10,** 1157–1163 (1977).
50. F. S. Bates, G. D. Wignall, and w. C. Koehler, *Phys. Rev. Lett.* **55,** 2425 (1985).

51. P. T. Gilmore, R. F. Alabella, and R. L. Laurence, *Macromolecues* **13,** 880–883 (1980).
52. P. G. de Gennes, *J. Chem. Phys.* **72,** 4756–4762 (1980).
53. R. J. Composto, J. W. Mayer, E. J. Kramer, and D. M. White, *Phys. Rev. Lett.* **57,** 1312 (1986).
54. P. G. de Gennes, Invited Lecture, Cornell (1984).
55. S. Wu, H. K. Chuang, and C. D. Han, *J. Polym. Sci., Polym. Phys. Ed.* **24,** 143–156 (1986).
56. F. Brochard and P. G. de Gennes, *Europhys. Lett.* **1,** 221–224 (1986); F. Brochard and P. G. de Gennes, *PCH Physico Chemical Hydrodynamics* **4,** 313–322 (1983).
57. K. Binder, private communication.
58. F. Brochard, *C.R. Acad. Sci. Paris,* **B305,** 657 (1987).
59. E. A. Jordan, R. C. Ball, A. M. Donald, L. J. Fetters, R. A. L. Jones, and J. Klein, *Macromolecules* **21,** 235 (1988).
60. H. Sillescu, *Makromol. Chem. Rapid Commun.* (to appear).
61. P. G. de Gennes, *J. Phys.* (*Paris*) **50,** 255 (1989).
62. F. Brochard-Wyart and P. G. de Gennes, Symposium volume of *Makromol. Chem.*, Macromolecular Symposium of Networks 90, Jerusalem (to be published).

7

Welding, Tack, and Green Strength of Polymers

Richard P. Wool

1. INTRODUCTION

In recent years, we have considered the problem of strength development at polymer–polymer interfaces in terms of the static and dynamic properties of random-coil chains.[1–4] When two pieces of solid polymer are brought into contact, wetting or close molecular contact (van der Waals) first occurs followed by interdiffusion of chain segments back and forth across the wetted interface. After a contact time t, we inquire as to the energy G required to separate the two pieces as a function of time, temperature T, contact pressure P, and molecular weight M of the linear random-coil chains, i.e.,

$$G = f(t,T,P,M) \tag{1}$$

where f is the welding function to be determined. Solutions to this problem have application to similar situations involving polymer–polymer interfaces such as processing of powder and pellet resin, internal weld lines of polymer melts during compression and injection molding, welding of surfaces, lamination of composites, coextrusion, blends, adhesion, fracture mechanics, tack and green strength of uncured linear elastomers.

In this chapter, we focus on problems of welding, tack, and green strength with emphasis on symmetric interfaces where the same amorphous polymer exists on both sides of the interface. Consideration is also given to the important case of incompatible asymmetric amorphous and crystalline interfaces where strength develops between two immiscible polymers. We begin by introducing the reptation model[5,6] for diffusion at a symmetric interface, determine a molecular description of the interface structure as a function of the variables t, T, P, and M, and relate this structure to the mechanical properties via a set of microscopic deformation mechanisms. The latter involve disentanglement by chain pullout, bond rupture, and connectivity between chains. The predictions for welding, tack, and green strength are compared with experiments of many investigators. Comparisons are also made with theoretical predictions of several investigators.[7–11]

Richard P. Wool • Department of Materials Science and Engineering, University of Illinois, Urbana, Illinois 61801.

2. THEORY

2.1. Molecular Dynamics of Random-Coil Chains

The motion of individual chains in amorphous bulk materials or concentrated solutions of linear random-coil (Gaussian) chains has been modeled by the reptation theory of de Gennes(5) and Edwards.(6) In this model the chain is confined to a tube having a similar shape as the random-coil configuration of the chain. The tube represents topological constraints to lateral motion of monomers imposed by the neighboring chains and predominantly restricts motion of the chain to that along the curvilinear length of the tube. Our use of this model is illustrated in Figure 1. At $t = 0$, the chain (solid line) is in its initial tube indicated by the dashed lines. The chain exhibits Brownian motion back and forth in the tube and, because the chain ends are free to pick any direction of motion away from the tube, the memory of the initial tube position in space is gradually lost as shown at times t_2 and t_3. However, the chain being in a dense system is always in a new tube but only parts of the chain, usually toward the center, retain memory of the initial tube for the relaxation time T_r. The length of chain $l(t)$ which "escapes" from the initial tube is also a random-coil chain obeying Gaussian statistics and we call it the minor chain.(2) As the minor chain increases in length we can represent its static dimensions by a "most probable" spherical envelope as shown in Figure 1. Finally at $t = T_r$, the reptation time, the chain has escaped or forgotten its original configuration. Details of the molecular dynamics of the reptation model have been discussed elsewhere in this book (see Chapter 6 by F. Brochard-Wyart). The following relations summarize the description of the motion of a linear polymer chain in an engangled melt:

The mean-square escape length of the minor chains, $\langle l^2 \rangle$:

$$\langle l^2 \rangle = 16\, D_1\, t/\pi \tag{2}$$

The curvilinear one-dimensional diffusion coefficient, D_1:

$$D_1 \approx M^{-1} \tag{3}$$

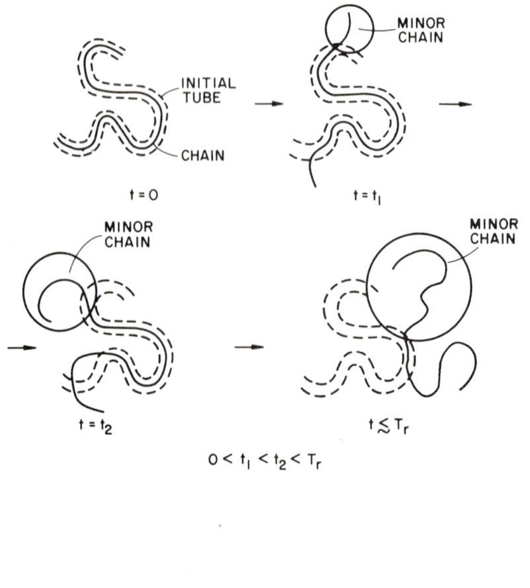

FIGURE 1. The reptation model for a random-coil chain in an entangled melt shows the chain relaxing from its original configuration or tube at times $0 < t_1 < t_2 < T_r$, where T_r is the tube disengagement or relaxation time. The minor chain is the portion of the chain which has relaxed from its original configuration and is shown contained in its random-coil most probable spherical envelope. The dotted tube represents the topological constraints to motion imposed by the neighboring chains in the melt. (From Reference 2.)

The mean-square monomer displacement of the minor chains:

$$\langle X^2 \rangle \approx \langle l^2 \rangle^{1/2} \tag{4}$$

The mean-square center-of-mass diffusion distance, $\langle X^2{}_{cm} \rangle$:

$$\langle X^2_{cm} \rangle \approx 2Dt \tag{5}$$

The center-of-mass self-diffusion coefficient, D:

$$D \approx M^{-2} \tag{6}$$

The reptation time, T_r:

$$T_r \approx M^3 \tag{7}$$

The fraction of chain remaining in the tube, $F(t)$:

$$F(t) = 1 - 4/\pi^{3/2} \quad (t/T_r)^{1/2} \tag{8}$$

where $t < T_r$. For $t > T_r$, $F(t)$ can be well approximated by[5,12]

$$F(t) \approx 8/\pi^2 \exp(-t/T_r) \tag{9}$$

Equations (8) and (9) form the basis for describing the viscoelastic shear relaxation modulus of the Doi–Edwards[12] constitutive theory of entangled linear polymer melts. Considerable experimental support exists for the molecular-weight dependence of the self-diffusion coefficient, Eq. (5), and is reviewed both in Chapter 6 by Brochard-Wyart and elsewhere by Tirrell.[13] Support for the minor chain model shown in Figure 1 has been obtained by Lee and Wool.[14] Using FTIR techniques, they showed that the orientation relaxation of step-strained centrally deuteriated polystyrene chains in a higher-molecular-weight melt first occurred by relaxation of the protonated chain ends while the center of the chain retained its orientation. The centrally deuteriated part of the triblock chain then relaxed at a predicted time and rate consistent with the reptation theory. In a homopolymer melt, the ends relax at the same rate as the center. The relaxation time, $T_r \approx M^3$, is also important in determining healing or welding times at polymer interfaces. In the next section we use the minor chain model to develop a time and molecular-weight-dependent molecular description of a polymer interface. Effects of temperature and pressure will also be discussed.

2.2. Molecular Description of a Polymer–Polymer Interface

The most important interface to consider is that formed by a symmetric A/A polymer pair consisting of linear monodisperse molecular weight polymers whose chain ends are uniformly distributed in the bulk melt material. Polydisperse symmetric A/A pairs and some asymmetric A/B pairs can also be investigated using methods presented in this section. Assuming instantaneous wetting (time-dependent wetting is considered later) of the interface we proceed to analyze the molecular aspects of the interface arising from interdiffusion. This problem is simplified further as shown in Figure 2. At $t = 0$, molecular contact is achieved at the interface (only the molecules on one side of the interface are shown for clarity). The chains begin to reptate via their chain ends which are shown initially as dots randomly distributed in space. As time proceeds, only that part of the

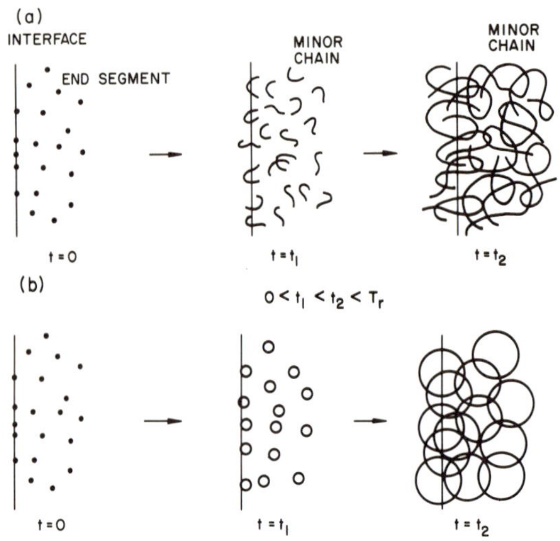

FIGURE 2. The interdiffusion process at a polymer–polymer interface in terms of the behavior of (a) the minor chains and (b) the minor chain spherical envelopes. Only one side of the interface is shown for convenience. Instantaneous molecular contact or wetting is assumed at $t = 0$. (From Reference 2.)

chain which has escaped its original configuration, i.e., the minor chain, can contribute to interdiffusion across the interface. Furthermore, we do not have to consider the exact location of each segment of the minor chain but only the spherical envelope to which it belongs, as shown in Figure 2. Thus, as time approaches the reptation time T_r, interpenetration and reentanglement of chains are achieved and the molecular properties of the virgin state are obtained at the interface plane. From this minor chain model, many molecular aspects of the interdiffusion process can be determined, as presented below. Solutions are given for $t < T_r$ (healing) and $t > T_r$ (virgin state). Details of the calculations are given in References 2, 3, 15, and 16.

2.2.1. Number of Chains Intersecting the Interface, *n*(*t*)

The number of random-coil chains intersecting unit area of the interface as a function of contact time t and molecular weight M behaves as

$$n(t) \approx t^{1/4} M^{-5/4} \qquad (t < T_r) \tag{10}$$

$$n_\infty \approx M^{-1/2} \qquad (t > T_r) \tag{11}$$

A chain is considered to intersect the interface when one of its ends has crossed the interface contact plane defined at $t = 0$. Function $n(t)$ also scales as X/M where X is the average monomer segment interpenetration distance defined by Eq. (4). Equation (10) is derived on the basis that all chains whose ends are within a distance of X from the interface will have intersected the interface at time t, and that the number of chains per unit volume varies as M^{-1}. The time dependence is atypically $t^{1/4}$ rather than $t^{1/2}$ as expected for usual atomic diffusion processes.

We note that the static or virgin state solution at $t > T_r$ is obtained by substituting for the reptation time $T_r \approx M^3$ in the dynamic solution. The static solutions are independent of the molecular dynamics model and can be derived from other considerations. Equation (11) was derived previously by Wool and Rockhill.[17] Thus, the dynamic solutions give the static soutions simply by substituting for the molecular-weight dependence of the relaxation time. This is an important feature in evaluating molecular aspects of strength development.

When a molecular property is suggested as controlling the time dependence of welding, that property can be used to simultaneously predict both the molecular-weight dependence of welding and the molecular-weight dependence of fracture in the virgin state. This three-way concurrence is useful in evaluating potential solutions to welding problems. For example, it is known from the experiments of Kausch *et al.*[18–20] and Wool *et al.*[1,21,22] that the fracture stress during crack healing and welding increases as $\sigma \approx t^{1/4}$. It could be hypothesized upon inspection of Eq. (10) that the number of chains intersecting the interface, $n(t) \approx t^{1/4}$, is responsible for the strength development. However, this hypothesis can be rejected by examining the virgin state property, Eq. (11), since the fracture stress of polymers does not decrease with increasing molecular weight as $\sigma_\infty \approx M^{-1/2}$. Further experimental study would show that the healing rate does not depend on $M^{-5/4}$. Conversely, molecular properties used to describe the virgin state strength can also be evaluated by their ability to describe the time and molecular-weight dependence of healing. This argument applies when the same microscopic deformation mechanisms are operative during welding and fracture of the virgin state.

2.2.2. Average Interpenetration Contour Length, $l(t)$

The average contour length of chain segments or cilia which have diffused across the interface is obtained from the minor chain model as

$$l(t) \approx t^{1/2} M^{-1/2} \tag{12}$$

$$l_\infty \approx M \tag{13}$$

This property has the same scaling relation as the minor chain length, $\langle l \rangle$. It will be shown later that $l(t)$ is important in determining mechanical properties where disentanglement and chain pullout are the dominant microscopic are deformation mechanisms.

2.2.3. Number of Bridges Crossing an Interface, $p(t)$

Each time a piece of interdiffusing chain crosses back and forth through the interface, it creates a molecular bridge. Intuitively, this may be visualized as a "sewing up" of the interface. Function $p(t)$ is determined by $p(t) = n(t)\sqrt{l(t)}$ where $\sqrt{l(t)}$ is the number of bridges created by each minor chain. Thus we obtain

$$p(t) \approx t^{1/2} M^{-3/2} \tag{14}$$

$$P_\infty \approx M^0 \tag{15}$$

This result was first obtained by de Gennes[9] and is equivalent to the "crossing density" calculated by Prager and Tirrell.[7] It is noteworthy that the virgin state number of bridges is independent of molecular weight. Thus, bridges alone cannot be used to describe strength development during welding in cases where the virgin strength exhibits a molecular-weight dependence over some range of molecular weight as observed for most polymers. Bridges are also useful in the development of entanglement models and for estimating the number of broken bonds in brittle fracture.

2.2.4. Average Monomer Interpenetration Depth, $X(t)$

The average monomer interpenetration depth of segments on the interdiffusing chains, $X(t)$, is obtained from an integration over the Gaussian segment density profiles of the interdiffused minor chain population,(2) as shown in Figure 3, such that

$$X(t) \approx t^{1/4} M^{-1/4} \tag{16}$$

$$X_\infty \approx M^{1/2} \tag{17}$$

The $t^{1/4}$ dependence is unique to the correlated motion of the chains for times less than the reptation time. This result can be understood by considering the Brownian motion and displacement s of an individual monomer along the contour of the tube as $s \approx t^{1/2}$ for $t < T_r$. However, the curvilinear motion is related to translational motion via $X \approx s^{1/2}$ because the chain is a random coil. For times greater than T_r, when the original tube has evaporated the motion of the monomer becomes uncorrelated since s is greater than the chain length and its displacement becomes similar to the center-of-mass displacement X_{cm},

$$X_{cm} \approx t^{1/2} M^{-1} \tag{18}$$

The crossover from $t^{1/4}$ to $t^{1/2}$ behavior has been observed by secondary ion mass spectroscopy,(23) which probes the concentration profile at a polymer interface for $t < T_r$ and distances $X < R_g$. Center-of-mass motion for $t > T_r$ and $X \gg R_g$ is well established from self-diffusion studies and found to be in agreement with Eq. (18). For chains in the bulk, the monomer motion also proceeds as $t^{1/4}$ for $t < T_r$ while the center of mass proceeds as $t^{1/2}$ at all times.

Equation (18) cannot be used to describe center-of-mass motion for chains whose random-coil configurations are altered by the interface as shown in Figure 4. The chains initially have a non-Gaussian configuration due to the reflecting boundary condition at the surface. As the minor chains interdiffuse, the configurations of the affected chains in the surface layer relax to Gaussian configurations. In Figure 4, the average shape and size of a chain at the interface (solid vertical line) before and after relaxation are shown. The distance η from the location of the highest segment density of the chain to the interface remains constant during the relaxation. Since the center of mass of each chain is close to the peak of the segment density profile, it can be seen that very little motion of the center

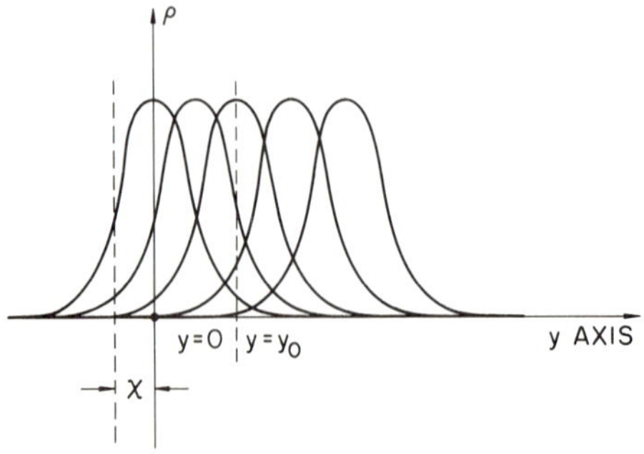

FIGURE 3. Segment density ρ of minor chains and the average interpenetration distance x. (From Refernece 2.)

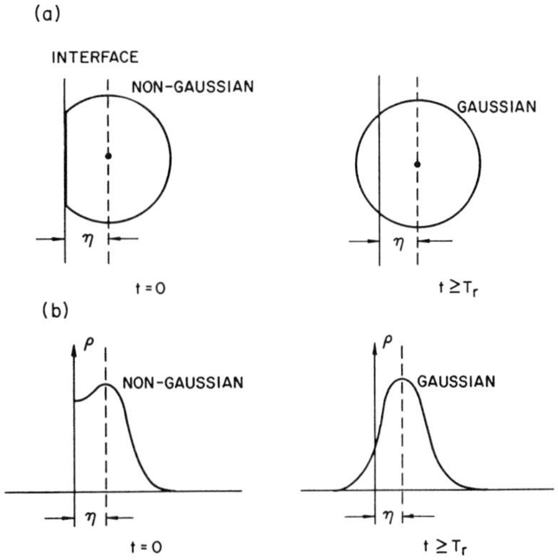

FIGURE 4. (a) Average shape and size of a chain at the interface before and after healing. (b) Segment density ρ of the chain in (a) before and after healing. (From Reference 2.)

of mass occurs for $t < T_r$. For a chain in the bulk, the center of mass is displaced by a distance approximately equal to its end-to-end vector **R** during the same relaxation time. Since $\mathbf{R} > \eta$, the "apparent center-of-mass self-diffusion coefficient" of chains affected by the interface would be smaller than that of chains further away from the interface in the bulk. However, Rouse relaxation of chain segments at $t << T_r$ give anomolously high diffusion coefficients.

The nonequilibrium configuration of the chains near the interface plays an important role in determining entanglement density and the structure of asymmetric incompatible amorphous interfaces. The reduced entropy of the confined chain configurations shown in Figure 4 acts as a driving force to achieve limited diffusion in the interface of immiscible polymers. The increase in entropy is counterbalanced by a positive enthalpy of mixing and results in an equilibrium thickness,[24] $d_\infty \approx \chi^{-1/2}$, where χ is the Flory–Huggins interaction parameter.[25] The short range diffusion (to about 50 Å) is dominated by Rouse motion of chain segments.[15]

2.2.5. Number of Monomers Crossing an Interface, $L_0(t)$

The number of monomers which have crossed from one side of the interface to the other is equivalent to the total interpenetration contour length, $L_0(t) = n(t)l(t)$. Summing over the segment density profiles of the interdiffused minor chains, we obtain[15]

$$L_0(t) \approx t^{3/4} M^{-7/4} \tag{19}$$

$$L_{0\infty} \approx M^{1/2} \tag{20}$$

The unusual $t^{3/4}$ dependence is again due to correlated motion effects and reverts to a normal $t^{1/2}$ diffusion dependence for times greater than T_r. This result can also be explored by experimental techniques discussed in Reference 23 where L_0 is determined from the integral of the concentration profile. Whitlow and Wool[23] observed the crossover in a PS/PS interface using SIMS analysis with high molecular weight (10^6) PS.

2.3. Concentration Profile

Integrating over the population of interdiffused minor-chain Gaussian segment density profiles shown in Figures 2 and 3, the normalized monomer concentration profile $\Phi(x,t)$ for the symmetric polymer–polymer interface was obtained by Zhang and Wool[15] for $t < T_r$ in the form

$$\Phi(x,t) = \{[l(t)/L + x^2/bL] \operatorname{erfc}[x/(b\sqrt{2n})] - \sqrt{(2n/\pi)}\,(x/L) \exp[-x^2/(2nb^2)]\} \tag{21}$$

where b is the bond length, n the number of monomers in the minor chain of length $l(t)$, x the diffusion depth such that $x = 0$ at the interface, and L is the contour length of the entire chain. The time dependence of the concentration profile in Eq. (21) derives from the time dependence of the minor chains via $l(t) = \sqrt{(16\,D_1\,t/\pi)}$.

Computed concentration profiles are shown in Figure 5 at fractional reptation times and diffusion depths normalized with respect to the radius of gyration of the Gaussian chains.[15] Figure 5 clearly shows that the concentration at the initially joined interface at $x = 0$ is discontinuous when the welding time is shorter than the reptation time. For normal Fickian diffusion, the atomic or monomer concentration profile at the origin immediately attains a value of 0.5 (assuming an initially sharp interface at $t = 0$) since the Brownian motion of the atoms is isotropic. For polymers, however, the majority of monomers in contact with the initial plane exhibit anisotropic motion due to their connectivity. They are not free to interdiffuse on the time scale of monomer motion but await the diffusion of the much smaller number of minor chain ends. Thus, $\Phi(x,t)$ contains a discontinuity at $x = 0$ which is an interesting peculiarity of the reptation model and persists for at least one reptation time. For $t > T_r$, the concentration profile returns to the classical Fickian profile given by

$$\Phi(x) = \operatorname{erfc}(x/L_d) \tag{22}$$

where L_d is the Einstein diffusion length given by $L_d = 2\sqrt{(Dt)}$. The non-Fickian characteristics of the concentration profile due to correlated motion effects in the reptation model have been discussed by de Gennes,[9] Tirrell *et al.*,[26] and Zhang and Wool.[15] Equation

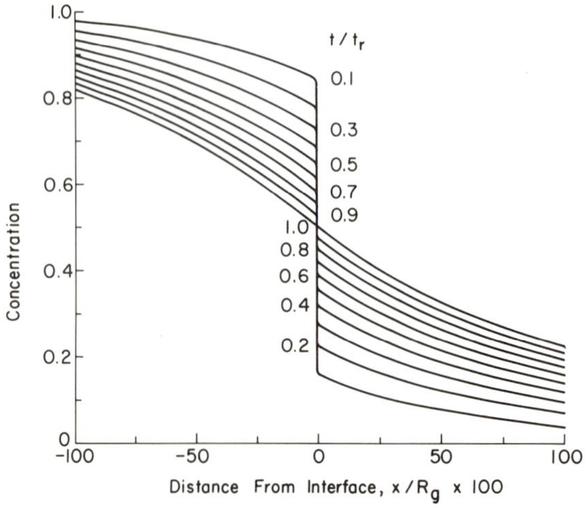

FIGURE 5. Concentration profile of a polymer–polymer interface at times t less than the reptation time T_r. The interdiffusion depth x is normalized with respect to the radius of gyration R_g. (From Reference 15.)

(21) has also been used to calculate the molecular properties of the interface presented in the last section and summarized in Table 1.[15]

2.4. Scaling Laws for a Polymer–Polymer Interface

The molecular properties described above and others listed in Table 1 have a convenient common scaling law which relates the dynamic properties $H(t)$ to the static properties H_∞ via the reduced time t/T_r by[4,15]

$$H(t) = H_\infty \, (t/T_r)^{r/4} \tag{23}$$

$$H_\infty \approx M^{(3r-s)/4} \tag{24}$$

where $r,s = 1,2,3,\ldots$

Table 1 summarizes the average dynamic and static molecular properties of polymer–polymer interface and lists values of r and s for individual properties. This table can be used to determine other properties and evaluate suggestions and theories relating interface structure to mechanical properties. Several of these properties have also been derived by de Gennes[9,27,28] and by Prager, Tirrell, and Adolf.[7,26,29]

If a mechanical property such as the fracture energy G is controlled by one of the properties $H(t)$, then the time dependence of welding is given by $G \approx t^{r/4}$ and the molecular-weight dependence by $M^{-s/4}$, where r and s are given in Table 1. Welding experiments of Wool *et al.* indicate that for the fracture energy, $r = 2$ and $s \approx 2$, which suggests that the interpenetration contour length $l(t)$ is controlling the strength development (see Table 1). However, considering the complexity of the fracture mechanics and deformation mechanisms, it may be too simplistic an argument to suggest that a single molecular property defined typically over a few hundred angstrom thickness of the interface is responsible for the energy dissipation and behavior of a much larger deformation zone at the crack tip.

Table 1.
Molecular Aspects of Interdiffusion at a Polymer–Polymer Interface

Molecular aspect	Symbol	Dynamic relation H(t)	Static relation H_∞	r	s
General property	$H(t)$	$t^{r/4}$ $M^{-s/4}$	$M^{(3r-s)/4}$	r	s
No. of chains	$n(t)$	$t^{1/4}$ $M^{-5/4}$	$M^{-1/2}$	1	5
No. of bridges	$p(t)$	$t^{1/2}$ $M^{-3/2}$	M^0	2	6
Average monomer depth	$X(t)$	$t^{1/4}$ $M^{-1/4}$	$M^{1/2}$	1	1
Total monomer depth	$X_o(t)$	$t^{1/2}$ $M^{-3/2}$	M^0	2	6
Average contour length	$l(t)$	$t^{1/2}$ $M^{-1/2}$	M	2	2
Total contour length, no. of monomers, N	$L_o(t)$	$t^{3/4}$ $M^{-7/4}$	$M^{1/2}$	3	7
Average bridge length	$l_p(t)$	$t^{1/4}$ $M^{-1/4}$	$M^{1/2}$	1	1
Center of mass depth[a]	X_{cm}	$t^{1/2}$ M^{-1}	$M^{1/2}$	2	4
Diffusion front length	N_f	$t^{1/2}$ $M^{-3/2}$	M^0	2	6

[a] This equation applies to chains in the bulk and does not apply to chains whose center of mass is within a radius of gyration of the surface.

3. FRACTAL NATURE OF A DIFFUSION FRONT

3.1. Atomic Diffusion Front

While the concentration profile due to diffusion varies smoothly with depth as shown in Figure 5, the diffusion field when viewed in two or three dimensions has a very rough nature. Consider the two-dimensional lattice diffusion of A atoms (shown in Figure 6) into B atoms (not shown) as discussed by Sapoval *et al.*[30–32] The concentration profile of A atoms is given by the Fickian result, $\Theta(x) = \mathrm{erfc}(x/L_d)$, where $L_d = 2(Dt)^{1/2}$ is the Einstein diffusion length at time t with diffusion coefficient D. The diffusion "front" (heavy line in Figure 6) is defined by those A atoms which are connected to the diffusion source at $x = 0$ via other A atoms and have an empty first or second B neighbor, which is itself connected to the source of B atoms. The interface with a diffusion front is seen to consist of solid regions with lakes or holes leading down to the "seashore" and unconnected islands of A atoms in the B "sea." The presence of holes and the connectedness between atoms in the interface is important in determining electrical and mechanical properties of the interface.

Sapoval *et al.* determined the frontier to have fractal properties involving self-similarity (scale invariance) and incompleteness of space filling (holes). The fractal geometry of the frontier is determined by the mass-to-radius relation[33]

$$M \approx R^{D_f} \tag{25}$$

where D_f is the fractal dimension of the diffusion front. The fractal geometry exists only over the width σ_f of the front as shown in Figure 6. In two dimensions, $D_f = 1.76$. The

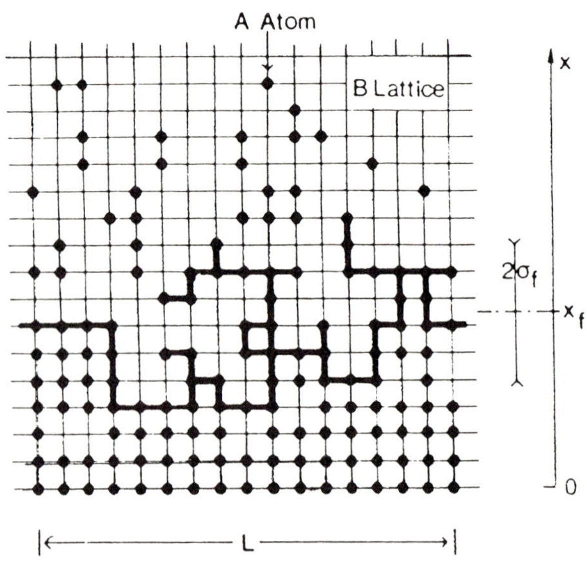

FIGURE 6. Schematic of a two-dimensional diffusion field at a given time. The heavy line represents the "diffusion front" determined by the leading edge of A atoms connected to their diffusion source at $x = 0$ and having at least one B neighbor connected to its source. The position of the center of the front occurs at X_f and the width of the front is $2\sigma_f$. (From Reference 30.)

number of particles per unit width of the front, N_f, increases with the diffusion length according to

$$N_f = 0.96\ L_d^{\alpha} \tag{26}$$

where the exponent α is given by

$$\alpha = (D_f - 1)\nu/(1+\nu) \tag{27}$$

in which $\nu = 4/3$ is the critical exponent for the percolation coherence length in two dimensions, resulting in $\alpha = 0.43$. The latter result is supported by computer simulation studies of Wool and Long.[34] The mean position of the frontier can also be shown to occur at a concentration corresponding to the percolation threshold, p_c.[30] The three-dimensional analysis of diffusion fronts is more complex.[31]

Equation (26) indicates how the interface "roughness" increases with diffusion length such that $N_f \approx L_d^{0.43}$. Thus, even though the average concentration profile varies smoothly, the diffusion front can have considerable lateral roughness as shown in Figure 6. Experimental examples of this behavior are described in Reference 34 based on studies of electrochemical reduction of metal ions diffusing in polymers by Mazur and Reich.[35] Figure 7 shows a polymer metal interface produced by electrochemically depositing silver atoms in a polyimide film. Atoms not in contact with the source (bottom of figure) were removed from the electron micrograph to reveal the highly ramified nature of the inter-

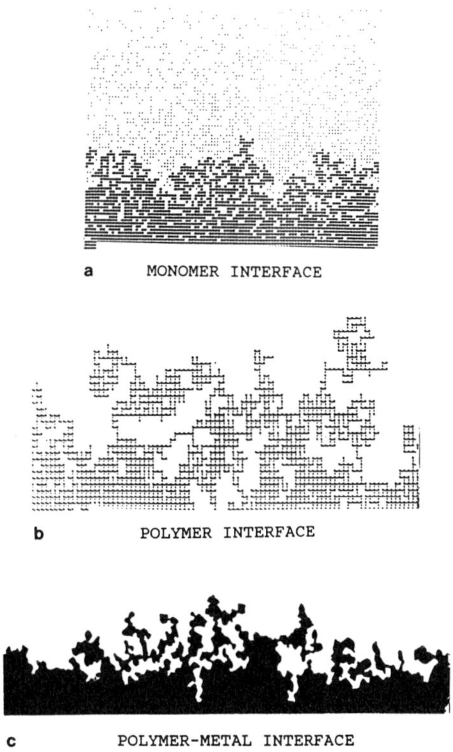

FIGURE 7. Computer simulation of (a) monomer interface, (b) polymer interface, and (c) the connected region of a silver/polyimide interface. (From References 34 and 36.)

face. The fractal dimension of the thinly sliced interface diffusion front was determined to be about 7/4, in agreement with Sapoval's theory. Details of this analysis are given elsewhere.[34,36] The ability to control the interface roughness and molecular connectivity has interesting implications for the mechanical and electrical property development of interfaces. Roughness promotes mechanical interlocking, but fractal characteristics with holes reduce electrical conductance and reduce signal speed in the metal layer by increasing the dielectric constant of the polymer.

3.2. Computer Analysis of Polymer–Polymer Interdiffusion

A computer simulation of reptating random-coil chains diffusing on a square lattice across an interface plane was conducted.[34] An IBM-AT microcomputer attached to a Cray XMP-48 was used to simulate the Brownian motion of the interpenetrating chains. The purpose of this study was to investigate the scaling laws derived for polymer interdiffusion shown in Table 1, to evaluate the non-Fickian concentration profiles for polymer diffusion at times less than the reptation time, and to examine fractal characteristics of polymer diffusion fronts.

Brownian motion was accomplished by a reptation algorithm which randomly chose a chain end and added a new monomer in one of three possible new directions and subtracted a monomer from the other end. The process was then repeated for each chain. The chain-end distribution function was uniform behind the interface plane. The basic reptation dynamic results were checked for consistency, namely, that (1) the relaxation time behaved as $T_r \approx M^3$, (2) the center-of-mass diffusion coefficient behaved as $D \approx M^{-2}$, and (3) that the monomer displacement behaved as $\{\langle r(t) - r(0)\rangle^2\} \approx t^{1/2}$. The chain density was chosen to be inversely proportional to the molecular weight and the monomer density in the virgin state was chosen to consist of an average of five monomers per lattice site due to interpenetration of random-coil chain statistical segments. At $t = 0$, the chains were allowed to diffuse across the interface plane and formed ramified structures as shown in Figure 7. The polymer interface is seen to consist of features resembling lakes, islands, and seashore similar to the previous atomic diffusion case. However, many of the holes near the $x = 0$ plane are not due to the fractal nature of diffusion but rather to the correlated motion effect producing discontinuities in the interface as shown in Figure 5. The computer simulation gave further support to the scaling laws shown in Table 1 and the concentration profile, Eq. (21). Fractal properties of polymer interfaces are discussed in the next section.

3.2.1. Fractal Polymer Diffusion Fronts

The analysis of the fractal nature of polymer diffusion fronts[34] is complicated by the interpenetrated nature of the random-coil chains allowing several statistical segments or monomers to occupy a single lattice site. Figure 7 shows typical diffusion fronts evolving with time. The diffusion front was obtained in the computer simulation using the same technique employed by Sapoval *et al.*[30] The front is potentially more ramified initially due to the connectivity of monomers within a given chain. Figure 8 shows the number of monomers on the diffusion front, N_m as a function of the average interdiffusion distance, $\langle X\rangle$, for random-coil chains of molecular weights ranging from 60 to 100 statistical units. Two regions of behavior are observed.

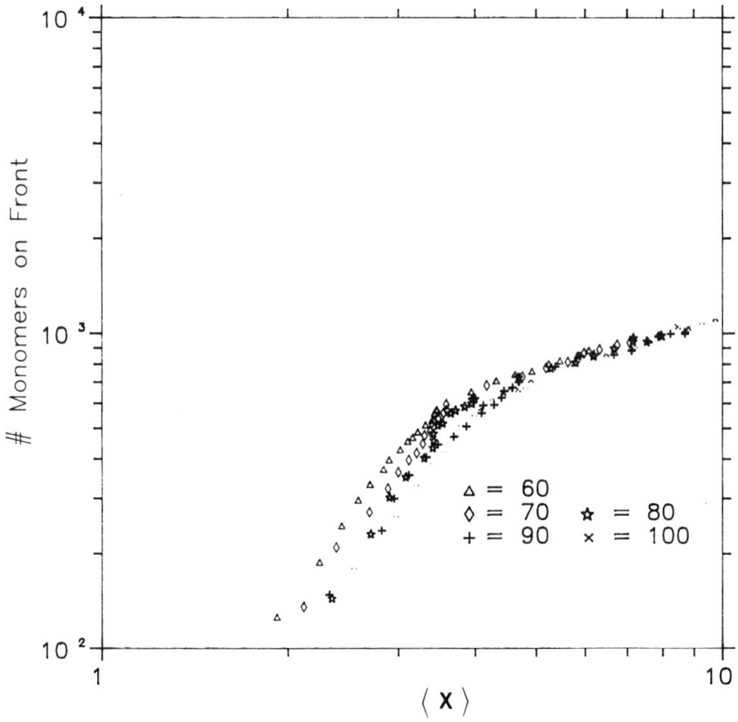

FIGURE 8. The number of monomers on a polymer interface diffusion front is plotted vs. the average monomer interpenetration depth $\langle x \rangle$ for molecular lengths ranging from 60 to 100 times the critical entanglement length. The reptation algorithm was used to simulate Brownian motion of linear flexible chains. (From References 34 and 36.)

For $t < T_r$, the front is discontinuous for reasons discussed above and can be approximated by summing contributions from single chains $\sqrt{l(t)}$ as

$$N_m \approx n(t)\sqrt{l(t)} \tag{28}$$

where $n(t)$ is the number of chains crossing unit area (or unit length) of the initial interface plane. From Table 1, it can be seen that $n(t)$ scales as $n(t) \approx X(t)/M$ where $X(t)$ is the average monomer diffusion distance (equivalent to $\langle X \rangle$ used in Figure 8). The contribution to the diffusion front from a single minor chain is given by the square root of the contour length which is proportional to $X(t)$. Substitution in Eq. (28) yields

$$N_m \approx X(t)^2/M \tag{29}$$

Figure 8 shows that the plot of log N_m vs. log $X(t)$ has a slope of 2 for all molecular weights and that at constant $X(t)$, N_m varies approximately as the inverse of M. However, for times less than the reptation time, the diffusion front is not considered to be fractal since the criterion of self-similarity is not obeyed due to discontinuities in the interface plane.

At $t = T_r$, since $X_\infty^2 \approx M$, $N_m(T_r)$ is independent of molecular weight. The time-dependent scaling law for N_m at $t < T_r$ is seen from Table 1 to be identical to that for bridges, $p(t)$.

When $t > T_r$, the number of monomers on the diffusion front behaves as

$$N_m \approx X^{0.43} \tag{30}$$

as shown by the convergence to a slope of 0.43 for all molecular weights in Figure 8. The number of lattice sites, N_L, on the polymer diffusion front was found to behave similarly to the number of monomers, N_m. A mass-to-radius analysis of the diffusion front geometry gave a fractal dimension of approximately 1.7. These results are in agreement with the atomic diffusion fractal analysis in two dimensions. Thus, the polymer diffusion front becomes fractal at times greater than the reptation time and at diffusion distances greater than the end-to-end vector of the chains. When we consider Rouse relaxation and segmental motion at $t << T_r$, the discontinuity or "gap" at $X = 0$ disappears and the polymer diffusion front becomes fractal even at very short times.[15]

4. MICROSTRUCTURAL FRACTURE CRITERIA

4.1. Chain Pullout

The Griffith approach to brittle fracture is given by[37]

$$(\delta U/\delta a)_\delta \geq 2\Gamma \tag{31}$$

where U is the stored strain energy consumed at constant displacement δ to advance a crack by the increment δa and create fracture surface energy $\Gamma\delta a$. The fracture energy, or the critical strain energy release rate, G_{Ic}, is related to Γ by

$$G_{Ic} = 2\Gamma \tag{32}$$

In principle, if one independently knows the fracture surface energy Γ of a solid, one should be able to predict its fracture energy using Eq. (32). However, for high-molecular-weight entangled polymers the experimental values are typically of the order $G_{Ic} \approx 1000$ J/m^2 and $\Gamma \approx 10^{-2}$ J/m^2. Closer agreement with the Griffith theory is obtained for brittle fracture of very low-molecular-weight unentangled polymers.[38,39]

The Griffith idea has considerable value if one can determine how stored energy is consumed in forming the fracture surface. To address welding and healing problems a stored strain energy approach to fracture was adopted which considers both chain pullout

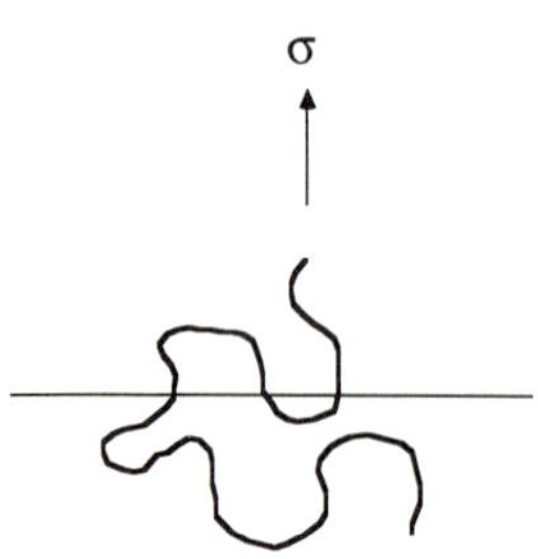

FIGURE 9. A segment of a stressed chain at an interface.

via disentanglement and chain fracture mechanisms at the interface. Let us consider a volume element containing a small area of interface as shown in Figure 9 subjected to normal stresses σ_{ij} and shear stresses τ_{ij}. The strain energy density U of the element is given by

$$\begin{aligned} U = &(\sigma_{xx}^2 + \sigma_{yy}^2 + \sigma_{zz}^2)/2E \\ &- (\sigma_{xx}^2\sigma_{yy}^2 + \sigma_{yy}^2\sigma_{zz}^2 + \sigma_{xx}^2\sigma_{zz}^2) \\ &\nu/E \\ &+ (\tau_{xy}^2 + \tau_{xz}^2 + \tau_{yz}^2)/2\mu \end{aligned} \tag{33}$$

where E, ν, and μ are Young's modulus, Poisson's ratio, and shear modulus, respectively.

We first consider a uniaxial stress field with $\sigma = \sigma_{xx}$ and all other stress terms equal to zero. In that case, the strain energy density is

$$U = \sigma^2/2E \tag{34}$$

We have shown theoretically and experimentally that the modulus behaves as[1,40]

$$E \approx t^0 M^0 \tag{35}$$

$$E_\infty \approx M^0 \tag{36}$$

such that E can be assumed to remain constant during healing and fracture.

The number of chains per unit volume, N_v, is given by

$$N_v = \rho\, N_a/M \tag{37}$$

where ρ is density and N_a is Avogadro's number. The strain energy per chain, U_c, is therefore obtained in the form

$$U_c = U/N_v = \sigma^2 M/(2E\, \rho\, N_a) \tag{38}$$

The strain energy associated with a segment of chain of length l is similarly derived as

$$U_c \approx \sigma^2 l \tag{39}$$

The stored elastic energy can be used to either pull the chain out from the interface or fracture the chain. A fracture criterion for chain pullout can be written as

$$U_c \geq U_p \tag{40}$$

where U_p is the energy to pull a chain out of its tube with a force f and velocity dl/dt.

The pullout force is given by $f = \mu\, dl/dt$, where μ is the friction coefficient for the chain segment of length l. The friction coefficient for the segment is related to the monomer friction coefficient μ_0 via $\mu = \mu_0 l$ such that the force may be written as

$$f = \mu_0\, l\, dl/dt \tag{41}$$

The energy required to pull a segment out at constant velocity is determined from

$$U_p = \mu_0 \int_{l=0}^{l=L} l(dl/dt)\, dl \tag{42}$$

as[41]

$$U_p = \mu_0 \, (dl/dt) \, l^2/2 \tag{43}$$

or

$$U_p \approx l^2$$

A similar result is obtained if the chain is pulled out at constant force rather than at constant velocity V. Equation (43) has also been derived by Prentice[42] and Evans.[43]

Substituting Eqs. (39) and (43) in Eq. (40) yields the critical fracture stress for chain disentanglement (and tack) in the form

$$\sigma \approx (\mu_0 V \, E \, b \, \rho \, N_a/M_0)^{1/2} \, l(t)^{1/2} \tag{44}$$

or

$$\sigma \approx X(t) \tag{45}$$

where $X(t) \sim \sqrt{l(t)}$ is the average monomer segment interpenetration distance, M_0 is the monomer molecular weight, and b is the bond length. Thus from Table 1, the fracture stress or tack in uniaxial tension should behave as

$$\sigma \approx t^{1/4} \, M^{-1/4} \tag{46}$$

and the virgin (green) strength as

$$\sigma_\infty \approx M^{1/2} \tag{47}$$

The velocity dependence, $V = dl/dt$, of tack and green strength during peel adhesion tests from Eq. (44) should obey

$$\sigma \approx V^{1/2} \tag{48}$$

Fracture by the chain pullout mechanism is favored at low molecular weight [small $l(t)$], low deformation rates (small V), high temperature (low μ_0), short healing times [small $l(t)$], and plasticizing agents (low μ_0).

4.2. Fracture Mechanics of Welding

The Dugdale model[44] was used as a first approximation to determine the molecular aspects of the fracture mechanics. In this model, shown in Figure 10, a crack of initial length a_0 propagates through a plastic line zone or craze of length r_p, ahead of the crack tip. The plane stress field near the crack for a Cartesian volume element xyz and angle α at distance r is given by Rice[45] as

$$\begin{aligned} \sigma_{xx} &= K_1/\sqrt{2r} \cos \alpha/2 \, (1\text{-}\sin \alpha/2 \, \sin 3\alpha/2) + \ldots \\ \sigma_{yy} &= K_1/\sqrt{2r} \cos \alpha/2 \, (1+\sin \alpha/2 \, \sin 3\alpha/2) + \ldots \\ \sigma_{xy} &= K_1/\sqrt{2r} \cos \alpha/2 + \ldots \end{aligned} \tag{49}$$

where K_I is the stress intensity factor. The length of the plastic zone is determined by

$$r_p = (\pi/8) \, K_I^2/\sigma_c^2 \tag{50}$$

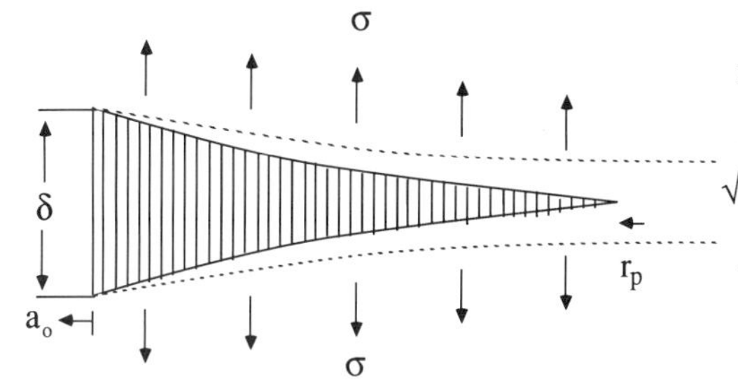

FIGURE 10. The Dugdale model for a crack propagating through a welded interface of thickness $X \sim l^{1/2}$. The stress σ acting on the plastic zone of length r_p draws the interface into craze-like fibrils of constant draw ratio λ which break down to produce the crack-opening displacement δ. (From References 40 and 41.)

where σ_c is the critical stress to initiate fracture.

The critical crack-opening displacement is given by

$$\delta = K_I^2/\sigma_c E \tag{51}$$

where E is the tensile modulus corrected for plane stress or plane strain.

Using the approximation that σ_c remains constant within the plastic zone, the critical strain energy release rate G_{Ic} is determined by

$$G_{Ic} = \sigma_c \delta \tag{52}$$

such that the fracture energy is determined as a critical force acting over a critical distance.

Figure 11 shows the essential features of the Dugdale model for a crack propagating through a line deformation zone in styrene–isoprene–styrene (SIS) block copolymer.[46] The microstructure elucidated by transmission electron microscopy (TEM) consists of styrene spheres (white) with a diameter of ca. 250 Å in an isoprene matrix (stained black with osmium tetroxide). For the craze zone, $r_p \approx 3$ μm and $\delta \approx 0.2$ μm. The craze propagates in the stress field applied normal to the craze by incorporating a single styrene sphere into the tip of the zone. As the zone propagates it thickens from 250 Å to 2000 Å by drawing further material into the zone and elongates by extending the material already in the zone. At the crack tip, many styrene spheres have been drawn into the zone and deformed nonuniformly such that the crack tends to propagate through the center of the zone. Similar mechanisms have been identified for crazes propagating through pure amorphous polystyrene.[47,48] Thus, a critical stress nucleates the line zone at distance r_p from the crack tip and the crack-opening displacement is determined by the breakdown of the material drawn into the zone.

To evaluate the molecular aspects of the welding problem we first calculate σ_c using the strain energy approach such that the stored energy per molecule in an xy-plane stress field is

$$U_c = (\sigma_{xx}^2/2E + \sigma_{yy}^2/2E + \sigma_{xy}^2/2\mu)/N_v \tag{53}$$

On substituting for the stress terms from Eq. (49) in Eq. (53), U_c is found to be dominated by the σ_{yy} term and can be well approximated by

$$U_c \approx \sigma_{yy}^2\ l(t) \tag{54}$$

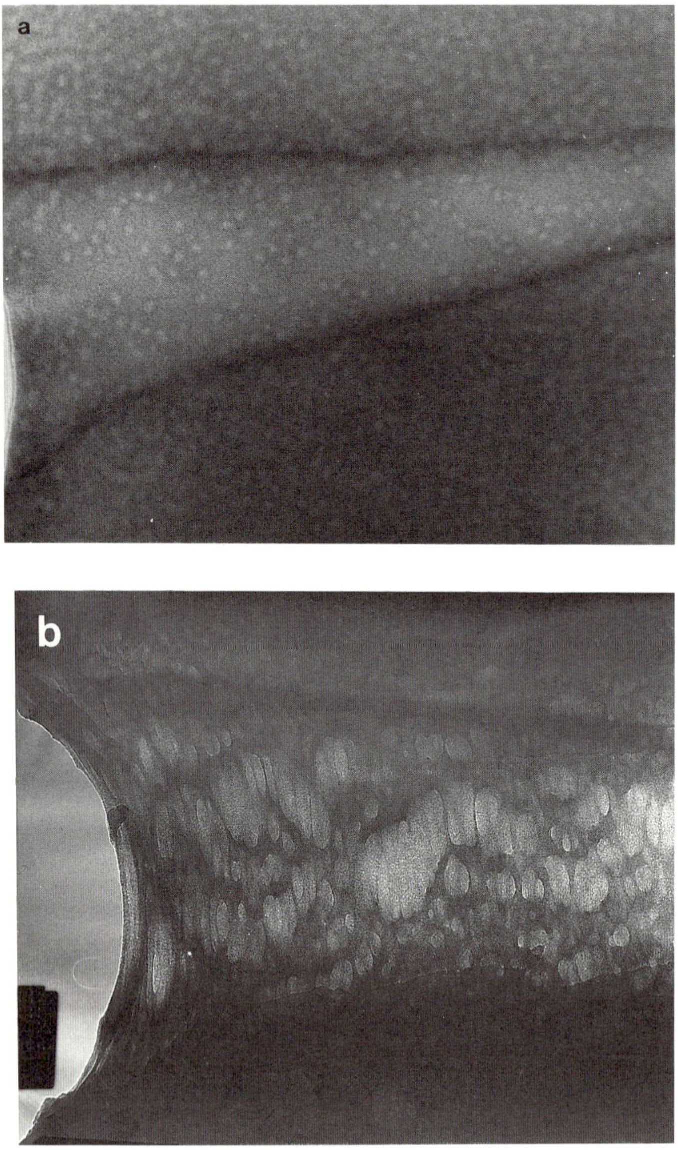

FIGURE 11. (a) Electron micrograph of a plastic zone in a mechanically stressed styrene–isoprene–styrene (SIS) thin film. The white spheres (styrene) have a diameter of about 250 Å. The number of spheres in the zone increases toward the crack tip due to zone thickening. (b) A lower magnification of (a) above shows the crack tip (left) and plastic deformation zone (craze) extending into the undeformed SIS. 1 cm = 770 Å. (From Reference 46.)

where $l(t)$ is again the average interpenetration length. Using the same criterion for chain pullout, Eq. (43), and letting $\sigma_{yy} = \sigma_c$ at critical conditions, we obtain

$$\sigma_c \approx l(t)^{1/2} \tag{55}$$

which is similar to the result obtained for the uniaxial case. For glassy polymers, σ_c can be related to the craze stress which differs from the shear yielding stress σ_Y.

The critical crack-opening displacement δ is determined by the breakdown of the entanglement network in the deformation zone and the extent to which the zone can thicken, as depicted in Figure 10. During welding, the zone consists of interdiffused short-chain segments mixed with longer chains possessing nonequilibrium configurations, and the entangled structure is complex. The crack-opening displacement consists of two components: a component from the virgin material, δ_z, and a component from the diffuse interface, δ_i, such that $\delta = \delta_z + \delta_i$. At low welding strengths the contribution from the weaker interface dominates, since the energy required to thicken and propagate the zone into the virgin material is too high. The crack-opening displacement is then determined by the breakdown of the entanglement network in the interface by the following approximation[49]:

$$\delta \approx \lambda_f l(t)^{1/2} \tag{56}$$

where λ_f is the constant draw ratio to tighten the slack between entanglements. Entanglement models[49,50] predict that $\lambda_f \approx 4$, which is in agreement with TEM studies of craze fibrils by Donald and Kramer.[50] The thickening velocity v_i of the craze zone in virgin glassy polymers has been studied by Kramer *et al.*[51] as a function of stress and molecular weight. For a constant stress acting normal to the craze for a time t, they find for polystyrene that $v_i \sim M^{-5/2}$ which suggests $\delta_z \sim M^{-5/2}\, t\lambda_f$. The breakdown of the craze by disentanglement and bond rupture is complex and depends on the rate of loading and temperature.

In the virgin state, the critical draw ratio for disentanglement, λ_c, is given (with the approximation $\lambda^2 \gg 1/\lambda$) by[49]

$$\lambda_c \simeq (2M/M_c)^{1/2} \tag{57}$$

where M_c is the critical entanglement molecular weight. If further disentanglement cannot occur following strain hardening at $\lambda_f = 4$, then the maximum molecular weight M^* for disentanglement is given by

$$M^* \approx 8M_c \tag{58}$$

Fracture by disentanglement of a welded interface is considered to occur in the following steps: (1) The entangled chains in the interface of thickness $X(t)$ are stretched to a constant draw ratio, $\lambda_f \approx 4$, in the deformation zone, as shown in Figure 10. The stretch is accomplished by taking up the "slack" between entanglements but produces little initial loss of the entanglement density at times less than the Rouse relaxation time T_{RO}, of the chains. (2) The zone thickens to consume the interface and incorporate all the minor chain segments into the stretched deformation zone. The zone consists of the shorter minor chains intermixed with the longer virgin chains. (3) Rouse motion of the stretched segments at $\lambda = 4$ causes retraction of the minor chains along their contour length and a loss of entanglements ensues. (4) The loss of entanglements occurs both perpendicular to the draw direction to create a partially fibrillated structure and parallel to the fibril axis to eventually produce fracture. (5) If the draw ratio of the minor chains in the fibrils λ_f, is greater than the critical draw ratio, $\lambda_c = (2l(t)/l_c)^{1/2}$, to produce a critically connected entangled network [similar to Eq. (57)], then the fibrils in the deformation zone will fracture by loss of minor chain entanglements at the Rouse relaxation time. (6) Fracture occurs by a percolating instability which develops when about half the entanglements in the stretched network are lost.

If fracture does not occur at T_{RO}, the deformation zone may thicken further into the virgin material at depths past the original interface thickness $X(t)$, and the crack opening displacement will be larger. Chain rupture will speed the disentanglement process by lowering the molecular weight and λ_c.

Substitution of Eqs. (55) and (56) in Eq. (52) enables the critical strain energy release rate to be determined as

$$G_{Ic} \approx t^{1/2} M^{-1/2} \tag{59}$$

$$G_{\infty} \approx M \qquad (M < M^*) \tag{60}$$

The critical stress intensity factor K_{Ic} is related to the fracture energy via

$$G_{Ic} = K_{Ic}^2/E \tag{61}$$

and is expected to increase with welding time as

$$K_{Ic} \approx t^{1/4} M^{-1/4} \tag{62}$$

$$K_{Ic\infty} \approx M^{1/2} \tag{63}$$

The length of the plastic zone in the Dugdale model from Eq. (50) should behave as

$$r_p \approx t^0 M^0 \tag{64}$$

since K_{Ic} and σ_c cancel in Eq. (50).

If the chains in the deformation zone cannot disentangle completely and bond rupture occurs, then the scaling laws become more complicated. For welding of glassy polymers, fracture involves a mixture of disentanglement and bond rupture mechanisms. De Gennes[(27)] has suggested that molecular bridges could control the strength, which is similar to the crossing density suggested by Prager and Tirrell,[(7)] and which is similar to the total number of entanglements per chain per unit area suggested by Kausch.[(52)] In that case, the fracture energy is given as

$$G_{Ic} \approx t^{1/2} M^{-3/2} \tag{65}$$

$$G_{Ic\infty} \approx M^0 \tag{66}$$

From the latter relations further deductions on plastic zone lengths, crack-opening displacements, etc., cannot be inferred directly. As will be seen later, bond rupture contributes about 0.01% to the fracture energy of glassy polymers and, even at very low temperatures, stress-activated plastic and viscoelastic processes are the dominant contribution to the fracture energy. The above relations for welding and fracture will be explored in the experimental section.

4.3. Stages of Healing

Healing of polymer interfaces has been described by Wool and O'Connor in terms of several stages involving[1] (a) surface rearrangement, (b) surface approach, (c) wetting, (d) diffusion, and (e) randomization. Most of the above discussion focused on the influence of diffusion on mechanical properties. However, several other factors can affect the welding process and are discussed below as the "stages of healing."

(a) Before the surfaces are contacted, one should consider the roughness or topography of the surface and how it changes with time, temperature, and pressure following contact. In fractured polymers, rearrangement of fibrillar morphology, etc., can affect the rate of crack healing. Chain end distributions near the surface can change as molecules diffuse back into the bulk polymer. Spatial changes of the molecular-weight distribution may also occur, e.g., where the low-molecular-weight species migrate to the surface. Chemical reactions, such as oxidation and crosslinking, can occur on the surface and complicate the molecular dynamics of diffusion. Many other processes, e.g., molecular orientation changes, can contribute to the stage of surface rearrangement. Each material and experimental technique usually possesses unique surface rearrangement processes which may need to be quantified. O'Connor[53] and Wool[21] found significant effects of surface rearrangement on the healing rate of lightly crosslinked polybutadiene such that the fracture energy at constant healing time decreased asymptotically with increasing surface rearrangement time (prior to contact). This was interpreted as the chain ends diffusing back into the bulk such that they were not as readily available for diffusion during healing.

(b) Surface approach considers time-dependent contact of the different parts of the surfaces to create the interface. For example, in crack,[1,18] craze,[54,55] and void[56,57] healing, contact may be achieved at different locations at different times in the interface depending on the closure mode.[1] For example, slow closure of a double cantilever beam crack would result in different extents of healing along the closed crack. This stage typically contributes as a boundary value problem to the other stages of wetting and diffusion.

(c) Wetting can occur in a time-dependent fashion at the interface. The dynamics of wetting are discussed in detail in Chapter 5 by Professor de Gennes and in several other chapters in this book. For our purposes we provide a brief phenomenological description of wetting to illustrate potential problems in evaluating the time dependence of welding. Figure 12 shows a schematic region of the plane of contact of a polymer interface.[1] Due to surface roughness, etc., good contact and wetting are not achieved instantaneously at all locations. Typically, wetted "pools" are nucleated at random locations at the interface and propagate radially until coalescence and complete wetting are obtained. This problem has been treated phenomenologically as a two-dimensional nucleation and growth process such that the fractional wetted area, $\phi(t)$, is given as[1]

$$\phi(t) = 1 - e^{-kt^m} \tag{67}$$

where k and m are constants depending on the nucleation function and radial spreading rates.

(d) The diffusion stage has been discussed above with respect to the instantaneous

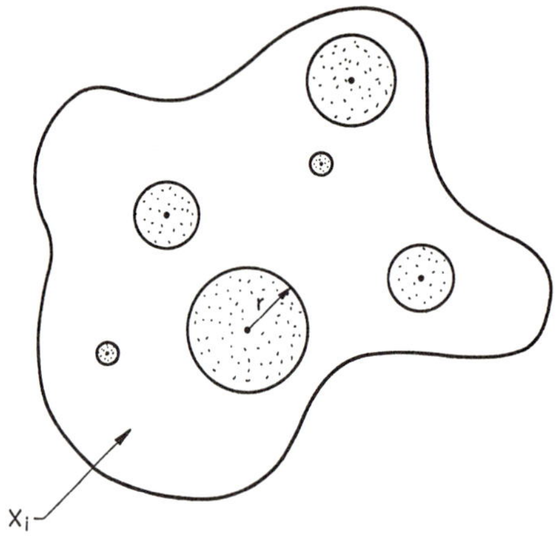

FIGURE 12. A schematic view of the stages of healing in an interface. Contacted regions of radius r grow and coalesce. Interdiffusion occurs only in the wetted regions while surface rearrangement occurs in the nonwetted areas. (From Reference 1.)

wetting condition. However, in the presence of a time-dependent wetting function, Eq. (67), we see from Figure 12 that diffusion will have progressed to different extents in different areas of the interface. If the intrinsic diffusion function $H(t)$ as given by Eq. (23) does not change its nature with time due to the other stages, then the net diffusion $H'(t)$ can be expressed as the convolution product

$$H'(t) = \int_{-\infty}^{\tau} H(t - \tau) \frac{\delta\Phi(\tau)}{\delta\tau} \delta\tau \tag{68}$$

where τ is a "dummy" variable on the time axis. The net result of the convolution process for typical $\Phi(t)$ functions is to mask the time dependence of the intrinsic diffusion function and related mechanical properties. This part of the problem can be solved mathematically by letting the wetting function be a Dirac delta function,

$$\frac{d\Phi(t)}{dt} = \delta(\tau) \tag{69}$$

such that Eq. (68) reduces to $H'(t) = H(t)$. This can be attempted experimentally by obtaining instantaneous wetting of atomically smooth surfaces under moderate contact pressure at temperatures above T_g.

(e) The randomization stage refers to the equilibriation of the nonequilibrium conformations of the chains near the surfaces and, in the case of crack healing and processing, the restoration of the molecular-weight distribution and random orientation of chain segments near the interface. The conformational relaxation is of particular importance in the strength development at incompatible interfaces.

The stages of crack healing can have interactive time-dependent functions such that the welding and tack problems can consist of processes involving five-way interconvoluted functions resulting in mechanical properties whose time dependence may not be readily interpretable. Thus, great care must be taken in designing welding experiments which attempt to critically address molecular theories of strength development.

5. HEALING EXPERIMENTS

5.1. Tack and Green Strength

In this section we examine interface structure and strength relations for tack and green strength. Several reviews are available.[11,58–60] The critical experiment is to show that the tack or stress at fracture of uncured linear elastomers behaves as $\sigma \approx t^{1/4}M^{-1/4}V^{1/2}$ and that the green or virgin strength behaves as $\sigma_\infty \approx (MV)^{1/2}$. In addition, the self-diffusion coefficient should be measurable as $D = A/M^2$ where A is a constant derivable from mechanical data. The concurrence of predictions derives from the molecular dynamics analysis of tack and green strength. These predictions apply to all amorphous linear polymers whose chain configurations can be described by Gaussian statistics and whose fracture behavior is dominated by chain disentanglement mechanisms.

Figure 13 shows tack data obtained by Skewis[61] for several polymer–polymer pairs, where we have replotted the tack (units of force applied to a cross-sectional area of 0.4 cm^2) vs. $t^{1/4}$. The original data were nonlinear when plotted on a $t^{1/2}$ scale to investigate diffusion mechanisms. However, the fracture energy behaves like $G = t^{1/2}$ while the tack behaves like $\sigma \approx t^{1/4}$, as evidenced by the linear plot in Figure 13. The SBR/butyl pair are incompatible and the long-time equilibrium strength is considerably weaker than either of the symmetric pairs. Similar $t^{1/4}$ results for tack have been reported by Wool and O'Connor[1,40] for polybutadiene, and by Voyutskii[11] for polyisobutylene.

The molecular-weight dependence of tack and green strength was investigated by Forbes and McLeod[62] for fractionated samples of Hevea. Their results are shown in Figure 14 in which the upper line represents the green strength and the bottom line represents the tack stress evaluated at a constant contact time of $t_c = 30$ sec for each molecular weight. The green strength curve was obtained theoretically using[1]

$$\sigma_\infty = k_1(M^{1/2} - M_i^{1/2}) \tag{70}$$

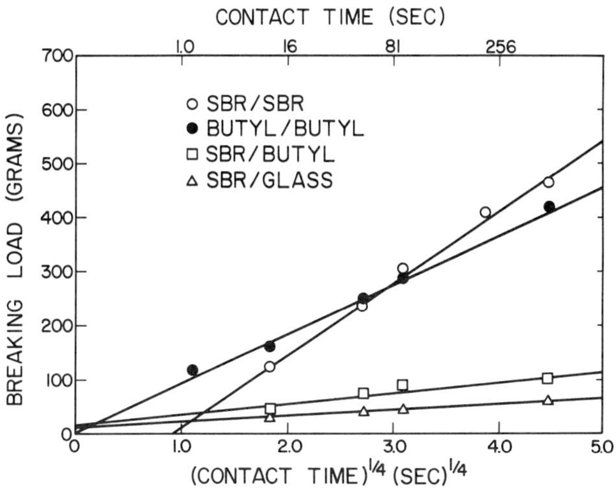

FIGURE 13. The tack or breaking load vs. $t^{1/4}$ for several polymer–polymer uncured pairs as investigated by Skewis.[61] Tack measurements were done with a contact load of 100 g and a contact area of 0.40 cm^2. The viscosity-average molecular weight M_v of the SBR was 260,000 and 225,000 for the butyl rubber.

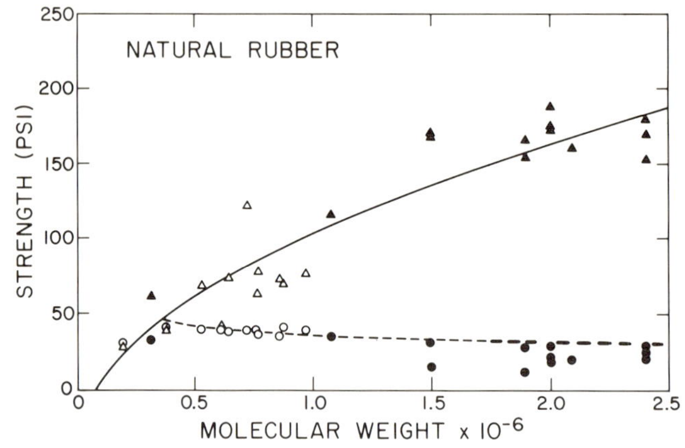

FIGURE 14. Tack (circles) and green strength (triangles) as a function of viscosity-average molecular weight for fractionated samples of natural rubber.[62] The solid line for green strength was obtained theoretically using Eq. (70) and the dashed line for tack using Eq. (71). The green strength was evaluated at a uniaxial test speed of 26.67 cm/min at 25 °C, and the tack was evaluated at a constant contact time of 30 sec for each sample.

where $k_1 = 0.144$ psi/$M^{1/2}$ and $M_i^{1/2} = 284.1$ is the intercept on the $\sigma_\infty = 0$ axis of the plot of σ_∞ vs. $M^{1/2}$. The curve through the tack data in Figure 14 was determined by theory as

$$\sigma = \sigma_0 + \alpha M^{-1/4} \tag{71}$$

where $\alpha = 767.2$ psi$\cdot M^{1/4}$ and $\sigma_0 = 11.68$ psi is a constant dependent on the initial wetting conditions of the interface. When $\sigma = k_2 t^{1/4}$ as in Figure 13, the constant k_2 is related to the constant α in Eq. (71) by

$$k_2 = \alpha/(M^{1/4}\, t_c^{1/4}) \tag{72}$$

where in this case $t_c = 30$ sec. Considering the difficulties in performing such experiments, the agreement between theory and experiment in Figure 14 is satisfactory. If the strength is controlled by bridges, then $\sigma \approx M^{-3/4}$ at constant t_c and $\sigma_\infty \approx M^0$. These predictions are inconsistent with the data in Figure 14. It has been suggested that the tack stress is independent of molecular weight, but in fact it behaves approximately as $M^{-1/4}$ in close agreement with disentanglement theories. This is a weak dependence on M, but related mechanical properties such as fracture work G should behave with a stronger molecular-weight dependence, as $G \approx t^{1/2} M^{-1/2}$.

In Figure 14 the green strength is seen to rise steadily up to $M \approx 1.5\times10^6$, at which point it remains fairly constant and bond rupture is expected to dominate the fracture mechanism by regulating the extent of disentanglement. Given sufficient time, the tack stress will increase to the green strength value by continued diffusion. Using tracer diffusion methods, Skewis[63] measured the self-diffusion coefficient of natural rubber at 25 °C as $D = 3\times10^{-14}$ cm^2/sec using samples with a viscosity-average molecular weight of $M_v = 234{,}000$. The polydispersity of his samples was not known. The self-diffusion coefficient from healing data is given by[1–3]

$$D = (k_2/k_1)^4\, b^2\, C_\infty/6MM_0 \tag{73}$$

where C_∞ is the characteristic ratio for the dimensions of a random coil chain. Letting $M = 234{,}000$, $M_0 = 68$ for polyisoprene, $b = 4.05$ Å from the c-axis of the *cis*-1,4-polyisoprene unit cell, and $k_2 = 14.9$ psi/sec$^{1/4}$ from Eq. (72), we obtain $D \approx 10^{-14}$ cm^2/sec which is of the same order of magnitude as the value obtained by Skewis.

The reptation time T_r is obtained from the self-diffusion coefficient in the form

$$T_r = \mathbf{R}^2/(3\pi^2 D) \tag{74}$$

where $\mathbf{R}^2$ is the end-to-end vector given by $\mathbf{R}^2 = M/M_0\, C_\infty\, b^2$. For natural rubber the healing time is approximated by

$$T_r = 2.48\times 10^{-15}\, M^3 \text{ sec} \tag{75}$$

When $M = 234{,}000$, $T_r = 32$ sec and, in Figure 14, this material should be in the green state at $t_c = 30$ sec as observed. The healing time required for the samples with $M = 1{,}000{,}000$ is about 40 min. These results differ with those of Roland and Bohm,[64] who found that the time required to achieve maximum autoadhesive strength of polybutadiene was orders of magnitude longer than the time required for the chains to diffuse a distance approximately equal to the end-to-end vector, **R**. Chain branching and nonuniform wetting at the interface may have contributed to this difference.

Pressure affects diffusion by decreasing the free volume, which slows the hopping process necessary to achieve Brownian motion and subsequently decreases the self-diffusion coefficient. This effect is more pronounced near the glass transition temperature and is important in polymer melt processing, where large hydrostatic pressures can be encountered. It is not very important in normal tack experiments where the contact pressures are usually much less than a kilobar. Thus, tack measurements should be dependent on pressure to some degree at short contact times, but should be largely independent of pressure at longer contact times when interdiffusion is occurring. Expressions for the pressure dependence of T_r and D have been given.[1,3]

The effect of strain rate and peeling velocity on tack has been examined by many investigators. Figure 15 shows data obtained by Hamed and Shieh[65] for the cohesive strength of an SBR elastomer (FRS-146) as a function of peel rate using a T-peel test configuration. The prediction that $\sigma \approx V^{1/2}$ is observed to be in excellent agreement for the high-temperature data (where chain pullout is expected to dominate) as judged by the slope of 1/2 for the plot of log σ vs. log rate. As the temperature approaches $T_g = -50$ °C, chain fracture begins to contribute to the deformation mechanism and the $V^{1/2}$ behavior is lost as predicted. The prediction that $\sigma \approx V^{1/2}$ is also supported by data of Tsuji *et al.*[66] for T-peel tests of polyisobutylene and by data of Bhowmick *et al.*[67] for adhesive tack and green strength of EPDM rubber. The dependence of the fracture energy G on the rate of peel is also $G \approx V^{1/2}$, since the extent of disentanglement and not the rate determines the crack-opening displacement for the chain pullout mechanism.

For peel adhesion, the pullout mechanism predicts the fracture energy to behave as

$$G \approx (t\ V/M)^{1/2} \tag{76}$$

$$G_\infty \approx M\ V^{1/2} \tag{77}$$

Several other theories of tack and adhesion have been presented. Bister *et al.*[68] predict that the peel energy behaves as

$$G \approx M^{2/3}\, (Dt)^{1/2} \tag{78}$$

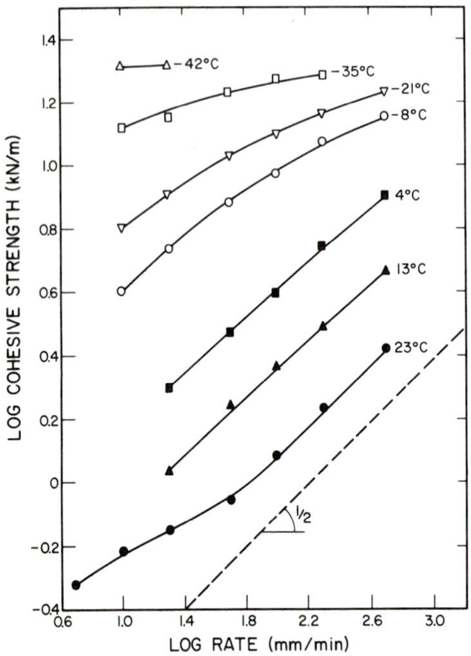

FIGURE 15. Cohesive tear strength of SBR as a function of peeling rate at several constant temperatures.[65] The dashed line represents the theoretical slope of 1/2 and was added for comparison purposes. The SBR was a cold emulsion styrene–butadiene random copolymer containing 23% bound styrene and had a Mooney viscosity (ML-4, 100 °C) of 40.

Since $D \approx M^{-2}$ and $T_r \approx M^3$, then Eq. (78) predicts that $G \approx t^{1/2} M^{-5/3}$ and $G_\infty \approx M^{-1/6}$. However, this molecular-weight dependence cannot be supported by data reported here or elsewhere. Most theories predict the $t^{1/2}$ dependence but, as seen from Table 1, this is not a unique function.

Wu *et al.*[69] predict that the adhesive strength is determined by $G \approx n(t)X(t)$ which behaves in an identical manner to bridges (see Table 1) as $G \approx t^{1/2} M^{-3/2}$ and $G_\infty \approx M^0$. The latter relations are not supported by experiments. However, Wu *et al.*[69] also suggest that the number of chains, $n(t)$, saturates quickly and that G depends on center-of-mass motion over larger distances as $G \approx t^{1/2} M^{-1}$. We disagree with the latter suggestion since $n(t)$ increases simultaneously with the strength for times $t < T_r$.

Vasenin's[70] kinetic theory of adhesion suggests that the tack stress behaves as

$$\sigma \approx t^{1/4} M^{-5/3} V \tag{79}$$

While the time dependence is supported by experiment, the molecular weight and rate dependences are at variance with the body of data. Many other theories of tack and adhesion are discussed in References 10 and 71. Kausch[18] also presents a review of crack healing studies.

5.1.1. Summary of Tack and Green Strength

Strength development at a polymer–polymer interface was analyzed in terms of the dynamic and static properties of random coil chains. Interdiffusion of chain segments across the interface was considered to be the controlling factor for tack and green strength of uncured linear elastomers. This concept is similar to that proposed by Voyutskii but differs markedly from contact theories as proposed by Anand.[72] In our approach, time-dependent wetting first occurs followed by interdiffusion. Increasing contact pressure and temperature should promote the establishment of molecular contact (wetting) at the interface up to a saturation point of complete wetting. However, interdiffusion is retarded by increased hydrostatic pressure and enhanced by temperature in the usual thermally activated manner. The effect of pressure on diffusion is to reduce the volume available (depending on the compressibility) for segmental motion and subsequently decrease the diffusion coefficient.

Increasing the tack test temperature increases the average interdiffusion chain segment length, as $l(T) \approx \exp(-Q/2kT)$, but decreases the stress required to pull the segment out, as $\sigma \approx \exp(Q/4kT)$, where Q is the activation energy. Therefore, the tack evaluated at constant contact time will decrease with increasing temperature for the interdiffusion controlled process.

The effect of molecular weight on tack at constant contact time t_c is to increase the tack according to $\sigma \approx M^{1/2}$ for those samples whose relaxation time $T_r < t_c$, and decrease the tack as $\sigma \approx M^{-1/4}$ for those molecular weights for which $T_r > t_c$. The tack should reach a maximum at a molecular weight corresponding to $T_r = t_c$. The latter value can be used to determine the self-diffusion coefficient of the chains.[1] At small contact times the results might be complicated by wetting processes. The position of the maximum in a plot of σ vs. M is relatively insensitive to the contact time, since $T_r \approx M^3$ and M at the maximum will consequently increase as $t_c^{1/3}$. These predictions appear to be in agreement with much experimental data reviewed by Hamed[58] and Rhee and Andries[59] but differ in some respects from their own interpretations of the same data.

The above results are not unique to elastomers and many of the predictions will be seen to describe welding and fracture of glassy polymers in the next section.

6. WELDING OF POLYMER INTERFACES

6.1. Wedge Cleavage Experiments

6.1.1. Symmetric Amorphous Interfaces

Figure 15 shows the welding method used to evaluate healing relations for polymer interfaces.[73,74] Monodisperse molecular weight films (or other materials) are bonded to substrates, usually composed of the same polydisperse molecular weight material, and the film surfaces are then welded together. The film surfaces to be contacted are first molded against highly polished metal plates and annealed in vacuum to assure maximum smoothness. To approximate the instantaneous wetting condition necessary to evaluate the effect of diffusion on welding, the surfaces are wetted at a higher temperature for a short time $t \ll T_r$, and subsequently healed for long times at a constant lower temperature. Under these conditions, the long-time healing data extrapolates to zero time.

The stress distribution and the fracture mechanics of the wedge cleavage method have been examined.[75,76] The wedge is driven into the interface, spreading the arms apart until the crack of length a begins to grow at a critical value of d as shown in Figure 16. By modeling the specimen as an elastic double cantilever beam with each beam partly

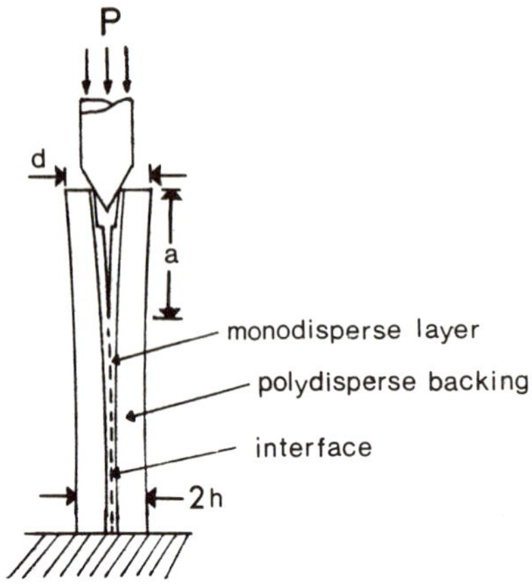

FIGURE 16. Wedge cleavage method of testing strength vs. welding time in polymers. Monodisperse molecular-weight films are attached to polymer substrates prior to thermal welding. (From Reference 1.)

supported by an elastic foundation, the critical stress intensity factor K_{Ic} is obtained at constant displacement as[75,76]

$$K_{Ic} = \frac{3}{4} Eh^{3/2} (d - 2h) f(h/a)/a^2 \cdot Q \tag{80}$$

where h is the half-width of the specimen (see Figure 16) and d is the separation of the specimen at the upper end. The function $f(h/a)$ is given by

$$f(h/a) = \frac{1 + 0.64\,(h/a)}{1 + 1.92(h/a) + 1.22(h/a)^2 + 0.39(h/a)^3} \tag{81}$$

and the factor Q is a correction for the extra compliance due to the presence of the end-slot,[76]

$$Q = 1 + [(h/h')^3 - 1]\,(l_s/a)^3 \tag{82}$$

where l_s is the length and h' is the half-width of the slot, respectively. The fracture energy is obtained from Eq. (80) via the relation $G_{Ic} = K_{Ic}^2/E$. The advantage of the wedge cleavage technique is that the crack propagates in a stable manner such that multiple data points can be obtained from a single sample, thus providing good statistics. It is possible to obtain the weld strength as a function of welding time by alternating crack propagation with healing treatments such that each measurement corresponds to crack advance into a previously unfractured portion of the interface.

Figure 17 shows results obtained by O'Connor and Wool (corrected for initial wetting) for welding monodisperse molecular weight polystyrene interfaces at 120 °C. The best fit to the data gives

$$K_{Ic} \approx t^{1/4} M^{-x} \tag{83}$$

where x is in the range 0.31 to 0.37. The latter value is closer to the chain pullout prediction of $x = 1/4$ than to the bridge fracture prediction of $x = 3/4$.

Figure 18 shows G_{Ic} welding data at 115 °C by McGarel and Wool and co-workers[77,78] for wedge cleavage welding of polystyrene. The data are plotted as G_{Ic} vs.

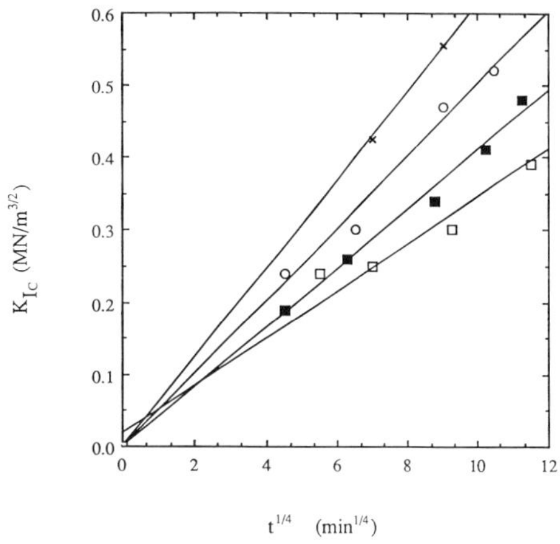

FIGURE 17. K_{Ic} vs. $t^{1/4}$ for welding of polystyrene of varying molecular weights using the wedge cleave technique.[1] Molecular weight key: crosses, 142,000; open circles, 217,000; closed squares 330,070; open squares 524,220.

$t^{1/2}M^{-1/2}$ (contour length) and show no tendency to superimpose when plotted with respect to the bridge fracture model; they exhibit some agreement with chain disentanglement theories. In accordance with the scaling laws for healing, if $G=l(t)$ then the virgin state should support $G_{Ic\infty} \approx M$ for some range of molecular weight.

Figure 19 shows the fracture energy G_{Ic} of polystyrene as a function of molecular weight (monodisperse fractions).[49,73,77] An abrupt change in fracture behavior is observed in the vicinity of $M_c \approx 32{,}000$ which parallels changes in the zero-shear viscosity behavior. For $M < M_c$, the data obtained by Robertson[79] show that the material is extremely fragile and the fracture energy well described by the linear equation, $G = 0.23 + 3.44\times10^{-5}M$ (J/m²). As M decreases, the fracture energy approaches the Griffith limit, $G = 2\Gamma$, where Γ is the surface energy. For PS at room temperature, $2\Gamma = 0.08$ J/m², which is smaller than the extrapolated limit of 0.23 J/m² for the fracture energy in Figure 19. However, this difference may be reasonable when consideration is given to the surface obtained by fracture compared with the flat surface normally used in surface

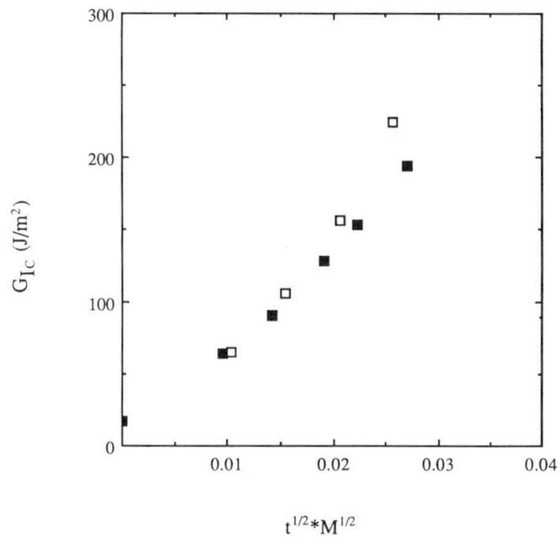

FIGURE 18. Fracture energy G_{Ic} vs $(t/M)^{1/2}$ for welding of polystyrene of molecular weight 152,000 (open squares) and 400,000 (closed squares).

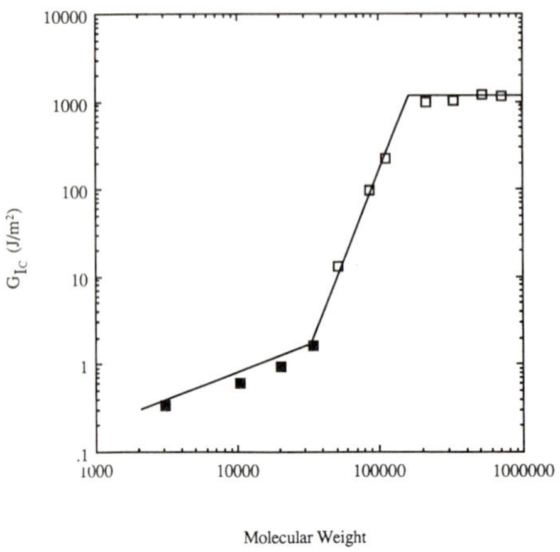

FIGURE 19. G_{Ic} vs. molecular weight for polystyrene in the virgin state. Data of Robertson(79) (closed squares) using wedge cleavage and O'Connor and Wool(73) (open squares) using compact tension fracture methods.

tension measurements. The same data can also be described by a power law, $G = 4.67\times10^{-3}M^{0.53}$ J/m^2. It has been suggested by Kramer(80) that $G \approx M^{1/2}$ based on the random coil radius $\langle S^2\rangle^{1/2} \approx M^{1/2}$ controlling the crack-opening displacement at constant craze stress.

In the molecular-weight range $M_c < M < M^*$ where $M^* \approx 8M_c \approx 250{,}000$, G_{Ic} increases from 1 to 1000 J/m^2 as shown in Figure 18 and can be described approximately by

$$G \approx (M^{1/2}-M_c^{1/2})^2 \tag{84}$$

When $M \gg M_c$ but less than M^*, then the relation $G \approx M$ is realized. The uniaxial fracture stress σ and the critical stress intensity factor K both behave as $\sigma, K \approx (M^{1/2}-M_c^{1/2})$ in this range.(1) Entanglement connectivity arguments based on percolation(81,82) have been presented to explain the role of M_c in fracture of glassy polymers.(4,49) When $M > M_c$, chain pullout and disentanglement are dominant mechanisms of fracture in glassy polymers. The fracture surface morphology consists of fractured remnants of craze fibrillar material and considerable energy is expended in the formation and disentanglement of the craze microfibrils. The critical energy to propagate a crack through a craze zone is given by $G = \sigma_c\delta$, where σ_c is the stress necessary to form the craze and δ is the critical crack-opening displacement. Both σ_c and δ should show a dependence on M and M_c which behaves approximately as $\sigma_c \approx \delta \approx (M^{1/2}-M_c^{1/2})$. Some support for this prediction is provided by fracture data on polycarbonate obtained by Pitman and Ward.(83)

When $M > M^*$, the fracture parameters become independent of molecular weight as shown in Figure 19. The number of broken bonds during fracture of PS in this molecular-weight range was determined at room temperature as 7×10^{17}/m^2.(17,84,85) Bond rupture contributes about 0.0001 kJ/m^2 to the fracture energy of 1 kJ/m^2 and is essentially negligible. However, the number of broken bonds roughly corresponds to the number of bridges crossing the fracture plane and suggests that bond rupture controls the extent of chain disentanglement at $M > M^*$. Thus, in fracture of glassy polymers, viscous and plastic deformation processes dominate the fracture energy dissipation while bond fracture contributes mechanistically to this dissipation by controlling their extent. As the fracture temperature of PS is increased to T_g, the number of broken bonds decreases[84,85] and chain pullout is favored.

Prentice[42] and Evans[43] have argued that the fracture energy should depend on molecular weight via relations similar to Eq. (43) for chain pullout such that $G \approx n_\infty L^2$. While they incorrectly assume that $n_\infty \approx M^0$ ($n_\infty \approx M^{-1/2}$) they suggest that $G \approx L^2 \approx M^2$ and show from PMMA fracture data similar to that shown in Figure 19 that a slope of 2 is obtained in the range $M_c < M < M^*$. A power law fit of the data in Figure 19 indicates that $G \approx M^{3.4}$ but the slope of 3.4 is due to the inhomogeneous nature of G vs. M and the data are more justifiably represented by relations similar to Eq. (83) above. If the relation $G \approx M^2$ were correct, then one could expect for welding that $G(t) \approx t/M$, and this is not consistent with welding data.

6.1.2. Incompatible Amorphous Interfaces

With immiscible, incompatible, amorphous polymers, the equilibrium structure of the interface is determined thermodynamically by the balance of two forces. The favorable decrease of free energy due to entropic relaxation of the surface chains' "compressed" configurations is balanced by the unfavorable enthalpy of mixing immiscible monomers in a layer of characteristic thickness d_∞ (see Section 2.2.4). Since $d_\infty \simeq \chi^{-1/2}$ according to Helfand, *et al.*,[24,86,87] and the Flory–Huggins interaction parameter χ behaves with temperature approximately as $\chi = a/T + b$, where a and b are constants, we expect for fracture, dominated by chain pullout, that $G \approx l_\infty \approx d_\infty^2$ and that the equilibrium strength behaves with welding temperature as[22,77]

$$G_\infty \approx T \qquad (a/T \gg b) \tag{85a}$$

and

$$G_\infty \approx T^0 \qquad (a/T \ll b) \tag{85b}$$

Figure 20 shows the equilibrium fracture energy as a function of welding temperature for the incompatible PS/PMMA interface obtained by Foster and Wool[77] using wedge

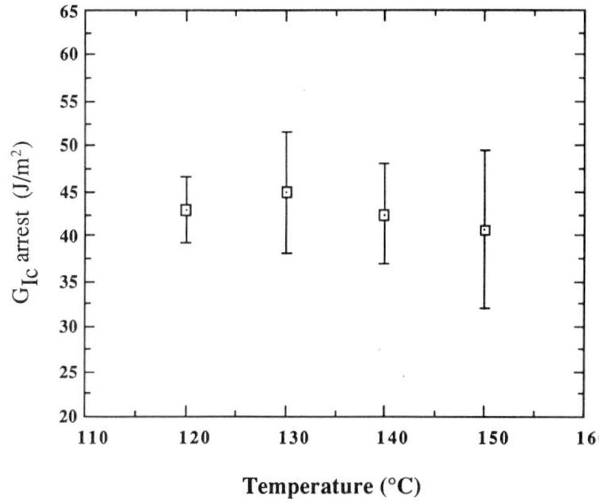

FIGURE 20. Fracture energy vs. temperature for a PS/PMMA welded interface. Each point represents a minimum of 20 data points and bars represent standard deviation. G_{Ic} shows little change with temperature in agreement with measurements of the Flory–Huggins interaction parameter by Russell *et al.* in this temperature range.

cleavage methods. The following molecular weights were used; M_w(PS)=302,000 with a polydispersity of 2.36 and M_w (PMMA)=120,000 with a polydispersity of 1.94. The fracture energy was found to be constant (42.5 J/m^2) and independent of temperature in the range 120–150 °C. Studies by Russell *et al.* of the temperature dependence of χ in this temperature range for PS/PMMA gave the constants for the temperature dependence of χ as a=3.906 and b=0.0284. Thus, χ changes by only 1.0% in this range. The experimental result in Figure 20 is consistent with the prediction in Eq. (85b) where χ has little dependence on temperature.

The fracture energy of 42.5 J/m^2 is greater than the initial wetted energy of about 5 J/m^2 obtained at short contact times and much greater than the combined surface energies Γ, of PS and PMMA, with $\Gamma \approx 0.1$ J/m^2. Comparing the strength of the PS/PMMA interface in Figure 20 with the virgin strengths of PS in Figure 19, we obtain the same fracture energy of $G \approx 42$ J/m^2 at M=70,000. This molecular weight has an end-to-end vector of magnitude R=178 Å for the random coil chains in the melt. The equilibrium thickness for the PS/PMMA interface is determined from the Helfand theory as $d_\infty = 2b/\sqrt{(6\chi)} \approx 30$ Å, where the statistical bond length, $b = 6.5$ Å. Thus, the theoretical incompatible interface thickness is much smaller than the end-to-end vector of the virgin material with the same strength. The theoretical value of $d_\infty \approx 30$ Å is in closer agreement with experimental thickness values, $d_\infty \approx 50$ Å, of a PS/PMMA interface, as measured by Neutron Reflection (Russell *et al.*, private communication). The incompatible interface is stronger than the virgin material with comparable interpenetration at M = 70,000 because of the higher molecular weight of the PS/PMMA materials and we conclude that the analogy is poor.

We should expect a better agreement by comparing the strength of the incompatible PS/PMMA interface with that of a partially healed symmetric PS/PS interface of the same interpenetration distance X. In this example, both interfaces are weak with low interpenetration depths but have high molecular weight virgin materials ($M \approx 200{,}000$) on both sides of the interface. With a healing interface, we have the following relations for the fracture energy, $G(t)/G_\infty = (t/T_r)^{1/2}$, where G_∞ is the fracture energy of the healed material; the average monomer interpenetration distance, $X(t)/X_\infty = (t/T_r)^{1/4}$ from Table 1; and at $t = T_r$, $X_\infty = 0.33$ R, as determined by Zhang and Wool. When $d_\infty = 2X(t')$ for some time t', and equating $G(t')$ with the fracture energy G of the PS/PMMA interface, we obtain,

$$d_\infty = 0.66\, R\, (G/G_\infty)^{1/2} \tag{86}$$

For example, from Figure 19, at M = 250,000, $G_\infty \approx 1000$ J/m^2, the end-to-end vector magnitude R = 338 Å, G = 42.5 J/m^2 and we predict from Eq. (88) that $d_\infty \approx 46$ Å. Examining another point from Figure 19, at M = 100,000 $G_\infty \approx 200$ J/m^2, R = 214 Å and we predict that $d_\infty \approx 65$ Å. The healing interface analogy gives much better agreement with both theory and experiment than the previous virgin state analogy. However, considering the complexity of the micromechanics and entanglement structure of the incompatible interface deformation zone, the comparison with the partially healed symmetric interface must be considered a crude approximation.

An interesting problem to explore with incompatible interfaces is the role of Rouse dynamics permitting loops rather than chain ends to be the dominant contribution to the interdiffused layer and entanglement process. Within the framework of the reptation theory, lateral diffusion of monomers occurs in the "tube" via Rouse relaxation of entanglement segments and can result in diffusion distances up to about 50 Å. Mechanical strength can develop by loop entanglement and loop "trapping." The effectiveness of the loop entanglements and the number of trapped loops remains to be determined for the amorphous symmetric interface.

6.1.3. Incompatible Crystallizable Interfaces

Strength development at incompatible, crystallizable interfaces such as polyethylene with polypropylene, PE/PP, can occur by different mechanisms than those discussed above. In the melt, limited diffusion to a depth d_∞ first occurs. When the interface cools, crystallization occurs first in the PP side due to the difference in melting points. Random nucleation of spherulites near the interface coupled with the volume contraction due to crystallization breaks up the interface. PE melt riverlets are drawn into the PP side and subsequently crystallize and, effectively, mechanically interlock the interface. This mechanism has been discussed by Galeski and co-workers[88,89] in terms of "local crystallization" creating "influxes" across the interface and has been investigated by Yuan, Wool, and co-workers[77,90]

Figure 21 shows a PE/PP interface during fracture by wedge cleavage methods. The interface was crystallized in a manner to optimize the mechanical interlocking effect. The fibrillar structures in the fractured interface are due to plastic deformation of the PE influxes. The fracture stress was measured as $\sigma \approx 10$ MPa, which is more than half the virgin strength of PE. In the absence of influxes, the interface is extremely brittle with

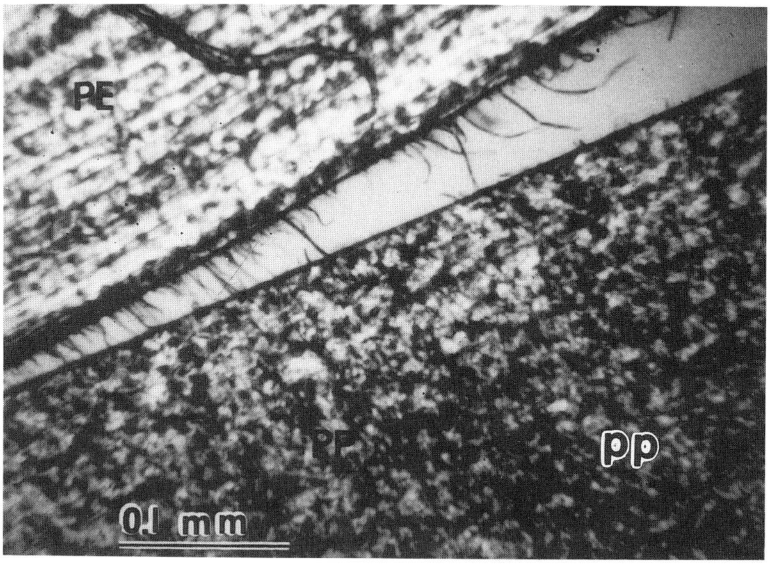

FIGURE 21. Fracture of a PP/LLDPE incompatible semicrystalline interface. The fibrils on the PE surface are due to pullout of PE material which was incorporated into the PP side during crystallization. (From Reference 90.)

$\sigma < 1$ MPa. Figure 22 shows a magnified view of the PP side obtained by SEM.[90] The PE fibrillar remnants are clearly seen in the interstices of the PP spherulites. In addition to the influxes, PP spherulites which nucleate on the interface plane entrap PE melt on a finer scale between the crystalline lamellar structures as seen by the "white decoration" of plastically deformed PE on several PP spherulites.

Thus for an incompatible, crystallizable interface, while melt diffusion produces an equilibrium interface of thickness 50 Å, the localized crystallization mechanism produces

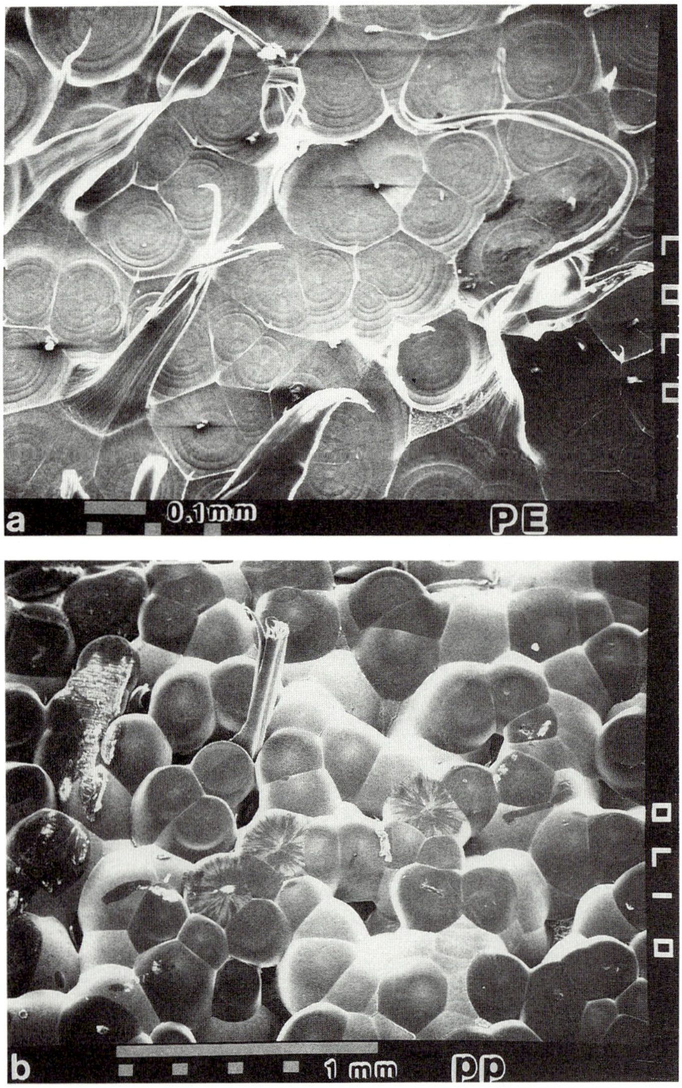

FIGURE 22. (a) Electron micrograph of a PE fracture surface from a PP/LDPE interface. Plastically deformed PE fibrils are observed in addition to circular imprints of PP spherulites on the PE side. (b) PP fracture surface from a PP/LDPE interface shows a fractured PE fibril remnant and PP spherulites (From Reference 90.)

influxes of several micron depth that impart significant strength to the interface. Delicate control of the processing history is necessary to optimize this effect as it is easy to make a weak interface with incompatible polymers. Methods of optimizing the strength have been discussed.[90] The same mechanism also exists at the interface between symmetric crystallizable interfaces such as PE/PE welding but plays a secondary role to diffusion and homogeneous crystallization.

7. LAP SHEAR WELDING

7.1. Polystyrene Welding

A lap shear joint method as shown in Figure 23 was used to study strength development during welding of PS interfaces by Kline and Wool.[91] The surfaces had previously not been in contact and care was taken to ensure rapid wetting of the interface. Lap shear samples were welded for a given time at constant temperatures above T_g and fractured at room temperature. Figure 24 shows a plot of the (average of 15 samples) shear failure stress τ_f vs. time to the fourth power at each constant temperature. The linear response of τ_f with $t^{1/4}$ is seen at each temperature and provides further support for diffusion-controlled strength at polymer–polymer interfaces.

The change in slope with temperature was found to be describable by the William–Landel–Ferry[92] theory of thermal activation near T_g. At T = 113.5 °C, the activation energy E_a = 93.2 kcal/mol is predicted from a WLF analysis using the parameters for PS, c_1 = 13.7, c_2 = 50, and T_g = 100 °C, which compares with E_a = 96.1 kcal/mol obtained from a plot of the superposition shift factor vs. $1/T$ in Figure 24.

The time to achieve complete healing, T_h, can be estimated from the upper plateau in Figure 24. At 391.3 K (118.1 °C), $T_h \approx$ 256 min for the polydisperse PS samples with M_n = 142,000 and M_w = 262,000. This time can be compared with related times measured

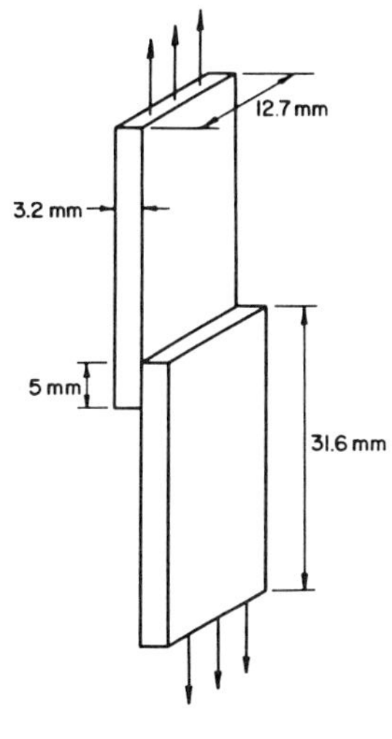

FIGURE 23. Lap shear joint method used to test weld strength of polymers. (From Reference 91.)

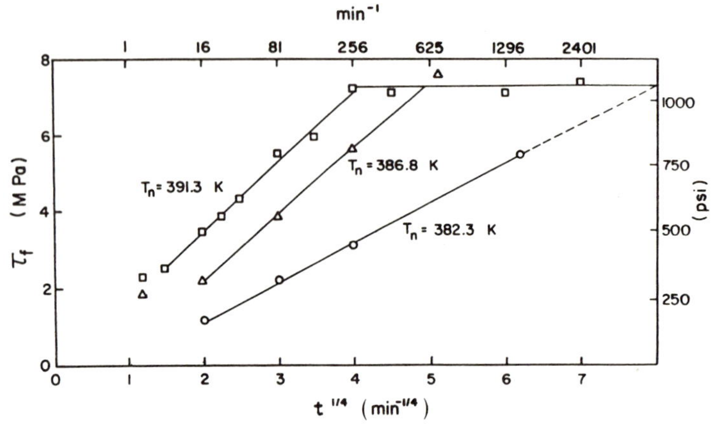

FIGURE 24. Shear stress vs. $t^{1/4}$ for lap welding of polystyrene at several temperatures. (From Reference 91.)

by other techniques. Lee and Wool,[(14)] using FTIR dichroism techniques, measured the relaxation time T_r of uniaxially oriented monodisperse PS with $M_n = 233{,}000$ at $T_r = 215$ min. Kramer *et al.*[(93)] report a self-diffusion coefficient for monodisperse PS with $M_n = 255{,}000$ as $D = 5.8\times10^{-17}$ cm²/sec at 125 °C. Shifting to 118 °C using the relation[(93)]

$$\log D/T = A - B/(T-T_\infty) \tag{87}$$

where $B = 710$, $T_\infty = 49$ °C, and $A = -9.49$, then $D_{118\ °\mathrm{C}} = 6.42\times10^{-18}$ cm²/sec. Using Eq. (74), the time required for the center of mass to diffuse a distance equal to the end-to-end vector is obtained as $T_r = 1000$ min, which is about four times longer than the time required to achieve complete strength of the lap shear joint with $M_w = 262{,}000$. On the other hand, using $T_r \approx M^3$ and $D \approx M^{-2}$, the relaxation time for the chains with $M_n = 143{,}000$ at 118 °C is about 180 min, which is in close agreement with the lap shear result. Owing to the ramified nature of polymer interfaces with holes and statistical irregularities as shown in Figure 7, the mechanical healing times are expected to be a little longer (but of the same order of magnitude) than the relaxation times determined from molecular dynamics and diffusion studies.

8. POLYMER PROCESSING WELD LINES

8.1. Compression Molding

As a final example of welding we consider the compression molding of resin pellets where the interfaces and weld lines form when the pellet surfaces contact in the mold, as shown in Figure 25. With increasing time and pressure, the interfaces wet, diffusion occurs, and the strength of the sample increases to its virgin strength provided that sufficient time for diffusion is allowed.

Wool and O'Connor[(1)] studied the time dependence of compression molding using PS pellets of the same molecular weight as described above in the lap shear experiments. PS plates, 6×6 in., were compression molded in a Carver Press using a window frame 1/8 in.

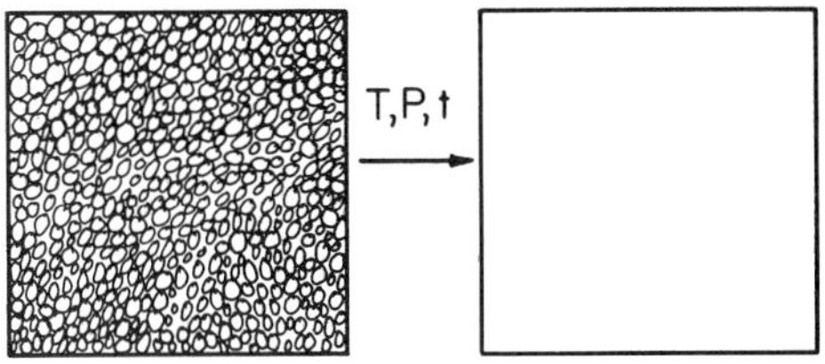

FIGURE 25. Schematic of powder/pellet interface welding during polymer melt processing. With time, temperature, and pressure, the pellet interfaces heal to form the plastic product.

thick. Each plate was molded at 127 °C ± 6 °C, with pressures of 800 psi for varying processing times t_p. The plate was withdrawn from the mold, quenched, cut into tensile "dogbone" samples, and fractured at room temperature in uniaxial tension. Weak interfaces within the sample provided a source of crack initiation. Figure 26 shows the fracture stress (average of 6 samples) plotted vs. the fourth power of the processing time. The prediction $\sigma(t) \approx t^{1/4}$ is again noted to be satisfactory despite the crudeness of the experiment. The processing time to reach the virgin state near $\sigma_\infty \approx 6000$ psi was between 10 and 20 min judging by the saturation of the strength at longer times. Using Kramer's diffusion data at 125 °C, the reptation time given by Eq. (74) for pellets with $M_n = 143{,}000$ is about 20 min, which is in reasonable agreement with the observed processing time.

The molecular-weight dependence of the virgin strength σ_∞ of compression-molded PS pellets has been determined by McCormick *et al.*[94] using similar sample preparation and fracture conditions. Figure 27[1] shows σ_∞ vs. $M^{1/2}$ for monodisperse molecular-weight PS using data from Reference 94. The data are described by

$$\sigma_\infty = 33.4\,(M^{1/2} - M_i^{1/2})\text{ psi} \qquad (88)$$

where the intercept at $\sigma = 0$ is $\sqrt{M_i} = 245$. The strength is seen to increase linearly in accordance with virgin state predictions up to $M \approx 160{,}000$ after which it decreases slightly with increasing M. The data were obtained from samples compression-molded at a constant time of 1 min at 170 °C for all materials. We have suggested that, under the processing conditions used, samples with $M < 160{,}000$ had reached their virgin state at

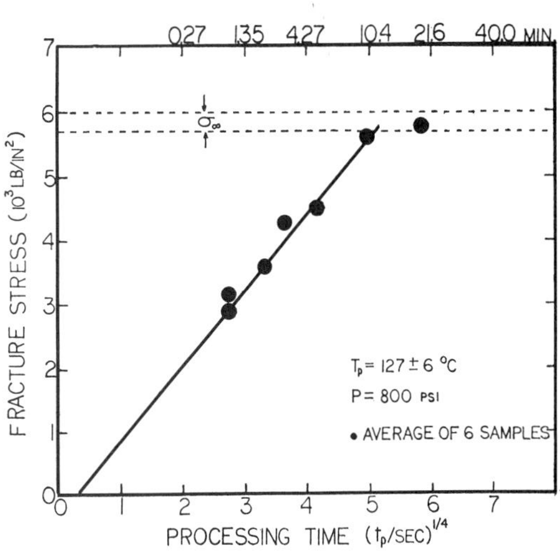

FIGURE 26. Fracture stress vs. processing time ($t^{1/4}$) for compression molding of polystyrene pellets. Dogbone samples were prepared from compression-molded plates (Wool and O'Connor).

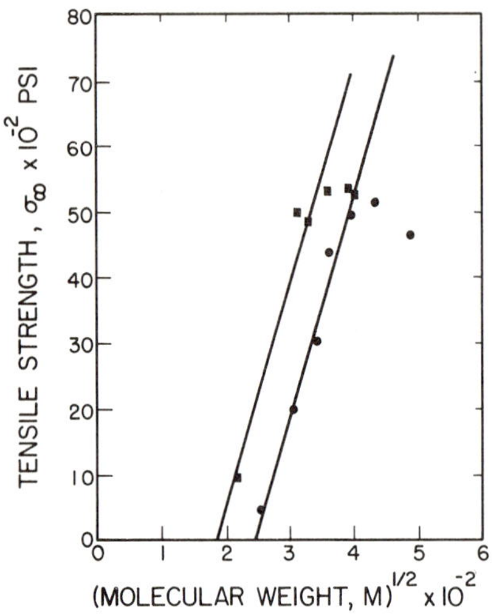

FIGURE 27. Tensile strength vs. square-root molecular weight of compression-molded polystyrene (pellets). Molding occurred at 170 °C for 1 min. Monodisperse molecular-weight pellets were used. Data of McCormick *et al.*

t = 1 min but samples with $M > 160{,}000$ were not completely healed at 1 min and the fracture stress should decrease as observed according to $\sigma \approx M^{-1/4}$. The reptation time for $M = 160{,}000$ is about 1 sec at 170 °C and the time of 1 min used in this experiment presumably includes warm-up time in the mold in addition to diffusion times. However, the maximum in strength with increasing M at constant contact time is expected and is similar to the tack data presented in Figure 14.

Strength development at internal weldlines in injection-molded specimens are expected to behave similarly, although some complications are expected with interfaces involving oriented molecules at the weldline. However, it is interesting to note that orientation relaxation occurs at about the same rate as normal diffusion.[(14)] Orientation relaxation may contribute to interdiffusion at the weldline in a similar manner as for unoriented surfaces, but problems are anticipated with "barrier" effects produced by oriented layers which could result in anisotropic diffusion rates.

ACKNOWLEDGMENTS

The author is grateful to the following agencies for providing financial and equipment support of this work: the Army Research Office, Durham N.C., Grant DAAL03-86-K-0034; the National Science Foundation, Grant DMR86-11551; the Materials Research Laboratory at the University of Illinois, Grant DMR86-12860; IBM Corporation; and Dow Chemical Company. Particular thanks are extended to my students who have contributed significantly to this work.

NOMENCLATURE

A	Self-diffusion constant
a	Crack length

α	Proportionality constant
b	Bond length
D	Diffusion coefficient
D_{f}	Fractal dimension
d_∞	Equilibrium interpenetration depth
δ	Crack-opening displacement
E	Tensile modulus
$F(t)$	Fraction of chain remaining in tube at time t
f	Function
$\varphi(t)$	Wetting function
$\varphi(x,t)$	Concentration profile
G	Fracture energy
G_{Ic}	Critical strain energy release rate
Γ	Fracture surface energy
$H(t)$	Molecular property of interface
H_∞	Equilibrium molecular property
h	Sample dimension
K_{Ic}	Critical stress intensity factor
k	Avrami constant, Boltzmann constant
L	Contour length of chain
L_{d}	Einstein diffusion length
L_0	Total number of monomers diffused
$l(t)$	Contour length of minor chain
λ_{c}	Critical Draw ratio for disentanglement
λ_{f}	Craze fibril draw ratio
M	Molecular weight
M_{c}	Critical entanglement molecular weight
M_0	Monomer molecular weight
M^*	Threshold molecular weight for disentanglement by Rouse relaxation
m	Avrami constant
μ	Chain friction coefficient
μ_0	Monomer friction coefficient
N_{a}	Avogadro's number
N_{f}	Number of particles on diffusion front
N_{m}	Number of monomers in polymer diffusion front
N_{v}	Number of chains per unit volume
n	Number of segments in minor chain
$n(t)$	Number of chains intersecting interface
P	Hydrostatic pressure
ρ_{c}	Percolation threshold
$\rho(\mathrm{t})$	Number of bridges intersecting interface
Q	A constant in wedge cleavage test
$\mathbf{R}$	End-to-end vector of Gaussian chain
R_{g}	Radius of gyration
r	Constant
r_{p}	Length of plastic zone in the Dugdale model

ρ	Density
S	Entropy
s	Constant
σ	Stress
T	Temperature
T_g	Glass transition temperature
T_r	Reptation time
t	Time
τ	Shear stress, convolution variable
U	Strain energy
U_c	Elastic chain energy
U_p	Energy to pull a chain out
V	Volume, deformation velocity
V_i	Craze-bulk interface velocity
ν	Percolation exponent
X	Monomer diffusion depth
xyz	Cartesian coordinates
χ	Flory–Huggins interaction parameter

REFERENCES

1. R. P. Wool and K. M. O'Connor, *J. Appl. Phys.* **52**(10), 5953–5962 (1981).
2. Y.-H. Kim and R. P. Wool, *Macromolecules* **16**, 1115–1120 (1983).
3. R. P. Wool, *Rubber Chem. Technol.* **57**(2), 307–318 (1984).
4. R. P. Wool, *J. Elastomers Plast.* **17**, 107–118 (1985).
5. P.-G. de Gennes, *J. Chem. Phys.* **55**, 572–579 (1971).
6. S. F. Edwards, *Proc. Phys. Soc. London* **92**, 9 (1967).
7. S. Prager and M. Tirrell, *J. Chem Phys.* **75**(10), 5194–5198 (1981).
8. K. Jud, J. G. Williams, and H. H. Kausch, *J. Mater. Sci.* **16**, 204 (1981); H. H. Kausch, *Pure Appl. Chem.* **55**, 833 (1983).
9. P.-G. de Gennes, *C. R. Acad. Sci. Paris* **292**(2), 1505 (1981).
10. S. Wu, *Polymer Interfaces and Adhesion,* Dekker, New York, (1982).
11. S. S. Voyutskii, *Autoadhesion and Adhesion of Polymers,* in: *Polym. Rev.* **4**(H. F. Mark and E. H. Immergut, eds.), Wiley–Interscience, New York (1963).
12. M. Doi and S. F. Edwards, *J. Chem. Soc., Faraday Trans. 2,* 1789 (1978).
13. M. Tirrell, *Rubber Chem. Technol.* **57**(3), 523–556 (1984).
14. A. Lee and R. P. Wool, *Macromolecules* **19**, 1063–1068, (1986); *Macromolecules* **20**, 1924 (1987).
15. Huanzhi Zhang and R. P. Wool, *Macromolecules,* **22**, 3018 (1989); and in press.
16. R. P. Wool, *Proc. IX Intl. Congress on Rheology* **3**, pp. 573–580, Acapulco, Mexico (1984).
17. R. P. Wool and A. T. Rockhill, *J. Macromol. Sci., Phys.* **B20**(1), 85–99 (1981).
18. H. H. Kausch, *Polymer Fracture,* 2nd edn., Chapter 10, Springer-Verlag, Heidelberg (1987).
19. K. Jud and H. H. Kausch, *Polym. Bull.* **1**, 697 (1979).
20. H. H. Kausch and K. Jud, *Plast. Rubber Proc. Appl.* **2**, 265 (1982).
21. R. P. Wool and K. M. O'Connor, *J. Polym. Sci., Polym. Lett. Ed.* **20**, 7 (1982).
22. J. L. Willett, K. M. O'Connor, and R. P. Wool *ACS Polym. Prepr.* **26**(2), 123–125 (1985); K. L. Foster and R. P. Wool, *Macromolecules,* in press.
23. S. J. Whitlow and R. P. Wool, *Macromolecules,* **22**, 2648 (1989); and in press; R. P. Wool, *Bull. Am. Phys. Soc.* **32**, (1987).
24. E. Helfand and Y. Tagami, *J. Chem. Phys.* **56**, 3592 (1972).
25. P. J. Flory, *Principles of Polymer Chemistry,* Cornell University Press, Ithaca, New York (1953).
26. M. Tirrell, D. Adolf, and S. Prager, *IMA Vol. Math. Appl.* **5**, Springer-Verlag, Berlin (1986).

27. P. G. de Gennes, *C. R. Acad. Sci. Paris, Ser. B* **291,** 219–221 (1980).
28. P. G. de Gennes and L. Leger, *Ann. Rev. Phys. Chem.* **33,** 49–61 (1982).
29. D. Adolf, M. Tirrell, and S. Prager, *J. Polym. Sci., Polym. Phys. Ed.* **23,** 413–427 (1985).
30. B. Sapoval, M. Rosso, and J. F. Gouyet, *J. Phys. (Paris), Lett.* **46,** 149–156 (1985).
31. M. Rosso, J. F. Gouyet, and B. Sapoval, *Phys. Rev. Lett.* **57**(25), 3195–3198 (1986).
32. M. Kolb, J. F. Gouyet, and B. Sapoval, *Europhys. Lett.* **3**(1), 33–38 (1987).
33. B. B. Mandelbrot, *The Fractal Geometry of Nature,* Freeman and Co., San Francisco (1982).
34. R. P. Wool and J. M. Long, *Proc. US–Germany Polymer Science Symposium,* Napa Valley, CA (Sept. 1987).
35. S. Mazur and S. Reich, *J. Phys. Chem.* **90,** 1365–1372 (1986).
36. R. P. Wool and J. M. Long (to appear).
37. A. A. Griffith, *Phil. Trans. R. Soc.* A**221,** 163 (1920).
38. R. E. Robertson, *Toughness and Brittleness in Plastics,* in: *Adv. Chem. Ser.*, No. 154, p. 89, American Chemical Society, Washington, D.C. (1976).
39. E. J. Kramer, *J. Mater. Sci.* **14,** 1381–1388 (1978).
40. R. P. Wool, *ACS Polym. Prepr.* **23**(2), 62 (1982).
41. R. P. Wool, Invited Lecture, *Bull. Am. Phys. Soc.* **27,** 297 (1982).
42. P. Prentice, *Polymer* **24,** 344–350 (1983).
43. K. E. Evans, *J. Polym. Sci., Polym. Phys. Ed.* **25,** 353–368 (1987).
44. D. S. Dugdale, *J. Mech. Phys. Solids* **8,** 100–104 (1960).
45. J. R. Rice, in: *Fracture* **2**(3) (H. Liebowitz, ed.), Academic Press, New York (1968).
46. L. C. Dolmon, I. M. Robertson, and R. P. Wool, *Bull. Am. Phys. Soc.* **30**(3), 488 (1985); L. C. Dolmon, M.S. Thesis, University of Illinois (1985).
47. B. D. Lauterwasser and E. J. Kramer, *Philos. Mag. A* **39**(4) 469–495 (1979).
48. E. Paredes and E. W. Fischer, *J. Polym. Sci., Polym. Phys. Ed.* **20,** 929 (1982).
49. R. P. Wool, in: *Amorphous Polymers and Non-Newtonian Fluids, IMA Vol. Math. Appl.* **6** (C. Dafermos, J. L. Ericksen, and D. Kinderlehrer, Eds.), pp. 169–187, Springer-Verlag, Berlin (1987); R. P. Wool, B.-L. Yuan, and O. J. McGarel, *Polym. Eng. Sci.* **29,** 1340 (1989).
50. A. M. Donald and E. J. Kramer, *J. Polym. Sci., Polym. Phys. Ed.* **20,** 899–909 (1982).
51. L. Burger and E. J. Kramer, *Macromolecules* **20,** 1980 (1987).
52. H. H. Kausch, Lecture presented to Materials Research Society, Boston (Dec. 1987).
53. K. M. O'Connor, Ph.D. Thesis Dissertation, University of Illinois, Urbana (1984).
54. R. P. Wool and K. M. O'Connor, *Polym. Eng. Sci.* **21,** 970–977 (1981).
55. O. J. McGarel and R. P. Wool, *J. Polym. Sci., Polym. Phys. Ed.* **25,** 2541 (1987).
56. K. M. O'Connor and R. P. Wool, *J. Appl. Phys.* **51,** 5075–5079 (1980).
57. R. P. Wool, in: *Adhesion and Adsorption of Polymers* (L.-H. Lee, ed.), Part A, pp. 341–362, Plenum Press, New York (1980).
58. G. R. Hamed, *Rubber Chem. Technol.* **54,** 576–595 (1981).
59. C. K. Rhee and J. C. Andries, *Rubber Chem. Technol.* **54,** 101–114 (1981).
60. R.-J. Chang and A. N. Gent, *J. Polym. Sci., Polym. Phys. Ed.* **19,** 1619–1633 (1981).
61. J. D. Skewis, *Rubber Chem. Technol.* **39,** 217 (1966).
62. W. G. Forbes and L. A. McLeod, *Trans. Inst. Rubber Ind.* **30**(5), 154–174 (1958).
63. J. D. Skewis, private communication and Reference 61.
64. C. M. Roland and G. G. A. Bohm, *Macromolecules* **18,** 1310–1314 (1985).
65. G. R. Hamed and C.-H. Shieh, *J. Polym. Sci., Polym. Phys. Ed.* **21,** 1415 (1983).
66. T. Tsuji, M. Masuoko, and K. Nakao, in Reference 71.
67. A. K. Bhowmick, P. P. De, and A. K. Bhattacharyya, *Polym. Eng. Sci.,* in press (1987).
68. E. Bister, W. Borchand, and G. Rehage, *Kautsch. Gummi, Kunstst.* **29,** 527 (1976).
69. S. Wu, H. K. Chuang, and C. D. Han, *J. Polym. Sci., Polym. Phys. Ed.* **24,** 143–156 (1986).
70. R. M. Vasenin *Adhes. Age* **8**(5), 18 (1965).
71. L.-H. Lee (ed.), *Adhesion and Adsorption of Polymers,* Vols. 12A and 12B, Plenum Press, New York (1980).
72. J. N. Anand, *J. Adhes.* **1,** 31 (1969).
73. K. M. O'Connor and R. P. Wool, *Bull. Am. Phys. Soc.* **30**(3), 389 (1985).
74. R. P. Wool, Final Report, "Crack Healing In Polymers," U.S. Army Research Office Grant DAAG-29-C-0143, Durham, NC (1983).

75. R. Guernsey and J. J. Gilman, *Exp. Mech.* **1,** 50 (1961).
76. M. F. Kanninen, *Int. J. Fract.* **9,** 83 (1973).
77. K. L. Foster and R. P. Wool, *Macromolecules,* in press.
78. O. J. McGarel, Ph.D. Thesis (to appear).
79. R. E. Robertson used the wedge cleavage method in Reference 38.
80. E. J. Kramer derived $G_{Ic} = \sigma_c (\lambda - 1)R^2 \sim M^{1/2}$ in Reference 39.
81. S. Feng and P. N. Sen, *Phys. Rev. Lett.* **52**(3), 216–219 (1984).
82. H. He and M. F. Thorpe, *Phys. Rev. Lett.* **54**(19), 2107–2109 (1985).
83. G. L. Pitman and I. M. Ward, *Polymer* **20,** 897–902 (1970).
84. J. L. Willett, K. M. O'Connor, and R. P. Wool, *J. Polym. Sci., Polym. Phys. Ed.* **24,** 2583–2589 (1986).
85. K. P. Paulson and R. P. Wool (to appear).
86. E. Helfand, in: *Polymer Compatibility and Incompatibility* (M. Solc, ed.), MMI Press, Midland, MI (1983).
87. T. A. Weber and E. Helfand, *Macromolecules* **9,** 311 (1976).
88. Z. Bartzak and A. Galeski, *Polymer* **27,** 544–548 (1986).
89. A. Galeski and E. Piorkowska, *J. Polym. Sci., Polym. Phys. Ed.* **21,** 1299 (1983).
90. B.-L. Yuan and R. P. Wool, *Polym. Eng. and Sci.,* **30** (1990).
91. D. B. Kline and R. P. Wool, *Polym. Eng. Sci.* **28,** 52 (1988).
92. J. D. Ferry, *Viscoelastic Properties of Polymers,* 3rd edn., Wiley, New York (1980).
93. P. F. Green and E. J. Kramer, *Macromolecules* **19,** 1108 (1986); E. J. Kramer, private communication.
94. H. W. McCormick, F. M. Brower, and L. Kin, *J. Polym. Sci.* **39,** 87 (1959).

8

Role of Electrostatics in Adhesion

Dan A. Hays

1. INTRODUCTION

The phenomenon of adhesion is the attraction of one material to another in either a gaseous or liquid medium. There are many mechanisms which can contribute to the adhesion between materials. The discussion in this chapter is limited to adhesion mechanisms due to electrostatic forces between dissimilar solid materials in a gaseous medium. The effect of electrostatic forces on adhesion is widely known and observed not only today but as early as 500 BC when the Greeks attracted pieces of straw to amber that had been charged by rubbing with another material. In spite of early observations and the ubiquitous nature of adhesion and triboelectric charging between materials, our present understanding is not satisfactory. Our understanding of electrostatic attraction is satisfactory when materials are separated. But when the materials are in intimate contact, the importance of electrostatic effects relative to other adhesion mechanisms becomes uncertain.

If two materials are charged through an imbalance of electrons or ions, the electrostatic force between the materials can act over a distance on the order of centimeters which is much greater than the range of any other mechanism of adhesion, except for gravitational attraction which can be neglected here. Progress in understanding the long-range electrostatic force between charged materials did not come about until the eighteenth century when Coulomb formulated a force law. Coulomb's law provided the mathematical foundation for describing electrostatic forces between objects with different material properties and geometries.[(1)]

When materials are brought into contact, the significance of the electrostatic force relative to other mechanisms of adhesion such as the van der Waals forces is unclear. Deryagin and colleagues[(2)] have emphasized the dominance of an electrostatic component of adhesion due to charge exchange between donor–acceptor molecular groups at an interface. The counter viewpoint is that electrostatic effects on adhesion are the exception rather than the general rule.[(3)] The argument for this opinion is that when different material combinations are contacted and separated, the charge exchange is strongly dependent on the material and surface conditions while the adhesion measurements are more weakly dependent on materials. Since the van der Waals attractive forces depend on bulk material properties and do not differ widely among a class of materials such as polymers,

Dan A. Hays • Webster Research Center, Xerox Corporation, Webster, New York 14580.

it is natural to conclude that adhesion is dominated by the van der Waals forces in most instances. Furthermore, measurements of the van der Waals forces over a separation range of several hundred angstroms are consistent with measured adhesive forces in many instances.[4]

The various mechanisms of adhesion are operational over different ranges of separation between materials. The dominant adhesion mechanism for a particular separation will depend not only on the combination of materials but also on the separation distance. Most adhesion mechanisms, such as interface diffusion, are significant only when the distance of separation is not much more than atomic dimensions. The van der Waals forces have a much larger range extending to several hundred angstroms. If the materials have acquired a net charge, an electrostatic Coulomb force acts over an even larger range of centimeters. When the materials are oppositely charged, the long-range attractive Coulomb force is the dominant adhesive force. Since insulative materials have a propensity to acquire and retain a net charge, long-range attraction due to electrostatic forces is a widely observed phenomenon.

In describing the adhesion between materials, it is useful to distinguish between the mechanisms which cause attachment of materials from mechanisms which oppose detachment or separation. The total forces which oppose detachment generally exceed the total attachment forces. The attachment forces are nevertheless important since they serve to bring the materials into intimate contact. Some adhesion mechanisms such as van der Waals or electrostatic forces contribute equally to both the attachment and detachment forces if the materials are brought near contact without touching and separated. Other mechanisms such as interface diffusion or charge exchange during contact only contribute to the forces which oppose detachment.

Electrostatic effects in adhesion are generally manifest when one or both materials acquire a net charge through an imbalance of electrons or ions. To establish the framework for subsequent description of adhesive forces due to a charge imbalance between materials, we begin with a brief review of relevant electrostatic concepts. We then describe the long-range Coulomb attractive force acting on a point charge in close proximity to a material, followed by a discussion of attraction between charged planar materials. Limitations on charge density imposed by ionization of the gaseous medium are described.

When planar materials are brought into intimate contact, charge transfer at the interface increases the adhesive force. We consider the dependence of the adhesive force on the charge exchange density and average distance of separation between the donor and acceptor charge exchange groups associated with each material. To assess the magnitude of the charge exchange, we review the recent experimental and theoretical literature describing charge exchange between metal–metal, metal–insulator, and insulator–insulator contacts.

For insulator charging, we describe several experiments designed to measure the charge penetration depth. Upon separation of the materials, the charge transferred during contact can flow back by one or more mechanisms. During separation, electron, ion, and photon emissions are observed which imply a high charge exchange density at the interface during contact. The difficulty in quantifying the charge exchange density while the materials are in contact makes an assessment of electrostatic effects on adhesion a difficult problem. Even if there is no net charge transfer, polarization effects due to charge displacement at the interface can in principle contribute to adhesion.

Although adhesion between planar materials is of practical significance in many situations, the material is oftentimes in a different geometrical form such as particles. The adhesion of particles to surfaces is either beneficial or a nuisance in many industrial processes. Since particles are usually charged, an electrostatic Coulomb force contributes to the particle adhesion. Technologies based on electrostatic adhesion of particles include xerography, electrostatic powder coating, electrostatic flocculation, and separation of dissimilar materials.[1] In xerography, charged particles referred to as toner are attracted to an electrostatic image on a photoconductor and subsequently electrostatically transferred to paper. The adhesion properties of toner are important in the development and transfer steps as well as in cleaning residual toner from the photoconductor prior to the next xerographic cycle. A recent example of an important industrial problem caused by particle adhesion involves particle contamination that limits production yields of photolithographically fabricated microelectronic circuits.[5] We review recent theoretical concepts and experimental observations concerning the role of electrostatics in the adhesion of charged particles to planar surfaces.

2. *ELECTROSTATIC CONCEPTS*

Detailed descriptions of electrostatic concepts are available from a variety of reference materials including Moore,[1] Smythe,[6] Crowley,[7] and Jackson.[8] Here we present a synopsis of key electrostatic concepts. Readers who are familiar with electrostatics are encouraged to bypass this section.

Coulomb's law discovered around 1785 states that the magnitude of the electrostatic force acting on point charges of magnitude Q_1 and Q_2 separated by a distance r is given by the expression (SI units)

$$F = Q_1Q_2/(4\pi\epsilon r^2) \tag{1}$$

The quantity ϵ is the permittivity which characterizes the polarization of the medium. For vacuum, the permittivity is a constant $\epsilon_0 = 8.85 \times 10^{-12}$ $C^2/N\text{-}m^2$. For a polarizable medium, $\epsilon = \kappa\epsilon_0$ where κ is the relative permittivity or dielectric constant of the material which can range from values near 1 for gases to 1200 for barium titanate. For common insulating materials such as polymers, the dielectric constants range from 2 to 10. If Q_1 and Q_2 are of opposite sign, the electrostatic force is attractive and directed along the line which joins the charges. When other point charges are present, the net force acting on Q_1 is the vector sum of the forces from the other charges. If Q_1 is a small test charge, the electrostatic force per unit test charge is defined as the electric field $\mathbf{E}$ at that point. The electric field is a vector since $\mathbf{F}$ has both magnitude and direction. If the dielectric polarization of the medium is linear in the electric field, the Maxwell displacement vector $\mathbf{D}$ is defined as

$$\mathbf{D} = \epsilon\mathbf{E} \tag{2}$$

Gauss's law is a more general form of Coulomb's law which relates the total charge Q enclosed within an arbitrarily shaped volume to a surface integral (summation) of the normal component of the displacement vector over the boundary of the enclosed region. If the surface is divided into infinitesimally small elements of area, **dS,** at points on the surface where the displacement is **D,** the integral form of Gauss's law can be expressed as

$$\oint_S \mathbf{D} \cdot \mathbf{dS} = Q \tag{3}$$

where the vector dot product $\mathbf{D} \cdot \mathbf{dS}$ is $D\mathrm{d}S\cos\Theta$ and Θ is the angle between the vector **D** and the vector **dS** represented by a magnitude of dS and direction perpendicular to the area and outwardly directed.

The integral form of Gauss's law is particularly useful in calculating the electric field for situations where the geometries of the charge distribution and materials possess symmetry. For example, if the boundary between two materials with dielectric constants κ_1 and κ_2 contains a surface charge density σ, one can calculate the difference in the electric field across the boundary by forming a Gaussian "pillbox" which overlaps the boundary with an infinitesimal height and top and bottom parallel to the boundary. For this situation, Gauss's law becomes

$$\kappa_1 E_1 - \kappa_2 E_2 = \sigma/\epsilon_o \tag{4}$$

where the positive electric field direction is from material 2 to 1.

The work required to move a test charge from a reference point A to another point B in a medium with an electric field is the line integral (summation) of the product of an elemental distance of displacement dl and the component of the electrostatic force vector **F** in the direction of the vector **dl**. The work done per unit test charge is given by the expression

$$W = -\int_A^B \mathbf{E} \cdot \mathbf{dl} \tag{5}$$

(The negative sign is required, since a force of opposite sign is applied to overcome the electrostatic force acting on the test charge.) If the electric field vector is expressed as the negative gradient of a scalar (no direction), i.e., $\mathbf{E} = -\mathbf{grad}\ V$, where **grad** is the gradient vector operator, then the right-hand side of Eq. (5) is equal to the difference in V between points B and A, i.e., $V_B - V_A$. In general, V is referred to as the electric potential at a point relative to a reference point. The potential energy of a charge Q at a point where the potential is V is equal to QV. The potential energy of a point charge Q_2 at a distance r from a point charge Q_1 is

$$\mathrm{PE} = Q_1 Q_2/(4\pi\epsilon r) \tag{6}$$

where the reference potential is zero at infinity.

The electrostatic force acting on a charge can be expressed as the product of the charge times the electric field. In general, the source of the electric field at a point is from a distribution of charges. If two charges of equal magnitude Q but opposite sign are separated by a small vector distance **dl,** the net electrostatic force acting on the dipole of dipole moment Q**dl** is zero if the electric field is uniform over the separation distance. But if the field is nonuniform, an electrostatic force acts on the dipole. For a cartesian coordinate system, the force in the z direction is

$$F_z = p_x\, \mathrm{d}E_z/\mathrm{d}x + p_y\, \mathrm{d}E_z/\mathrm{d}y + p_z\, \mathrm{d}E_z/\mathrm{d}z \tag{7}$$

where quantities p_i are the dipole moments in the i directions. With similar expressions for the F_x and F_y components, the electrostatic force vector acting on a dipole in a non-uniform electric field is

$$\mathbf{F}_\mathrm{d} = (\mathbf{p} \cdot \mathbf{grad})\mathbf{E} = \mathbf{grad}(\mathbf{p} \cdot \mathbf{E}) \tag{8}$$

The work done in moving a dipole in an electric field from position A where the electric field is $\mathbf{E}_\mathrm{A}$ to position B where the electric field is $\mathbf{E}_\mathrm{B}$ is

$$W_\mathrm{D} = -\int_{\mathbf{E}_\mathrm{A}}^{\mathbf{E}_\mathrm{B}} \mathbf{p} \cdot d\mathbf{E} \tag{9}$$

3. ELECTROSTATIC FORCE BETWEEN A POINT CHARGE AND PLANAR MATERIAL

Figure 1 illustrates a point charge Q in a region 1 with dielectric constant κ_1 at a distance $z = h$ from another planar material occupying region 2 with a dielectric constant κ_2. We assume the dimensions of the materials are much larger than h. We wish to consider the electrostatic force acting on a point charge adjacent to the interface of planar materials with different electrical conductivities and dielectric constants. For multiple charges near the interface, the calculated electrostatic force for each charge is valid provided the spacing between charges is much larger than the maximum distance between a charge and the interface. The case where multiple charges are closer to each other compared to the interface is discussed in the next section.

If the material in region 2 is a conductor, its surface will form the boundary of an equipotential volume since the internal electric field is zero in the presence of an external electric field. We arbitrarily assume the potential of the conductor to be zero (grounded). To obtain the force acting on a point charge in a vacuum (or gaseous) medium at a distance h from the conductor, it is convenient to use the method of images to calculate the electrostatic image force.[6] If in the absence of the conductor a point charge of opposite sign is placed at an equal distance h from the boundary in region 2, symmetry dictates that the net potential at the interface is zero as required for a grounded conductor. The electric field in region 1 is equivalent to replacing the conductor with an image charge in region 2. Thus, the magnitude of the electrostatic image force acting on a point charge is given by the expression

$$F_\mathrm{I} = Q^2/(16\pi\epsilon_0 h^2) \tag{10}$$

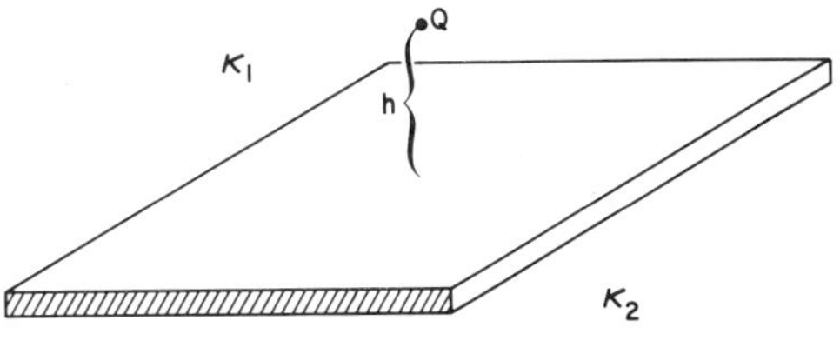

FIGURE 1. Point charge near a planar material.

For insulative materials in which the dielectric constants of materials 1 and 2 are κ_1 and κ_2, respectively, the magnitude of the image force for $\kappa_2 > \kappa_1$ is[6]

$$F_I = (Q^2/16\pi\kappa_1\epsilon_0 h^2)(\kappa_2 - \kappa_1)/(\kappa_2 + \kappa_1) \tag{11}$$

[Equation (11) is equivalent to Eq. (10) when $\kappa_1 = 1$ and κ_2 is infinite which is the effective dielectric constant for a conductor.] The magnitude of the potential energy, which is also referred to as the adhesive energy (the work done to move the point charge from h to infinity), is

$$W_E = (Q^2/16\pi\kappa_1\epsilon_0 h)(\kappa_2 - \kappa_1)/(\kappa_2 + \kappa_1) \tag{12}$$

Stoneham and Tasker[9] suggest that many interfacial phenomena at a metal/ dielectric interface can be interpreted in terms of an electrostatic image interaction due to charge trapped in the dielectric near the interface. For a metal/insulator interface with an electron trapped in the insulator with a dielectric constant of 10 at a distance of 0.1 to 0.2 nm from the interface, the electrostatic adhesive energy as calculated from Eq. (12) ranges from 0.18 to 0.36 eV. If there are 10^{19} electronic charges per square meter, the total adhesive energy is 0.29 to 0.58 J/m^2 which is comparable to measured adhesive energies. However, the validity of this calculation is based on the assumption that a monolayer of electronic charges is energetically favorable near an interface.

4. ELECTROSTATIC FORCE BETWEEN CHARGED PLANAR MATERIALS

Many materials are in a planar or sheet form such as plastic film. Since insulative materials tend to acquire and retain charge by frictional or contact charging, they adhere to other materials by an electrostatic force which acts over large separation distances. The adhesion of charged films is viewed as a nuisance in most instances. In fact, an antistatic agent is oftentimes added to dissipate charge. In other instances the electrostatic adhesion of film is desired. A recent example is a new food wrap in which the adhesive property is attributed to an electrostatic force from internally trapped charge.[10]

Figure 2 illustrates planar materials 1 and 2 separated by a medium of dielectric constant κ. We consider several cases for different materials. Let us first consider material

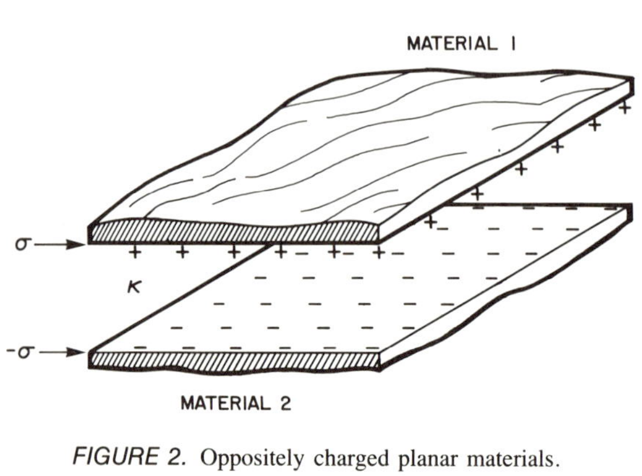

FIGURE 2. Oppositely charged planar materials.

1 to be a dielectric with a uniform surface charge density σ facing material 2, which is a grounded conductor. The charge density induces an equal and opposite charge density in the surface of the conductor. From Gauss's law [Eq. (4)] with $E_2 = 0$ in the conductor, the electric field in the intervening medium is

$$E = \sigma/(\kappa\epsilon_0) \tag{13}$$

To calculate the electrostatic force between the two materials, we first recognize that in the absence of the conductor the electric field from a sheet of charge in a medium of dielectric constant κ extends outward on each side of the sheet with a magnitude of

$$E = \sigma/(2\kappa\epsilon_0) \tag{14}$$

If the electric fields from the positive and negative charge densities illustrated in Figure 2 are added vectorially, we obtain Eq. (13) for the electric field in the intervening medium and $E = 0$ in regions 1 and 2. The magnitude of the electrostatic force per unit area is calculated by taking the electric field from the charge density at the surface of medium 1 [Eq. (14)] times the magnitude of the charge density at the surface of the conductor. The electrostatic force per unit area, which is also referred to as the electrostatic adhesive pressure, is then

$$P_A = \sigma^2/(2\kappa\epsilon_0) \tag{15}$$

We note that P_A is independent of the separation unlike the electrostatic force between a point charge and planar material. Since the adhesive force is independent of separation for charged planar materials, the adhesive energy per unit area is P_A times the separation distance.

We now want to consider typical values for the charge density and the corresponding magnitude of the electrostatic adhesive pressure. For a gaseous medium, the maximum electric field which the gas can support is limited by electrical breakdown. For planar electrodes, the breakdown electric field depends on the gas composition, gas pressure, and gap spacing over which the electric field is applied.[11] Figure 3 shows the Paschen curve for the maximum electric field that air can support at standard pressure as a function of the gap between planar conducting electrodes. Even though the Paschen curve for air

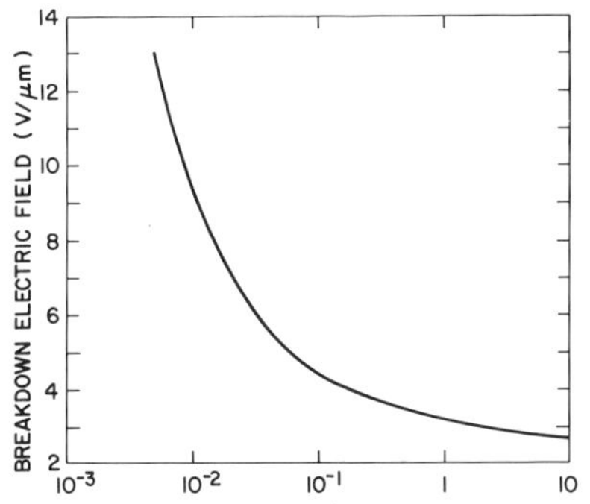

FIGURE 3. Paschen curve for air breakdown.

breakdown is obtained by applying a voltage between conducting electrodes, the curve is applicable to situations where the electric field in an air gap is supplied by a uniform charge density on a dielectric. Since materials which acquire a net charge by contact with other materials will usually undergo a separation greater than a centimeter, the maximum electric field which air can support is on the order of 3 V/μm. The charge density corresponding to this electric field is 30 $\mu C/m^2$, according to Eq. (13). From Eq. (15), the electrostatic force per unit area is 5×10^{-5} MPa. This pressure is 20 times larger than the weight per unit area of a 250-μm-thick polymeric film. It is clear that an electrostatic force due to a net charge on planar materials is effective in bringing objects into intimate contact.

We have assumed that the charge density on planar materials is uniformly distributed. For this situation, the electrostatic pressure is independent of the separation between the charged sheet and planar conductor. If, however, the uniformity of the charge density varies over an average distance of λ which is much less than the separation distance, the electrostatic pressure is given by Eq. (15) where σ is the average charge density. When λ is greater than the separation distance, the electrostatic pressure is given by an average of the quantity σ^2 in Eq. (15). For example, if the sheet contains equal amounts of positive and negative charged regions, the adhesion is zero for large separations but increases for distances less than λ. Thus, the electrostatic pressure for a planar geometry depends on the separation distance if the charge is nonuniform. This is often the case, since the charge on insulative materials is limited by air breakdown which deposits a nonuniform charge pattern of opposite charged ions over the surface.

We now want to consider the role of electrostatics in the adhesion of planar materials after they are brought into intimate contact by long-range attachment forces such as electrostatic or van der Waals forces and mechanical pressure. We consider three types of material combinations: metal–metal, metal–insulator, and insulator–insulator.

Many materials acquire an appreciable charge when brought into contact with each other. This is evident when the materials are separated. This phenomenon is variously referred to as contact charge exchange, contact electrification, triboelectricity, or frictional charging when the materials are specifically rubbed together. Following the work of Deryagin and Smilga[12] and Deryagin, Krotova, and Smilga,[2] we consider contact charge exchange between insulating materials 1 and 2 where material 1 is electron accepting and material 2 is electron donating, as illustrated in Figure 4. We assume the average distance between opposite charges due to electron transfer from the donor to the acceptor groups is r. The average separation between the donor–acceptor pairs for which transfer has occurred is a.

We will first consider the case when $r \ll a$. From Eq. (1), the electrostatic attractive force between an electron donor (cation) and electron acceptor (anion) pair is

$$F_e = e^2/(4\pi\kappa\epsilon_0 r^2) \tag{16}$$

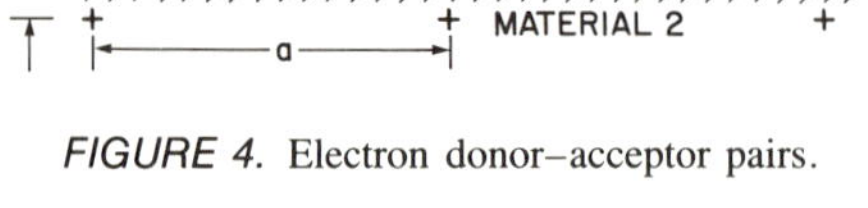

FIGURE 4. Electron donor–acceptor pairs.

where e is the electronic charge, 1.6×10^{-19} C. Here κ is viewed as an average interfacial dielectric constant for the two materials. If the number of cation–anion pairs per unit area is N, the adhesive force per unit area is

$$P_{r \ll a} = NF_e = Ne^2/(4\pi\kappa\epsilon_0 r^2) \tag{17}$$

If the average distance between cation–anion pairs is comparable to the cation–anion separation, the attractive force between each cation and anion increases due to the electric field from neighboring dipoles. Let us consider the electrostatic force acting on an anion from neighboring dipoles.[13] If the dipoles are distributed over a uniform lattice, the electrostatic force due to the neighboring dipoles is zero by symmetry for a direction parallel to the interface. In the perpendicular direction, the electrostatic force from the neighboring anions is also zero by symmetry. But the electric field from the neighboring cations increases the electrostatic force since the electrostatic force vector from each cation has a component along the perpendicular direction. Let us assume the position of a neighboring dipole to be $r_{ij} = ia\mathbf{u} + ja\mathbf{v}$, where a is a lattice spacing and $\mathbf{u}$ and $\mathbf{v}$ are unit vectors with directions that differ by the angle Θ. The component of the electrostatic force perpendicular to the interface due to the cation–anion pair at position i,j [normalized to the electrostatic force [Eq. (16)] at $i,j = 0,0$] is $r^3/(r^2 + r_{ij}^2)^{3/2}$ or $[1 + (r_{ij}/r)^2]^{-3/2}$. The total dipole electrostatic force acting on an anion is obtained by summing over all lattice sites. For a square lattice in which $\mathbf{u}$ and $\mathbf{v}$ are perpendicular ($\Theta = \pi/2$), $r_{ij}^2 = a^2(i^2 + j^2)$. The total adhesive force per unit area is equal to the density of cation–anion pairs times the sum of the attractive forces between an anion and the adjacent cation [Eq. (16)] and the neighboring cation–anion dipoles. Table 1 displays the dependence of the total adhesive pressure on the density N and average separation r between the cations and anions for an interfacial dielectric constant of 3. The density of cation–anion pairs for a square lattice is $N = 1/a^2$ where a is the average distance between the cation–anion pairs. The electrostatic adhesive pressure is in units of MPa. [For a hexagonal lattice, $r_{ij}^2 = a^2(i^2 + j^2 + ij)$ and $N = 2/(3^{1/2}a^2)$. Calculations of the dipole force for a hexagonal lattice are not significantly different from a square lattice.] In the limit when $r \gg a$, the calculation of the adhesive pressure is simplified. The charge density in materials 1 and 2 is $\sigma_1 = -eN$ and $\sigma_2 = eN$,

TABLE 1

Electrical Component of Adhesive Pressure in Units of MPa

$P_{r \ll a} = 7.6 \times 10^{-17}(N/r^2)$ $\qquad$ $P_{r \gg a} = 4.6 \times 10^{-34}(N^2)$

N (m^{-2})	a (nm)	r (nm) 0.3	0.5	1	10	100
10^{14}	100	8.4×10^{-2}	3.0×10^{-2}	7.6×10^{-3}	7.7×10^{-5}	4.8×10^{-6}
10^{15}	31	8.4×10^{-1}	3.0×10^{-1}	7.6×10^{-2}	9.5×10^{-4}	4.7×10^{-4}
10^{16}	10	8.4	3.0	7.7×10^{-1}	4.8×10^{-2}	4.6×10^{-2}
10^{17}	3.1	85	31	9.5	4.7	4.6
10^{18}	1.0	1.0×10^{3}	5.8×10^{2}	4.8×10^{2}	4.6×10^{2}	4.6×10^{2}
10^{19}	0.31	4.8×10^{4}	4.7×10^{4}	4.7×10^{4}	4.6×10^{4}	4.6×10^{4}

respectively. For $r \gg a$, we can approximate each layer as a uniform charge density. From Eq. (15), the adhesive force per unit area is

$$P_{r \gg a} = e^2 N^2 / (2\kappa\epsilon_0) \tag{18}$$

This simplified formula gives the same adhesive pressure as calculated by summing dipole fields when $r \gg a$. Although Eq. (18) is oftentimes used in describing the electrostatic component of adhesion, it is only valid when $r \gg a$. If it is used to describe the case when $r \ll a$, the adhesive pressure is underestimated as emphasized by Derjagin *et al.*[(2)]

Let us compare the electrostatic component of the adhesive pressure with the theoretical van der Waals pressure between two planar materials.[(14)] The pressure due to the van der Waals adhesive forces per unit area between two planar materials is

$$P_{vdW} = A / (6\pi Z^3) \tag{19}$$

where A is the materials-dependent Hamaker constant and Z is a separation distance of atomic dimensions. Assuming $A = 1 \times 10^{-19}$ J and $Z = 0.4$ nm, then $P_{vdW} = 83$ MPa. It is clear from Table 1 that the electrostatic component of adhesion is comparable to or greater than the van der Waals component when the density of cation–anion pairs is approximately $10^{17}/m^2$ or greater. A surface density of $10^{17}/m^2$ corresponds to a charge density of 1.6×10^{-2} C/m^2 or 16,000 μC/m^2. Since a monolayer of atoms is approximately $10^{19}/m^2$, the electrostatic component of adhesion is significant when the density of cation–anion pairs is greater than 1% of a monolayer.

5. CHARGE EXCHANGE PROPERTIES OF MATERIALS

To assess the role of electrostatics in the adhesion of materials in contact, we need to consider the following question. Is there experimental evidence that the charge exchange density during contact, σ_c, is greater than 16,000 μC/m^2? The difficulty in answering this question is that no one has identified a satisfactory method for measuring the charge exchange density while materials are in contact. The only way one can measure the charge exchange density is to separate the materials. But as the materials are separated, some of the charge will flow back by various mechanisms such as electron tunneling. We expect the charge exchange density upon separation, σ_s, will in general be less than σ_c. There is always the possibility that the charge density during contact is much greater than the separation charge exchange density. The interface electric field for a charge exchange density of 16,000 μC/m^2 is very high. Equation (13) gives the expression for the interface electric field when $r \gg a$. For $\sigma_c = 16{,}000$ μC/m^2 and $\kappa = 3$, $E = 600$ MV/m.

In this section we review a variety of charge exchange measurements and describe the effect of material properties and contact conditions on the sign and magnitude of the charge exchange. A book by Harper, *Contact and Frictional Electrification*,[(15)] and a review article by Lowell and Rose-Innes[(16)] represent useful reference materials on the subject.

Studies of contact charging between solid materials can be grouped into three categories of material pairs: metal–metal, metal–insulator, and insulator–insulator. For each material combination, we summarize the current experimental and theoretical understanding with particular emphasis on those studies which provide a quantitative measure of the charge exchange density.

5.1. Metal–Metal Charging

Metals are characterized by a high electrical conductivity due to free valence electrons which can move about the positive charged ion cores. Figure 5a illustrates a simplified electronic energy level diagram for the surface of a metal. For a temperature near 0 K, the free electrons occupy energy levels up to the Fermi energy, U_F. For a higher temperature T, the probability that an energy level U near U_F is occupied is given by the Fermi distribution function

$$f(U) = [1 + \exp{(U - U_F)/kT}]^{-1} \tag{20}$$

where k is the Boltzmann constant. Electrons at the Fermi energy are bound to the metal by an energy ϕ_m, the energy required to remove an electron from inside the metal to just outside. If the metal has a net charge and therefore a surface potential V_s relative to zero at infinity, the energy required to remove the electron to infinity is $\phi_m + eV_s$ where e is the electronic charge (1.6×10^{-19} C). The energy of an electron at the Fermi energy relative to infinity is referred to as the electrochemical potential and is equal to $-\phi_m - eV_s$.

The electronic properties of free electrons in a potential well formed by the boundaries of a metal are described by quantum mechanics. Since electrons have wavelike properties, the electron wave functions at the surface extend beyond the ion cores by a distance on the order of 1 to 3 nm. Within the bulk of the metal the volume charge density of the electrons is equal and opposite to the ion core density but at the surface a space

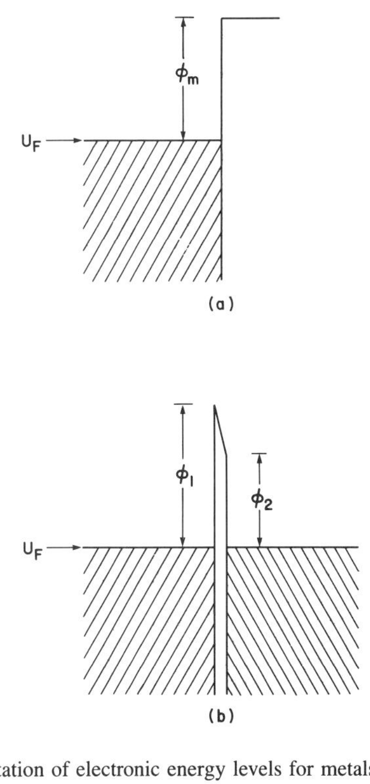

FIGURE 5. Schematic representation of electronic energy levels for metals: (a) work function of a metal; (b) metals in thermodynamic equilibrium.

charge density is formed due to electron tunneling beyond the ion cores. A high electric field is associated with the space charge density at the surface.[17]

When two dissimilar metals are brought near each other, electrons will tunnel from one metal to the other when the separation is less than the tunneling distance. The transfer brings the materials into thermodynamic equilibrium which is equivalent to the electrochemical potential being equal in each metal, as illustrated in Figure 5b. The difference in the surface potentials of the metals, $V_2 - V_1$, also referred to as the contact potential, is given by

$$V_c = (\Phi_1 - \Phi_2)/e \tag{21}$$

Since work function values for a variety of metals range from 3 to 5 eV, the contact potential ranges from 0 to 2 V. The charge density associated with the contact potential is confined to atomic dimensions at each metal surface since the high density of free electrons shields the bulk of the metal. Assuming the average distance over which the surface potential changes to be d and the electric field at the interface to be V_c/d, then from Eqs. (13) and (15) the electrostatic adhesive pressure is

$$P_A = \epsilon_0 V_c^2/(2d^2) \tag{22}$$

where we let $\kappa = 1$. For $V_c = 1$ V and $d = 0.5$ nm, the electrostatic adhesive pressure is 20 MPa. Since the van der Waals adhesive pressure for metals is typically 100 MPa, it would seem that the van der Waals force dominates the adhesion between most metals in contact.

To test the validity of the electrostatic analysis, one can measure the charge transfer when metals are separated. Measurements of charge transfer between spherical and planar metals have been described by Harper,[18] Bredov and Kshemianskaia,[19] and Lowell.[20] A linear dependence of the charge transfer on contact potential is observed as predicted. However, the magnitude of the charge transfer is less than expected on the basis that thermodynamic equilibrium is maintained until the tunneling cutoff distance is on the order of 1 nm. Lowell pointed out that the discrepancy can be understood if one considers the fact that most metal surfaces are not atomically smooth but rather consist of asperities which space the metals apart. Since the asperities represent a small fraction of the apparent contact area, the average separation between the metals when the last asperity makes contact with the other metal is typically 30 to 50 nm. Under these conditions both the electrostatic and van der Waals forces are reduced considerably. Thus, since metals are not generally smooth, the adhesion due to electrostatic and van der Waals forces is reduced considerably by asperities.

The basic physics of metal–seminconductor contacts is understood quite well in view of many detailed studies driven by the technological significance of semiconductors.[21,22] Since the density of free electrons (and holes) in undoped or doped semiconductors is considerably less than metals, the charge penetration depth into the semiconductor is on the order of a micrometer. This implies that the charge exchange density for metal–semiconductor contacts is less than that for metal–metal (assuming smooth) contacts unless the semiconductor has a high density of surface states. Measurements of contact charge exchange between separable metal–seminconductor contacts are limited. Dervos and Truscott[23] recently described charge exchange measurements between p- or n-type doped silicon and freshly evaporated metals in ultrahigh vacuum. Measurements obtained while the materials are in contact allow one to calculate the contact area, barrier height, and surface state density. The measured charge exchange shows an approximately linear dependence on the metal work function but the charge density is much less than expected from theory. The measured charge exchange densities were approximately two orders of magnitude less than the expected values.

5.2. Metal–Insulator Charging

The contact charge exchange between metals and insulators has been studied extensively, since the electronic properties of metals are well understood and insulators such as polymeric materials are of considerable commercial importance. However, the understanding of electronic mechanisms for metal–polymer contact charging remains unsatisfactory. This situation is due to a number of experimental difficulties including the need to be able to characterize a surface to a level of 10^{-4} of a monolayer. A test of the theoretical concepts requires measurements of both the charge transfer while the materials are in contact and the actual contact area. But in measuring the charge transfer upon separating the materials, there is the possibility of charge back-flow during separation. For measurements of the contact area, one can expect the actual area of contact to be much less than the apparent area due to surface irregularities and undulations on a microscopic scale. Microscopic roughness of surfaces precludes intimate contact unless the materials are viscoelastic, in which case the microscopic contact area increases with time. If f is the fractional area actually making contact, the *apparent* charge exchange density upon separation is

$$\sigma_a = f\,\sigma_s \tag{23}$$

The value of f is highly variable since it depends on surface topography, hardness, and contact conditions. If one of the materials is highly elastic or not too viscous, the value of f will approach 1. Since many polymeric materials are viscoelastic, intimate contact is achieved with sufficient interfacial pressure. For polymer–metal contacts when sufficient pressure is applied to cause indentation, f is estimated to be greater than 0.5.[(16)] Roberts[(24)] assumes f to be near 1 for rubber–glass contacts.

Several types of contact between metals and polymers have been used to measure contact charge exchange properties. It is well known that frictional contacts can generate appreciable charge accumulation on a polymer. However, for many situations it is unclear if the charge accumulation results from increased contact area or if the relative motion promotes charge transfer. Since we are only interested in the role of electrostatics in the adhesion of materials brought into contact, we limit our discussion to contacts with no deliberate relative motion between the materials. For this situation, we discuss rolling metal–polymer contacts, repeated metal–polymer contacts, and mercury–polymer contacts.

5.2.1. Rolling Metal–Polymer Contacts

Davies[(25)] has described contact charge exchange measurements of polymers brought into repeated rolling contact with metals of different work functions. A linear relationship between the charge exchange density and metal work function is observed. Other investigators have not observed a work function dependence for some materials like polytetrafluoroethylene.[15,26] If the polymer charging depends on the metal work function, an effective work function for the polymer can be obtained by determining the metal work function which does not charge the polymer. Equilibrium charge densities as high as 2500 $\mu C/m^2$ have been observed under vacuum conditions for some metal–polymer combinations. The charge exchange can be described by a model in which one assumes the charging to be proportional to the difference in the work function of the metal and the effective work function of the polymer and inversely proportional to an average charge penetration depth into the polymer which is a constant independent of the metal work function.[25]

5.2.2. Repeated Metal–Polymer Contacts

Fabish *et al.*[27] and Cottrell *et al.*[28] have described repeated contact charging measurements between various metals and polymer films. The charging typically continues to increase even after thousands of repeated contacts within nominally the same contact area. One possible explanation for the accumulation could be that the actual contact area increases. However, observations by Lowell[29] and Cunningham and Goodings[30] indicate that the charging continues to increase after there could be no possible effect of plastic deformation on contact area. As discussed by Lowell, other explanations include charge conduction away from the surface when the materials are separated or mechanical stirring of the polymer surface by interfacial stress that creates new charge exchange sites. A satisfactory understanding of charge accumulation by repeated contacts is lacking.

There are several conceptual difficulties as to why insulators such as polymers should accept or donate any charge when brought into contact with metals. The reason for this is illustrated in Figure 6a, which displays the relevant energy levels for a polymer such as polyethylene in contact with a metal. The highest occupied molecular orbitals (HOMO) are about 9 eV below the vacuum energy level while the lowest unoccupied molecular orbitals (LUMO) are only about 1 eV below the vacuum level. Since the Fermi energy of metals is typically 3 to 5 eV below the vacuum level, there are no electronic states in the polymer that can either accept or donate electrons at that energy level. Furthermore, as Bauser *et al.*[31] have emphasized, the probability for electron or hole injection into the LUMO and HOMO levels, respectively, is extremely low for a reasonable contact time. The fact that charge exchange does not depend on dwell time (in contrast to repeated contacts) led Bauser to suggest that the charge transfer is likely controlled by surface states which have energy levels near the middle of the polymer energy gap, as illustrated in Figure 6b. The polymer work function Φ_p corresponds to an energy level between electron donor and acceptor states. It is presumed that the surface states represent electron accepting or donating molecular groups such as oxidative groups or impurities. To account for a linear dependence of contact charging on the metal work function, the surface states density must be uniformly distributed in energy.

The surface states model of contact charging implies contact charging properties that are dominated by chemical species extrinsic to the chemical composition of the polymer. Although it is well known that charge exchange of polymers can be readily modified by surface modification such as corona treatment, it is also a fact that for many polymers the

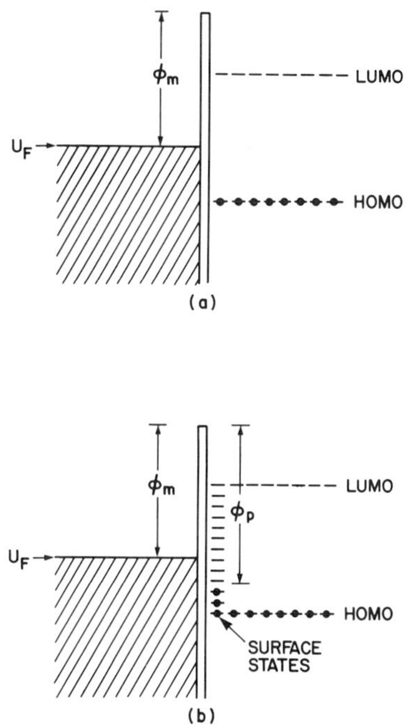

FIGURE 6. Schematic representation of electronic energy levels for metal–insulator contacts: (a) metal–insulator; (b) metal–insulator with surface states.

intrinsic chemical properties control the contact charging properties. Because the bulk properties are important, it is possible to order materials such as polymers into a triboelectric series in which a material higher in the list will charge positively when contacted or rubbed with a material lower in the list.

Duke and Fabish[32–36] have proposed a molecular-ion model of charge injection between metals and intrinsic electronic states associated with pendant molecular groups in polymers. The model assumes that when electrons are injected into LUMO levels as illustrated in Figure 6a, the energy is lowered by several electron volts through intermolecular and intramolecular polarizations. The position and energy distribution of the localized acceptor states are inhomogeneously broadened. When holes are injected into the HOMO levels, polarization shifts the molecular-ion donor state energy to higher energies by several electron volts.

Figure 7 illustrates the energetics for electron injection from a metal into acceptor states of a polymer with pendant molecular groups. Since the electron tunneling probability falls off rapidly for electron energies below the Fermi energy U_F, the electron injection occurs within a window of width $\Delta U \simeq 0.4$ eV. (The "window" theory has been contested based on contrary observations reported by Cottrell *et al.*[28]) The general expression for the polymer volume charge exchange density is

$$q_{MI} = \int_{U_F}^{U_F+\Delta U} \rho_D(U) f(U)\,dU - \int_{U_F-\Delta U}^{U_F} \rho_A(U)[1 - f(U)]\,dU \tag{24}$$

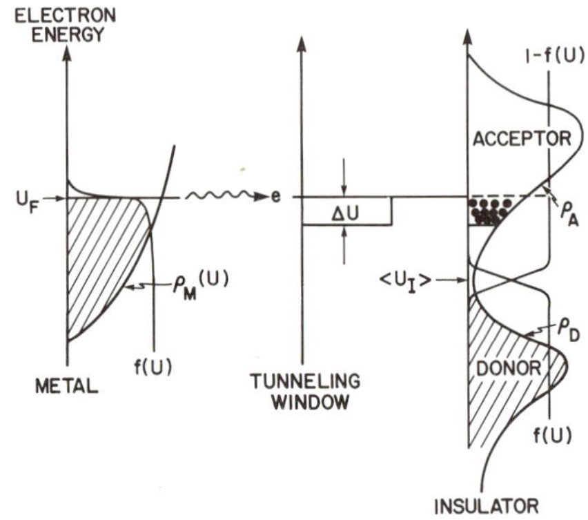

FIGURE 7. Schematic representation of metal–insulator contact for electron injection into insulator acceptor states.

where ρ_D and ρ_A are the density of donor and acceptor states, respectively, and $f(U)$ is the probability that a state is occupied as expressed by Eq. (20). The dependence of the density of states on energy is determined by contact charging measurements with metals of different work functions. The Gaussian distributed density of states has a width of several electron volts and a peak density of $2 \times 10^{21}/\text{m}^3\text{-eV}$. It is assumed that the electronic states are occupied over a depth of 1 to 2 μm for steady-state conditions. The charge exchange per unit area is obtained by multiplying Eq. (24) by 1 to 2 μm. For the case of electron injection as illustrated in Figure 7, Eq. (24) can be simplified to

$$q_{\mathrm{MI}} = -\rho_A(U_F)\,\Delta U \tag{25}$$

Since steady-state contact charging to a depth of 1 to 2 μm is achieved after thousands of repeated contacts, we assume that back-flow due to electron tunneling is unlikely under these conditions. The maximum contact charge exchange density obtained by Fabish *et al.* was only 70 μC/m^2.

For polystyrene with phenyl pendant groups, the charge exchange density for electron injection into acceptor groups is the highest for a metal with a work function of 3 eV and decreases to near zero for 5 eV. For poly(2-vinyl pyridine), which is a polymer similar to polystyrene except nitrogen is incorporated in the phenyl group, the charge exchange density for hole injection into the donor pendant groups is the lowest for a metal work function of 4 eV and the highest for 5.5 eV. Thus, there is a range of metal work functions for which the contact charging of polystyrene decreases and poly(2-vinyl pyridine) increases. It follows that if the electrostatic contribution to adhesion is significant relative to the van der Waals forces, the adhesion of polystyrene and poly(2-vinyl pyridine) should show opposite trends when separated from metals with different work functions. Preliminary adhesion measurements by Fabish[(37)] on samples formed by heat fusion to metals are suggestive of this trend.

5.2.3. Liquid Metal–Polymer Contacts

Since the actual contact area for metal–polymer contacts is uncertain, it is difficult to obtain a reliable measure of the separation charge exchange density. It is for this reason that a number of researchers have chosen a liquid metal such as mercury for polymer contact charging studies. Furthermore, liquid metal contacts enable a nonfrictional contact without mechanical disturbance of the polymer surface. A nondisruptive contact between materials represents a model system for adhesion studies since most adhesive contacts are of this kind. It is for these reasons that mercury with a work function of approximately 4.5 eV represents an ideal metal for contact charging studies. One can argue that the vapor pressure of mercury confounds the interpretation of measurements due to complications of material transfer. Although this is a possibility, one can nevertheless learn a great deal about electronic properties of polymer surfaces through systematic studies.

We assume the fractional contact area between mercury and polymers to be near one. This is only valid if the radius of curvature of surface undulations is greater than the ratio of the mercury surface tension (0.48 N/m) to interfacial pressure. For a pressure equal to the van der Waals forces per unit area (about 100 MPa), the critical radius of curvature is 10 nm. The apparent polymer contact area is easily determined by developing the charge pattern with charged particles such as xerographic toner.

Mercury–polymer contacts have provided the highest charge exchange densities for single contact events. If repeated contacts are made, the contact charging does not increase in the manner observed for solid metal–polymer contacts. Medley[(38)] reported charge densities of 1600 $\mu C/m^2$ for single mercury–polymer contacts obtained in vacuum to circumvent air breakdown limitations. For measurements in air, a grounded conductor provided behind a thin polymer film reduces the electric field in the air gap as the mercury is separated. For a 25-μm-thick polymer film, a charge density of 500 $\mu C/m^2$ can be measured in air without air breakdown. This is comparable to the maximum charge densities reported by Donald[(39)] for mercury–polymer contacts in air.

Mercury–polymer contacts have been used to study the effect of polymer surface modification on contact charging. Mercury contact charging of many polymers such as freshly cast polyethylene films is only 20 $\mu C/m^2$, but increases an order of magnitude when exposed to ozone from either the ambient or a generator.[(40)] The increase in the negative charging of polyethylene exposed to the ambient is attributed to surface oxidation by ozone (1 part in 10^7 molecules) reacting with unsaturated carbon–carbon (double) bonds (1 part in 10^4 bonds). Similar charging behavior is observed for polystyrene films exposed to ozone or ultraviolet radiation.[(41)]

El-Kazzaz and Rose-Innes[(42)] compared the contact charging of polymers by metals in the solid and liquid states. The sign of the contact charging was the same when either solid or liquid mercury, gallium, and indium were contacted to various insulators. The maximum contact charge exchange density was 300 $\mu C/m^2$ for the liquid metals. Essentially no charge accumulation was observed for repeated contacts with liquid metals, in contrast to solid metals. This observation tends to suggest that charge accumulation on insulators upon repeated contacts with solid metals is due to surface deformation.

5.3. *Insulator–Insulator Charging*

Contact charging is observed for a wide range of insulative materials. A variety of natural and synthetic insulative materials have been ordered into various triboelectric series. As a general rule, polymers containing halogens tend to charge negatively while polymers containing nitrogen groups tend to charge positively.

A theory of the electrostatic component of adhesion between amorphous solids described by Deryagin *et al.*[2] is based on the surface of one material containing electron donor groups and the other surface containing electron acceptor groups where in general there is a difference in the energy levels of the two molecular species. If the acceptor groups are at a lower energy, electron transfer from the donor groups will increase the interfacial electric field and, consequently, the potential energy of the acceptor groups. The equilibrium charge transfer is determined by minimizing the free energy of the system as a function of the number of reacted pairs, taking into account an increase in entropy. If we assume the density of acceptors to be much greater than the donor density of 10^{17} to $10^{18}/m^2$ (0.01 to 0.1 of a monolayer), then the charge transfer density is 10^{15} to $10^{17}/m^2$ for a donor–acceptor energy level difference of 0.5 to 1 eV, donor–acceptor separation of 0.5 to 1.0 nm, and interfacial dielectric constant of 10. The adhesive pressure for the higher density is comparable to the van der Waals contribution.

Fowkes[43,44] characterizes the donor–acceptor groups in terms of basic–acidic sites, respectively. The surfaces of metals generally contain metal oxides. The adhesion between polymers and metal oxides is enhanced by the acid–base interaction. Acidic sites of polymers are often hydrogens. All cations such as aluminum, magnesium, etc. on oxide surfaces are electron acceptors (acidic). Many oxide surfaces have interacted with water to produce surface hydroxyl sites which can be either basic or acidic. The adhesion of oxygen-containing polymers that are bases, such as polyesters, polyethers, polyamides, or polycarbonates, is enhanced by acidic surface metal hydroxide groups. Polymers with basic nitrogen or sulfur groups provide strong bonding to some metal cations.

There are many factors which can affect the contact charging of insulative materials. Not only is the chemical composition of the bulk important, but additives in the bulk such as plasticizer, catalyst, colorant, conductive agent, and antioxidant can affect the contact charging properties. Since contact charging is an interfacial phenomenon, the chemical composition of the surface and modification by oxidation, for example, is very important.

6. CHARGE PENETRATION DEPTH IN INSULATORS

The appropriate formula for calculating the adhesive pressure due to charge exchange depends on the average charge penetration depth in the insulator compared to the average lateral spacing between charges. It is clear from the above discussion that there are different views on the charge penetration depth into insulators. If insulator surface deformation occurs upon repeated contacts with solids, it is conceivable that the charge penetration is on the order of 1 μm. On the other hand, if the insulator contacts are gentle as with liquid metals, one might expect a charge penetration depth no greater than the tunneling distance of 1 nm.

Several approaches have been described which enable indirect estimates of the charge penetration depth. In the work of Fabish *et al.*[27] involving repeated metal contacts to thin polymer films, the steady-state surface potential was found to be a function of film thickness. The surface potential increased quadratically for thin films (micrometers) and linearly for thicker films. The quadratic dependence was attributed to charge injection, which formed a uniform volume density of acceptor or donor groups for the thin films less than 2 μm.

Nordhage and Backstrom[45] have measured the dependence of polyethylene contact charging on an applied electric field and work function of various metals. If the average charge penetration depth in the polymer is L, the electric field E_A shifts the energy by an amounts eE_AL. The ratio of the change in the charge exchange per unit electric field to that per unit work function is equal to eL. A value of $L = 40$ nm was calculated from this ratio. This value is much larger than the tunneling distance, but it should be pointed out that the measurements were obtained with sliding contacts which could induce surface deformation.

Hays[46] and Homewood[47] have calculated the charge penetration depth from measurements associated with gentle contacts between mercury and polymers. When a mercury–polymer contact is established, one expects a potential change at the interface which is proportional to the contact charge exchange density (electric field) and average penetration depth L. The change in the surface potential can be measured with millivolt precision. But one must be careful not to assume the change in surface potential to be entirely due to charge exchange. Hays[46] has pointed out that an appreciable potential change (+0.25 V for polystyrene) is due to polarization of molecular groups at the interface of dissimilar materials. After correcting for the surface potential change due to polarization, a potential drop of −10 mV is measured for a separation contact charge density of $-200\ \mu C/m^2$ on the polymer. The calculated maximum charge penetration depth is 1 nm, which is on the order of the tunneling distance. Homewood[47] calculated a depth of 0.4 nm for mercury contacts to polymethylmethacrylate.

Possart and Roder[48] have recently described a method for using a potential contrast scanning electron microscope to measure the potential distribution near an aluminum–polyethylene interface. The polyethylene film was solvent cast onto anodized aluminum and annealed at 363 K for several hours. The composite was cut perpendicular to the polymer–metal interface and overcoated with a thin layer of graphite for SEM examination. By choosing the proper primary electron energy, one can map the potential variation at the interface by measuring the energy shift of the secondary electrons. An energy shift of 0.8 V was observed over a distance of 10 μm in the polyethylene. Thus, it would seem that an aluminum–polyethylene interface brought near thermodynamic equilibrium by a combination of solvent and heat has the electrical characteristics of a metal–semiconductor interface.

Summarizing the above discussion, the charge penetration depth into polymers from metal contacts seems to depend on the contact conditions. The penetration depth can be several microns for repeated contacts, 40 nm for sliding contacts and 1 nm for gentle contacts. Since gentle contacts seem to be the appropriate contact condition for adhesive contacts between materials, one expects a charge penetration depth on the order of 1 nm unless the polymer is either solvent or heat cast against the metal, in which case thermodynamic equilibrium increases the charge penetration depth. We recall that the magnitude of the average charge separation or penetration depth r relative to the average distance a between charges determines the appropriate formula to use in calculating the electrical component of the adhesive pressure. In reference to Table 1, if r is typically 1.0 nm or less, the appropriate adhesion formula is given by Eq. (17). Most discussions of the electrostatic component of adhesion utilize Eq. (18), however. If r is 0.3 to 0.5 nm, this formula underestimates the electrostatic adhesive pressure by an order of magnitude for $N = 10^{17}/m^2$. On the other hand, it seems that if a polymer is solvent or heat cast against another material, the charge penetration depth could be microns, in which case the appropriate adhesion formula is Eq. (18).

7. CHARGE BACK-FLOW DURING SEPARATION

It is rather remarkable that the maximum charge exchange density measured upon separating materials is on the order of 1600 $\mu C/m^2$, which is only 10^{16} electronic charges/m^2 or approximately one thousandth of a monolayer ($10^{19}/m^2$). One might suppose that there are some materials that would have a higher density of charge exchange sites. Assuming this to be the case, one can argue there must be some fundamental mechanisms which either limit the higher charge transfer while the materials are in contact or provide significant charge back-flow during materials separation. It is the latter possibility that we wish to consider in this section. We will only discuss metal–insulator separation in vacuum, since much of the relevant data have been obtained under this condition to circumvent charge back-flow due to air breakdown.

Lowell[49] has discussed the role of tunneling in the back-flow charge when metals and insulators are separated. For purposes of discussion, we assume the insulator to have electron accepting groups on the surface at an energy Δ (about 0.5 eV) below the Fermi energy of the metal. When the materials are in contact, the number density of occupied acceptor states is N_0. When the materials are separated, the potential energy of the occupied states increases. If the increase in the potential energy makes the acceptor energy level greater than the Fermi energy, electrons will tunnel back to the metal if the separation is less than the tunneling cutoff distance of $z_c \simeq 2$ nm (see discussion in Section 5.1). We would like to determine how large N_0 can be before back-flow by tunneling reduces the charge exchange during separation. We assume that N_0 is sufficiently high so that the average lateral distance between charges is less than 2 nm. For this condition we can use Eq. (13) to calculate the electric field in the gap. The potential change across the gap at cutoff is

$$V = eN_0 z_c/\epsilon_0 \tag{26}$$

When the potential energy change, eV, is equal to Δ at the separation cutoff distance of 2 nm, one can calculate the maximum separation charge exchange density σ_m from the expression

$$\sigma_m = \epsilon_0 \Delta / e z_c \tag{27}$$

For $\Delta = 0.5$ eV and $z_c = 2$ nm, the maximum calculated separation charge exchange density is 2300 $\mu C/m^2$. This density is comparable to the maximum measured separation charge exchange density. Furthermore, since Δ depends on the difference in energy between the energy of the acceptor groups and the work function of the metal, σ_m is linear in the metal work function as observed by a number of researchers. (This interpretation of the linear dependence of insulator contact charging on the metal work function is in contrast with the model proposed by Bauser *et al.*[31] based on the assumption of a density of states uniformly distributed in energy.) On the other hand, if the density of charge exchange sites is such that σ is less than σ_m, the separation charge exchange density depends only on the density of acceptors and is independent of the metal work function.

This situation represents a possible interpretation of polytetraflouroethylene contact charging data that is independent of the metal work function.[15,26]

For tunneling-limited charging and separations greater than the tunneling cutoff distance of about 2 nm, the potential of the insulator is proportional to the separation. Since the electric field at the cutoff point in the above example is on the order of 2.5×10^8 V/m or 250 V/μm, the potential is 25,000 V for a separation distance of only 100 μm. If the separation occurs in a gaseous medium, charge back-flow will occur as a consequence of air breakdown according to the Paschen curve, shown in Figure 3. Emission of light and radio waves accompany the air breakdown. When materials are separated in vacuum, Deryagin and colleagues[2] in the early 1950s reported high-energy electron emission with an energy on the order of ten thousand electron volts. The electron emission was measured from one of the materials after separation. The emission continues over a time interval on the order of tens of minutes. This implies that the initial electron intensity and energy is much higher during initial separation. The electron emission is attributed to initial charge exchange densities of 3000 to 30,000 $\mu C/m^2$ in the areas where the materials were in intimate contact. The high-energy electrons produce radiation in the visible[50] and X-ray[51] parts of the spectrum due to interaction with the opposing surface.

Dickinson *et al.*[52] also report that during fracture of adhesive bonds in vacuum a variety of emissions including electrons, ions, neutral molecules, and photons in the form of light and radio waves accompany large charge separations between materials. However, Dickinson *et al.* are not of the opinion that the charge separation induces field emission of electrons which then induce secondary emissions as described by Derjaguin *et al.*[53] Rather they are of the opinion that a high potential from the charge exchange induces breakdown of gas desorbed during separation.[54]

Considering the various mechanisms for the back-flow of charge when materials are separated, it seems that it is difficult to measure a separation charge exchange density much greater than approximately 3000 $\mu C/m^2$.

8. INTERFACIAL POLARIZATION

It is generally observed that when molecules of a gas or liquid are adsorbed onto a metal surface, the contact potential of the metal changes by an amount up to 1 V.[55] The mechanisms for the potential change are attributed to either charge exchange between the adsorbate and metal as discussed above or interfacial polarization by induction or orientation of permanent dipole moments associated with the adsorbate. The adhesion of the adsorbate is increased by these mechanisms. Adhesion due to the formation of a dipole layer at an interface represents another mechanism which in principle can be important.

There are two viewpoints on why a dipole layer is formed at a metal–adsorbate interface. One model[56] is based on the fact that the surface of a metal has a very high electric field due to electron wave functions extending beyond the ion cores. When a molecule comes into close proximity with the surface, the high electric field polarizes the covalent bonds of atoms adjacent to the metal to form dipoles. The other model[57] assumes that as an atom approaches a metal surface, the image charges associated with the valence electrons are closer to the metal on the average than the image charge from the ion core. Since the average image force for the electrons is greater than the image force for the ion core, a dipole moment is formed by an electron distribution skewed toward the metal.

Smith[58] has measured contact potential changes for a variety of monomolecular films adsorbed on a mercury surface. If the number density of molecular groups is N, the potential change is

$$\Delta V = pN/\kappa\epsilon_0 \tag{28}$$

where $p = \alpha_p E_S + p_0$ is the dipole moment. The quantity α_p is the polarizability of a molecular group, E_S is the natural electric field at the metal surface, and p_0 is the component of a permanent dipole moment normal to the surface. The dielectric constant κ increases from 1 to approximately 2 for increasing N up to a monolayer. By measuring the initial slope for the dependence of the potential change on surface coverage, the dipole moment can be calculated. For the adsorption of alkanes on mercury, the dipole moment per C—H bond is 0.03 debye (1 debye = 3.3×10^{-30} C-m). With a bond polarizability of 7×10^{-41} C-m^2/V, the natural electric field is 1.4×10^9 V/m. The maximum observed contact potential change for increasing surface coverage was + 0.18 V. This is comparable to the contact potential change of + 0.25 V measured for a separable contact between polystyrene and mercury.[46]

Since contact potential changes on the order of 0.25 V are observed when liquids are adsorbed on metal surfaces,[59] one can expect a similar dipole layer formation for contacts between polymers and metals. The importance of an interfacial energy shift due to dipole layer formation at a metal–polymer contact has been emphasized by Hays[46] and Pong *et al.*[60]

Although interfacial polarization can affect the electronic structure for contact charge exchange, we want to consider the effect of induced dipoles on the adhesive force and adhesive energy between materials. We first consider the polarization model based on a surface electric field inducing the polarization. To obtain an order-of-magnitude calculation based on an approximate description, we use Eq. (7) which expresses the force acting on a dipole in a gradient electric field. We assume that the electric field gradient is the surface field, E_s, divided by the distance the electron wave functions extend beyond the ion cores, t_e, on the order of 0.5 nm. The adhesive force per unit area due to interfacial polarization is approximately

$$P_p = N\, p\, E_s/t_e \tag{29}$$

For $N = 10^{19}/\text{m}^2$, $p = 0.03$ debye, $E_s = 1 \times 10^9$ V/m, and $t_e = 0.5$ nm, the adhesive pressure from polarization is on the order of 2 MPa which is small compared to the van der Waals forces. The dipole adhesive energy is the work done to move the dipole from the metal surface to infinity where the electric field is zero. Using Eq. (9) with $p = \alpha_p E$, the adhesive energy per unit area is given by the expression

$$U_p = N\, \alpha_p E_s^2/2 \tag{30}$$

which is equal to 3.5×10^{-4} J/m^2.

9. CHARGED PARTICLE ADHESION

We have described the role of electrostatics in adhesion between materials with a planar interface. Although this is a common geometry between contacting materials, often one or both materials are not planar but rather are of different radii of curvature at the interface. If the adhesion mechanism acts over a range greater than atomic dimensions, such as the van der Waals or Coulomb electrostatic forces, the adhesive force will depend on the shape of the materials. The adhesion of spherically shaped particles to a planar material is of considerable interest since particle adhesion is important in many technological processes. A book by Zimon, *Adhesion of Dust and Powder,* represents a good general reference on the subject.(61)

In the zone where a particle and substrate are in intimate contact, the contributions to the adhesive force are similar to the mechanisms described for a planar interface. But in the region outside of the contact zone, the contribution from the longer-range forces decreases with increasing separation between the materials. For a perfectly rigid spherical particle of radius b resting on a rigid planar substrate, the force due to the van der Waals interactions is(14)

$$F_{\mathrm{vdW}} = Ab/(6Z^2) \tag{31}$$

where the A and Z parameters are similar to those discussed in connection with Eq. (19). If the spherical particle is insulating with a net charge of Q, a Coulomb electrostatic image force contributes to the particle adhesion. Assuming the charge on a particle in contact with a conductive planar substrate to be uniformly distributed on the surface as illustrated in Figure 8, the Coulomb electrostatic image force is(62,36)

$$F_{\mathrm{I}} = \alpha\, Q^2/(16\pi\epsilon_0 b^2) \tag{32}$$

where α equals 1 for a particle dielectric constant $\kappa_p = 1$ and 1.9 for $\kappa_p = 4$. The electrostatic image force for a uniformly charged sphere with $\kappa_p = 1$ is equivalent to the image force obtained when all of the charges are located at the center of the particle. For $\kappa_p > 1$, polarization of the particle by the electric field from the image charges enhances the adhesive force.

For a rigid insulating sphere with a charge uniformly distributed on the surface and in contact with a rigid conductive planar substrate in a dry ambient, one would expect adhesive forces due to the van der Waals and image force components equal to

$$F_{\mathrm{AD}} = F_{\mathrm{vdW}} + F_{\mathrm{I}} = Ab/(6Z^2) + \alpha\, Q^2/(16\pi\epsilon_0 b^2) \tag{33}$$

The magnitudes of the adhesive force components have been discussed by Bowling.(63) Assuming a 5-μm-radius particle and a charge of 1×10^{-14} C corresponding to a surface charge density of 20 μC/m^2, the force due to the van der Waals interactions is 5×10^{-7} N ($A = 1 \times 10^{-19}$ J and $Z = 0.4$ nm) while the electrostatic image force is only 2×10^{-8}

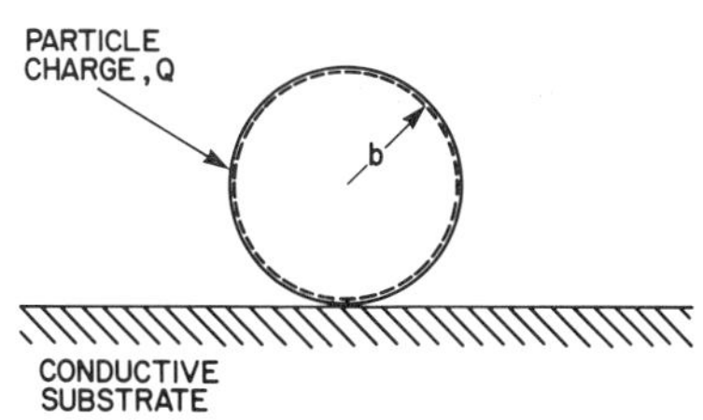

FIGURE 8. Uniformly charged particle on a conductive substrate.

N. Since the charge density of 30 $\mu C/m^2$ is typical of xerographic toner that is deliberately charged by triboelectricity, one can expect the net charge on particles in an uncontrolled environment to be generally less. The air breakdown limitation on the maximum surface charge density of isolated spheres increases with decreasing particle radius as expressed by the formula $30 \times b^{-0.3}$ $\mu C/m^2$ where b is in units of centimeters.(15) For a 5-μm-radius particle, the maximum charge density is 300 $\mu C/m^2$. Although small particles can acquire and retain higher charge densities, the typical charge level on particles is much lower, which implies that the van der Waals forces will dominate the adhesion.

Interest in the adhesive properties of charged particles has been stimulated by the significance of the phenomenon in the xerographic process. Although electric fields are used in the xerographic process to detach and transfer toner particles from one surface to another, most measurements of toner particle adhesion have been obtained by applying acceleration forces supplied by either a centrifuge or impact. Centrifuge adhesion measurements on xerographic toner have been reported in a number of studies with conflicting conclusions. Krupp(14) was first to report the dominant role of the van der Waals forces as implied by the discussion of Eq. (33). Donald(64) observed a significant Coulomb electrostatic image force contribution while Mastrangelo(65) concluded that the van der Waals forces were dominant. More recently, Lee and Ayala(66) have presented centrifuge adhesion data that show a strong dependence on toner charge. Measurements of toner adhesion by electric field detachment have been limited since knowledge of the detachment force is more complicated in comparison to the centrifugal force. Goel and Spencer(13) employed both electric field and centrifuge measurements to conclude that for large toner particles the image force is important while for small toner, the van der Waals forces are dominant. Hays(67) obtained simultaneous measurements of the particle charge and adhesion to conclude that the adhesive force was greater than what one expects for an electrostatic image force model of a uniformly charged spherical particle. The conflicting data suggest that both the van der Waals and electrostatic image forces can be significant depending on the material, particle size, and measurement conditions.

Most particles such as xerographic toner are not perfect spheres but are highly irregular in shape. The irregular shape implies the charge distribution on insulative particles can be nonuniform. Since Eq. (33) assumes spherical uniformly charged particles, one must reexamine the validity of the assumptions in describing the adhesion of irregularly shaped particles. Lee and Jaffe(68) discuss this point in interpreting their centrifuge adhesion measurements, which exhibit a strong dependence on toner charge. They argue that the surface roughness due to asperities on particles reduces the van der Waals forces since only a few small regions actually contact the substrate. To explain the enhanced adhesion due to the Coulomb electrostatic image force, they show that the particles tend to orient to bring a patch of charges adjacent to the substrate. By assuming a few percent of the particle area to be charged, the magnitude of the adhesion due to a Coulomb electrostatic force is comparable to the centrifuge adhesion measurements.

If an external electric field E_A is applied to a spherical charged insulative particle on a conductive substrate, the applied force F_A is(62,67)

$$F_A = \beta QE_A + 4\pi\gamma\epsilon_0 b^2 E_A{}^2 \tag{34}$$

The coefficients β and γ for the electrostatic force and polarization force, respectively, depend on κ_p. For a range in κ_p from 1 to 4, β varies from 1 to 1.6 and γ varies from 0 to

0.25. The electrostatic force is either directed toward or away from the substrate depending on both the sign of the charge and direction of the applied field. The polarization force is independent of the field direction and is always directed toward the substrate.

Since the adhesion of polymeric particles can increase due to time-dependent interfacial deformation, it is of interest to measure the adhesion for short contact times such as in the attachment of a particle to a substrate. From an analysis of spherical bouncing charged particles, one can measure the adhesive energy between a particle and substrate for an interaction time on the order of microseconds.[69] Figure 9 shows the charge induced in one electrode of a cavity when a 12-μm-diameter spherical charged particle is injected into the cavity and bounces back and forth. As the particle loses energy on each bounce, it is captured on one of the electrodes by the Coulomb electrostatic image force. The particle continues to lose energy and finally attaches to the electrode by a short-range force. By measuring the time between bounces, one can calculate the average speed of the particle and the kinetic energy. By determining the minimum kinetic energy before the particle is attached to an electrode, one can calculate the adhesive energies due to the Coulomb electrostatic image force and short-range force. The electrostatic adhesive energy for a uniformly charged spherical dielectric particle is given by the expression

$$E_e = Q^2/(16\pi\epsilon_0 b) \tag{35}$$

where we have neglected a polarization correction of approximately 10%. The electrostatic adhesive energy is 1.4×10^{-13} J for a particle charge of 1.9×10^{-14} C which is a typical xerographic charge. The short-range adhesive energy is also 1.4×10^{-13} J, which is 500 times larger than the value calculated for the van der Waals adhesive energy

$$E_s = Ab/(6Z) \tag{36}$$

Dahneke has also measured anomalously high adhesive energies for uncharged micrometer-diameter polystyrene latex spheres.[70] The unexpected large adhesive energy was attributed to an increase in the van der Waals forces by an elastic flattening theory[71] but this model has been questioned.[72]

Derjaguin and coworkers have published extensive theoretical and experimental studies on the adhesion of elastic particles to rigid substrates. They show that when a stress is applied to an elastic spherical particle on a rigid substrate, the flattening of the particle at

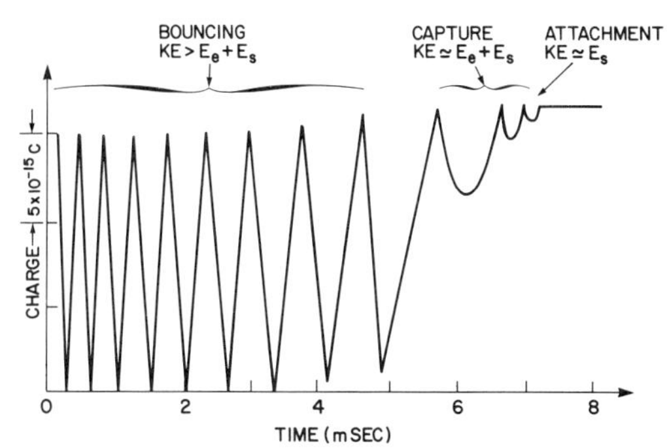

FIGURE 9. Charge induced in a cavity electrode due to a bouncing spherical charged particle.

the contact point decreases the van der Waals forces when considering the contributions from both inside and outside of the contact zone.[73] The maximum force required to detach a particle that has been elastically flattened corresponds to the instant when the particle makes a point contact. The adhesive force and adhesive energy are given by Eqs. (31) and (36), respectively, for a rigid spherical particle.

Derjaguin *et al.*[74] have made simultaneous measurements of the adhesive force and charge on particles removed from a substrate by impact acceleration. To obtain sufficient acceleration to detach the particles, a bullet with different energies is directed onto the backside of a target. The impact creates a high positive acceleration which presses the particle onto the substrate followed by a high negative acceleration which enables particle detachment. Measurements of the particle charge show that the detached particles have a much larger charge than initially. The increased particle charge is attributed to contact charge exchange in the contact zone formed by the positive acceleration. As the particle is separated from the substrate, an adhesive force given by Eq. (15) for the contact area must be overcome. From an analysis of the contact area based on a Hertz theory describing the dependence of the particle shape on stress, one would expect the dependence of the adhesive force on particle size to be near-quadratic as experimentally observed.[74] From a calculation based on measurements of the adhesive force and particle charge, the magnitude of the contact charge exchange density in the contact area is on the order of 300 to 30,000 $\mu C/m^2$. These observations corroborate the role of electrostatics in the adhesion of elastic particles detached by impact acceleration. Derjaguin *et al.*[75] reported similar effects for an elastic particle impinging on a surface.

To separate out the contributions of the van der Waals from the electrostatic force due to charge exchange at an interface, one would like to measure the adhesion as a function of the charge exchange density. One approach for varying the charge exchange density between metals and insulators is to measure the dependence of the adhesion on the metal work function since the van der Waals forces should not be strongly dependent on the metal type. Another approach is to use a photoconductor for the insulator since the contact charging can be modulated by light exposure at the interface. This approach has been described by Krupp and Schnabel[76] and Derjaguin *et al.*[77] Krupp and Schnabel obtained centrifuge adhesion measurements on 3 to 5 μm zirconium coated gold spheres dusted on photoconductive cadmium sulfide. Light exposure changed the adhesive force by a factor of up to three which demonstrates unambiguously that for some material combinations the electrostatic force exceeds the van der Waals forces. In the studies reported by Derjaguin *et al.*, impact acceleration adhesion measures were obtained for 5 to 30 μm polymeric particles on photoconducting selenium. The adhesive properties were largely influenced by donor–acceptor interactions between functional groups of the polymers and selenium surface but a dependence on light exposure was also observed which reflected the spectral response of the photoconductor. These observations are consistent with a view that electrostatic forces dominate in the adhesion of particles to surfaces.

A variety of methods have been used for supplying detachment forces to measure the adhesion between particles and other materials.[61] The more widely used methods are based on acceleration since, if the mass of the particle is known, one can readily calculate the applied force. Examples of applied forces supplied by acceleration include centrifuge, impact, and oscillatory vibration. Examples of nonacceleration methods for supplying detachment forces include electric field detachment, air drag, and cantilever beam. Although adhesion measurements of single particles with a cantilever beam are tedious, the method enables simultaneous measurements of the adhesion and charge exchange. Uber *et al.*[78] have described measurements on particles attached to a thin quartz cantilevered beam. When another material is contacted with the side of the particle, the particle will adhere since the beam deflects when the contacting material is retracted. By observing the maximum deflection at the moment of separation, one can calculate the adhesive force from a knowledge of the mechanical and physical properties of the beam. Furthermore, the particle contact charging can be determined by applying an electric field normal to the beam and observing the deflection. No correlation was observed between the adhesive force and charge transfer, however.

10. SUMMARY

The electrical component of the adhesive force between the planar surfaces of solids becomes important if the charge exchange density corresponds to 10^{17} or greater electronic charges per square meter, about 1% of the surface atoms. This charge transfer corresponds to 16,000 $\mu C/m^2$. The maximum density measured upon separation of materials is 1600 $\mu C/m^2$, an order of magnitude less. Since considerable charge back-flow can occur upon separation, we cannot rule out the possibility that the charge exchange density during contact is much greater. This is, in fact, the viewpoint of Deryaguin and collaborators from various observations including high-energy emissions during separation.

The adhesion of charged particles in a dry ambient is set by the Coulomb electrostatic image force and the van der Waals forces. The electrostatic image force is enhanced if the charge distribution on the particle is nonuniform. The adhesive energy due to the electrostatic image force for a typical xerographic particle charge is comparable to the adhesive energy due to a short-range force.

NOMENCLATURE

A	Hamaker constant
α	Polarizability
$\mathbf{D}$	Electric displacement
Δ	Energy
e	Electronic charge
$\mathbf{E}$	Electric field
ϵ_0	Permittivity of free space
ϵ	Permittivity
f	Fraction
$f(U)$	Fermi function
$\mathbf{F}$	Force
k	Boltzmann constant
κ	Dielectric constant
N	Number density
p	Dipole

P	Pressure
ϕ	Contact potential
PE	Potential energy
q	Volume charge density
Q	Charge
ρ	Density of states
S	Surface area
σ	Surface charge density
T	Temperature
Θ	Angle
U	Energy
U_F	Fermi energy
V	Voltage
W	Work

REFERENCES

1. A. D. Moore (Ed.), *Electrostatics and Its Applications,* Wiley, New York (1973).
2. B. V. Deryagin [Derjaguin], N. A. Krotova, and V. P. Smilga, *Adhesion of Solids,* Consultants Bureau, New York (1978).
3. A. J. Kinloch, Review: The science of adhesion, Part 1, Surface and interfacial aspects, *J. Mater. Sci.* **15,** 2141–2166 (1980).
4. J. N. Israelachvili and D. Tabor, The measurements of van der Waals dispersion forces in the range 1.5 to 130 nm, *Proc. R. Soc. London, Ser. A* **331,** 19–39 (1972).
5. K. L. Mittal (ed.), *Surface Contamination: Genesis, Detection and Control,* Vol. I and II, Plenum Press, New York (1979).
6. W. R. Smythe, *Static and Dynamic Electricity,* McGraw-Hill, New York (1939).
7. J. M. Crowley, *Fundamentals of Applied Electrostatics,* Wiley, New York (1986).
8. J. D. Jackson, *Classical Electrodynamics,* Wiley, New York (1962).
9. A. M. Stoneham and P. W. Tasker, Metal–nonmetal and other interfaces: the role of image interactions, *J. Phys. C* **18,** L543–L548 (1985).
10. Sticky film uses static electricity, *The Japan Economic Journal* (Mar. 15, 1986).
11. J. D. Cobine, *Gaseous Conductors,* Dover Publications, New York (1958).
12. B. V. Deryagin [Derjaguin] and V. P. Smilga, Electronic theory of adhesion, *J. Appl. Phys.* **38**(12), 4609–4616 (1967).
13. N. S. Goel and P. R. Spencer, in: *Adhesion Science and Technology* (L-H. Lee, ed.), Vol. 9B, pp. 763–829, Plenum Press, New York (1975).
14. H. Krupp, Particle adhesion theory and experiment, *Adv. Colloid Interface Sci.* **1,** 111–239 (1966).
15. W. R. Harper, *Contact and Frictional Electrification,* Oxford University Press (1967).
16. J. Lowell and A. C. Rose-Innes, Contact electrification, *Adv. in Phys.* **29**(6), 947–1023 (1980).
17. J. R. Smith, Self-consistent many-electron theory of electron work functions and surface potential characteristics for selected metals, *Phys. Rev.* **181**(2), 522–529 (1969).
18. W. R. Harper, The Volta effect as a cause of static electrification, *Proc. R. Soc. London, Ser. A* **205,** 83–103 (1951).
19. M. M. Bredov and I. Z. Kshemianskaia, The electrification observed after the contact of two bodies, *Sov. Phys. Tech. Phys.* **2,** 844–850 (1957).
20. J. Lowell, Contact electrification of metals, *J. Phys. D.* **8,** 53–63 (1975).
21. S. M. Sze, *Physics of Semiconducting Devices,* Wiley, New York (1981).
22. C. B. Duke and C. Mailhiot, A microscopic model of metal–semiconductor contacts, *J. Vac. Sci. Technol. B* **3**(4), 1170–1177 (1985).

23. C. Dervos and W. S. Truscott, Physical processes for contact charge transfer, *J. Electrost.* **16,** 137–146 (1985).
24. A. D. Roberts, Surface charge contribution in rubber adhesion and friction, *J. Phys. D.* **10,** 1801–1819 (1977).
25. D. K. Davies, Charge generation on dielectric surfaces, *Br. J. Appl. Phys.* **2,** ser. 2, 1533–1537 (1969).
26. A. R. Akande and J. Lowell, Contact electrication of polymers by metals, *J. Electrost.* **16,** 147–156 (1985).
27. T. J. Fabish, H. M. Saltsburg, and M. L. Hair, Charge transfer in metal/atactic polystyrene contacts, *J. Appl. Phys.* **47**(3), 930–939 (1976); The distribution of localized electronic states in atactic polystyrene, *J. Appl. Phys.* **47,** 940–948 (1976).
28. G. A. Cottrell, J. Lowell, and A. C. Rose-Innes, Charge transfer in metal-polymer contacts and the validity of "contact charge spectroscopy," *J. Appl. Phys.* **50**(1), 374–376 (1979).
29. J. Lowell, Charge accumulation by repeated contacts of metals to insulators, *J. Phys. D* **17,** 1859–1870 (1984).
30. S. Cunningham and A. Goodings, Plastic deformation and the transfer of electrostatic charge between metals and polymers during repeated-contact experiments, *J. Electrost.* **18,** 103–108 (1986).
31. H. Bauser, W. Klopffer, and H. Rabenhorst, in: *Advances in Static Electricity,* Vol. 1, pp. 2–9, Auxilia S.A., Burssels (1970).
32. C. B. Duke and T. J. Fabish, Charge-induced relaxation in polymers, *Phys. Rev. Lett.* **37,** 1075–1078 (1976).
33. C. B. Duke, T. J. Fabish, and A. Paton, Influence of polarization fluctuations on the electronic structure of molecular solids, *Chem. Phys. Lett.* **49**(1), 133–136 (1977).
34. C. B. Duke and T. J. Fabish, Molecular charge states and contact charge exchange in polymers, *J. Appl. Phys.* **48**(10), 4256–4266 (1977).
35. C. B. Duke, W. R. Salaneck, T. J. Fabish, J. J. Ritsko, H. R. Thomas, and A. Paton, Electronic structure of pendant-group polymers: molecular-ion states and dielectric properties of poly(2-vinyl pyridine), *Phys. Rev. B* **18**(10), 5717–5739 (1978).
36. T. J. Fabish, Electronic structure of polymers, *CRC Crit. Rev. Solid State Sci.* **8**(4), 383–420 (1979).
37. T. J. Fabish, private communication.
38. J. A. Medley, The electrostatic charging of some polymers by mercury, *Br. J. Appl. Phys.* **4** Suppl. 2, S28–S30 (1953).
39. D. K. Donald, Contact electrification of insulators and its relevance to electrets, *J. Electrochem. Soc.* **115,** 270–272 (1968).
40. D. A. Hays, Contact electrification between mercury and polyethylene: effect of surface oxidation, *J. Chem. Phys.* **61,** 1455–1462 (1974).
41. D. A. Hays and P. K. Watson, Contact charging of polymers, in: *Second International Conference on Electrophotography* D. R. White, ed., pp. 108–112, Soc. Phot. Sci. Eng. (1974).
42. A. El-Kazzaz and A. C. Rose-Innes, Contact charging of insulators by liquid metals, *J. Electrost.* **16,** 157–163 (1985).
43. F. M. Fowkes, in: *Materials Research Society Symposium Proceedings* E. A. Giess, K.-N. Tu, and D. R. Uhlmann, eds., Vol. 40, pp. 239–250, Materials Research Society, Pittsburgh, Pennsylvania (1985).
44. F. M. Fowkes, in: *Surface and Interfacial Aspects of Biomedical Polymers,* J. D. Andrade, ed., Vol. 1, pp. 337–372, Plenum Press, New York (1985).
45. F. Nordhage and G. Backstrom, Electrification in an electric field as a test of the theory of contact charging, *Static Electrification,* Inst. of Phys. Conf. Ser. No. 27, P. 84, The Institute of Physics, London and Bristol (1975).
46. D. A. Hays, The contact potential and charge exchange at a mercury–polymer interface, *Static Electrification,* Inst. Phys. Conf. Ser. No. 48, p. 265, The Institute of Physics, London and Bristol (1979).
47. K. P. Homewood, An experimental investigation of the depth of penetration of charge into insulators contacted by a metal, *J. Phys. D.* **17,** 1255–1263 (1984).
48. W. Possart and A. Roder, Measurement of electrical potential distribution in a polymer near the contact to a metal by means of scanning electron microscopy, *Phys. Status Solidi* **84,** 319–325 (1984).
49. J. Lowell, Tunnelling between metals and insulators and its role in contact electrification, *J. Phys. D.* **12,** 1541–1554 (1979).
50. V. A. Klyuev, E. S. Revina, V. I. Anisimova, Yu, A. Khrustalev, and Yu, P. Toporov, Light emission in breaking adhesive bonds under high vacuum, *Kolloidn. Zh.* **41**(2), 348–351 (1979).

51. V. I. Anisimova, V. A. Klyuev, T. N. Vladykina, N. A. Krotova, Yu. P. Toporov, and B. V. Derjaguin, Parameters of x-ray radiation during breakdown of adhesive contact under vacuum, *Dokl. Phys. Chem.* **233**(1–3), 239–242 (1977).
52. J. T. Dickinson, L. C. Jensen, and A. Jahan-Latibari, Fracto-emission: the role of charge separation, *J. Vac. Sci. Technol. A* **2**(2), 1112–1116 (1984).
53. B. V. Derjaguin and Yu. P. Toporov, Comments on "The question of emission of charged particles in failure of solids," *J. Mater. Sci.* **19,** 2423–2426 (1984).
54. J. T. Dickinson, Reply to 'Comments on "On the question of emission of charged particles in the case of failure of solids," ' *J. Mater. Sci.* **19,** 2426–2430 (1984).
55. D. M. Young and A. D. Crowell, *Physical Adsorption of Gases,* Butterworths, Washington (1962).
56. F. C. Tompkins, in: *The Solid–Gas Interface* (E. A. Flood, ed.), pp. 765–786, Marcel Dekker, New York (1967).
57. P. R. Antoniewicz, Surface-induced dipole moments of adsorbed atoms, *Phys. Rev. Lett.* **32**(25), 1424–1425 (1974).
58. T. Smith, Monomolecular films on mercury, *Adv. Colloid Interface Sci.* **3,** 161–221 (1972).
59. K. W. Bewig and W. A. Zisman, Surface potentials and induced polarization in nonpolar liquids adsorbed on metals, *J. Phys. Chem.* **68**(7), 1804–1813 (1964).
60. W. Pong, d. Brandt, Z. X. He, and W. Imaino, Contact charging of insulating polymers, *J. Appl. Phys.* **58**(2), 896–901 (1985).
61. A. D. Zimon, *Adhesion of Dust and Powder,* Consultants Bureau, New York (1982).
62. M. H. Davis, Electrostatic field and force on a dielectric sphere near a conducting plane—A note on the application of electrostatic theory to water droplets, *Am. J. Phys.* **37**(1), 26–29 (1969).
63. R. A. Bowling, An analysis of particle adhesion on semiconductor surfaces, *J. Electrochem. Soc.: Solid State Science and Technology* **132**(9), 2208–2214 (1985).
64. D. K. Donald, in: *Recent Advances in Adehsion* L. H. Lee, ed., pp. 129–143, Gordon and Breach, New York (1973).
65. C. J. Mastrangelo, The effects of charge, size and shape on toner photoconductor adhesion in electrophotographic systems, *Photogr. Sci. Eng.* **26**(4), 194–197 (1982).
66. M. H. Lee and J. Ayala, Adhesion of toner to photoconductor, *J. Imaging Technol.* **11**(6), 279–284 (1985).
67. D. A. Hays, Electric field detachment of toner, *Photogr. Sci. Eng.* **22**(4), 232–235 (1978).
68. M. H. Lee and A. B. Jaffe, Toner adhesion in electrophotography, to appear in Symposium Proceedings of Particles on Surfaces: Detection, Adhesion and Removal, 17th Annual Meeting of the Fine Particle Society, San Francisco (July 28–Aug. 2, 1986).
69. D. A. Hays and W. H. Wayman, Adhesion of charged particles, *Static Electrification, Inst.* Phys. Conf. Ser. No. 66. p. 237, The Institute of Physics, London and Bristol (1983).
70. B. Dahneke, Further measurements of the bouncing of small latex spheres, *J. Colloid Interface Sci.* **51**(1), 58–65 (1975).
71. B. Dahneke, The influence of flattening in the adhesion of particles, *J. Colloid Interface Sci.* **40**(1), 1–13 (1972).
72. B. V. Derjaguin, V. M. Muller, and Yu. P. Toporov, Effect of contact deformation on the adhesion of particles, *J. Colloid Interface Sci.* **53**(2), 314–326 (1975).
73. B. V. Derjaguin, V. M. Muller, Yu P. Toporov, and I. N. Aleinikova, The role of the pressing-on in the adhesion of elastic particles, *Powder Technol.* **37**, 87–93 (1984).
74. B. V. Derjaguin, I. N. Aleinikova, and Yu P. Toporov, On the role of electrostatic forces in the adhesion of polymer particles to solid surfaces, *Powder Technol.* **2**, 154–158 (1968/69).
75. B. V. Derjaguin, V. M. Muller, Yu P. Toporov, and N. S. Mikhovich, Influence of the contact electrification on the recoil and adhesion of particles impinging on a surface, Abstracts of Particles on Surfaces: Detection, Adhesion and Removal symposium, p. 39, 17th Annual Meeting of the Fine Particle Society, San Francisco (July 28–Aug. 2, 1986).
76. H. Krupp and W. Schnabel, Light-modulated electrostatic double layer adhesion, *J. Adhes.* **5**, 269–277 (1973).
77. B. V. Derjaguin, Yu P. Toporov, and I. N. Aleinikova, Electroadhesive properties of polymer particles in contact with the photoconductor surface, *J. Colloid Interface Sci.* **54**(1), 59–68 (1976).
78. A. E. Uber, J. F. Hoburg, and G. W. Penney, Experimental investigation of electrostatic effects in the adhesion of dry dusts, *IEEE Trans. Ind. Appl.* **IA-20**(1), 148–154 (1984).

9

Adhesion Through Silane Coupling Agents

Edwin P. Plueddemann

1. INTRODUCTION

A high school teacher invited four men from the community to meet with his class on creativity. He showed them an assortment of rocks on a table and asked for their comments.

The poet composed a short poem describing rocks as potential parts of a building or as missiles for harm. The artist discussed artistic arrangements, and the potential for sculpture that an artist can see in an individual rock. The biologist contemplated a life-form based on siloxane polymers rather than carbon chains.

The chemist mused on the water resistance of rocks, and how their chemistry could be used as a model for developing water resistant bonds in organic–inorganic composites. Silica and certain metal silicates withstand the action of water for hundreds of years even though it is known that individual siloxane bonds and metal oxane bonds are rather easily hydrolyzed. All of these bonds have a relatively high degree of ionic character, suggesting that bond making and breaking are controlled more by equilibrium constants than by rate constants. Although the bonds hydrolyze, they also reform rather easily. The rock structure has such good water resistance because each atom has multiple oxane bonds to its neighbor. Under equilibrium conditions it is highly improbable that three or four bonds will be hydrolyzed at a given instant if the equilibrium constant is at all favorable for bond formation.

Silane coupling agents provide a hybrid link between an organic matrix and inorganic reinforcements. They form stable covalent organic bonds with matrix polymers with high thermodynamic stability, and multiple inorganic bonds to the reinforcement that have favorable equilibrium constants for bond retention. Since the oxane bonds to the mineral surface are most subject to failure by hydrolysis, improvement in water resistance of composites may be expected if certain equilibrium requirements are satisfied:

- Drive the initial oxane bond reaction to completion.
- Provide an interphase region with minimum equilibrium concentration of water, i.e., make it hydrophobic.

Edwin P. Plueddemann • Dow Corning Corporation, Midland, Michigan 48640.

- Provide a high degree of crosslinking in the interphase region. Resistance to hydrolysis increases with an increasing degree of crosslinking.

Progress has been made in recent years in improving the performance of composites modified with silane coupling agents by applying these three principles. (See also the following chapter by Miller and Ishida on the interface of composites.)

2. GENERAL EQUILIBRIUM CONDITIONS

All silane coupling agents are of the general structure $X_3Si—R—Y$, where Y is an organofunctional group selected for bonding to organic polymers while X is a hydrolyzable group on silicon that forms Si—OH groups in water or may react directly with mineral hydroxide groups M—OH to form oxane bonds M—O—Si.[1] Y groups are attached to silicon through stable C—Si bonds. Other proposed commercial coupling agents (Table 1) have organofunctional groups bonded to a central atom[2] through oxygen. The resulting

TABLE 1
Representative Commercial Coupling Agents

Trade name	Organofunctional group	Chemical structure
A. Q9-6300	Vinyl	$(CH_3O)_3SiCH{=}CH_2$
B. Z-6076	Chloropropyl	$(CH_3O)_3SiCH_2CH_2CH_2Cl$
C. Z-6040	Epoxy	$(CH_3O)_3SiCH_2CH_2CH_2OCH_2\underset{\diagdown O \diagup}{CH—CH_2}$
D. Z-6030	Methacrylate	$(CH_3O)_3SiCH_2CH_2CH_2OOC(CH_3){=}CH_2$
E. A-1100 (UCC)	Primary amine	$(C_2H_3O)_3SiCH_2CH_2CH_2NH_2$
F. Z-6020	Diamine	$(CH_3O)_3SiCH_2CH_2CH_2NHCH_2CH_2NH_2$
G. Z-6062	Mercapto	$(CH_3O)SiCH_2CH_2CH_2SH$
H. Z-6032	Cationic styryl	H $(CH_3O)_3SiCH_2CH_2CH_2NHCH_2CH_2H \cdot HCl$ CH_2 $C_6H_4CH{=}CH_2$
I. S-3076S (Hercules)	Azide	$(CH_3O)_3Si—R—SO_2N_3$
J. Volan-A (DuPont)	Methacrylatochrome	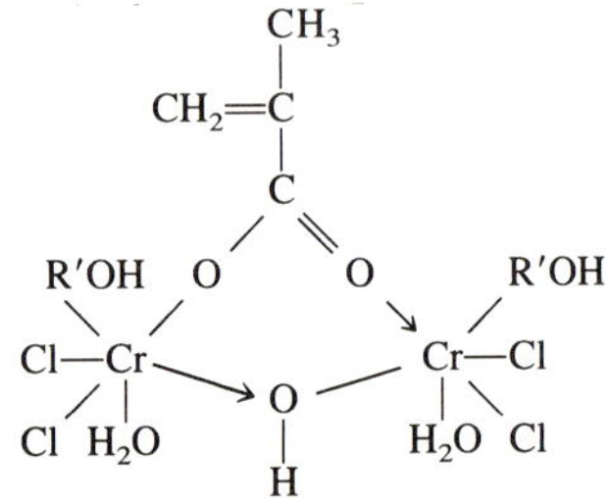
K. TTM-33 (Kenrich)	Methacrylate-titanate	$(CH_2{=}C(CH_3)COO)_3TiOCH(CH_3)_2$
L. Caveco-Mod (Cavedon)	Methacrylate-Al-zirconate	Undisclosed

Z—O—R—Y compounds have much poorer hydrolytic stability than comparable Si—R—Y structures, and therefore do not provide the degree of water-resistant "coupling" that is possible with silanes. Many of these materials are very surface active and may be effective dispersion aids, but a coupling agent should be defined as a material that provides a water-resistant bond between an organic matrix and an inorganic filler or reinforcement.

2.1. Reaction with the Mineral

Silane coupling agents are commonly applied to reinforcements (e.g., fiberglass) from aqueous solutions at a pH of about 4. Under these conditions the alkoxy groups hydrolyze rapidly to give solutions of the silane–triols $RSi(OH)_3$. When applied to a mineral substrate, these triols first hydrogen-bond to the hydroxyl surface and condense when dried to a polysiloxane structure having oxane bonds between the silane and mineral M—O—Si.[(2)] The rate of condensation of silanols is faster on E-glass than on silica or in the bulk.[(3)] Condensation of coupling agents with filler may be accelerated by adding a catalyst[(4)] or by drying the treated mineral at a moderately elevated temperature.[(5)] The catalyst may be an amine, an alkyl titanate, or an amino-organofunctional silane such as H of Table 1. The performance of some polyester/glass bead composites with various coupling agents are given in Table 2.

It has generally been observed that silanes on particulate or fibrous minerals are most effective if applied as monomers, but are more effective as primers on massive surfaces if applied as siloxane oligomers.[(6)] Condensed siloxanes derived from monomers are more highly crosslinked and provide better water resistance to particulate composites. Siloxane primers derived from oligomers are less highly crosslinked and may supply a tougher interphase region, which is especially important in bonding metal panels or other massive surfaces.

2.2. Hydrophobic Interphase Region

Hydrophobic resins with hydrophobic coupling agents produce composites with maximum water resistance. Vanderbilt and Jaruzelski[(7)] compared an epoxy resin having

TABLE 2
Glass Microbead[a] Filled Polyester Castings
(0.5% Silane as Integral Additive)

Coupling agent on glass	Reference to Table 1	Flexural dry	Strength of castings (MPa) 24-hour boil
Untreated		72	33
Vinyl silane	A	82	41
Methacrylate silane	D	145	67
Vinylbenzyl silane	H	154	62
9/1 mix	D/H	155	82
D + 1% BDMA[b]		143	85

[a]60% Potters 3000 microbeads in "Paraplex" P-43 Polyester.
[b]BDMA = benzyldimethylamine.

3% water absorption with a hydrocarbon resin (Buton) having 0.3% water absorption. After seven days in boiling water, a quartz-filled epoxy casting retained 80% of its dry strength, while a quartz-filled Buton casting retained 96% of its original strength. Silicones also may retain very good wet adhesion to silica even though silicones have a very high rate of water diffusion. The equilibrium amount of water at the interface is more important than the rate of diffusion of water in the polymer.

Some coupling agents, such as aminosilanes E and F are themselves hydrophilic, and any unreacted silane at the interface provides a weakness against hydrolysis. The aminosilane may be applied at very low concentrations or the treated surface may be rinsed with water to remove excess silane for optimum performance in adhesion.[8] The interphase region derived from amino-functional silanes may also be made more hydrophobic by mixing the aminosilane with relatively large proportions of hydrophobic silanes, such as those with methyl, phenyl, vinyl, or chloropropyl groups on silicon. Such mixed silanes should be useful with condensation polymers (urethanes, epoxies, phenolics, polyimides, etc.) where there are a limited number of functional groups in the resin to react with amino groups on the filler. Performance of mixed silanes in bonding two reactive urethane polymers to glass are shown in Table 3. As much as 98% inert hydrophobic silane could be used in the mixture of silanes to get optimum water resistance of the urethane bond to glass. Mixtures of neutral, hydrophobic silanes with unsaturated silanes in unsaturated polyester composites showed no advantage over the reactive silanes (D and H) themselves. Any increase in hydrophobicity contributed by the hydrophobic silane was counterbalanced by a decrease in reactivity with the resin.[9]

2.3. Optimum Structure in the Interphase Region

In 1965, Vanderbilt and Clayton[10] reported that the bond of vulcanized rubber to glass could be improved markedly by using a primer containing a reactive silane and a difunctional monomer. The reactive silane alone, or the silane with a monofunctional monomer, was much less effective. Crosslinking at the interface seemed to be a requirement for good adhesion.

The effect of polymer structure on adhesion was illustrated in a study of acrylic copolymers with constant silane modification but varying glass transition temperatures.[11] A series of copolymers of methyl methacrylate (MMA) and ethyl acrylate (EA) was prepared having glass transition temperatures between those of the homopolymerss: MMA = 105 °C and EA = −23 °C. Molecular weights were maintained at about 10,000 by including 1.5% mercaptopropyltrimethoxysilane as chain transfer agent. Additional silane modification was contributed by 1% methacryloxypropyltrimethoxysilane in each composition.

Films on glass microscope slides were dried for 30 minutes at 100 °C and examined for adhesion both dry and after soaking for one day in water. All films showed good dry adhesion, but copolymer films of intermediate composition that were flexible or rubbery showed poorest wet adhesion (Table 4). Similar acrylic copolymers without silane modification generally had very poor wet adhesion to glass.

Retention of adhesion across the interface is best if the silane-modified polymer is either tarry (has viscous flow) or rigid. It is virtually impossible to obtain a direct water-resistant bond between a flexible, or elastomeric, polymer and a hydrophilic mineral.

TABLE 3
Bonding of Flexible Two-Component Urethane to Glass
(Cure 2 h at 80°C)

$RSi(OMe)_3$ in 9/1 Mix with Silane F in Primer	Peel Strength (N/cm)		
	Dry	2 h boil H_2O	5 h boil H_2O
Unprimed control	0.5	nil	nil
Methyl silane G	1.6	0.1	nil
Vinyl silane A	1.8	0.6	nil
Phenyl silane I	C	C	C
Chloropropylsilane B	C	C	C

C = Cohesive failure in urethane at >10 N/cm

Bonding of RIM Urethane to Glass
(15 min at 80°C)

Primer on glass 30% IPA solution of silanes	Peel strength after boiling in water (N/cm)	
	2 h	5 h
Unprimed control	nil	nil
Vinylsilane A	0.1	nil
Aminosilane F	0.5	nil
50/50 A/F	1.1	nil
70/30 A/F	3.4	1.3
80/20 A/F	C	0.1
90/10 A/F	C	C
98/2 A/F	C	C

C = Cohesive failure in polymer at >30 N/cm

TABLE 4
Methyl Methacrylate (MMA), Ethyl Acrylate (EA) Copolymers

Copolymer composition	A	B	C	D	E	F
MMA	100	70	50	40	30	20
EA	—	30	50	60	70	80
Mercaptosilane	1.5	1.5	1.5	1.5	1.5	1.5
Methacrylate silane	1.0	1.0	1.0	1.0	1.0	1.0
Film adhesion[a] to glass	Very hard	Hard	Tough	Rubber	Soft	Tarry
dry	10	10	10	10	10	10
wet (1 day)	10	10	9	5	8	9

[a]Rate of adhesion: 0 = nil to 10 = excellent.

Water-resistant adhesion may be obtained by modifying the interface through application of a primer or by surface reactions of a silane-modified polymer. The modified interface layer may be very thin (a few hundred angstroms) without altering the flexibility of the polymer itself. In general, thermoplastic elastomers are best bonded through a tarry primer, while vulcanized elastomers are best bonded through a crosslinked, resinous interphase region.

These observations along with many others are illustrated graphically in Figure 1. Structural requirements of the interphase region may be understood in terms of equilibrium bonding across the interface through oxane bonds in the presence of water. It is possible to obtain water-resistant composites if equilibrium conditions can be maintained between a silane-modified polymer and the mineral surface. Equilibrium conditions are lost if silanol groups, resulting from hydrolysis at the interface, are removed physically from the interface. The various structures in the interphase region are all assumed to have optimum silane modification to form oxane bonds —M—O—Si— to a hydrophilic mineral surface M—O—H.

- Monomers and oils may be removed physically from the interface as individual links to the surface hydrolyze, and they are therefore lost for rebonding.
- Tarry polymers with viscous flow are self-healing. As silanol bonds are formed by hydrolysis at the interface, they flow to an active site and re-form oxane bonds. Failure is generally cohesive within the polymer, but polymer strength is not very great.
- At a rubbery interface, silanol groups formed by hydrolysis retract from the surface, removing themselves physically from any possibility of reforming oxane bonds. Adjacent silanols may also condense to siloxanes that are not useful for adhesion (Figure 2). Failure is always adhesively to the substrate.
- Silanols formed at a rigid interface cannot move beyond the dimensions of molecule segments and so are in a position to re-form bonds with the original or adjacent active sites on the surface. Excessive crosslinking at the interface will provide good chemical resistance of bonding, but may lead to brittle mechanical failure.

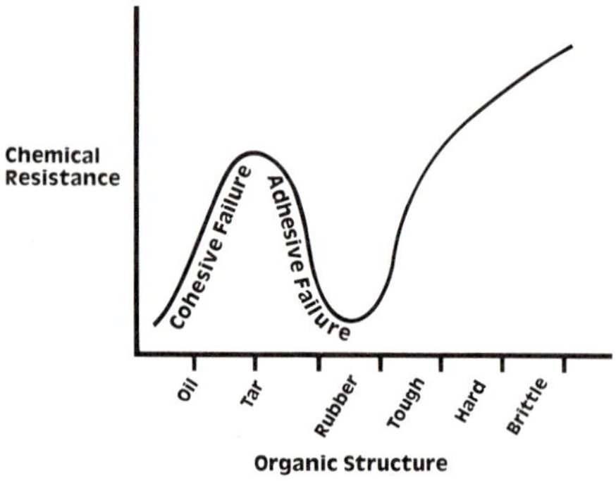

FIGURE 1. Relation of structure of the interphase region with chemical resistance of composites.

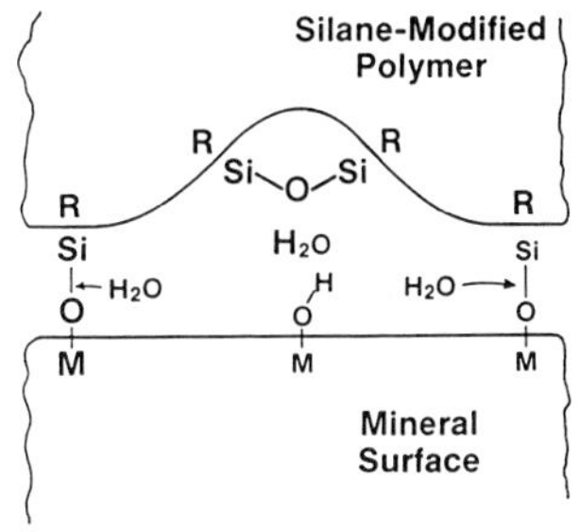

FIGURE 2. Reasons for optimum structure requirements.

3. CROSSLINKING THE INTERPHASE REGION

High-temperature epoxies, phenolics, and polyimides are very rigid at room temperature and so have favorable structures for water resistant bonds to glass, metals, and other mineral surfaces. Almost any compatible silane coupling agent will be effective in developing very water-resistant composites. It is often desired to prepare water-resistant composite from less rigid resins. In this case, it is important to generate a maximum degree of crosslinking in the interphase region while retaining the toughness of the resin itself.

3.1. Inhibition of Cure by Fillers

Mineral fillers and reinforcements have catalytic effects on thermosetting resins that generally inhibit their cure. Such inhibition has been observed especially with unsaturated polyesters cured with peroxides and with epoxies cured with amines. In both polyesters and epoxies, it was observed that silane treatments of fillers overcome cure inhibition as measured by cure exotherms (Table 5). Silanes that allowed maximum exotherms, i.e., that were most effective in overcoming surface inhibition, were generally most effective coupling agents, as indicated by chemical resistance of the composites.[12]

TABLE 5
Resin Exotherms with Silane-Treated Glass Filler
(5-g E-glass in 30-g Resin)[a]

Organofunctional silane 0.2% on glass	Polyester-Bz_2O_2 ΔT(°C)	Epoxy-"Z" ΔT(°C)
Untreated	−22	−17
Epoxy	−20	−8
Diamine	−15	−1
Mercaptan	−11	−8
Cationic styryl	−10	0
Phenyl	−7	−15
Methacrylate ester	−3	—

[a]Exotherms in oil bath at 115 °C (epoxy) or water bath at 90 °C.

3.2. Inhibition of Cure by Silanes

Aminofunctional silanes are generally not the best coupling agents to use on reinforcements in anhydride-cured epoxy composites. Chiang and Koenig[13] demonstrated by FTIR studies of these systems that the anhydride reacted with primary amine groups of the silane to form cyclic imides. Thus, although complete reaction of epoxy and anhydride were obtained, some of the potential crosslinking in the interphase region was lost. Silanes B, C, or H of Table 1 are recommended rather than silanes E or F for composites with anhydride-cured epoxies.

General-purpose polyesters are often cured with a ketone peroxide with cobalt naphthenate promoter and a tertiary amine accelerator. It has been observed that carboxy-functional silanes are not effective in composites of these resins while cationic silanes are most effective. This is not surprising, since carboxyl groups would remove the amine accelerator from the interphase region, while a cationic silane (H of Table 1) would provide additional acceleration to obtain maximum cure in the interphase region.

3.3. Additional Crosslinking Through the Coupling Agent

It was observed some time ago that the relative performance of a series of unsaturated silane coupling agents in styrene-diluted polyester composites was directly related to their relative reactivity in copolymerizing with styrene[14] (Table 6). A methacrylate-functional silane (D of Table 1) has been the world standard for coupling unsaturated polyesters since 1962.

Poly-acrylic-functional silanes were tested as coupling agents in polyester composites as a means of obtaining additional crosslinking in the interphase region.[15] Poly-functionality, in itself, was not sufficient to provide better performance than the standard methacrylate-functional silane. Bis-methacrylate or acrylate functional silanes attached to

TABLE 6
Relation of Reactivity with Styrene to Effectiveness of Silane Finishes

Silane finish reactive group	Relative reactivity with styrene	Order of wet strength of polyester laminates
Fumarates	3.3	2
Methacrylates	1.9	1
Acrylates	1.34	3
Isoprene	1.3	9
Styrene	1.0	5
Vinyl sulfides	0.2	6
Maleates	0.15	4
Crotonates	0.05	11
Vinyl Si (vinyl Cl)	0.05	7
Allyl Si (allyl Cl)	0.03	8
Allyl esters	0.011	10
Vinyl ethers	0.01	12

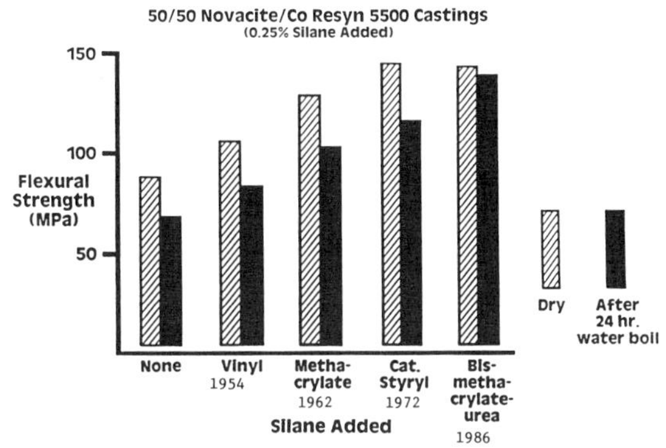

FIGURE 3. History of silane coupling agents.

silicon through flexible branched chain groups probably undergo a high degree of cyclocopolymerization which does not contribute more crosslinking than a monofunctional monomer. Esters prepared by reacting isocyanotoethyl methacrylate with polyamine-functional silanes have unique performance in developing water-resistant composites, possibly because rigid urea linkages in the molecule allow a higher degree of crosslinking. Performance of Novacite silica-filled polyester casting with various silanes as integral additives are shown in Figure 3. The new bis-methacrylate functional silane appears to be a significant improvement over silane D (Table 1) in such composites, especially in resistance to deterioration by water.

3.4. Additional Crosslinking Through Siliconates

There have been some efforts to mix orthosilicates (e.g., tetraethyl silicate) with standard commercial silane coupling agents.[16] Mixtures with ethyl silicate were of lower cost, but generally not as good coupling agents as the pure silanes themselves. Methyl silicate might be more effective than ethyl silicate, but it has not been considered seriously because of its toxicity. Compounds with the general structure $(CH_3O)_3Si—R—Si(OCH_3)/3$ (Table 7) were compared as potential crosslinkable adhesion enhancers[17] with ethyl silicate and phenyltrimethoxy silane in mixtures with Z-6032 coupling agent as primers for thermoplastics elastomers on glass. All of the additives improved the performance of Z-6032 as a primer, but additives 1 and 4 were most universally effective (Table 8). The same adhesion enhancers were tested with Z-6030 and with Z-6032 as additives to novacite-filled unsaturated polyester castings as summarized in Table 9. Compounds 1–4 of Table 7 were especially promising, with compound 4 being the best.

Mixtures of 9/1 by weight of Z-6020 and crosslinkers of Table 7 were used as sole curing agents for a bisphenol diglycidyl ether (Dow DER-330). Amine equivalent weights of the mixtures were calculated as 82 for mixing with stoichiometric epoxy. Films of these mixtures were cured on glass and aluminum for four days at room temperature. Coated coupons were boiled for six hours in water and allowed to stand two days more in water at room temperature. Adhesion of the films were rated qualitatively by attempting to loosen

TABLE 7

Crosslinkable (Potential) Adhesion Enhancers

1. $(MeO)_3SiCH_2CH_2Si(OMe)_3$
2. $(MeO)_3SiCH_2CH_2CH_2Si(OMe)_3$
3. $(MeO)_3Si(CH_2)_6Si(OMe)_3$
4. 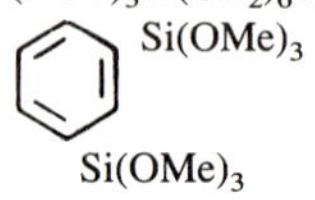 also *para* isomer
5. $PhSi(OMe)_3$
6. $Si(OEt)_4$
7. $(MeO)_3SiCH_2CH_2\underset{Me_2}{Si}—O—\underset{Me_2}{Si}CH_2CH_2Si(OMe)_3$

TABLE 8

Wet Peel Strength of Thermoplastic Elastomers to Primed Glass (N/cm)

9/1 Mix of Z-6032/additive in primer (cf. Table 7)	CXA 2022[a] in 70 °C water 1 h	Kraton 1102[b] in 100 °C water 6 h
Z-6032 (no additive)	12	nil
1[c]	24	19
2	17	14
3	9	1
4	16	32
5	17	1
6	5	11
7	8	19

[a]CXA 2022 = Ethylene-vinylacetate terpolymer of undisclosed composition (DuPont)
[b]Kraton 1102 = Styrene-butadiene block copolymer (Shell)
[c]Nos. 1–7 refer to the entries in Table 7.

TABLE 9

50/50 Novacite-filled Polyester Castings (Co-Resyn 5500) (Add 0.5% Silane Based on Filler)

Coupling agents as additives	Flexural Strengths of Castings (N/cm)	
	Dry	After 24 h water boil
None	88	67
Q9-6300 (vinyl)	107	82
Z-6030 (methacrylate)	130	101
Z-6030 9/1 mix with (1)[a]	160	126
Z-6032 (Cationic Styryl)	143	114
Z-6032 9/1 mix with (1)	146	125
Z-6032 9/1 mix with (2)	130	128
Z-6032 9/1 mix with (3)	138	128
Z-6032 9/1 mix with (4)	145	137
Z-6032 9/1 mix with (5)	124	118
Z-6032 9/1 mix with (6)	136	121
Z-6032 9/1 mix with (7)	123	119

[a]Nos. 1–7 refer to entries in Table 7.

TABLE 10
9/1 of Z-6020/Crosslinker as Curing Agents for DER-331
(Cured 4 Days at Room Temperature)

Silane mix as curing agent	Wet adhesion of film[a]	
	Glass	Aluminum
Z-6020 alone	fair	fair
9/1 Z-6020/(1)[b]	exc.	good
9/1 Z-6020/(2)	fair	fair
9/1 Z-6020/(3)	fair	poor
9/1 Z-6020/(4)	exc.	good
9/1 Z-6020/(5)	good	poor
9/1 Z-6020/(7)	good	poor

[a]Wet = 6 hours in boiling water plus 2 days in room-temperature water.
[b]Nos. 1–7 refer to entries in Table 7.

with a razor blade (Table 10). Again, compounds 1 and 4 of Table 7 seem to be outstanding as adhesion enhancers when used with Z-6020 silane.

Performance of adhesion enhancers for standard silane coupling agents is related to the degree of crosslinking provided by the enhancer. Bis-trimethoxysilyl compounds of structure $(MeO)_3Si—R—Si(OMe)_3$ are more effective than $PhSi(OMe)_3$ or $Si(OEt)_4$. Bis-trimethoxysilyl compounds with small —R— groups are generally better than those with long-chain —R— groups. *Meta*-bis-trimethoxysilyl benzene (compound 4 of Table 7) is uniformly the best of the series. Siloxanes derived from this compound must be highly crosslinked owing to spatial orientation of silyl groups that prevents initial formation of cyclic siloxanes.

4. CONCLUSION

The general concept of equilibrium bonding through oxane bonds in the presence of water has led to the development of greatly improved adhesion promoters for organic polymer to mineral substrates. This should lead to great improvement in corrosion resistance of polymer composites, and should lead to polymer-bonded metal structures with resistance to moisture comparable to that of silicate rocks. A new class of crosslinkable adhesion enhancers is now available that should be to standard silane coupling agents what octane enhancers are to straight-run gasoline.

REFERENCES

1. E. P. Plueddemann, *Silane Coupling Agents,* Plenum Press, New York (1982).
2. H. Ishida and J. L. Koenig, *J. Colloid Interface Sci.* **64**(3), 565 (1978).
3. C. H. Chiang, H. Ishida, and J. L. Koenig, *J. Colloid Interface Sci.* **74**(2), 396 (1980).
4. R. L. Kaas and J. L. Kardos, *SPE, 32nd ANTEC,* Paper 22 (1976).
5. H. A. Clark and E. P. Plueddemann, *Mod. Plast.* **40**(6), 133 (1963).
6. E. P. Plueddemann, *SPI, 39th Ann. Tech. Conf. Reinf. Plast.,* 4-C (1984).
7. B. M. Vanderbilt and J. J. Jaruzelski, *SPI, 17th Ann. Tech. Conf. Reinf. Plast.,* 10-D (1962).

8. A. G. Goury, A. H. Jerne, and P. T. Calahan, *Adhesion Aspect of Coatings* (K. L. Mittal, ed.), p. 170, Plenum Press, New York (1981).
9. E. P. Plueddemann and P. G. Pape, *Mod. Plast.* **63**(7), 78 (1986).
10. B. M. Vanderbilt and R. E. Clayton, *Ind. Eng. Chem. Prod. Res. Dev.* **4**(1), 16 (1965).
11. E. P. Plueddemann, *Adhes. Age* **18**(6), 36 (1975).
12. E. P. Plueddemann and G. L. Stark, *SPI, 35th Ann. Tech. Conf. Reinf. Plast.*, 20-B (1980).
13. C. H. Chiang and J. L. Koenig, *SPI, 35th Ann. Tech. Conf. Reinf. Plast.*, 23-D (1980).
14. E. P. Plueddemann, H. A. Clark, L. E. Nelson, and K. R. Hoffman, *Mod. Plast.* **39,** 135 (1962).
15. E. P. Plueddemann, P. G. Pape, and H. M. Bank, *41st Ann. Tech. Conf. Reinf. Plast.*, 2-D (1986).
16. P. C. Yates, U.S. Patent (to DuPont) 3,816,152 (1974).
17. E. P. Plueddemann, presented to 42nd Annual Tech. Conf. on Reinforced Plastics SPI, Cincinnati, Ohio (1987).

10

Adhesive–Adherend Interface and Interphase

James D. Miller and Hatsuo Ishida

1. INTRODUCTION

The science of adhesion is multidisciplinary in scope. Both chemical and physical forces contribute to adhesion and lack of adhesion. It is impossible to understand adhesive forces without an appreciation of both chemical and physical interactions. It is a goal of adhesion research to separate and measure the individual factors which lead to adhesion. This commendable goal has been attempted by some and achieved by few. Two difficulties have always hampered such a correlation. First, it is difficult to develop physical test methods which scientifically measure adhesive forces. Second, it has required decades for the development of analytical techniques which can begin to provide researchers with information about the nature of the chemical and physical forces at work at an interface or in an interphase.

It is necessary at this point to define what is meant by interface and interphase. An interface is a surface which is formed by the common boundary of two homogeneous phases in contact. By definition an interface has area but no volume. With respect to adhesion, the interface is the boundary between adherend and adhesive. Depending on one's outlook, the interface can be described on a subatomic, atomic, molecular, or more gross level. None of these descriptions is wrong when used in context, however, some confusion exists because the world can be viewed on different scales.

For example, one might be able to examine the difference in electron density with distance from a metal surface or interface. Such a detailed analysis may even make it impossible to specify the exact position of the surface due to gradients involved (e.g., electron spillout). Even though such a precise analysis may show that it is difficult to locate the exact position of the surface, one could paint the metal without expressing any difficulty about locating the surface.

The boundary between two homogeneous phases is rarely an interface but is often comprised of a third phase or interphase which has a characteristic thickness and proper-

---

James D. Miller and Hatsuo Ishida • Department of Macromolecular Science, Case Western Reserve University, Cleveland, Ohio 44106. *Present address of James D. Miller*: Amoco Performance Products, Inc., Bound Brook, New Jersey 08805.

ties which differ from those of either of the continuous homogeneous phases. Depending on the system it is usually possible to describe an interface between one or both of the homogeneous phases and the interphase.

Such a detailed explanation of terms would not normally be necessary. However, it is the interface and interphase regions which determine adhesion properties. The remainder of the text will be dedicated to research findings which describe the interface and interphase, how they relate to adhesion, and how they can be modified to improve the situation.

2. INTERFACE

Very few, if any, analytical techniques are capable of actually examining the interface between homogeneous phases *in situ*. This is due, in part, to interferences of the homogeneous phases. Therefore, most interfacial information is generated by artificially creating a surface which models the interface of interest. Then, highly sophisticated techniques which are capable of distinguishing the surface of a substrate from its bulk properties can be employed.

Typical of this type of study is the characterization of the chemical composition of surfaces which often differ from their bulk structure. The analysis of surfaces is important because it is this region of a substrate or adherend which will participate in the adhesion process. Although there is interest in promoting adhesion to as many types of surfaces as exist, we will restrict our survey to a limited number of examples which illustrate the utility of surface analysis.

Many factors must be considered when analyzing surface properties. These include, but are not limited to, chemical composition, processing variables, contaminants, surface treatments, and exposure to environmental conditions. An example is the analysis of commercially important carbon fibers used in high performance composites. The necessity of surface and interfacial analysis is illustrated by Larsen,[1] who stated that the greatest single problem associated with carbon and graphite fibers has been the interface area and the development of strong bonds between the fiber and matrix resin.

Carbon fibers are manufactured by a carbonization of polyacrylonitrile (PAN) or other starting materials. The surface of carbon fibers is believed to consist of a number of functional groups and residual elements. X-ray photoelectron spectroscopy (XPS) has been used to detail the surface functionality of these fibers.[2–7] Takahagi and Ishitani[6] have used XPS to show that oxygen-containing functionality is produced on the fiber surface as a result of the processing conditions where free radicals produced in the inert atmosphere of the furnace at high temperature are quenched on exposure to air. This oxygen-containing functionality is located in a thin layer near the surface since it is easily removed by 0.5 h of argon ion etching.

An important component of the carbon fiber surface is impurities such as alkali metal ions which may concentrate at the fiber surface. These are especially important as they have been correlated with the thermo-oxidative stability of the fibers and their composites. Scola and Laube[8] used ion scattering spectroscopy (ISS) and secondary ion mass spectroscopy (SIMS) to identify the surface elemental composition of carbon fibers from various suppliers. All the fibers examined showed the presence of carbon, oxygen, sodium, and potassium within 0.5–1 nm of the surface. Depending on the manufacturer, calcium, chromium, iron, and fluorine were also detected at the surface. The amount of each of these surface components also varied to a great extent depending on the manufacturer. The researchers also demonstrated that the surface composition was different from a region approximately 10 nm below the surface obtained by sputtering. Attempts at a correlation between surface chemistry and composite properties showed that the thermo-oxidative stability of polyimide (PMR-15) composites was poor with carbon fibers having high concentrations of sodium and potassium. It was suggested that this could indicate that the impurities accelerate the thermo-oxidation of the resin and fiber in the interface region or that the thermal-oxidative aging generates poor bonding at the fiber/resin interface.

Additional processing steps are applied to carbon fiber surfaces after they are formed to provide increased resin compatability and enhanced composite properties. These are usually oxidative treatments which produce functional groups on the surface and ultimately at the interface in the composite. XPS has also been used to identify these functional groups.[6,9–11] Takahagi and Ishitani[6] used digital subtraction of the C_{1s} spectra of unmodified carbon fibers from the spectra of moderately and extensively oxidized fibers. The difference spectra are illustrated in Figure 1. The oxidative treatment modified the surface to a great extent. Moderate oxidation left the surface with a composition of 73% hydroxyl, 17% carboxyl, and 10% carbonyl functionality. Extensive oxidation increased the concentration of acid groups producing 54% carboxyl, 24% hydroxyl, and 22% carbonyl functionality. Oxidation and other treatments are applied to carbon fiber surfaces in order to facilitate interfacial adhesion and thus increase composite properties. A good example of the correlation between interfacial adhesion, measurable surface properties, and composite performance is seen in Figure 2 using the carbon fiber specimens observed in Figure 1. They[6] demonstrate that the interlaminar shear strength of composites prepared from these fibers and an epoxy resin varies linearly with the oxygen/carbon ratio on the fiber surface as measured by XPS.

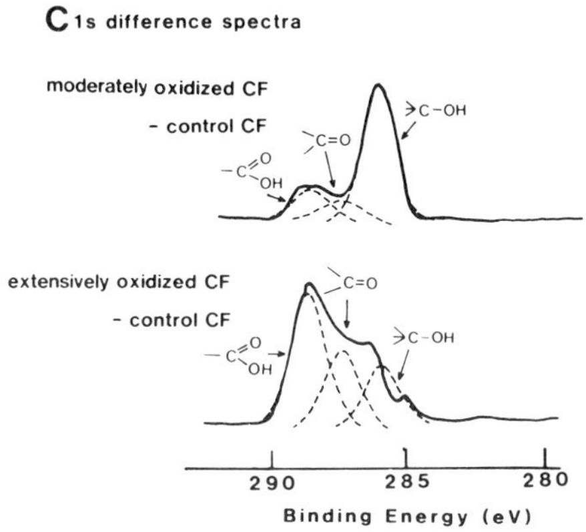

FIGURE 1. XPS digital difference spectra of surface oxidized carbon fibers. [From A. Ishitani, in: *Molecular Characterization of Composite Interfaces* (H. Ishida and G. Kumar, eds.), p. 321; Figure 8, p. 328, Plenum Press, New York (1985).]

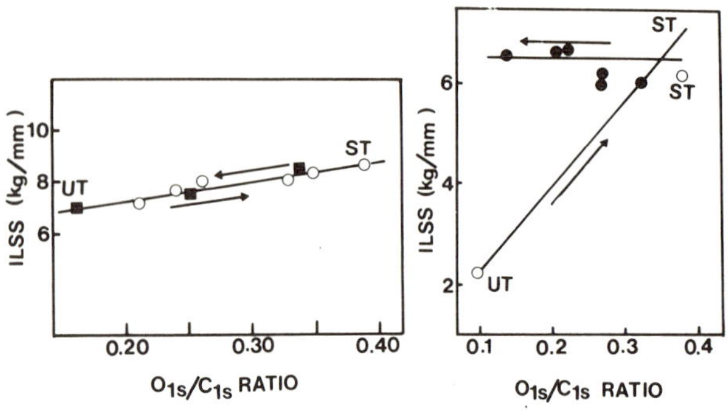

FIGURE 2. Relationship between interlaminar shear strength and the O_{1s}/C_{1s} ratio in carbon fiber (left) and graphite fiber (right) reinforced epoxy composites: UT—control, ST—surface oxidized fibers, closed circles, squares—oxidized fibers heated in vacuum. [From A. Ishitani, in: *Molecular Characterization of Composite Interfaces* (H. Ishida and G. Kumar, eds.), p. 321; Figure 9, p. 329, Plenum Press, New York (1985).]

There are numerous chemical methods which have been applied to modify carbon fiber surfaces and thereby increase interfacial adhesion. Many of these have had limited to moderate success and will only be listed here. They include liquid and gaseous oxidizing agents,(12–18) plasmas,(19,20) chemical solutions,(1) chemical coupling agents,(21–23) vapor deposition,(24,25) and electropolymerization.(26)

In addition to the chemical influence on interfacial adhesion, there may well be physical effects. Tuinstra and Koenig(27) showed that the diameter of crystallites on the graphite fiber surface measured by Raman spectroscopy was related to the degree of interaction at the fiber/resin interface and the shear strength of the composite. Since the smaller diameter crystallites gave the higher shear strengths, it is also possible that the improved properties resulted from the increased chemical functionality produced at crystallite boundaries and defects.

It is clear that all steps (e.g., starting materials, process conditions, post-treatments, etc.) required to produce a fiber or any other material influence surface properties. These surface properties, in turn, determine the adhesion properties and the ultimate performance of the material.

In general, the interface between any two substances effects their potential for adhesion. Carbon fibers are only one example where modifications have been made in an effort to improve the situation. Surface modifications have been attempted on other fibers, both inorganic (e.g., glass, ceramic) and organic (e.g., nylon, polyaramid, PET), particulates (e.g., clays, carbonates, silicas, carbon black), polymers, metals, woods, and other materials.

It is generally accepted that surface modifications improve adhesion either by increasing wetting or by forming chemical bonds across the interface. The relative importance of these two mechanisms depends to a great extent on the sample and conditions, and it is fair to say that both conditions are generally desirable.

Any surface treatment which is to be considered interfacial must be based at least in part on the chemistry of the substrate(s). A good example is the concept behind the use of so-called "coupling agents." These are bifunctional molecules which have one functionality suitable for reaction with the surface and a second functionality capable of reaction with another material. The result is, theoretically, the promotion of adhesion between the two materials.

For instance, Nakatsuka *et al.*[28,29] prepared several dihydrogen phosphates of functional alcohols and proved their significance as coupling agents in calcium-carbonate-filled vulcanizates. The dihydrogen phosphate portion of the molecule complexes with the calcium carbonate surface leaving an organofunctionally modified surface. SEM micrographs of the calcium carbonate surface showed a textured rather than smooth surface as a result of the modification. When the phosphate is prepared with methacryloxy and mercaptohexyl functional groups, the properties of the filled, peroxide-cured EDPM and sulfur-cured SBR, respectively, were increased dramatically. For instance, the tensile strengths were increased over 50% with little or no decrease in elongation at break.

The role of coupling agents is crucial in modifying interfaces to promote dry and wet adhesion to a variety of surfaces. Ideally, the coupling agent aids the formation of chemical bonds between the reinforcement and the matrix. Ahagon and Gent[30] demonstrated the effect of the number of interfacial chemical bonds between a glass surface and an elastomer on the failure strength. Glass surfaces were modified with a combination of a reactive vinylsilane and a nonreactive ethylsilane. The modified glass was coated with a layer of polybutadiene and crosslinked *in situ* by a free radical process. Under near equilibrium conditions at low rates of detachment and high temperature, the strength of adhesion was found to increase proportionally with the fraction of reactive vinylsilane. This trend is illustrated in Figure 3. There was a 35-fold increase in strength for the vinylsilane modification relative to the ethylsilane treatment. This difference is in rough agreement with the relative magnitudes of the strengths of covalent and dispersion bonds.[30]

One of the more common and useful methods for studying interfacial adhesion in multicomponent systems is to examine fracture surfaces with scanning electron microscopy (SEM). Figures 4 and 5 are typical of the results of many researchers. Figure 4 is the micrograph of the fracture surface of a glass-sphere-filled high-density polyethylene. The fracture surface shows what is commonly referred to as interfacial failure. This consists of

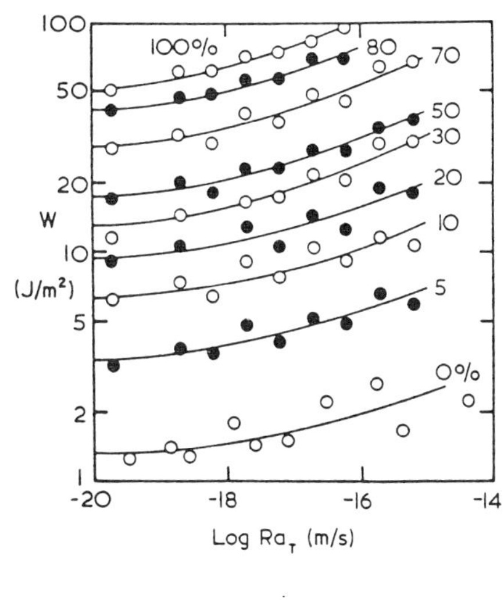

FIGURE 3. Adhesion energy as a function of the effective rate of detachment for polybutadiene crosslinked with 0.2% dicumyl peroxide against glass surfaces treated with increasing proportions of vinylsilane in admixture with ethylsilane. [From A. Ahagon and A. N. Gent, *J. Polym. Sci., Polym. Phys. Ed.* **13,** 1285; Figure 6, p. 1293 (1975).]

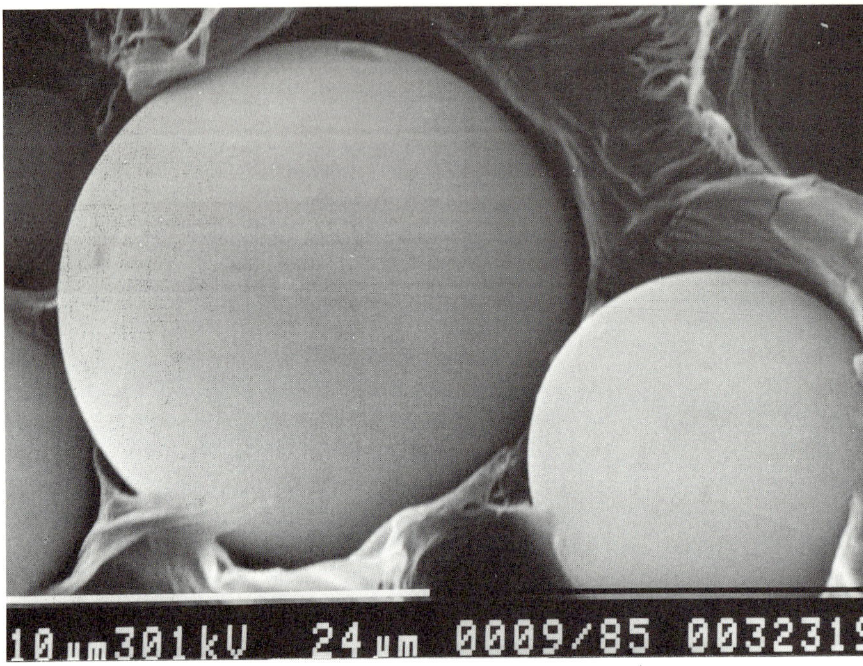

FIGURE 4. Scanning electron micrograph of the fracture surface of a glass-sphere-filled high-density polyethylene.

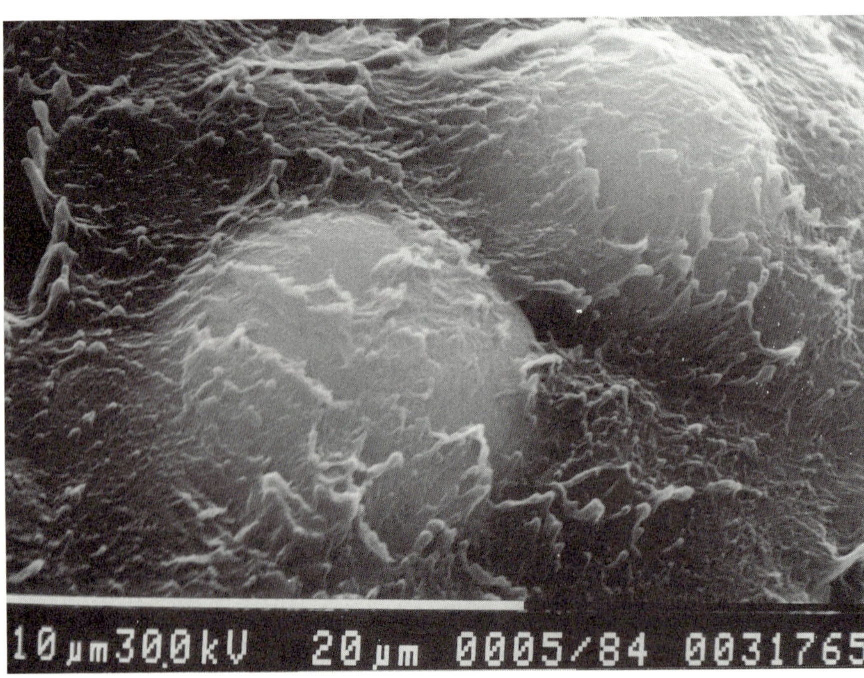

FIGURE 5. Scanning electron micrograph of the fracture surface of a glass-sphere-filled high-density polyethylene where the glass has been modified with an azide-functional silane.

a clean glass surface, dewetting at the interface, and fracture in the matrix near the equator of the sphere. Figure 5 is a similar system, however, the glass surface has been modified with an azide-functional alkoxysilane. In this case, there is adhesion of the matrix to the glass and the failure is referred to as cohesive (i.e., in the matrix). The fracture surface shows that the glass surface is completely covered with matrix material, there is no dewetting at the interface, and fracture in the matrix occurs near the poles of the spheres.

3. CHEMICAL MODIFIERS

Many approaches are currently available for improving adhesion between two incompatible surfaces. Most have been developed over the past thirty to forty years due to the realization of the advantages of multicomponent systems. For adhesion to a solid surface, the chemical modifiers are usually designed to react with the substrate surface and therefore impart a new surface usually comprised of a reactive functional group. This is the concept of the so-called "coupling agent."

The first inference to the coupling agent concept dates back to the testimony of Rochow in 1951 before the U.S. Patent Office when he referred to the experiments of O.K. Johannson in 1938 which involved treating glass with organochlorosilanes with the object of improving the adhesion of lacquers to glass.[31] Bjorksten and Yaeger reported in 1952 on the improved wet strength of polyester laminates. They had tried numerous compounds and determined that a nonaqueous treatment of a mixture of vinyl trichlorosilane and β-chloroalkyl alcohol performed the best. About the same time, Torrey[32] reported good wet strength retention using a methacrylic acid coordinated to two moles of trivalent chromium chloride and marketed as Volan® by du Pont. It is believed that in the presence of water the chromium chlorides hydrolyze and interact with surface silanols on glass. At the same time, coordination of the acid group to two chromium atoms is sufficiently stable and allows permanent functionalization of the surface with a reactive moiety attached to the acid.

The widest variety of organofunctional silanes have been investigated by E. P. Plueddemann and coworkers at Dow Corning Corporation. An early compilation of these experiments dealing with coupling agents for glass reinforced polyester and epoxy composites can be found in the literature.[33] More recently the chemistry, science, and use of silane coupling agents have been reviewed by Pleuddemann.[34] (See also Chapter 9 by Plueddemann.)

Development of silane coupling agents over the years has not departed from the concept of reactivity with both substrate and matrix materials; however, structural changes have been investigated in an effort to optimize performance. In 1962 Clark and Plueddemann[35] reported the synthesis and properties of a large variety of organofunctional, substituted silanes in polyester glass laminates. Silanes with various vinyl, hydroxy, thio, carboxy, amine, alkyl, and ester substitutions were evaluated. Out of twenty different ester-substituted silanes the propyl methacrylate substitution performed the best. This compound, $(CH_3O)_3Si(CH_2)_3OOC(CH_3)C{=}CH_2$ (γ-methacryloxypropyltrimethoxysilane), has gained wide commercial use in polyester systems today.

Just as the methacrylate substitution was the best performer in polyester resins, other substitutions have been determined appropriate for other resins. A wide variety of silanes with different organofunctional groups were developed and are now commercially available. These have reactive groups tailored to a resin chemistry of choice (e.g., vinyl, epoxy, amino, mercapto). The interaction, however, is not always covalent. For instance, aminopropyltriethoxysilane works well in a variety of resin systems including thermoplastics.

A representative list of current commercial organofunctional silanes and other coupling agents are listed in Table 1. Table 2 shows a comparative ranking of the wet adhesion performance of many commercial coupling agents in glass/polyester systems. The ranking emphasizes the need to match the chemistry of the organofunctional group to the matrix material. The effect of silane coupling agents on the dry and wet flexural strength of glass/polyester composites is seen in Table 3. The use of the coupling agent increases dry strength but, more importantly, causes a high percent of the dry strength to be maintained under wet conditions.

Structural changes beyond the organofunctional substitution have also been investi-

TABLE 1
Representative Commercial Coupling Agents[a]

Trade name	Organofunctional group	Chemical structure
A. Q9-6300	Vinyl	$(CH_3O)_3SiCH{=}CH_2$
B. Z-6076	Chloropropyl	$(CH_3O)_3SiCH_2CH_2CH_2Cl$
C. Z-6040	Epoxy	$(CH_3O)_3SiCH_2CH_2CH_2OCH_2\overset{O}{CH{-}CH_2}$ (epoxide ring: O bridging CH—CH_2)
D. Z-6030	Methacrylate	$(CH_3O)_3SiCH_2CH_2CH_2OOC(CH_3){=}CH_2$
E. A-1100 (UCC)	Primary amine	$(C_2H_5O)_3SiCH_2CH_2CH_2NH_2$
F. Z-6020	Diamine	$(CH_3O)_3SiCH_2CH_2CH_2NHCH_2CH_2NH_2$
G. Z-6062	Mercapto	$(CH_3O)SiCH_2CH_2CH_2SH$
H. Z-6032	Cationic styryl	$(CH_3O)_3SiCH_2CH_2CH_2NHCH_2CH_2H \cdot HCl$ CH_2 $C_6H_4CH{=}CH_2$
I. S-3076S (Hercules)	Azide	$(CH_3O)_3Si{-}R{-}SO_2N_3$
J. Volan-A (DuPont)	Methacrylatochrome	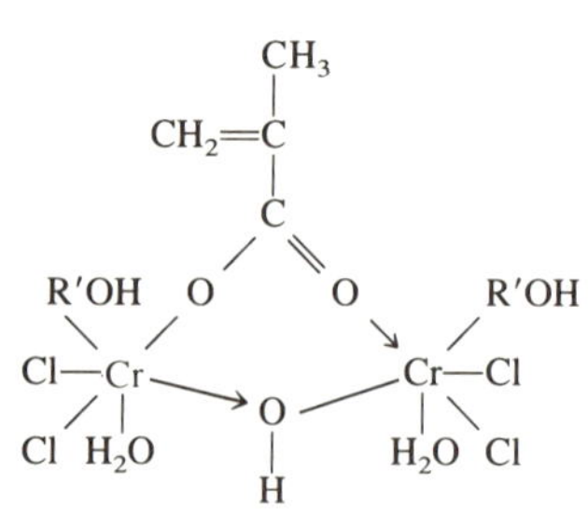
K. TTM-33 (Kenrich)	Methacrylate-titanate	$(CH_2{=}C(CH_3)COO)_3TiOCH(CH_3)_2$
L. Caveco-Mod	Methacrylate-Al-zirconate	undisclosed
M. X1-6100	90/10 mix $PhSi(OCH_3)_3$/Z-6020	
N. X1-6106	Z-6040-modified Cymel-303 melamine resin	
O. X1-6121	Product of Z-6020 with isocyanatoethylmethacrylate (IEMA)	

[a]From E. P. Plueddemann, Proc. 1st Tech. Conf. Am. Soc. for Comp., Dayton, OH, p. 264; Table I, p. 273 (Oct. 1986).

TABLE 2

Coupling Agents for Polyesters (Screening Tests on Microscope Slides)[a]

Commercial designation	Functional group of primer (on Si, Cr, etc.)	Order of wet adhesion to glass (1 = best) Bz_2O_2 at 130 °C	MEK peroxide at 70 °C
A-1100	Aminopropylsilane	11	9
Z-6020	Diamine	9	4
Z-6030	Methacrylate	2	5
Z-6031	Quat. amm. methacrylate	10	3
Z-6032	Vinylbenzyl cationic	8	1
Z-6032 + Mal. anhyd.	Vinylbenzyl zwitterion	3	10
Z-6040	Epoxy	5	12
Z-6062	Mercaptan	6	6
Z-6076	Chloropropyl	15	16
Q0-6300	Vinyl	12	11
Q1-6106	Epoxy-melamine resin	7	7
Z-6020 + 1 IEMA	Amino-ureamethacrylate	4	2
Z-6020 + 2 IEMA	Bis-ureamethacrylate	1	8
Volan-A	Methacrylate-chrome	14	13
KR-6	Methacrylate-titanate	13	12
Caveco Mod APG-2	Alumino-zirconate	16	15
None		17	17

[a]From E. P. Plueddemann, Proc. 1st Tech. Conf. Am. Soc. for Comp., Dayton, OH, p. 264; Table II, p. 274 (Oct. 1986).

gated. For instance, the hydrolyzable group on silicon which serves as an intermediate to the formation of silanols capable of reacting with hydroxyl-covered surfaces were varied. The best wet strength retention was obtained with three hydrolyzable groups on the silicon atom.[(33)] The nature of the hydrolyzable group on silicon (e.g., $—OCH_3$, $—OCH_2OCH_3$, —Cl, —Ac, $—OCH_2CH_2OCH_3$) does not make a great deal of difference in performance.[(33)] However, it should be noted that hydrolysis rates in aqueous solution can be controlled by the nature of this group. These hydrolysis rates have been studied by Plueddemann[(36)] and others.[(37,38)]

Functionalized siloxane polymers do not perform well as coupling agents. For in-

TABLE 3

Silane Treatments (0.2%) on Glass Roving in Pultrusion Rods[a]
[Paraplex P-43 Polyester Resin (Rohm and Haas) with 1% Bz_2O_2]

Silane treatment on glass	Flexural strength (MPa) Dry	Wet (4 h boil)	(24 h boil)
None	966	448	280
Z-6030	1145	952	883
Z-6031	904	632	531
Z-6032	1131	987	828

[a]From E. P. Plueddemann, Proc. 1st Tech. Conf. Am. Soc. for Comp., Dayton, OH, p. 264; Table V, p. 275 (Oct. 1986).

stance, Plueddemann *et al.*[33] and Clark and Plueddemann[35] treated glass surfaces with the methacrylate functionalized disiloxane, linear polysiloxane, and polysilsesquioxane (polymer with undefined structure having the generalized formula $RSiO_{3/2}$). None of these oligomeric or polymeric structures performed nearly as well as the monomeric species.

The chemical nature of the organic spacer between silicon and the reactive functionality was also investigated.[33] The general conclusions for a methacrylate substituted silane were that methylene, ether, or ester linkages of three to nine atoms in length gave equivalent properties. Deleterious effects were seen by incorporation of hydroxyl, nitrogen, or long polyglycol spacers. A study of the surface orientation and performance of a methacrylate functional silane with different length methylene spacers has recently been published.[39] Mechanical properties in silica/polyester castings decreased for the silanes with large and highly flexible methylene spacers.

Coupling agent molecules which are polyfunctional in either the silyl or reactive organic functionality have been synthesized and tested for performance in reinforced plastics. For instance, Plueddemann[33] synthesized a trimethoxysilyldimethacrylate [$(MeO)_3Si(CH_2)_3OCH_2CCH_3(CH_2$methacrylate$)_2$] and a highly methoxylated silylmonomethacrylate [$(MeO)_3SiOSi(OMe)_2(CH_2)_3$methacrylate]. Both of these molecules performed equivalently to the more conventional trimethoxysilylpropylmethacrylate.

Polyfunctional silane coupling agents have also been developed by researchers at Union Carbide. Marsden and Pepe[40] synthesized molecules which are (1) ionic or nonionic, (2) contain large organofunctional groups (C_{16}–C_{20}) with a multiplicity of carbon–carbon double bonds, and (3) contain more than one silyl group per molecule each of which can have up to three hydrolyzable substituents. None of these molecules performed as well in wet strength retention of polyester/glass pultruded rods as the standard γ-methacryloxypropyltrimethoxysilane. However, they gave substantial improvements in glass-fiber abrasion resistance and fiber stiffness over the methacryloxypropyltrimethoxysilane control. Abrasion resistance was aided the most by an ionic silane molecule having only one silyl group per molecule. Fiber stiffness measured in a three-point bend configuration was the highest for an ionic silane having more than one silyl group per molecule. These silane molecules incorporate some of the structural features of lubricants and film formers used in fiberglass sizes. Therefore, these silanes impart additional abrasion resistance and fiber stiffness over the conventional silane-only treatment.

Organofunctional silanes are not the only class of surface modifiers used to promote adhesion. However, the organofunctional silanes and the methacrylato-chromium complex are the only compounds widely accepted as true coupling agents, that which "improves the chemical resistance (especially to water) of the bond across the interface."[34]

Another class of surface modifiers are the organic titanates. Although claimed to promote adhesion, they are generally used to aid plasticization, dispersion, or lubricity by lowering the surface energy of the substrate. They are used in various industrial applications including coatings, inks, electronics, and cosmetics. Similar to the silanes, the titanates are designed to be bridging molecules which may react with an inorganic surface and a surrounding matrix material. While the mono-organofunctional trialkoxysilane has become the standard molecular design of the silane coupling agent, there is a greater structural diversity in the commercially available titanates. The basic forms of titanates are either chelates or tetraorganofunctional compounds using alkylate, carboxyl, sulfonyl, phenolic, phosphato, pyrophosphato, or phosphito groups to bind to the titanium center. Coordination-type titanates and quaternary-type titanates are also available.

An example of a tetraalkoxy titanates is tetraisopropyl titanate. These compounds react with water to form the hydrolyzed product which may proceed to a hydrous or anhydrous titanium dioxide depending on time and temperature conditions. The alkoxy titanates also react rapidly with carboxylic acids and acid anhydrides. They undergo alcoholysis with higher alcohols and ester interchange with carboxyl esters.

Reactive organofunctionality can be attached to the titanate for use as a surface modifier. For instance, an isopropyl trimethacryl titanate (e.g., KR 33CS, Kenrich Petrochemicals, Inc.) or an isopropyl tri(*N*-ethylamino-ethylamino) titanate (e.g., KR 44, Kenrich Petrochemicals, Inc.) would impart a reactive moiety to the substrate surface. The mechanism for this modification is the reaction of the monoalkoxy portion of the titanate molecule with water or free protons on the inorganic surface. Other organic functional titanates (e.g., acrylic, mercapto, hydroxyl) and oleophillic titanates (e.g., isopropyl triisostearoyl titanate, KR TTS Kenrich Petrochemicals, Inc.) are available. Titanate chelates have the general formula $(Ti(ORO)_2$. They are in general much less reactive than the alkoxide titanates. The chelating oxygens can also be replaced by nitrogens.

Through the proposed reaction at the surface of an inorganic substrate, the organic titanates are suggested to provide many property changes in a variety of systems. These effects include but are not limited to wet-out, dispersability, adhesion promotion, impact resistance, plasticization, processability, viscosity reduction, and deagglomeration. Debate continues over the success of organic titanates as adhesion promoters as there is little evidence that these compounds improve the chemical resistance of the bond across the interface usually measured by wet strength retention. It is certain, however, that the titanates play an important role in modification of the interface in a variety of chemical systems.

Similar to the way in which titanium is converted to the organofunctional titanates, other approaches have been made with aluminum and zirconium to give classes of organofunctional zirconates and aluminates. A series of organofunctional zircoaluminates are manufactured by Cavedon Chemical Company, Inc. The zircoaluminates were developed in the desire to obtain an optimum adhesion promoter for metallic substrates. They are low-molecular-weight compounds with two different types of organic ligands selectively bound to a zircoaluminate backbone. One ligand imparts hydroxylic and hydrolytic stability to the inorganic portion of the molecule while the second ligand contains the desired organofunctionality.(41) The combination of aluminum and zirconium was designed to include the advantages of both elements in a single organometallic complex. Aluminum has a high affinity for a variety of inorganic surfaces but forms hydrolytically unstable bonds with organofunctional ligands.(41) Thus, aluminum acylates bind to inorganic surfaces but do not function as well as coupling agents.(42) Zirconium has less propensity for inorganic surfaces but is tetravalent and forms more stable organofunctional complexes.(41) Therefore, a synergistic effect was the goal for combining the elements in one complex.

The general formula for the zircoaluminates available from Cavedon Chemical is $[Al_2(OR_1O)_a(OH)_bCl_c]_x[OC(R_2)O]_y[Zr(OH)_dCl_e]_z$. The subscripts a, b, c, d, and e must sum to appropriate values to fulfil stoichiometry requirements. The subscripts x, y, and z

determine the molecular weight of the product and can vary from 1 to 100.[43] The aluminum-containing segment has pairs of aluminum atoms joined by bidentate chelating ligands —OR_1O— (e.g., propylene glycol).[43] The zirconium-containing segment is a zirconium oxychloride moiety. The organofunctional ligand —$OC(R_2)O$— (e.g., methacrylato) bridges the chelated aluminum moiety to the zirconium oxychloride moiety through a complexed carboxylate. The zircoaluminates are available with a range of reactive organofunctionality including carboxy, methacryloxy, amino, and mercapto. It is reported that the attachment of the zircoaluminates to metal substrates is substantially irreversible[44,45] based on experiments of the long-term viscosity reduction of aqueous slurries of alumina trihydrate. Only the zircoaluminate and a silane provided long-term (i.e., after seven days) viscosity reduction while a surfactant and a titanate offered only short-term viscosity reduction. Physical properties of glass-fiber reinforced polyester laminates reacted with a zircoaluminate showed comparable dry and wet tensile and flexural strength to a chrome complex and a silane.[46] The zircoaluminates have been demonstrated to be effective-viscosity reducers for a wide range of substrate surfaces including clay, titanium dioxide, silica, calcium carbonate, aluminum trihydrate, carbon black, and organic pigments.[41]

Other examples of chemical surface modification used to promote adhesion include the bonding of polymers to hard tooth tissue or bone. For example, *N*-(2-hydroxy-3-methacryloxypropyl)-*N*-phenylglycine (NPG-GMA) was found to irreversibly adsorb on hydroxyapatite from solution.[47] Dental resins filled with synthetic hydroxyapatite modified with NPG-GMA had tensile strengths about 50% greater than that of the polymer filled with unmodified apatite.[47] Nakatsuka *et al.* prepared a variety of dihydrogen phosphates of functionalized alcohols and proved their significance as coupling agents for calcium carbonate in filled vulcanizates.[28,29] Organic borates have been synthesized as possible coupling agent type molecules.[48] Metallo-organic cobalt salts are used for the promotion of rubber adhesion to brass-plated steel tire cords[49] and further developments in this chemistry have led to improvements in initial and aged adhesion.[50]

4. EVIDENCE OF CHEMICAL BONDING

Covalent bond formation at interfaces is one of the oldest theories used to explain the promotion of adhesion. This theory is a particularly popular explanation of the role of coupling agents used in fibrous and particulate reinforced composites. The theory requires that covalent chemical bonds form between the two homogeneous phases creating a chemical linkage between the phases. In practice, attempts at this linkage are often made by the addition of a molecule or a phase which can presumably bond to both phases creating a bridge of primary bonds.

Hypothetically, it is not difficult to specify the chemical functional groups which might be reactive with each of the given homogeneous phases. For exampe, in a glass/epoxy composite a molecule with silanol groups to react with the glass and an amine to bond with the epoxy would meet this requirement. Similarly, if the homogeneous phases were both polymers then a different molecule with two organic functionalities might be envisioned. Surely, this was the idea behind the first coupling agent molecules.

There is a substantial amount of performance data indicating at least indirectly that this type of bonding occurs. However, it has been much more difficult to substantiate the theory with direct evidence that these bonds actually exist and are responsible for the improved performance. Most of the highly sophisticated state-of-the-art analytical techniques have been used in an attempt to identify interfacial bonds and the list is too long to detail here. However, as recently as 1982, Arthur Adamson stated that the subject of bond formation with a surface "has been hampered by the nearly total lack of any direct experimental information."(51) Adamson went on to point out that "no bond lengths are available, nor have any infrared absorptions assignable to any adsorption bond been observed."(51)

It is clear that direct and conclusive information from interfacial and surface regions is necessary before questions about bonding can be answered. In the past 3–4 years there has been a brief number of experimental findings which have been concluded to be the result of direct bonding between adjacent atoms or chemical groups of the adsorbate and adsorbent. These, hopefully, will begin to address the interfacial bonding issue.

Seyedmonir *et al.*(52) assigned an infrared absorption band at 925 cm^{-1} to the Mo—O—Si groups on a 700 m^2/g silica gel support impregnated with an aqueous solution of $(NH_4)_2Mo_2O_7$ (6 wt% Mo on SiO_2) and calcined at 500 °C. Kinney and Staley(53) reacted high surface area silica (400 m^2/g) with titanium tetrachloride ($TiCl_4$) and trimethylaluminum (Me_3Al) and observed their (FT-IR) photoacoustic spectra. They concluded the presence of Si—O—Al and Si—O—Ti species but assign only SiO, AlO, and TiO stretching vibrations from the observed spectra.

Naviroj *et al.*(54) used Fourier transform infrared (FT-IR) spectroscopy to study the surfaces of high surface area metal oxides modified with γ-aminopropyldimethylethoxysilane and isopropyltriisostearoyltitanate. Weak spectral bands near 963 and 950 cm^{-1} were attributed to Al—O—Si and Ti—O—Si linkages, respectively. Culler(55) looked at γ-aminopropyltriethoxysilane adsorbed from aqueous solution on silicon oxide by FT-IR. After digital subtraction of the substrate contribution it was observed that the bands in the region 1200–1000 cm^{-1} due to Si—O—Si bonds shifted in frequency as a function of concentration in solution. Bands at 1122 and 1032 cm^{-1} were attributed to intermolecular siloxane bonds in the adsorbed multilayers. At submonolayer coverages obtained from dilute solution, bands at 1111 and 1072 cm^{-1} were observed. The 1072 cm^{-1} band was also observed after extraction of the surface with water. As a result, it was suggested that the 1072 cm^{-1} band might be representative of a Si—O—Si bond with the substrate surface. It should be noted that it is extremely difficult to separate surface linkages from intermolecular siloxane formation using multialkoxy adsorbates. For instance, the peak location of the intermolecular siloxane bond shifts with surface concentration and structure at low surface coverage.(56) At submonolayer coverage on nickel oxide, it was found that the reaction product of hydrolyzed γ-methacryloxypropyltrimethoxysilane gives an intermolecular siloxane band at 1070 cm^{-1}(57) similar in position to the band observed by Culler.(55)

Covalent bond formation between organofunctional trialkoxysilane adsorbate molecules and a solid lead oxide surface was recently reported by Miller and Ishida(58) using diffuse reflectance FT-IR. The surface bond is the plumbosiloxane (Pb—O—Si) linkage resulting from the condensation reaction of hydrolyzed γ-methacryloxytrimethoxysilane and the surface hydroxyl groups. The adsorbate surface bond gives rise to an antisymmetric stretching vibration near 965 cm^{-1}.

Typically, adsorbate–substrate bands have been difficult to assign with any certainty due to the chemical similarity of these bonds to those existing in the bulk substrate. An example is the possible bond formed between hydrolyzed alkoxysilanes and siliceous surfaces. Many researchers over the years have sought for conclusive evidence demonstrating the presence and function of this bond. Infrared spectroscopy has been used for a number of these investigations because it offers a sufficient combination of sensitivity and selectivity to provide researchers with useful surface information. A problem that has plagued researchers in this field is that infrared bands resulting from siloxane condensates of various structures, adsorbate–surface siloxane bonds, and substrate silicon–oxygen networks all fall in a similar region of the spectrum generally between 1200 and 1000 cm^{-1}.

This problem can be overcome from an experimental standpoint by changing the M—O linkage in the substrate as demonstrated by Miller and Ishida[(58)] and Naviroj *et al.*[(54)] This allows the absorptions resulting from siloxane condensates, adsorbate–surface bonds, and substrate lattice bonds to be resolved in the infrared spectrum. For instance, Miller and Ishida[(59)] showed that at a concentration of two equivalent molecular layers of adsorbate molecules, polysiloxane condensates give infrared bands at 1120 and 1063 cm^{-1} while the adsorbate–substrate bond (Si—O—Pb) is shifted out of this region and absorbs at 967 cm^{-1}. Such an analysis allows the adsorbate–substrate bond to be isolated and quantified. Figure 6 shows an example illustrating the infrared band resulting from the bond between adsorbate and substrate as a function of surface coverage. Spectra A–F reflect increasing adsorbate concentrations. The number of surface bonds as reflected by the intensity of the band increases through spectrum C but plateaus at higher concentrations. Presumably, all the available surface sites become occupied above a certain concentration. Adsorption isotherms based on the number of bonds formed can also be drawn from this information as seen in Figure 7.[(59)]

Other examples of evidence for adsorbate–substrate bonds include Chiang *et al.*,[(60)] who used an infrared–emission technique to measure the linewidth of a molecule–substrate vibrational mode. Adsorption of CO on Ni(100) produced a C—Ni mode at 475 cm^{-1} with a linewidth of 15 cm^{-1}. Chabal[(61)] observed an overtone infrared absorption of the wagging mode of H on W(100) at saturation coverage using grazing-incidence reflection. Chabal[(62)] also observed infrared bands for Si—H and Si—D on Si(100).

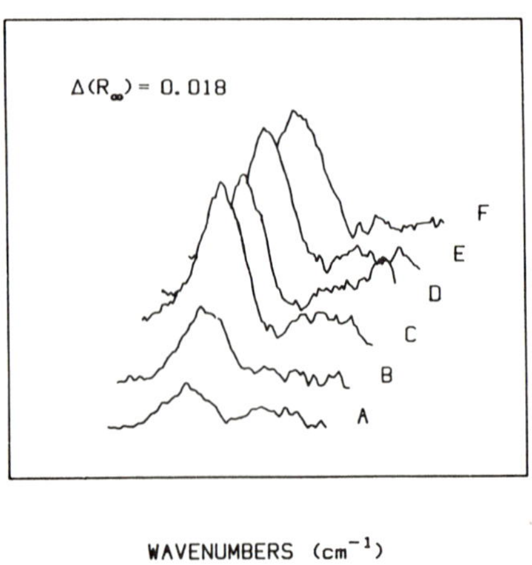

FIGURE 6. Diffuse reflectance infrared difference spectra (1000–910 cm^{-1}) of the plumbosiloxane (Pb—O—Si) bond (967 cm^{-1}) formed between adsorbate molecules and the solid substrate. Spectra A–F represent increasing adsorbate concentrations from 0.40 mg/m^2 to 6.00 mg/m^2. [From J. D. Miller and H. Ishida, *J. Chem. Phys.*, Figure 7 (in press).]

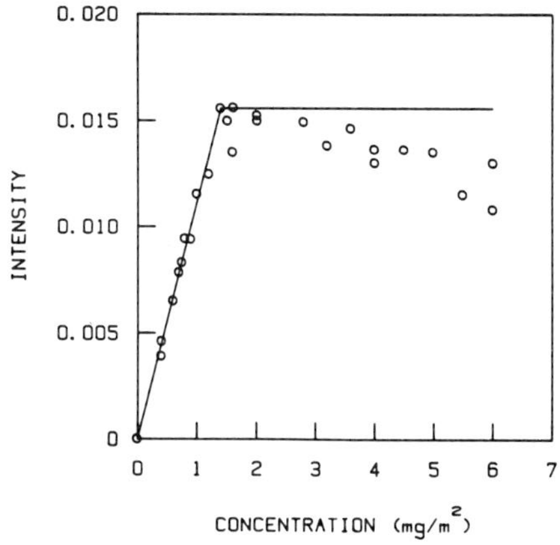

FIGURE 7. Adsorption isotherm based on the number of adsorbate to substrate covalent bonds for an organofunctional silane deposited on lead oxide. The solid line is a theoretical isotherm predicted from the data. [From J. D. Miller and H. Ishida, *J. Chem. Phys.*, Figure 8 (in press).]

One of the very promising techniques for examining the adsorbate–substrate bond is high-resolution solid-state nuclear magnetic resonance (NMR). NMR has superior selectivity and provides a wealth of chemical information, but has lagged behind in use for surface studies due to insufficient sensitivity. High-resolution solid-state ^{29}Si studies using magic angle spinning (MAS) and cross polarization (CP) have detailed the surface of high surface area silica gels. Resolution of resonance lines from interconnecting lattice sites (Si atom with four siloxane attachments), lone-hydroxyl silanol sites (Si atom with three siloxane bonds and one hydroxyl), and geminal or silanediol sites (silicon atom with two siloxane bonds and two hydroxyls) has been achieved.[(63,64)]

These studies have been extended to examine the surface of silicas modified with organofunctional silanes.[(63–67)] Since the NMR technique can discriminate between nuclei in different environments the ^{29}Si CP/MAS NMR experiments easily resolve the trimethylsilicon atom near 15 ppm attached to the silica gel surface after trimethylsilylation with trimethylchlorosilane. This is seen in Figure 8. It should also be noted that the silylation, in addition to attaching the trimethylsilicon, converts a lone-hydroxyl site to a fully siloxane bonded site and a geminal site to a lone-hydroxyl site. This causes a redistribution of intensity of the resonances between -80 and -120 ppm which result from nuclei of the substrate. Additional assignments can be made for the various levels of reaction achieved by multifunctional silanes. For instance, assignments have been made for the reaction products of methyltrichlorosilane which is seen in Figure 9. These include each step in the hydrolysis of the chlorine groups, as well as the various siloxane bonded combinations with unhydrolyzed or hydrolyzed chlorine groups present. One disadvantage of the ^{29}Si NMR technique is that it loses selectivity as the amount of polysiloxane structure increases. Therefore, structural differences in highly condensed polysiloxanes and polysilsesquioxanes become more difficult to determine. The application of ^{29}Si NMR to surfaces should increase rapidly as gains in sensitivity are achieved.

The preceding references reflect the majority of the direct spectroscopic evidence for the presence of adsorbate–surface covalent bonds. More numerous are examples of the effect of the adsorbate on the molecular structure of the adsorbent. Probably best known are the infrared absorption studies demonstrating the frequency shifts and intensity changes of surface hydroxyl stretching modes as a result of adsorbate presence.[(68–74)]

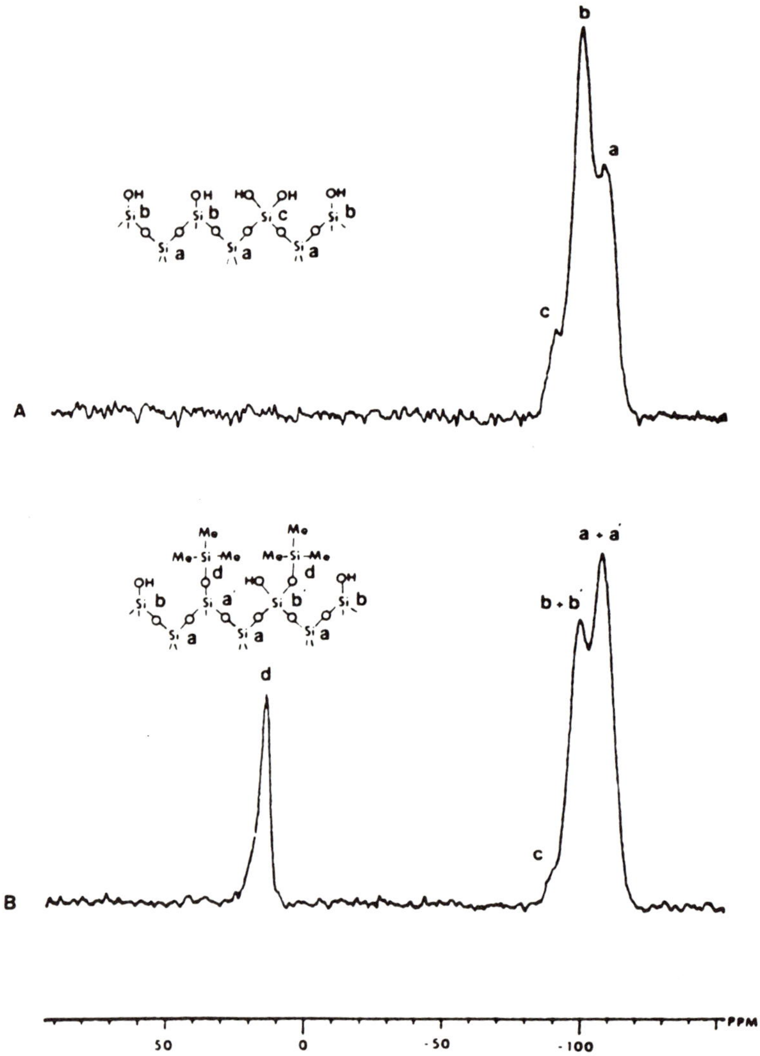

FIGURE 8. ^{29}Si NMR spectra of Fisher S-157 silica gel (A) before and (B) after reaction with trimethylchlorosilane. [From D. W. Sindorf and G. E. Maciel, *J. Am. Chem. Soc.* **103,** 4263; Figure 1, p. 4264 (1981).]

There are also many studies detailing perturbations of adsorbate molecules in the presence of substrate surfaces. Examples include orientation of adsorbate molecules, surface areas occupied by adsorbed molecules, and intermolecular adsorbate reactions.[(75–80)]

One technique, which to our knowledge has not currently demonstrated evidence of direct covalent bond formation at surfaces but shows a great deal of promise, is inelastic electron tunneling spectroscopy (IETS). Some examples in the literature have looked at the surface of aluminum oxide exposed to alkoxysilanes.[(72,81–85)] In none of these reports could an interfacial bond be detected. However, the signal-to-noise ratio of adsorbed

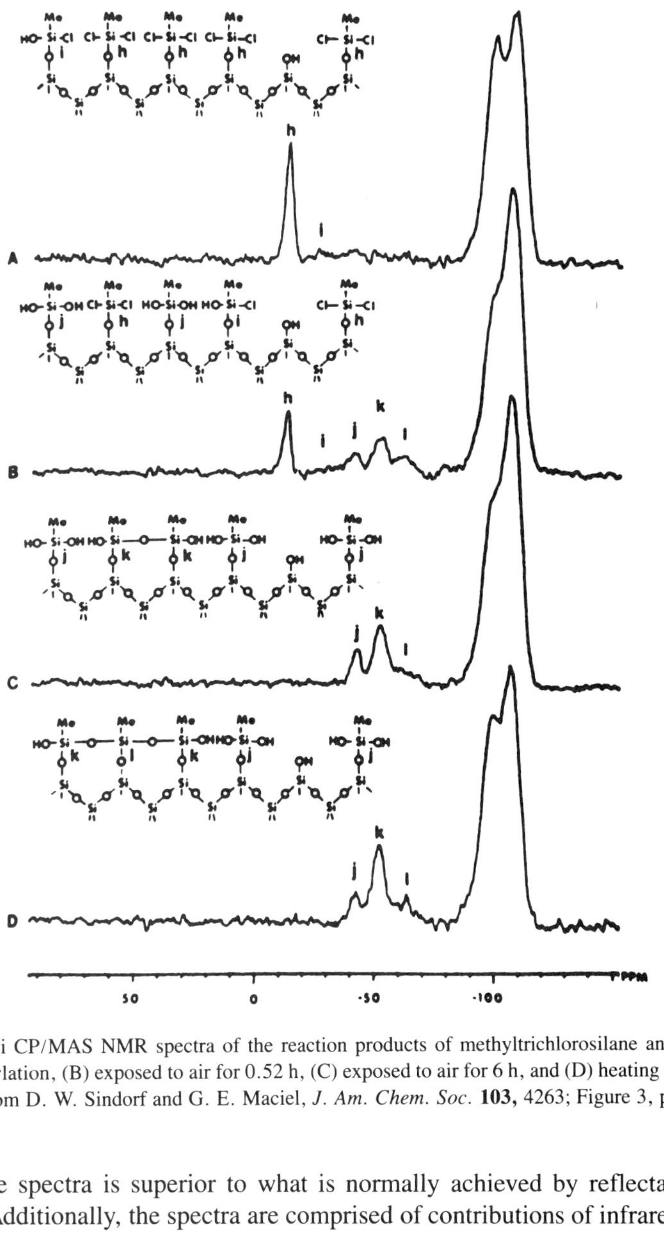

FIGURE 9. ^{29}Si CP/MAS NMR spectra of the reaction products of methyltrichlorosilane and silica gel (A) directly after silylation, (B) exposed to air for 0.52 h, (C) exposed to air for 6 h, and (D) heating sample C for 24 h at 200 °C. [From D. W. Sindorf and G. E. Maciel, *J. Am. Chem. Soc.* **103,** 4263; Figure 3, p. 4265 (1981).]

species in the spectra is superior to what is normally achieved by reflectance infrared techniques. Additionally, the spectra are comprised of contributions of infrared-active and Raman-active modes. These factors once better understood should aid in the understanding of adsorbate–surface reactions.

One might query the need for such precise information about surface bonds, especially in light of the difficulties encountered in acquiring this type of data. With respect to adhesion, it is possible that direct chemical bonds are responsible for a large degree of adhesive strength. Unless the bonds can be measured and quantified, it is impossible to prove or disprove this and other theories. Additionally, this type of information can also aid in the development of new chemical systems designed to take advantage of unknown or unrealized phenomena.

As a final note, it should be pointed out that many authors have explained behavior or even claimed to have shown evidence for the presence of covalent bonds between adsorbate molecules and a solid surface. Unfortunately, although there is a mass of indirect evidence supporting the theory of surface bonds, the actual direct evidence is very limited. The literature must be read carefully to separate conclusions based on direct evidence from those based on deductive or inductive reasoning.

5. INTERPHASE

5.1. Theories

An interphase is a region with finite volume and distinct physical properties or gradient in properties. An interphase may be formed but is not required when two homogeneous phases are brought into contact with each other. If either homogeneous phase influences a region of the other phase in a manner such as to change its chemical or physical structure, then this region of modified material can be termed an interphase. Therefore, an interphase has properties different from either homogeneous phase and the presence of an interphase requires an interface between the interphase and each homogeneous phase.

The characteristics and influence of an interphase can often be controlled by the nature of the materials and process. There are many examples of interphases in synthetic and natural materials. One can easily envision that when two metal surfaces are welded together, the weld line has different properties than either of the original materials. When polymers are coextruded there is a finite region which represents a miscible or nonmiscible mixture of the two materials. In more natural settings one often encounters the layer of sand near the edge of the tide which contains a great deal of water and is displaced by each footstep. This is an interphase between the two homogeneous phases, the densely packed sand found at greater depths, and the water.

The interphase is thought to play an important role in adhesive bonding in polymer composites and polymer adhesives. It is believed that many important factors in the performance of adhesive bonds depend on the nature of interphases. These include transfer of stress, moisture permeation, volume shrinkage or expansion, crystallization, etc. Since it is generally concluded that the interphase region is the most crucial in determining the effectiveness of adhesive bonds, a great deal of research effort has been made to understand, control, and modify its properties.

An often asked question is: what are the desired properties for an interphase region in an adhesive bond? Plueddemann[(34)] has addressed this question in reference to adhesive strength in the presence of water at the interface in reinforced polymer composites. He concludes that when using silane coupling agents, retention of adhesion across the interface is dependent on the morphology of the interphase material. Adhesion is maximized if the silane reacted polymer (i.e., interphase material) is either tarry (i.e., has viscous flow) or rigid. In this way, hydrolyzed silanols attached to tarry polymers can migrate to surface sites and reform oxane linkages. In the case of rigid polymers, silanols formed are held in molecular contact with surface sites and reform with the original or adjacent sites. Poor retention of adhesion occurs with rubbery polymers because they irreversibly retract from the reinforcement surface when the siloxane bonds are broken by water intrusion.

A good example of the influence of interphase morphology was demonstrated by Plueddemann(86) with a series of copolymers of ethyl acrylate and methyl methacrylate. A constant silane treatment was applied to glass microscope slides and the matrix comonomer ratio was varied, giving polymers with glass transition temperatures ranging from 105 to −23 °C. When the adhesion of the films to glass was measured under wet conditions, the copolymer films which were flexible or rubbery (e.g., 40:60 MMA/EA) gave the poorest adhesion. Rigid (e.g., 100:0 MMA/EA) and tarry (e.g., 20:80 MMA/EA) copolymers gave excellent wet adhesion.

In addition to the chemical nature of the interphase, many have hypothesized about the most desirable physical properties for an interphase region. In polymer composites, a stiff interphase generally gives the highest ultimate strength materials but with some reduction in ultimate strain. Composites with weak or debonded interphases give higher ultimate strains but at large decreases in strength.(87,88) For promoting toughness, a deformable or flexible interphase might be desirable. This type of interphase would be able to accommodate interfacial stresses due to differential thermal shrinkage. Broutman and Agarwal(89) estimated that maximum composite toughness should be realized when the modulus of the interphase is approximately 80 MPa. Erickson *et al.*(90) proposed that cure inhomogeneities near the surface of reinforcements might produce such a flexible or deformable interphase.

The concept of an interphase with a modulus intermediate to the reinforcement and matrix has been suggested. This is hoped to minimize stress concentrations which occur at interfaces between phases with drastically different moduli. Kardos *et al.*(91,92) tested this hypothesis by preferentially crystallizing the matrix in the vicinity of the reinforcing fiber surface in short graphite fiber/polycarbonate composites. According to Plueddemann,(34) performance data on reinforced composites suggest that a restrained layer (produced by a "tightening up" of the polymer structure in the interphase through the use of silane coupling agents) with higehr modulus than the matrix is required for maximum bonding and hydrothermal stability.

A second concept which has been suggested is an interphase which is ductile and can absorb energy.(93) Tryson and Kardos (94) investigated this concept by coating a glass fiber with a flexible epoxy which was then filament wound in a brittle epoxy matrix. Improvements in a number of properties were significant but there were other varibales beyond interphase properties (e.g., less fiber damage, fiber flaw healing) which may have contributed to the improvements.

The growing experimental evidence of the importance of the role of an interphase has led researchers to attempt to model the effect of this region on composite properties. Theoretical models are necessarily limited to the more ideal systems, but are likely to guide many of the future developments. For instance, Maurer(95) extended the van der Poel method(96–98) to calculate the complex shear and bulk moduli of particulate-filled composites with an interfacial layer. This strictly theoretical model is based on the description of a composite where each particle of the discontinuous phase is surrounded by a shell of interlayer or interphase, which is in turn covered with a shell of matrix material, and which is finally surrounded by composite material. Calculations based on this description and the conditions that displacements and radial and tangential stresses are continuous at the boundaries of the phases show near-exact correspondence with experimental data. For example, composites of glass-bead-filled styrene–acrylonitrile copolymer with an artificially applied ethylene–vinylacetate interfacial layer or interphase were prepared. From measurements of the properties of the individual phases it was possible to accurately predict the properties of the composite as seen for G' in Figure 10.

Such models can also be used to determine whether an interphase exists. For instance, the thermal transitions (e.g., T_g) of a polymeric third phase will be enhanced if it exists as an interphase around the solid inclusions (i.e., reinforcements) as opposed to intermixed with the matrix material.[99]

A large number of other models also exist that attempt to describe composite behavior. Theocaris *et al.*[100–102] have worked extensively on these models and dealt with the concept of the extent or efficiency of adhesion at the interface. The effect of interfacial adhesion on the mechanisms of craze and shear-band formation in polymers has also been addressed theoretically.[103,104] Finite-element analysis of the effects of the interface and interphase also exists.[103,105]

5.2. Properties

The interest in adhesion between solid surfaces and polymers has prompted numerous investigations into the physical and chemical responses of polymer molecules in contact with surfaces (i.e., interphase material). Changes in hydrogen bonding, orientation, mechanical properties, miscibility, heat of fusion, molecular weight, heat of adsorption, thermal transitions, and chain mobility are examples of the differences which have been noted between polymers near or in contact with solid surfaces and the bulk polymers. These changes in polymer properties in the interphase usually decrease with distance from the surface, but are often seen up to hundreds of microns in distance from the surface.

Lipatov[106] made some generalizations about the mobility of polymer molecules near surfaces that seem to hold true for most polymeric systems. First, surfaces influence

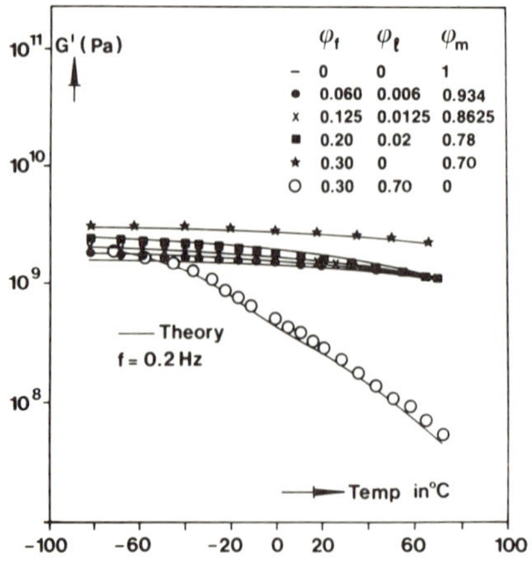

FIGURE 10. Theoretical and experimental storage moduli of glass-filled styrene–acrylonitrile copolymer with artificially applied ethylene–vinylacetate interfacial layer. [From F. H. J. Maurer, Ph.D. Thesis, Figure 45, p. 80, University Duisburg (1983).]

segmental mobility and relaxation in polymers due to the adsorption interaction and the conformation restriction imposed by the surface. This is demonstrated by numerous examples showing a substantial rise in the glass transition temperature of polymers near a surface compared to the bulk polymer.[107–109] The methods which have been used for these measurements include dilatometry, dynamic-mechanical analysis, calorimetry, dielectric relaxation, and NMR. Similarly, there are also data indicating an increase in the average relaxation time of polymers near a surface.[110–113] Second, the formation of polymer structure near a surface will be different from the bulk due to the changes in molecular mobility. For instance, the density of polymers near a surface is generally less than in the bulk, due to the retardation of relaxation processes. This has been shown by the increased sorption of filled polymers.[114] At the same time, there is usually a rise in packing density for polymer molecules very close to a solid surface, probably due to the orienting nature of the surface.[115,116] Lipatov and Babich[117] have also shown that the elastic modulus of a polymer increases with decreasing surface to volume ratio (i.e., thickness or distance from surface) as a result of the decreased molecular mobility.

Controlled polymer grafts on solid surface have been used to study the properties of the grafted chains and their interaction with the solid surface and with matrix polymer chains. For instance, Eastmond *et al.*[118,119] modified the surface of glass beads with trichloroacetyl groups which were used to generate radicals and initiate graft polymerization of methyl methacrylate. Composites fabricated with the grafted glass and using polymethylmethacrylate as a matrix had tensile properties identical to cmposites fabricated with the nongrafted glass. Examination of the fracture surfaces of these composites by scanning electron microscopy[120] showed that while grafts on adjacent beads are mutually miscible, grafts are immiscible with homopolymer chains at equilibrium. Such immiscibility, similar to that found between graft copolymers and the homopolymer of one of the comonomers, has been seen by other researchers.[121–123] Eastmond showed that entanglements between graft and homopolymer chains could be achieved by forming the homopolymer in the presence of the grafts and avoiding equilibrium conditions. Eastmond and Mucciariello[124] also photochemically detached and isolated polymer grafts on glass surfaces in order to characterize their properties and immiscibility with the homopolymer. The grafted polymer had a high proportion of high molecular weight (greater than 500,000 g/mol) chains and changing the reaction time effected the distribution.

Maurer *et al.*[125] analyzed the properties of bound polymer in polyethylene composites. The heat of fusion was found to be appreciably lower than that of the matrix. Additionally, the molecular mass distribution of the bound polymer had an average higher than that of the unbound fraction as predicted from theory.

Kessaissia *et al.*[126] studied the thermal transition of alkyl chains grafted onto silica as a model for reinforced polymers. Subambient-temperature inverse gas chromotography was used to show transitions at −20 and −75 °C common to all grafts and a third transition which was related to the chain length of the graft. Similarly, heat capacity data of silicas bonded with n-octyl, n-octadecyl, amino-butyl and 2,4-dinitroanilinobutyl groups was measured as a function of temperature by adiabatic calorimetry.[127] Heats of adsorption on modified silicas at approximately zero surface coverage have been measured using vapor-phase chromatography.[128] The results revealed the existence of long range interactions (more than one molecular diameter) between probe molecules and the silica surface. Vidal *et al.*[129] used modified silica to initiate homopolymerization of isobutylene and styrene as well as copolymerization of isobutylene-co-isoprene.

Thermostimulated creep measurements have been used to study the relaxation times in the matrix of a glass-reinforced epoxy.[130] The epoxy matrix exhibits two relaxations: the faster corresponds to the network and the slower is associated with the dangling chains. Incorporation of silicone in the interphase increases the proportion of dangling chains. Using a silane to modify the interphase produces two additional relaxations.

Polymer surface interactions have also been demonstrated by rheological measurements[95,131] and dynamic mechanical measurements.[95,132] A substantial amount of the initial research on composite interphases was carried out on filled rubber systems. There are numerous examples in the literature.[133–137]

The role of an interphase region should not be underestimated as often it can be the controlling factor in the performance of the composite. Maurer *et al.*[125] showed a strong relationship between tensile strength and weight fraction of adsorbed polymer in kaolin-filled polyethylene as seen in Figure 11. Numerous samples were compounded as a function of volume fraction of kaolin and processing conditions. The tensile strength as well as the fracture behavior (i.e., cohesive vs. adhesive failure and ductile vs. brittle matrix failure) correlated with the amount of bound polymer as measured by thermogravimetric analysis after dissolution of the unbound matrix polymer. The reduced mobility of the bound polyethylene resulted in a decrease in the heat of fusion from 225 to 120 J/g. The strong relationship between ultimate composite performance and identifiable characteristics of the interphase region has been seen in other systems. Maurer *et al.*[138] showed that there is a 3.6-nm-thick interphase of bound polyethylene in Aerosil-filled polyethylene. Similarly, Kendall and Sherliker[139,140] showed the presence of a bound polymer layer or interphase in various polyethylene–silicate composites having a thickness of approximately 2 nm.

Changes in polymer morphology in the interphase have been noted by a number of researchers. Dolakova and Hudecek[141] showed a residual interfacial fibrillar structure

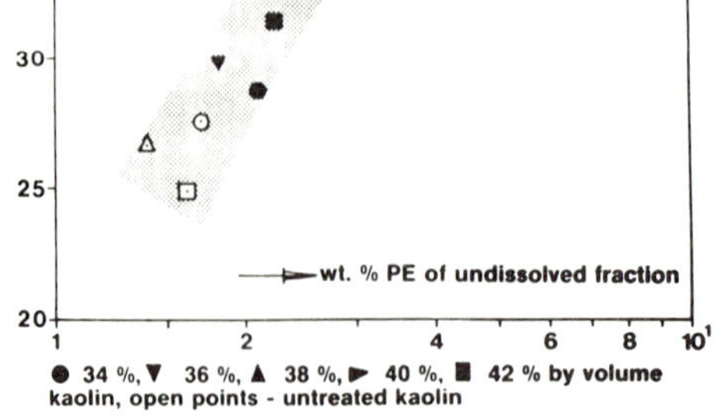

FIGURE 11. Relationship between the tensile strength of polyethylene–kaolin composites and the weight percent bound (undissolvable) polyethylene. [From F. H. J. Maurer, R. Kosfeld, and Th. Uhlenbroich, *Colloid Polym. Sci.* **263**(8), 624, Figure 10 (1985).]

after removing the matrix in glass-bead-, silica-, and kaolin-filled polyethylene. Chacko *et al.*[142] demonstrated in calcium-carbonate-filled polyethylene a disordered lamellar structure around the particles and a decreased crystallinity in the presence of the particle for low-molecular-weight polyethylene.

With regard to amorphous polymers few general conclusions can be drawn. However, there are examples in the literature of the influence of the solid surface on the interphase structure of amorphous polymers.[143,144] For instance, Lipatov *et al.*[145] showed that in polyurethane films on a steel surface there was a change in globule size with distance from the surface up to 160 microns. They concluded that globule size and packing density, which influence the real contact area with the surface, determine the number of functional groups interacting with the surface which, therefore, influences the adhesion bond strength.

Recently, there has been a great deal of interest in controlling morphologies near the interface in composites. This is especially true with thermoplastic matrices where the matrix and composite properties are determined to a great extent by the crystallization of the matrix. The interaction of polymer chains with a solid surface alters the kinetics of crystallization and may also effect nucleation. For example, Peacock *et al.*[146] examined the crystallization of poly(ether-ether-ketone) (PEEK) under the constraint of reinforcing fibers in actual composites. Using an etching technique, the researchers showed that nucleation from the fibers dominates the morphology and that discrete spherulitic growth emerges from the fiber surface before initiation takes place in the matrix. Lee and Porter[147] demonstrated that a transcrystalline interface could be developed in fiber-reinforced nylon 6 and PEEK. The transcrystalline interphase is more strongly developed if crystallization takes place at a high temperature or if the time allowed for melt-annealing is long.

Lipatov[132] concluded that strong adsorption interaction of a polymer with the surface of a solid slows down crystallization, while a very weak interaction has no influence on the crystallization, and moderate interaction leads to the surface acting as a center of formation of nuclei. These types of controlled crystallization are thought to be able to improve interfacial adhesion and other composite properties.[148] Probably the most extreme example of controlled morphology involving solid surfaces is epitaxial polymerization, where the polymer morphology is determined by the morphology of the solid surface on which the polymerization takes place.

Molecular level studies of the orientation of polymers at the reinforcement surface have also begun. Nguyen and Ishida[149] examined the molecular interactions between poly(ether-ether-ketone) and its model compounds and the basal plane of graphite. Controlled interfacial matrix morphology (i.e., interphase) is bound to be an important factor in the optimization of the properties of thermoplastic composites. This realization will have a great impact on the processing requirements of these materials. The quality of the composite will, to even a greater extent than in the past, be dependent on the way in which the materials are processed into usable parts.

While it is clear that an interphase region can develop due to morphological considerations in thermoplastics, there is a somewhat analogous situation in thermosetting materials. Solid surfaces can influence the nature of polymerization reactions in different ways. Changes in crosslink density, gel time, molecular weight, molecular weight distribution, and rate of polymerization have all been noted. Any changes in the cure conditions caused by the presence of reinforcement surfaces are bound to change the properties of the matrix in the vicinity of the reinforcement. Deviations in cure history including incomplete, retarded, accelerated, and crosslinking reactions have been suggested to produce an interphase in thermosetting composites. Garton and Daly[150,151] used a modification of internal reflection spectroscopy (IRS) in the mid-infrared region to model the curing reactions in anhydride-cured epoxies in the approximately 300 nm region adjacent to a silicon dioxide or aromatic polyamide surface. In both cases the yield of ester crosslinks was reduced about 22% compared to a freshly polished germanium surface. Similarly, Garton and Stevenson[152] showed that oxidative treatments to a carbon surface strongly influence the chemical state of adjacent anhydride-cured epoxy resin. A possible explanation is that small molecule curing agents are depleted in the interphase by adsorption on the reinforcement surface. Numerous other citings[153–155] detail curing inhomogeneities caused by the presence of reinforcement surfaces.

These types of curing inhomogeneities are likely to result in a more deformable interphase with a modulus less than that of the bulk matrix. A more restrained interphase can be produced in thermosetting composites with the use of coupling agents. The coupling agents are surface active molecules and will retard the adsorption of curing agents reducing any cure inhomogeneities as shown by Garton and Daly.[151] Coupling agent molecules also have reactive functionalities matched to the matrix chemistry and may help to bind polymer to the reinforcement surface or provide additional crosslinks in the interphase region.

5.3. Silane Interphase

The chemistry of organofunctional silanes in the interphase has gained attention due to numerous experimental results demonstrating that the beneficial effect of their usage cannot be explained by a simple or ideal concept of interfacial modification. In fact, usage of the commercially available organofunctional trialkoxysilanes does not result in a uniform monolayer coating of the reinforcement, but in nonuniform layers of a finite thickness and varying structure. Plueddemann has calculated, based on ignition loss, that commercial glass has an average coverage of 10 to 20 monolayers of silane coupling agent.[34] It has also been found empirically that increasing or decreasing the coverage from this level causes a decrease in composite performance. Other investigations using analytical techniques including ellipsometry, radioactive tracers, and Fourier transform infrared spectroscopy confirm silane depositions of greater than monolayer coverage.[156–160] Certainly a number of variables influence this thickness including the organofunctional group on silicon, the alkoxy group, the nature of the oxide surface, the availability of water, pH, concentration and age of the silane solution, drying conditions, etc.

Organofunctional silanes are thought of as bifunctional molecules because they have potential reactivity with both hydroxyl covered reinforcement surfaces and matrix chemistries. The commercial silanes are actually tetrafunctional due to the trialkoxy silane moiety in addition to the organofunctionality. The condensation reactions of the hydrolyzed trialkoxysilanes are responsible for many of the effects observed in the interphase. There are a number of variables which can be used to characterize the structure of silanes in the interphase. They include the amount or percent of silane, uniformity and thickness of the silane layers, orientation and organization of silane molecules, degree of siloxane formation, molecular weight and molecular weight distribution, the degree of available organofunctionality, solubility in the matrix, and the degree of chemically adsorbed and physically adsorbed silane.

Studies have been undertaken to determine the structure of the resulting silanes in this region. Schrader *et al.*(158,161,162) found that the adsorption of γ-aminopropyltriethoxysilane from aqueous solution resulted in a film which could be regarded as containing three fractions. Fraction 1 consisted of physically adsorbed silane hydrolyzate and represented 98% of the deposited silane. Fraction 2 was chemically adsorbed polymeric silane and had a thickness of about ten monolayer equivalents. Fraction 3 was a monolayer or less, tightly held to the surface, and could not be removed by a four-hour water boil.

Nakatsuka *et al.*(163) studied the molecular weight of extractable silane from the surfaces of calcium carbonate and clay using gel permeation chromatography. The molecular weight of extractable silane was less than 4000 and was higher on the clay surface than on calcium carbonate. Ishida and Miller(164) correlated the molecular weight of the physiosorbed siloxane with the acid–base nature of the reinforcement surface. Similar molecular weight distributions of the siloxane were seen on surfaces having similar aqueous slurry pH values regardless of their chemical, morphological, or topological makeup.

Attempts to control the siloxane structure on the solid surface with tetrafunctional silanes are possible, but only at much higher concentrations of the tetrafunctional additive than would be required in the bulk.(165) This is due to a cyclization tendency of the organofunctional silane on the solid surface. Miller and Ishida(56) were able to follow the formation of siloxane linkages as a function of adsorbate concentration on a silane-modified surface as illustrated in Figure 12. At coverage less than approximately 0.3 monolayer, no intermolecular siloxane bonds form. At higher concentrations, the number of siloxane linkages formed increased with increasing coverage and approached an asymptotic value. The asymptote represented a structure in which the yield of siloxane bond formation was 90%. The siloxane structure on the substrate surface is also influenced by the condensation reaction in solution. Aqueous condensation of hydrolyzed

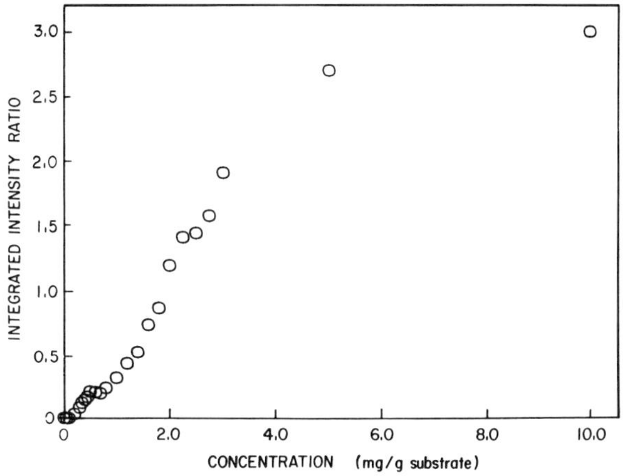

FIGURE 12. Integrated intensity ratio of the siloxane bands to an internal standard band as a function of the concentration of surface modification. [From J. D. Miller and H. Ishida, *Langmuir* **2**, 128, Figure 6, p. 129 (1986).]

alkyltrialkoxysilanes has been studied by Pohl and Osterholtz.(38) Figure 13 shows that the condensation reaction is slowest near p*D* 4.5.

Migration of an aminofunctional silane in the interphase of an Al_2O_3/polyethylene joint was demonstrated by Sung *et al.*(166) using X-ray microprobe analysis. A much broader distribution of the position of silane molecules with respect to distance from the surface was observed when the silane-modified aluminum adherend was dried at room temperature as opposed to drying at elevated temperatures.

The influence of organofunctional silanes in the interphase on the rheological properties of filled polymer melts and composite systems has been studied. A number of trends have been noted, including both increases and decreases in viscosity depending on the system. It can generally be concluded that when large decreases in viscosity are observed, the silane acts to better disperse solid particles in the matrix and decrease particle–particle interaction. Smaller contributions to a viscosity reduction might be explained by a plasticization or lubrication effect caused by the silane. Increases in viscosity are probably due to the presence of a bound layer of matrix material around the particles held by either chemical or physical forces and promoted by the presence of the silane.

Kulkarni *et al.*(167) showed large decreases in viscosity in dispersions of mineral fillers in nonaqueous low-viscosity fluids upon incorporation of a silane. Han *et al.*(168) demonstrated that titanates generally reduce the viscosity in calcium carbonate, talc, and fiberglass-filled polyolefins. Boaira and Chaffey(169) showed that a silane, *N*-(4-vinylphenyl)methyl-*N'*-(3-trimethoxysilylpropyl) ethylenediamine hydrochloride, greatly decreased viscosity in mica-filled polypropylene. Increases in viscosity were noted for glass-sphere-filled polymer melts when the glass surface was modified with silane and titanate treatments.(170,171)

It has also been shown that for a given system the viscosity can be either increased or decreased depending on the type of surface modification.(172,173) Figure 14 shows that for steel spheres dispersed in low-density polyethylene with modifiers added directly to the melt, the dynamic viscosity is raised for surface modification with an aminosilane and a titanate, lowered for a lubricant, and either raised or lowered depending on frequency for surface modification with a wetting agent.(172) The researchers also showed that rheo-

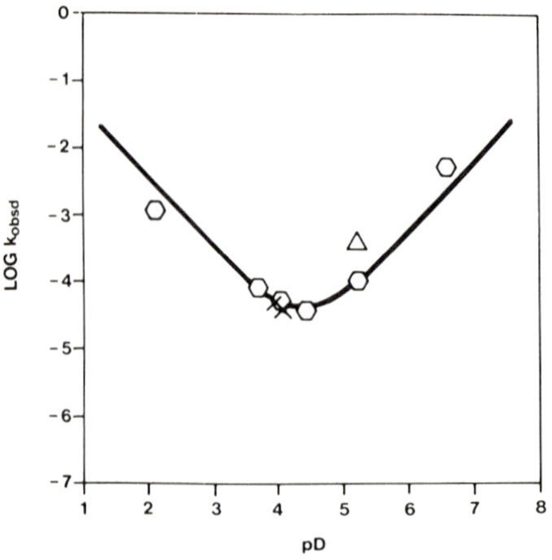

FIGURE 13. p*D*–rate profile for the condensation of γ-glycidoxypropylsilanetriol (hexagon) to the siloxane dimer in acetate buffer at 35 °C. Values for γ-aminopropylsilanetriol (triangle) and γ-methacryloxypropylsilanetriol (X) are also shown. [From E. R. Pohl and F. D. Osterholtz, in: *Molecular Characterization of Composite Interfaces* (H. Ishida and G. Kumar, eds.), p. 157; Figure 4, p. 167, Plenum Press, New York (1985).]

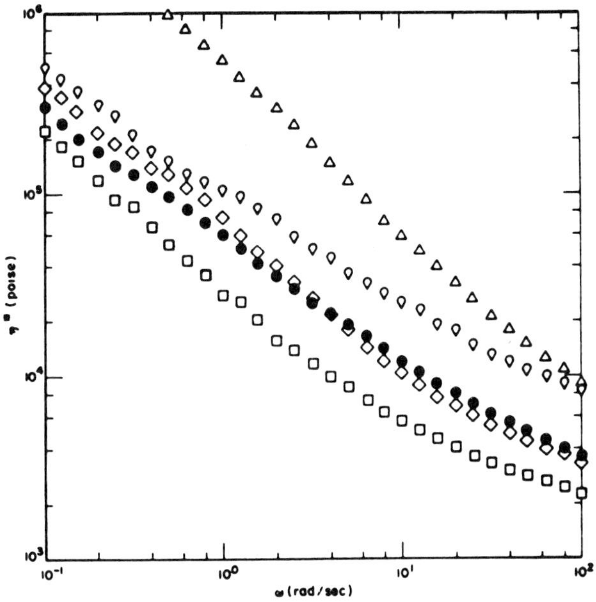

FIGURE 14. Effect of surface treatments added directly to the melt on the dynamic viscosity of steel-sphere-filled (60 volume percent) polyethylene at 200 °C: closed circles—untreated, triangles—aminosilane, inverted drops—titanate, diamonds—wetting agent, squares—lubricant. [From D. M. Bigg, *Polym. Eng. Sci.* **22**(8), 512, Figure 4, p. 515 (1982).]

logical properties were dependent on the method of application of the surface treatment (e.g., direct addition, pretreatment).

There are numerous reports of the effect of the presence of organofunctional silanes on the mechanical properties of filled polymers. However, most of these investigations deal with only surface-modified versus control systems. There are relatively few investigations which attempt to relate the structure of silanes in the interphase with properties. Schrader(174) made a correlation between the presence of the three silane fractions(158) and composite strength. It was concluded that fraction 1 did not contribute to bond strength and caused bond deterioration if present in excess. Fractions 2 and 3 were determined to provide the adhesive bond strength with joint life linearly dependent on the amount for fraction 2.

Emadipour *et al.*(175) used a single fiber pull-out test to investigate the effect of solution concentration of different aminofunctional silanes on the interfacial shear strength of glass-reinforced epoxies. The thinner coatings of the silane gave better properties in all cases. Graf *et al.*(176) made simultaneous mechanical and molecular characterization of a composite system of E-glass/polyester with a methacrylate functional silane. They showed that the silane structure consisted of physisorbed silane which was soluble in organic solvents, and chemisorbed silanes which could not be removed from the glass surface. The physisorbed silane reduced the strength of the composites. Kokubo *et al.*(177) made analogous observations that removal of physiosorbed silanes by extraction improved properties in mica-filled composites. Similarly, a number of researchers have noted an optimum concentration of silane which maximizes composite properties.[166,176–180]

Kaul *et al.*[181] attempted to correlate the peel strength of Al_2O_3/polyethylene joints with the aminosilane structure in the interphase. The silane structure was varied by changes in concentration, pH, and drying conditions and monitored by photoelectron spectroscopy and Fourier transform infrared spectroscopy. It was concluded that an interdiffusion between the silane and the polyethylene is necessary to optimize adhesion. Boerio and Gosselin[182] studied variables related to silane interphase structure, including pH and drying time, and attempted to correlate them with joint strength in metal/epoxy lap joints.

5.4. Other Additives

The interphase is a region which can be affected by a number of physical and chemical factors. Coupling agents are certainly one of the more important factors, but a number of others have appeared recently. Plueddemann[183] has shown that thin elastic interlayers which bond to the matrix resin by interdiffusion or other mechanisms can be used to generate tough composites having rigid matrices. The concept of interpenetrating polymer networks as a mechanism for adhesion promotion or reinforcement has been discussed.[34] Interpenetrating polymer networks have been observed in silane-modified polymer films deposited from a latex.[184] These films resulted in improved adhesion and solvent resistance without any chemical reaction between phases.

A highly chlorinated solid paraffin used with magnesium oxide to act as a HCl scavenger has been reported to improve properties in mica-reinforced polypropylene[185] when the polymer has no reactive groups and processing necessitates high-intensity mixing. Plueddemann and Stark[186] also showed that a combination of chlorinated paraffin, a trace of magnesium oxide, and a reactive amine-functional silane with dicumyl peroxide promotes adhesion in thermoplastics. The chemical and physical interactions in these systems have also been investigated. Garton *et al.*[187] showed that the chlorinated hydrocarbons and silanes, for that matter, modify the polymer morphology in the interphase in mica-reinforced polypropylene. Huang *et al.* demonstrated that magnesium oxide catalyzes the dehydrochlorination of chlorinated paraffins[188] and the resulting conjugated polyenes can graft to the polypropylene matrix during the extrusion process.[189]

Chun and Woodhams[190] showed that maleated polypropylene waxes prepared by the reaction of a polypropylene wax with maleic anhydride[191] can be used as an interfacial modifier or coupling agent in mica-filled polypropylene. The maleated polypropylene was particularly effective when used with mica modified with γ-aminopropyltriethoxysilane. For example, at 50 weight percent mica, the ultimate tensile strength of composites without modification or with aminosilane treatment only is 30 MPa. The maleated polypropylene and aminosilane combination gives an ultimate tensile strength of 43 MPa.

Grafting of polar monomers to polymers to improve adhesion has also become popular. For example, the grafting of acrylic acid to polypropylene in the presence of radical activators has been reported.[192,193] Plueddemann and Stark[194] used a silane-modified hexa(methoxymethyl)melamine resin primer to modify the interphase and improve adhesion of polar thermoplastics to glass. It was postulated that the primer could act as a solvent for the molten thermoplastic in the interphase and form an interpenetrating polymer network upon cooling.

6. FUTURE DIRECTIONS

It appears that the study of interfaces and interphases as they relate to adhesion will continue to be a discipline ripe for new discovery and for solving existing problems. Recently, an assessment of the science of interfaces and thin films was published as one of four reports requested by the science advisor of the President of the United States.[195] The reports were aimed at evaluating research opportunities in critical areas of science and technology.

Much of the advancement in understanding and controlling interfacial properties will come through developments in sophisticated analytical techniques that will allow researchers to probe and measure the chemistry and physics of interfaces. A good example of recent development in this area is scanning tunneling microscopy (STM), which can measure the position of individual atoms on surfaces.[196] The resolution achievable is as little as one angstrom, creating three-dimensional images that show actual atomic position and defects.

STM operates by measuring the current of tunneling electrons when a small voltage is applied between the surface and a very sharp needle positioned so close to the surface that the wavefunctions of electrons in the needle tip overlap wavefunctions of electrons in the surface.[197] Originally, the scanning tunneling microscope was limited to the surfaces of conducting surfaces in vacuum. However, the technique has been expanded to examine insulating surfaces by using a cantilevered beam to measure interatomic forces between the needle tip and the surface. Researchers have also expanded the technique to work in ambient conditions, without the requirement of high vacuum, and to systems where the surface is immersed in a liquid medium.

The development of nuclear magnetic resonance imaging is another technique which will provide new information about interface and interphase properties. The migration of water to the interface in composites is an example of the type of phenomenon that can be studied with the imaging technique.

A large number of modern analytical techniques have contributed to the current understanding of interfaces. These include X-ray photoelectron spectroscopy, scanning electron microscopy, reflectance infrared spectroscopy, Raman and surface-enhanced Raman spectroscopy, Auger spectroscopy, ion scattering spectroscopy, secondary ion mass spectroscopy, acoustic microscopy, low-energy electron diffraction, and extended X-ray absorption fine structure. More complete lists with description are available.[51,198]

The most important advances in adhesion research will come through the development of techniques which can probe interfaces and interphases *in situ*. This will require devices that can discriminate between interfacial and bulk regions. Gauging by past developments, such advances are likely and should shed great light in the area of adhesion mechanisms.

REFERENCES

1. J. V. Larsen, *Surface Treatments and Their Effects on Carbon Fibers,* U.S. Naval Ordnance Laboratory (1971).
2. F. Hophgarten, *Fibre Sci. Technol.* **11,** 67 (1978).
3. F. Hophgarten, *Fibre Sci. Technol.* **12,** 283 (1979).
4. K. Walterson, *Fibre Sci. Technol.* **17,** 289 (1982).
5. A. Ishitani, *Carbon* **19,** 269 (1981).
6. T. Takahagi and A. Ishitani, in: *Molecular Characterization of Composite Interfaces* (H. Ishida and G. Kumar, eds.), p. 321, Plenum Press, New York (1985).
7. W. Chen and J. P. Wightman, Fourteenth Biennial Conference on Carbon, American Carbon Committee, p. 107 (1979).
8. D. A. Scola and B. L. Laube, in: *Molecular Characterization of Composite Interfaces* (H. Ishida and G. Kumar, eds.), p. 423, Plenum Press, New York (1985).
9. A. Proctor and P. M. A. Sherwood, *Carbon* **21,** 53 (1983).
10. A. Proctor and P. M. A. Sherwood, *Surf. Interface Anal.* **4,** 212 (1982).
11. A. Proctor and P. M. A. Sherwood, *Carbon* **21,** 53 (1983).
12. J. Barr, U.S. Patent 3,791,840 (1974).
13. J. C. Goan, Graphite Fiber Oxidation, Naval Ordnance Laboratory TR-153 (1969).
14. J. Judd, British Patent 1,479,455 (1977).
15. R. J. Dauksys, Coupling of Epoxy Polymers to Graphite Fibers, A.F.M.L. 72–23 (May 1972).
16. H. Nyo, D. L. Heckler, and P. L. Hoernschemeyer, *SAMPE 24th Natl. Symp.* **24,** 51 (1979).
17. F. Molleyer and M. Bastick, Carbon '76, International Carbon Conference, 2nd Preprint, p. 500 (1976).
18. Kobetis, *Polym. Mech. (Engl. Transl.)* **4**(4) (1979).
19. N. J. DeLollis, *SAMPE J.* **15**(3), 10 (1979).
20. H. Yasuda and N. Morosoff, *Polym. Prepr.* **19**, 488 (1978).
21. J. C. Watts, U.S. Patent 3,367,812 (1968).
22. H. Lubowitz, French Patent 1,580,456 (1969).
23. T. Fujisawa, A. Tanaka, K. Hashimoto, and M. Kobayashi, *Polym. Prepr.* **20**(1), 853 (1979).
24. J. V. Milewski and J. V. Anderle, GTC Technical Report 127.7-1 (September 1967).
25. W. L. Lockman, R. Penty, and A. Jahn, U.S. Patent 3,894,864 (1975).
26. R. V. Subramanian and J. J. Jukubowski, *Polym. Eng. Sci.* **18**(7), 590 (1978).
27. F. Tuinstra and J. L. Koenig, *J. Comp. Mater.* **4,** 492 (1970).
28. T. Nakatsuka, H. Kawasaki, K. Itadani, and S. Yamashita, *J. Appl. Polym. Sci.* **27,** 259 (1982).
29. T. Nakatsuka, H. Kawasaki, K. Itadani, and S. Yamashita, *J. Colloid Interface Sci.* **82,** 298 (1981).
30. A. Ahagon and A. N. Gent, *J. Polym. Sci., Polym. Phys. Ed.* **13,** 1285 (1975).
31. E. G. Rochow, *An Introduction to the Chemistry of Silicones,* 2nd ed., Chapman and Hall, London (1951).
32. J. V. P. Torrey, *Mod. Plast.* **30,** 154 (1952).
33. E. P. Plueddemann, H. A. Clark, L. E. Nelson, and K. R. Hoffman, *Mod. Plast.* **39,** 135 (1962).
34. E. P. Plueddemann, *Silane Coupling Agents,* Plenum Press, New York (1982).
35. H. A. Clark and E. P. Plueddemann, *Mod. Plast.* **40**(6), 133 (1963).
36. E. P. Plueddemann, 24th Ann. Tech. Conf. Reinforced Plastics/Composites Inst., SPI, 19-A (1969).
37. H. Ishida and J. L. Koenig, *Appl. Spectrosc.* **32,** 462 (1978).
38. E. R. Pohl and F. D. Osterholtz, in: *Molecular Characterization of Composite Interfaces* (H. Ishida and G. Kumar, eds.), p. 157, Plenum Press, New York (1985).
39. J. Jang, H. Ishida, and E. P. Plueddemann, 41st Ann. Tech. Conf. Reinforced Plastics/Composites Inst., SPI, 2-C (1986).
40. J. G. Marsden and E. J. Pepe, 40th Ann. Tech. Conf. Reinforced Plastics/Composites Inst., SPI, 17-C (1985).
41. L. B. Cohen, *Plast. Eng.* **39**(11), 29 (1983).
42. D. H. Solomon and B. G. Hawthorne, *Chemistry of Pigments and Fillers,* Wiley, New York (1983).
43. L. B. Cohen, U.S. Patent 4,539,048 (Sep. 3, 1985).
44. L. B. Cohen, *High Solids Coatings,* **9**(3) (September, 1984).
45. L. B. Cohen, 39th Ann. Tech. Conf. Reinforced Plastics/Composites Inst., SPI, 19-E (1984).

46. L. B. Cohen, ANTEC '84, Society of Plastics Engineers, p. 639 (1984).
47. D. N. Misera and R. L. Bowen, *J. Colloid Interface Sci.* **61,** 14 (1977).
48. M. M. Fein, B. K. Patnaik, and F. K. Y. Chu, U.S. Patent 4,073,766 (1978).
49. W. J. van Ooij, *Rubber Chem. Technol.* **52,** 604 (1979).
50. W. J. van Ooij, *Rubber Chem. Technol.* **57,** 686 (1984).
51. A. W. Adamson, *Physical Chemistry of Surfaces,* Wiley, New York (1982).
52. S. R. Seyedmonir, S. Abdo, and R. F. Howe, *J. Phys. Chem.* **86,** 1233 (1982).
53. J. B. Kinney and R. H. Staley, *J. Phys. Chem.* **87,** 3735 (1983).
54. S. Naviroj, J. L. Koenig, and H. Ishida, ONR Tech. Report No. CWRU/DMS/TR-6 (December 1982).
55. S. Culler, Ph.D. Thesis, Case Western Reserve University (1984).
56. J. D. Miller and H. Ishida, *Langmuir* **2,** 127 (1986).
57. J. D. Miller and H. Ishida, unpublished data.
58. J. D. Miller and H. Ishida, in: *Chemically Modified Surfaces,* Vol. 1, *Silanes, Surfaces, and Interfaces* (D. E. Leyden, ed.), p. 525, Gordon and Breach, New York (1986).
59. J. D. Miller and H. Ishida, *J. Phys. Chem.* **86,** 1593 (1987).
60. S. Chiang, R. G. Tobin, P. L. Richard, and P. A. Thiel, *Phys. Rev. Lett.* **52,** 648 (1984).
61. Y. J. Chabal, *J. Vac. Sci. Technol.* **A3,** 1448 (1985).
62. Y. J. Chabal, *Phys. Rev. Lett.* **55,** 845 (1985).
63. G. E. Maciel and D. W. Sindorf, *J. Am. Chem. Soc.* **102,** 7607 (1980).
64. G. E. Maciel and D. W. Sindorf, *J. Am. Chem. Soc.* **103,** 4263 (1981).
65. D. W. Sindorf and G. E. Maciel, *J. Chromatogr.* **205,** 438 (1981).
66. D. W. Sindorf and G. E. Maciel, *J. Phys. Chem.* **86,** 5208 (1982).
67. G. S. Caravajal, D. E. Leyden, and G. E. Maciel, in: *Chemically Modified Surfaces, Vol. 1 Silanes, Surfaces, and Interfaces* (D. E. Leyden, ed.), p. 283, Gordon and Breach, New York (1986).
68. D. L. Allara, in: *Vibration Spectroscopies for Adsorbed Species* (A. T. Bell and M. L. Hair, eds.), p. 37, American Chemical Society, Washington D.C. (1980).
69. J. D. Miller and H. Ishida, *Surf. Sci.* **148,** 601 (1984).
70. T. Morimoto, H. Yanai, and M. Nagao, *J. Phys. Chem.* **80,** 471 (1976).
71. V. Yu. Borovkov, W. K. Hall, and V. B. Kazanski, *J. Catal.* **51,** 437 (1978).
72. P. N. T. Van Velzen, *Surf. Sci.* **140,** 437 (1984).
73. V. N. Abramov, A. V. Kiselev, and V. I. Lygin, *Russ. J. Phys. Chem.* **37,** 1507 (1963).
74. M. L. Hair, *Infrared Spectroscopy in Surface Chemistry,* Marcel Decker, New York (1967).
75. L. H. Little, *Infrared Spectra of Adsorbed Molecules,* Academic Press, London (1966).
76. W. D. Bascom, *J. Phys. Chem.* **76,** 3188 (1972).
77. R. S. McDonald, *J. Am. Chem. Soc.* **79,** 850 (1957).
78. J. G. Roth and M. J. Dignam, *Can. J. Chem.* **54,** 1388 (1976).
79. J. Chmielowiec and B. A. Morrow, *J. Colloid Interface Sci.* **94,** 319 (1982).
80. H. Knozinger and w. Stahlin, *Prog. Colloid Polym. Sci.* **67,** 33 (1980).
81. P. N. T. Van Velzen, *Surf. Sci.* **146,** 319 (1984).
82. P. N. T. Van Velzen, *Surf. Sci.* **161,** L609 (1985).
83. A. F. Diaz, V. Hetzler, and E. Kay, *J. Am. Chem. Soc.* **99,** 6780 (1977).
84. T. Furukawa, N. K. Eib, K. L. Mittal, and H. R. Anderson Jr., *Surf. Interface Anal.* **4**(6), 240 (1982).
85. D. M. Brewis, J. Comyn, D. P. Oxley, R. G. Pritchard, S. Reynolds, C. R. Werrett, and A. J. Kinloch, *Surf. Interface Anal.* **6**(1), 40 (1984).
86. E. P. Plueddemann, *Adhes. Age* June (1975), p. 36.
87. A. S. Kenyon and H. J. Duffy, *Polym. Eng. Sci.* **7,** 1 (1970).
88. A. S. Kenyon, *J. Colloid Interface Sci.* **27,** 761 (1968).
89. L. J. Broutman and B. D. Agarwal, 28th Ann. Tech. Conf., Reinf. Plast./Composites Inst., SPI, 5-B (1973).
90. P. W. Erickson, A. A. Volpe, and E. R. Cooper, 19th Ann. Tech. Conf., Reinf. Plast./Composites Inst., SPI, 21-A (1964).
91. F. S. Cheng, J. L. Kardos, and T. L. Tolbert, *SPE Journal* **26,** (1970).
92. J. L. Kardos, *J. Adhes.* **4,** 1 (1972).
93. R. E. Lavengood and M. J. Michno, Jr., Proc. Div. Techn. Conf., Engr. Prop. Struct. Div., SPE, p. 127 (1975).

94. L. D. Tryson and J. L. Kardos, 36th Ann. Tech. Conf., Reinforced Plastics/Composites Inst., SPI, 2-E (1981).
95. F. H. J. Maurer, Ph.D. Thesis, University Duisburg (1983).
96. C. van der Poel, *Rheol. Acta* **1,** 198 (1958).
97. F. R. Schwarzl and J. v. d. Eikhoff, Calculation of Elastic Properties of Composites by means of the van der Poel Theory, Central Lab. TNO, Delft, NL, Report No. CL 71/55 (1971).
98. J. C. Smith, *J. Res. Natl. Bur. Stand.* **78A,** 355 (1974).
99. C. Scott, H. Ishida, and F. H. J. Maurer, in: *Composite Interfaces* (H. Ishida and J. L. Koenig, eds.), p. 177, Elsevier, New York (1986).
100. G. C. Papanicolaou, P. S. Theocaris, and G. Spathis, *Colloid Polym. Sci.* **258,** 1231 (1980).
101. P. S. Theocaris, G. Spathis, and B. Kefalas, *Colloid Polym. Sci.* **260,** 837 (1982).
102. P. S. Theocaris and G. C. Papanicolaou, *Fibre Sci. Technol.* **12,** 421 (1979).
103. M. E. J. Dekkers and D. Heikens, *J. Mater. Sci.* **18,** 3281 (1983).
104. M. E. J. Dekkers and D. Heikens, *J. Mater. Sci.* **19,** 3271 (1984).
105. D. F. Adams, in: *Composite Interfaces* (H. Ishida and J. L. Koenig, eds.), p. 351, Elsevier, New York (1986).
106. Y. S. Lipatov, *Rubber Chem. Technol.* **49,** 1311 (1976).
107. Y. S. Lipatov and V. P. Privalko, *Vysokomol. Soedin.* **15B,** 749 (1973).
108. V. Solomkov, Thesis, Kiev State Univ. (1971).
109. G. Fritschy, E. Papirer, *J. Appl. Polym. Sci.* **25,** 1867 (1980).
110. Y. S. Lipatov and T. E. Geller, *Vysokomol. Soedin.* **9A,** 241 (1967).
111. Y. S. Lipatov, *Vysokomol. Soedin.* **10B,** 879 (1968).
112. Y. S. Lipatov, *Vysokomol. Soedin.* **7,** 1430 (1965).
113. G. M. Bartenev, B. I. Revyakin, N. M. Lyalina, and K. Boturov, *Polym. Sci. USSR* **25**(3), 595 (1983).
114. Y. S. Lipatov, *Trans. Inst. Plast. Ind.* **4,** 83 (1966).
115. T. K. Kwei, *J. Polym. Sci.,* Part A, **3,** 3229 (1965).
116. Y. S. Lipatov, *Polymer* **16,** 582 (1975).
117. Y. Lipatov and V. Babich, *Vysokomol Soedin.* **10,** 848 (1968).
118. G. C. Eastmond, C. Nguyen-Huis, and W. Piret, *Polymer* **21,** 598 (1980).
119. C. H. Bamford, G. C. Eastmond, and V. J. Robinson, *Trans. Faraday Soc.* **60,** 751 (1964).
120. G. C. Eastmond and G. Mucciariello, *Polymer* **23,** 164 (1982).
121. G. C. Eastmond and D. G. Phillips, *Polymer* **20,** 1501 (1979).
122. K. Kendall and F. R. Sherliker, *Br. Polym. J.* **12,** 85 (1980).
123. J. Clarke and B. Vincent, *J. Chem. Soc., Faraday Trans.* **77,** 1831 (1981).
124. G. C. Eastmond and G. Mucciariello, *Br. Polym. J.* **16,** 63 (1984).
125. F. H. J. Maurer, R. Kosfeld, and Th. Uhlenbroich, *Colloid Polym. Sci.* **263**(8), 624 (1985).
126. Z. Kessaissia, E. Papirer, and J. Donnet, *J. Colloid Interface Sci.* **82**(2), 526 (1981).
127. J. C. van Mittenburg and N. E. Hammers, *J. Chromatogr.* **268,** 149 (1983).
128. R. S. Chabal and L. E. St. Piere, *Macromolecules* **1,** 152 (1968).
129. A. Vidal, A. Guyot and J. P. Kennedy, *Polym. Bull.* **6,** 401 (1982).
130. J. P. Bayoux, C. Pillot, D. Chatain, and C. Lacabanne, in: *Molecular Characterization of Composite Interfaces* (H. Ishida and G. Kumar, eds.), p. 287, Plenum Press, New York (1985).
131. J. E. Stamhuis and J. P. A. Loppe, *Rheol. Acta* **21,** 103 (1982).
132. Y. Lipatov, *Physical Chemistry of Filled Polymers,* Rubber and Plastics Research Association of Great Britain, England (1979).
133. A. R. Payne, *J. Appl. Polym. Sci.* **6,** 21 (1962).
134. P. P. A. Smit and A. K. van der Vegt, *Kautsch. Gummi* **23,** 147 (1970).
135. I. Pilskin and N. Tokita, *J. Appl. Polym. Sci.* **16,** 473 (1972).
136. S. A. Paipetis, *Fibre Sci. Technol.* **21,** 107 (1984).
137. S. Kaufman, W. P. Slichter, and D. D. Davis, *J. Polym. Sci.,* Part A-2, **9,** 829 (1971).
138. F. H. J. Maurer, H. M. Schoffeleers, R. Kosfeld, Th. Uhlenbroich, in: *Progress in Science and Engineering of Composites* (T. Hayashi, K. Kawata, and S. Umekawa, eds.), ICCM-IV, p. 803, Tokyo (1982).
139. K. Kendall and F. R. Sheriker, *Br. Polym. J.* **12,** 85 (1980).
140. K. Kendall and F. R. Sheriker, *Br. Polym. J.* **12,** 111 (1980).
141. V. Dolakova and F. Hudeck, *J. Macromol. Sci.* **B15,** 337 (1978).

142. V. P. Chacko, F. E. Karasz, R. J. Farris, and E. L. Thomas, *J. Polym. Sci., Polym. Phys.* **20,** 2177 (1982).
143. A. N. Kuksin, *Vysokomol. Soedin.* **A12,** 2332 (1970).
144. L. Bozveliev, *J. Appl. Polym. Sci.* **17,** 1963 (1973).
145. Y. S. Lipatov, A. N. Kuskin, and L. M. Sergeeva, *J. Adhes.* **6,** 259 (1974).
146. J. A. Peacock, B. Fife, E. Nield, and R. A. Crick, in: *Composite Interfaces* (H. Ishida and J. L. Koenig, eds.), p. 299, Elsevier, New York (1986).
147. M. E. J. Dekkers and D. Heikens, in: *Composite Interfaces* (H. Ishida and J. L. Koenig, eds.), p. 161, Elsevier, New York (1986).
148. M. Ratzsch, in: *Interfaces in Polymer, Ceramic, and Metal Matrix Composites* (H. Ishida, ed.), p. 425, Elsevier, New York (1988).
149. H. X. Nguyen and H. Ishida, *Makromol. Chem.* **5,** 135 (1986).
150. A. Garton and J. H. Daly, *J. Polym. Sci., Polym. Chem.* **23,** 1031 (1985).
151. A. Garton and J. H. Daly, *Polm. Comput.* **6,** 195 (1985).
152. A. Garton and T. K. Stevenson, in: *Composite Interfaces* (H. Ishida and J. L. Koenig, eds.), p. 265, Elsevier, New York (1986).
153. J. L. Kardos, *Trans. N.Y. Acad. Sci.* **35,** 136 (1973).
154. A. S. Kenyon and L. E. Nielsen, *J. Macromol. Sci., Chem.* **A3,** 275 (1969).
155. R. E. Cuthrell, *J. Appl. Polym. Sci.* **12,** 1263 (1968).
156. D. J. Tutas, R. R. Stromberg, and E. Passaglia, *SPE Trans.* **4,** 256 (1964).
157. W. D. Bascom, *Adv. Chem. Ser.* **87,** 38 (1968).
158. M. E. Schrader, I. Lerner, and F. J. Doria, *Mod. Plast.* **45,** 195 (1967).
159. O. K. Johannson, F. O. Stark, G. E. Vogel, and R. M. Fleischmann, *J. Comput. Mater.* **1,** 278 (1967).
160. S. E. Berger and C. T. Desmond, *Plastics Compounding,* p. 55 (May/June 1983).
161. M. E. Schrader, *J. Adhes.* **2,** 202 (1970).
162. M. E. Schrader, *J. Polym. Sci.*, Part C, **34,** 281 (1971).
163. T. Nakatsuka, H. Kawasaki, K. Itadani, and S. Yamashita, *J. Appl. Polym. Sci.* **24,** 1985 (1979).
164. H. Ishida and J. D. Miller, *Macromolecules* **17,** 1659 (1984).
165. H. Ishida and J. D. Miller, *J. Polym. Sci., Polym. Phys. Ed.* **23,** 2227 (1985).
166. N. H. Sung, A. Kaul, I. Chin, and C. S. P. Sung, *Polym. Eng. Sci.* **22,** 637 (1982).
167. R. D. Kulkarni, P. S. Leung, and E. D. Goddard, *Colloids and Surfaces* **5,** 321 (1982).
168. C. D. Han, C. S. Sanford, and H. J. Yoo, *Polym. Eng. Sci.* **18,** 849 (1978).
169. M. S. Boaira and C. E. Chaffey, *Polym. Eng. Sci.* **17,** 715 (1977).
170. J. D. Miller, H. Ishida, and F. H. J. Maurer, *Rheol. Acta* **27,** 397 (1988).
171. R. E. S. Bretas and R. L. Powell, *Rheol. Acta* **24,** 69 (1985).
172. D. M. Bigg, *Polym. Eng. Sci.* **22,** 515 (1982).
173. C. D. Han, T. van den Weghe, P. Shete, and J. R. Haw, *Polym. Eng. Sci.* **21,** 196 (1981).
174. M. E. Schrader, 25th Ann. Tech. Conf., Reinforced Plastics/Composites Inst., SPI, 13-E (1970).
175. H. Emadipour, P. Chang, and J. L. Koenig, *Res. Mechanica* **5,** 165 (1982).
176. R. T. Graf, J. L. Koenig, and H. Ishida, *J. Adhes.* **16,** 97 (1983).
177. M. Kokubo, H. Inagawa, M. Kawahara, D. Terunuma, and H. Nohira, *Kobunshi Ronbunshu* **38,** 201 (1981).
178. O. K. Johnannson, F. O. Stark, and R. Baney, AFML-TRI-65-303, Part 1 (1965).
179. R. Wong, *J. Adhes.* **4,** 171 (1972).
180. Y. Eckstein and E. J. Berger, *Proc. Org. Coat. Appl. Polym. Sci.* **48,** 23 (1983).
181. A. Kaul, N. H. Sung, I. Chin, and C. S. P. Sung, *Polym. Eng. Sci.* **24,** 493 (1984).
182. F. J. Boerio and C. A. Gosselin, 36th Ann. Tech. Conf., Reinforced Plastics/Composites Inst., SPI, 2-G (1981).
183. E. P. Plueddemann, 29th Ann. Tech. Conf., Reinforced Plastics/Composites Inst., SPI, 24-A (1974).
184. M. A. Lutz and K. E. Polmanteer, *J. Coat. Technol.* **51**(652), 37 (1979).
185. F. J. Meyer and S. Newman, *Polym. Compos.* **1**(1), 37 (1980).
186. E. P. Plueddemann and G. L. Stark, 35th Ann. Tech. Conf., Reinforced Plastics/Composites Inst., SPI, 20-B (1980).
187. A. Garton, S. W. Kim, and D. M. Wiles, *J. Polym. Sci., Polym. Lett. Ed.* **20,** 273 (1982).
188. H. Ishida and Z. Huang, *Polm. Compos.* **6,** 34 (1985).
189. H. Ishida and Z. Huang, Antec '84, p. 205 (1984).

190. I. Chun and R. T. Woodhams, Antec '84, p. 132 (1984).
191. I. Chappell, F. R. Benn, and J. W. Dwyer, *J. Chem. Soc.* 533 (1977).
192. Eur. Pat. Appl., EP 41-214.
193. I. Sasaki and F. Ide, *Kobunshi Ronbunshu* **38**(2), 67 (1981).
194. E. P. Plueddemann and G. L. Stark, Proc. Adhesion Society Ann. Meeting, Savannah, GA (1980).
195. *Research Briefings 1986,* National Academy Press, Washington D.C. (1986).
196. G. Binnig and H. Rohrer, *Sci. Am.*, p. 50 (August 1985).
197. G. Binnig, C. F. Quate, and C. Gerber, *Phys. Rev. Lett.* **56,** 930 (1986).
198. *Opportunities in Chemistry,* p. 266, National Academy Press, Washington D.C. (1985).

11

Adhesion at Metal Interfaces

Amitava Banerjea, John Ferrante, and John R. Smith

1. INTRODUCTION

When two solid surfaces are brought into contact, strong adhesive bonding forces can develop at the interface formed by the solids. The magnitude of the forces will depend on the state of the surfaces, their cleanliness, and the fundamental properties of the two solids, both surface and bulk.

Both theoretical and experimental studies have been conducted on the bonding forces that develop at adhered interfaces. Theoretical studies, to date, have been directed at the bimetallic interface(1); experimental observations have included metals in contact with themselves, polymers, semiconductors, diamond, and ceramics.(2)

Surface properties which have been correlated experimentally with adhesion include atomic or molecular orientation, reconstruction, segregation, and the chemistry of the surface species. Where dissimilar solids are in contact, the contribution of each is considered in relation to the effect of the resultant interactive chemistry on bond strength.

Experimental or practical adhesion studies typically involve the testing of not only bond strengths at interfaces, but also the effects of elastic and plastic deformation.(3) If that were not complicated enough, experiments are also influenced by impurities absorbed from the environment(4) or segregated from the bulk,(5) by surface properties created by processing, by inherent strains in outer layers, and by the speed of the adhesion or the debonding process. An advantage of a theoretical approach is that these numerous variables can be separated. Currently, theorists have only considered the most ideal interfaces from first principles, i.e., perfect single crystals whose surfaces are free of both impurities and defects. The challenge for the future will be to include and assess the role of defects and impurities, while at the same time broadening the base of information on ideal interfaces.

These theoretical efforts would be greatly aided by experimental force measurements in adhesion. The experiments of Tabor and Winterton(6) on mica and of Israelachvili(7) on polymer interfaces provide the type of data needed, namely, adhesive force versus dis-

---

Amitava Banerjea • NASA Lewis Research Center, Cleveland, Ohio 44135; and Physics Department, Kent State University, Kent, Ohio 44242. *John Ferrante* • NASA Lewis Research Center, Cleveland, Ohio 44135. *John R. Smith* • Physics Department, General Motors Research Laboratories, Warren, Michigan 48090.

tance between surfaces.[6] For other types of interfaces, i.e., ionic, covalent, or metallic, such data are not available. For metal interfaces, however, Pashley *et al.*[3] have carried out some very careful measurements of the maximum adhesive force. More recently, the scanning-tunneling/atomic-force microscope (STM/AFM)[8,9] may be useful in providing such data.

We need to define a basic adhesion process which is an appropriate starting point for calculation as well as an appropriate ideal limit for the understanding of experimental or practical adhesion.[10] We bring two solids together at $T = 0$ K in an adiabatic and reversible manner. Then at each separation a between the two surfaces, the total energy $E(a)$ is a minimum functional of the electron density and atomic coordinates. The basic adhesive energy $E_{ad}(a)$ is then defined as

$$E_{ad}(a) = \frac{E(a) - E(\infty)}{2A} \tag{1}$$

where A is the cross-sectional area. It should be noted that, for practical adhesion, $E_{ad}(a)$ should include the effects of elastic deformation, impurity and surface roughness effects,

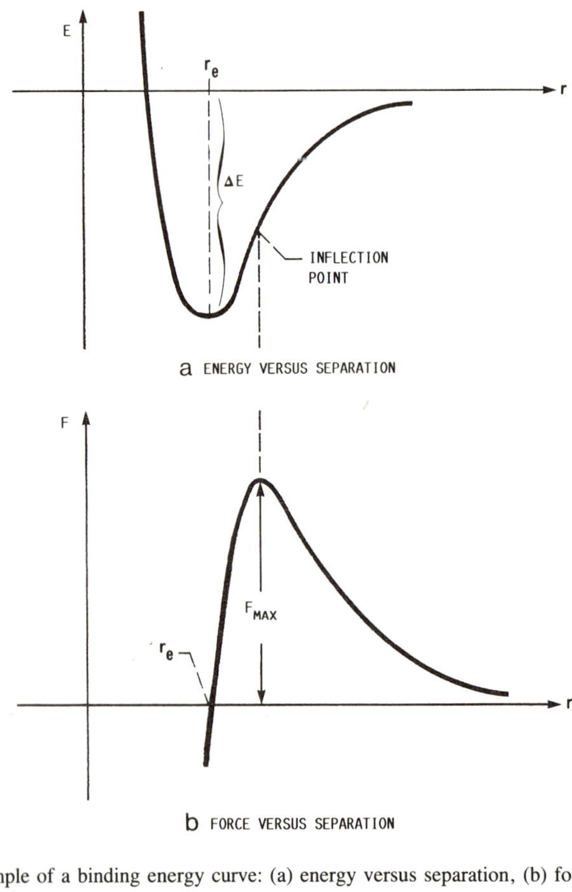

FIGURE 1. Example of a binding energy curve: (a) energy versus separation, (b) force versus separation.

as well as inelastic effects. Henceforth, when we use the word adhesion, we shall be referring to basic adhesion as distinguished from practical adhesion. The latter applies to real materials with the defects, impurities, and other considerations already mentioned, while the former applies to adhesion forces for ideal materials and calculated from first principles.

The objective of this chapter is to review briefly, by way of background, what has been accomplished in our understanding of the theory of those properties which influence metallic adhesion. The thrust will be to gain a better fundamental understanding of those properties or processes that affect metallic adhesive bond strength.

We now discuss the general properties expected for creating two free surfaces by motion, perpendicular to the surfaces. We shall refer to this process as rigid adhesion, which we define as the separation of two surfaces without deformation or relaxation. For these tensile forces, i.e., separation of the material normal to a plane, we expect the binding energy versus separation curve to have the shape shown in Figure 1a. The well depth represents the binding energy; this is the surface energy for perfectly matched, identical half-spaces. The rise in the curve for separations less than the equilibrium value represents compression. The derivative of this curve is the force (actually the tensile stress) necessary to separate the surfaces. The maximum in the force–distance curve is the "breaking force," and it occurs at the inflection point of the binding energy curve (Figure 1b). This is the simplest representation of what occurs experimentally and can be thought of as a rigid fracture curve. In ductile solids in tension, there would be a "stretching" or ductile extension before the brittle fracture. However, because of the complexity of a real tensile test, it is necessary to understand this rigid component of the process separately.

2. THEORETICAL CONSIDERATIONS

Now we present some theoretical approaches to the interface problem. The interface poses difficulties for the solid-state theorist. In order to obtain the correct energetics and electronics structure of the interface, one needs a fully quantum-mechanical approach. Consequently, it is necessary to solve the Schrödinger equation (or the Dirac equation for heavier metals where relativistic effects are important) for the large number (about 10^{23}) of electrons. In doing this, techniques in solid-state physics rely on the three-dimensional periodicity of the lattice to simplify the calculations.[11] This periodicity is lost (in one direction) at interfaces; consequently, special approaches are needed to handle this problem. The theoretical models used depend on the materials comprising the interface. For example, at free-electron metal interfaces, binding may not be localized, while at transition metal, semiconductor, or ceramic interfaces, we might expect localized bonding to be important. Pair potentials are often used to represent nonmetals.[12] In this method, the interaction between atoms is approximated by some two-body potential, e.g., a van der Waals potential, and the energetics are obtained by establishing the geometry of the defect and summing over two-body interactions. For metals, this approach is inadequate because of the mobility of the electrons which redistribute their positions in the vicinity of a defect. Smith and Ferrante[13] have estimated this electron redistribution contribution to the grain boundary energy for simple metals and have shown it to be large compared to pair potential contributions. There are attempts to include these volume-dependent contributions into a pair potential[14]; however, a question arises concerning how well these can represent complicated defect structures. Another fully quantum-mechanical approach used for these situations involves Green's function[15] or tight-binding methods.[16,17] These approaches have been used recently to calculate grain boundary energies in semiconductors.

2.1. First-Principle Calculations

In order to render the full quantum-mechanical problem tractable, a number of reasonable approximations are made. Usually, the nuclear and electronic contributions to the energy are separated using the Born–Oppenheimer approximation, which relies on the fact that the nuclei are much heavier than, and hence move much more slowly than, the electrons. The nuclei are then treated as stationary, and only the electronic problem remains. The solutions for the electron energies are still a formidable problem, and in order to solve it the solid-state theorist relies on the periodicity of a crystalline solid, which implies that the solutions need only be obtained within one unit cell instead of all space. The lattice periodicity is generally exploited by solving the problem in "reciprocal," or wave-vector, space within the unit cell of the reciprocal lattice. Further, the many-electron effects are approximated by using the Hohenberg–Kohn[18] density-functional theory within the local density approximation. One is then left to solve the Schrödinger-like Kohn–Sham equations[19] for the system of interest. In the case of a surface or interface, some of the simplifications just described are lost because the periodicity perpendicular to the surface/interface is lost. The periodicity parallel to the plane of the interface still exists, however, and can be exploited. This means that the wave vector perpendicular to the interface, $\mathbf{k}_\perp$, is no longer a meaningful quantity, although the wave vector parallel to the interface, $\mathbf{k}_\|$, still is.

Thus, in order to implement these goals, we begin by solving the Kohn–Sham one-electron equations for the geometries and materials involved at the interface.[19] In atomic units (In which $\hbar = c = m_e = e = 1$), the Kohn–Sham equations may be written as

$$[-\tfrac{1}{2}\nabla^2 + V(\mathbf{r})]\Psi_i(\mathbf{k}_\|,\mathbf{r}) = \epsilon_i(\mathbf{k}_\|)\Psi_i(\mathbf{k},\mathbf{r}) \tag{2}$$

where Ψ_i and ϵ_i are the one-electron wave functions and energies, respectively, and

$$V(\mathbf{r}) = \Phi(\mathbf{r}) + V_{xc}[\rho(\mathbf{r})] \tag{3}$$

$V_{xc}[\rho(\mathbf{r})]$ is the exchange and correlation energy, $\Phi(\mathbf{r})$ is the electrostatic energy given by

$$\Phi(\mathbf{r}) = \int d\mathbf{r}' \frac{\rho(\mathbf{r})}{|\mathbf{r} - \mathbf{r}'|} - \sum_j \frac{Z_j}{|\mathbf{r} - \mathbf{R}_j|} \tag{4}$$

and $\rho(\mathbf{r})$ is the electronic charge density, the first term is the electron–electron interaction and the second is the electron–ion interaction,

$$\rho(\mathbf{r}) = \sum_{\text{occ}} |\Psi_i(\mathbf{k}_\|,\mathbf{r})|^2 \tag{5}$$

Z_i is the nuclear charge of the atom at site $\mathbf{R}_i$; $\mathbf{k}_\|$ is the component of the Bloch vector parallel to the surface and occ refers to all the occupied states. This potential is treated typically in the local density approximation,[20] in which its value at any point $\mathbf{r}$ is given in terms of $\rho(\mathbf{r})$. The solutions of Eqs. (2)–(5) for a given defect geometry are obtained by guessing a potential and then iterating until the potential obtained from Eq. (3) gives back the wave functions in Eq. (2), and *vice versa*. This procedure is called self-consistency. Self-consistency is a necessity for computation of electronic structure. For example, the question of existence of surface states, much less their dependence on $\mathbf{k}_{\|}$ depends on the solution being fully self-consistent. Once these solutions are obtained, the total energy, which is a function of the electron density, is obtained as

$$E(\rho) = E_{ke}[\rho] + E_{es}[\rho] + E_{xc}[\rho]$$

where E_{ke} is the kinetic energy contribution, E_{es} is the electrostatic contribution, and E_{xc} is the exchange and correlation energy contribution to the total energy.

2.2. Jellium-Model Calculations

The first solutions of these equations for a surface (the solid vacuum interface) were obtained for a jellium model by Lang and Kohn.[20] Jellium refers to treating the positive charge as uniformly distributed rather than point charges. This approximation is most

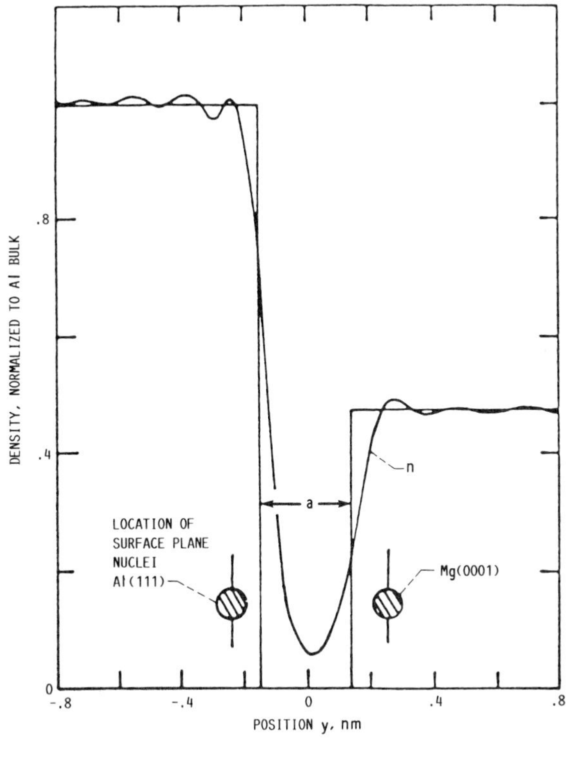

FIGURE 2. Electron number density n and jellium ion charge density n_+ for an Al–Mg contact. When $a = 0.0$, the distance between Mg and Al atomic planes is $(d_{Al} + d_{Mg})/2$, where d_{Al} and d_{Mg} are the respective bulk interplanar spacings.

appropriate for the simple metals, e.g., Al, Zn, Mg, and the alkali metals. We will consider these metals first. They are called simple metals because their conduction electrons are nearly free, i.e., interact very weakly with the lattice of ion cores and thus are well described for some purposes by a free-electron or jellium model. The first self-consistent, quantum-mechanical treatment of adhesion using the jellium model was obtained by Ferrante and Smith.(21) An example of an Al–Mg bimetallic interface showing both the electron and positive charge density distributions is shown in Figure 2.

For this model, we solve for the electronic charge distribution and potential via Eqs. (2)–(5); the $\mathbf{k}_{\|}$ component now is simply that for a free electron model. The electron density distribution, which only varies in the direction normal to the interface, is inhomogeneous. The discrete atomic nature of the lattice is treated by perturbation theory for the electron–ion interactions and lattice sums for the ion–ion interaction as described elsewhere.(22) We now describe some typical results for the electron density and potential distributions.

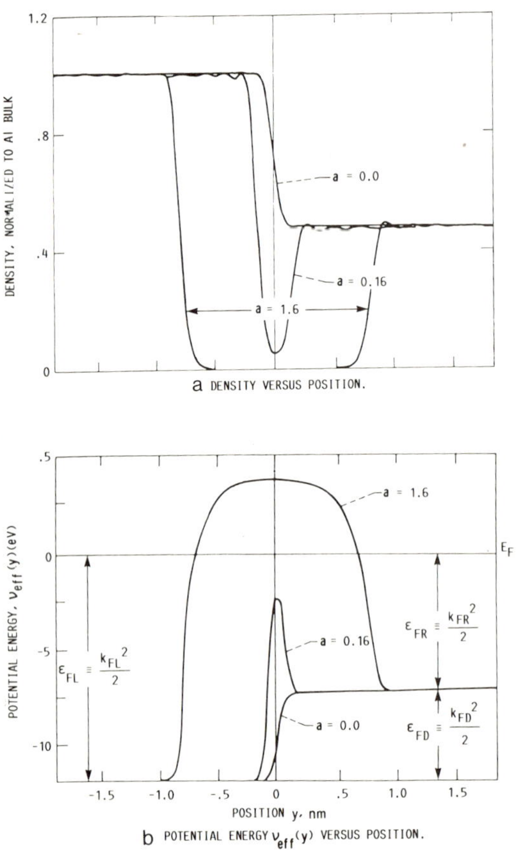

FIGURE 3. Electron density and potential energy for an Al–Mg contact (Al is on the left) at separations of 0, 0.16, and 1.6: (a) density versus position, (b) potential energy $v_{eff}(y)$ versus position.

The electron densities of an Al–Mg contact for separations of 0, 0.16, and 1.6 nm are shown in Figure 3. For $a = 0.0$, the electron density distribution falls smoothly between the Al and Mg. However, for a somewhat larger separation, $a = 0.16$ nm, a significant dip in the electron density is evident. Thus one would have to conclude that the electron density distribution in the interface is strongly sensitive to the interfacial spacing. For $a = 1.6$ nm, the electron density distribution is close to what one finds at a solid–vacuum interface.

For $a = 0.0$, the potential rises smoothly from the Al to the Mg. At $a = 0.16$ nm, there arises a significant "bump" in the potential over 5 eV in height. Again this shows the sensitivity of the interfacial electronic properties to interfacial spacing. For $a = 1.6$ nm, the potential distributions are again close to solid–vacuum distributions.

One might well ask the question: "How much does the electron density distribution differ from what one would obtain by simply overlapping solid–vacuum distributions?" These are compared for the Al–Na interface at $a = 0.0$ in Figure 4. It shows that simple overlap is a reasonably accurate procedure for the interfacial electron density distribution. But it is the interaction between the two metals, in which we are primarily interested, that causes the small difference between the self-consistent and the simple-overlap distributions. In Figure 5, we plot this difference, self-consistent minus simple-overlap, again for the Al–Na interface at $a = 0.0$. We see that the density difference shown in Figure 5a exhibits quite a bit of structure. The potential difference that corresponds to this density difference is shown in Figure 5b. The well-known contact potential ΔV is shown there to be a rather small part of the potential difference; $\Delta V = 0.81$ eV, which is small compared to the actual potential barriers shown in Figure 3. In fact, there is no obvious correlation between the contact potential and interfacial electronic barriers. We now increase the

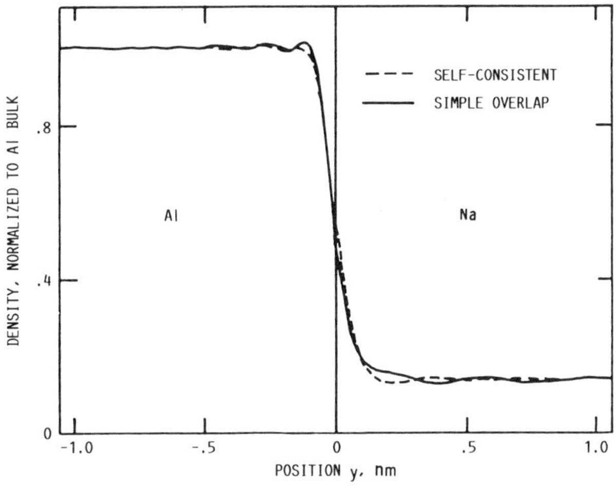

FIGURE 4. Electron density distributions in the Al–Na interface at zero separation ($a = 0.0$). The dashed curve is the fully self-consistent result, while the solid curve results from a simple overlap of the solid–vacuum results for each of the two metals.

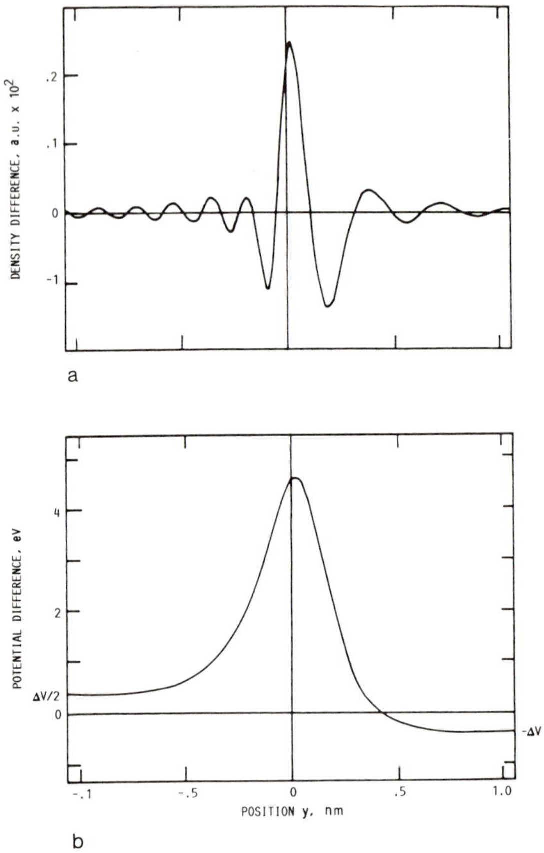

FIGURE 5. Difference electron density and potential energy at an Al–Na interface. (a) Self-consistent electron densities, $n(y,a)$, minus overlapped solid–vacuum electron densities for a = 0.0 in an Al–Na contact. (Al is on the left). This density difference plot shows the charge rearrangement due to interaction between the Al and Na surfaces. (b) Self-consistent electron potentials, $\upsilon_{eff}(n,y)$, minus overlapped solid–vacuum electron potentials, for a = 0.0 in an Al–Na contact. ΔV is the contact potential.

separation to a = 1.6 nm and again view the difference distributions for the Al–Na contact, as shown in Figure 6.

The distributions are now seen to look more like the classic ones sometimes discussed in texts.[11] We see that electrons are removed from the Na and deposited in the Al. One should expect that result, because the work function of the Na is smaller than that of the Al, and the charge redistribution will continue until the Fermi levels in the two metals are the same. We note that there is essentially no density difference in the vacuum region between the two metal surfaces. The potential difference plot reflects that fact with a linear potential in the vacuum region. Also, the contact potential has become the dominant barrier. It has not changed with spacing, only the scale has been changed.

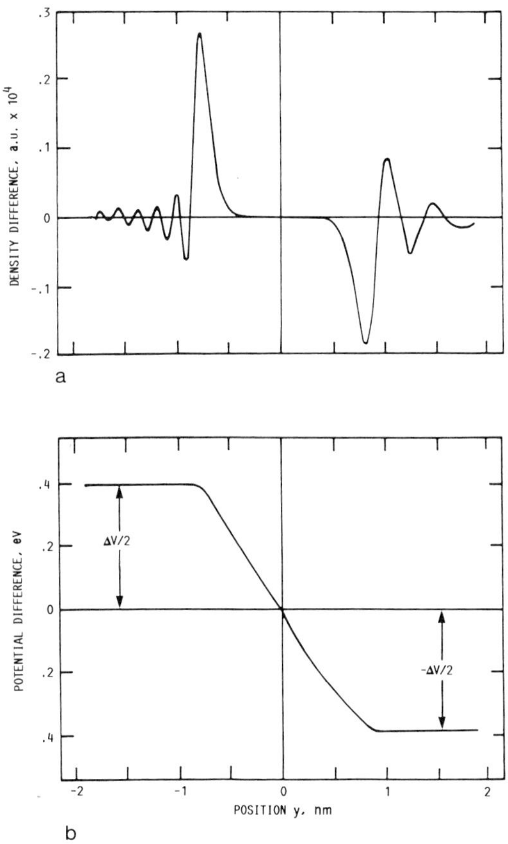

FIGURE 6. Difference electron density and potential energy at an Al–Na interface. (a) Self-consistent electron densities, $n(y,a)$, minus overlapped solid–vacuum densities, $n(y,a)$, for a = 1.59 nm in an Al–Na contact. (Al is on the left.) This shows clearly that electrons are transferred from the Na to the Al. (b) Self-consistent potentials, $\upsilon_{\rm eff}(n,y)$, minus overlapped solid–vacuum electron potentials for a = 1.59 nm in an Al–Na contact. ΔV is the contact potential resulting from the charge transfer shown in (a).

3. BINDING ENERGIES

Since we have now solved for the electronic structure, we are ready for the second step, which is the computation of the energies of adhesion as a function of the interfacial separation a, as discussed above. Actually, there has been a paucity of such calculations, i.e., binding energy versus separation. The only quantum-mechanical calculations of the binding energy [Eq. (1)] as a function of separation have been those performed by Ferrante and Smith[21] for simple metal contacts. As stated previously, the calculations used a one-dimensional jellium model corrected for three-dimensional effects. The results of this calculation for bimetallic contacts for the high-density surfaces of a number of simple metals [Al(111), Mg(0001), Zn(0001), and Na(110)] are shown in Figure 7.

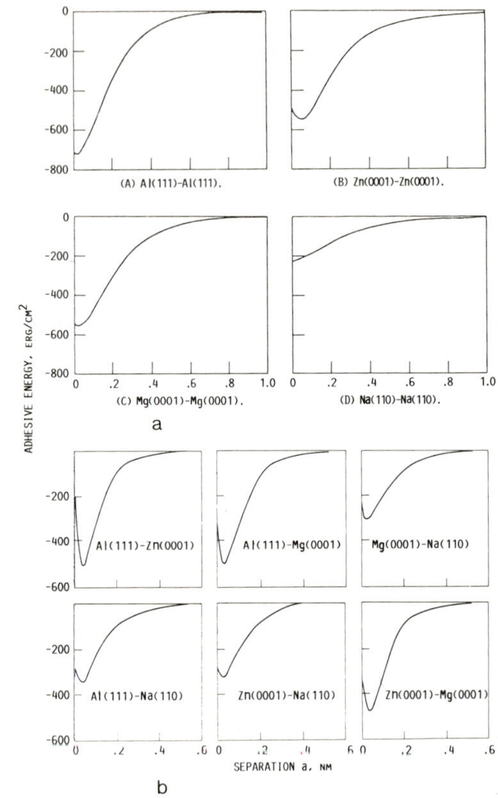

FIGURE 7. Adhesive binding energy versus separation *a:* (a) commensurate adhesion is assumed; (b) incommensurate adhesion is assumed.

TABLE 1
Binding Energy for Combinations of Simple Metals

Metal combination	Binding energy (erg cm^{-2}) Incommensurate	Binding energy (erg cm^{-2}) Perfect registry (commensurate)
Al(111)–Al(111)	490	715
Zn(0001)–Zn(0001)	505	545
Mg(0001)–Mg(0001)	460	550
Na(110)–Na(110)	195	230
Al(111)–Zn(0001)	520	
Al(111)–Mg(0001)	505	
Al(111)–Na(110)	345	
Zn(0001)–Mg(0001)	490	
Zn(0001)–Na(110)	325	
Mg(0001)–Na(110)	310	

Relaxation was not included in this calculation, with both materials frozen at their bulk structures (rigid adhesion). We see that the binding energy curves have the general shape expected, with equilibrium spacing close to $a = 0.0$. There are two important observations concerning these results: The strength of the interfacial bond is quite high, comparable in magnitude to surface energies of the bulk materials, and the range of the strong bonding is of the order of interplanar spacings, 0.2 nm. A summary of the same-metal and different-metal binding energies for all combinations of these simple metals is given in Table 1. It is noteworthy that when the two metals are the same, the binding energy is the surface energy. Therefore, electron sharing of the metal interface can lead to strong, short-range bonding forces. In the systematics of these cases, we will find an important energy relationship which will be discussed later.

3.1. Fully Three-Dimensional Calculations

Transition metals are more difficult to deal with than *sp* metals, primarily because the *d* electrons lead to a much more localized electronic behavior. Because of the localized nature of the *d* electrons, the electron density can vary by a factor of 10 in a distance of about 1 Å, as shown in the self-consistent charge contours of Figure 8, taken from work of Arlinghaus *et al.*[23] These exhibit the electron density distribution near a Ni(100) surface. As in the case of the simple metals, there is a charge rearrangement at the surface, but in this case it is a localized rearrangement (Figure 8). One could not conceive of modeling transition metal interfaces with a bijellium, perturbation-theory approach as we used for simple metal interfaces. A method based on localized basis functions appropriate for transition metals—the self-consistent local orbital (SCLO) method—is described at length elsewhere.[24]

In the SCLO method, the wave functions are expanded in terms of the localized orbitals $a_j(\mathbf{r} - \mathbf{R}_j)$:

$$\Psi_i(\mathbf{k}_{\|},\mathbf{r}) = \sum_{i,j} C_{ij}(\mathbf{k}_{\|},\mathbf{R}_j)a_j(\mathbf{r} - \mathbf{R}_j) \tag{6}$$

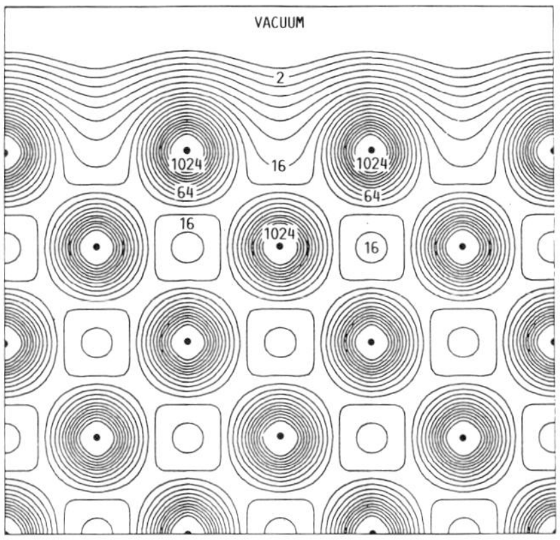

FIGURE 8. Electronic charge density contours at a Ni(100) surface. (From Reference 24.)

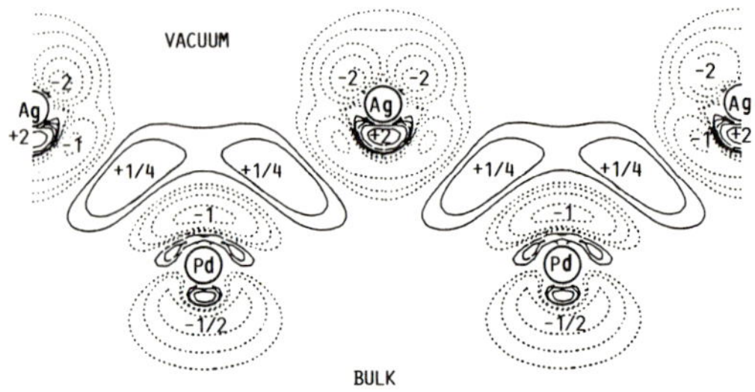

FIGURE 9. Charge transfer of the palladium [100] slab upon silver adsorption. The difference density ρ(Pd [100] + ρ(1 × 1)Ag)$_3$ − ρ(Pd [100] − ρ(Ag) is plotted along (110) plane in units of 10^{-2} e/a_0. Successive contours are separated by a factor of two. Solid (dashed) lines denote an accretion (depletion) of electrons after adsorption.

The $a_j(\mathbf{r} - \mathbf{R}_j)$ comprise all ground-state atomic orbitals (including the core), plus other localized orbitals. Matrix elements of the Hamiltonian are then determined between local basis functions, and the electronic structure is obtained by solving the corresponding matrix eigenvalue problem in order to solve the Kohn–Sham equations (2)–(5). An example of this localized binding is shown in Figure 9 for the Ag–Pd(100) interface.[25,26] Here, a monolayer of Ag was deposited on a Pd(100) surface epitaxially. This is a convenient configuration experimentally, because electrons can be photoemitted through the thin Ag layer to obtain information about the interface.[21] Given the short screening lengths possessed by metals, even a monolayer of Ag is representative of a thicker Ag film interface. In Figure 9, the sum of the isolated Ag monolayer electron density distribution and the clean Pd(100) surface distribution is subtracted from the distribution of the Ag monolayer chemisorbed on the Pd(100). This difference is then due to the charge transfer and/or rearrangement occurring upon interaction between the Ag and Pd(100). It is the transition metal analog to Figure 5a. Dashed contours refer to a charge depletion, and solid contours to an accumulation. Here we see that electrons have been "scooped up" from around both the Ag and Pd atoms and deposited in the interstitial region between the Ag and Pd layers. This is not unlike what one finds in a covalently bonded molecule, so as in simple bimetallic interfaces, we find a molecular analogy.

TABLE 2
Transition Metal Surface Energies

Metal	Surface energy (erg cm^{-2})	
	Theory	Experiment[a]
Cu	2300	2016
Ag	1650	1543
Ni	3050	2664
Fe	3100	2452

[a]H. Wawra, *Z. Metallkd.* **66**, 395 (1975).

This method has been used in computations of cleavage or surface energies for a series of transition metals.[25,26] The results for copper, silver, nickel, and iron are given in Table 2. Reasonable agreement was obtained with experiment. The calculations were conducted for (100) surfaces, while the experimental data pertain to polycrystalline metals, owing to lack of single-crystal data. The iron and nickel calculations were for the spin-polarized or ferromagnetic state.

3.2. Universality in the Shapes of the Binding Energy Curves

As stated earlier, Rose, Ferrante, and Smith[27] discovered that the binding energy curves of Ferrante and Smith scaled onto one binding energy curve and, in fact, that this scaling corresponded to a much wider class of bonding, which we will now discuss. This universality can be represented in the simple form

$$E_{ad}(a) = \Delta E \times E^*_{ad}(a^*) \tag{7}$$

where

$$a^* = (a - a_0)/l \tag{8}$$

$$l = \left\{ \frac{\Delta E}{[d^2E_{ad}(a)/da^2]_{a_0}} \right\}^{1/2} \tag{9}$$

a_0 is the equilibrium separation, ΔE is the equilibrium binding energy, and $E^*(a^*)$ is a universal energy relation. The scaling length l can equally well be chosen as the Thomas–Fermi screening length for simple metals in contact. If we have two different simple metals in contact, an average of the Thomas–Fermi screening lengths works as well.

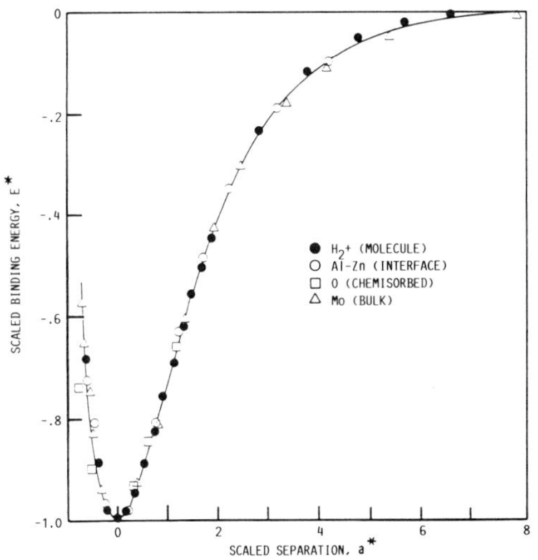

FIGURE 10. Binding energy as a function of interatomic separation for four systems, scaled as described in the text. The solid line is the fit to Eq. (10).

This universality was found to extend to other systems, besides adhesion, that could be thought of in terms of covalent binding, cohesion, chemisorption, and molecular binding. In Figure 10, we show an example of scaling the binding energy relations for all four systems by the prescription given above. Rose, Smith, and Ferrante found that this universal shape can be represented accurately by a simple functional form, the Rydberg function,

$$E^*(a^*) = -(1 + a^*)e^{-a^*} \tag{10}$$

A plot of this equation showing the fit for the four systems is the solid line in Figure 10. Use of this form led to defining an equation of state at zero temperature for all classes of solids.[28,29] We should emphasize that the universality is independent of the functional form used to represent it. In the next section, we will give some examples of the use of these results for the development of semiempirical techniques which can be applied to the adhesion problem.

4. SEMIEMPIRICAL TECHNIQUES

Now we address some of the issues involved in practical adhesion and some techniques that may be used to approach them. As indicated earlier, real interfaces are quite complex. In addition, fully quantum-mechanical approaches to modeling binding of real interfaces are also quite complex. We now present some simpler approximate techniques which show promise in modeling the complexities of real interfaces.

The difficulty in fully quantum-mechanical calculations involves loss of periodicity and inclusion of relaxation effects, i.e., finding the minimum energy configuration for a defect. The problem is that these difficult calculations must be performed for each possible configuration and the energy minimum found. The difficulty in performing such calculations where there is a low degree of symmetry is obvious. However, with recent advances using supercells, a way to attack the periodicity problem,[30,31] and a new technique for finding energy minima and inclusion of molecular dynamics,[32] fully quantum-mechanical calculations are feasible, but still quite difficult. For nonmetals, pair potentials may be a reasonable approach for some problems.[12] These have the advantage of making the search for energy minima much simpler, but do not properly take into account the many-body effects in metals.[14] A promising approach for metals involves some new semiempirical approaches—the embedded-atom method (EAM)[33] and the equivalent-crystal theory (ECT)[34,35]—which treat the many-body effects and allow for a simpler treatment of relaxation. These semiempirical methods, along with the fully quantum-mechanical methods, have the promise of attacking real material problems. We will now present a discussion of EAM and ECT and give the results of some adhesion calculations using ECT.

4.1. Embedded-Atom Method

The embedded-atom method of Foiles, Baskes, and Daw[33] is inspired by the effective-medium theory of Norskov and Lang,[36] and the quasi-atom method of Stott and

Zaremba.[37] In this procedure, the energy required to embed an ion in a jellium is considered. The total energy is written in the form

$$E = \sum_i F[\rho(\mathbf{R}_i)] + \frac{1}{2} \sum_{\substack{i,j \\ i \neq j}} Z_i(\mathbf{R}_{ij}) Z_j(\mathbf{R}_{ij}) / \mathbf{R}_{ij} \tag{11}$$

where $\mathbf{R}_{ij}$ is the interion distance, $F(\rho)$ is called the embedding energy and depends only on $\rho(\mathbf{R}_i)$, the electron density contributed by all of the atoms except the ith one at the site, the ith atom, while the second term represents the pair repulsion between ion cores where $Z(\mathbf{R}) = Z_0(1 + \beta \mathbf{R}^\nu) \exp(-\gamma \mathbf{R})$, Z_0 being a nuclear charge and β, ν, and γ determined from fitting to experiments such as vacancy energies or elastic constants. In its most recent form, the "universal" binding energy relation of Rose, Ferrante, and Smith[27] is used to define the embedding energy for the cohesive energy case. The electron density is obtained from overlapping atomic densities. Once the embedding function is obtained, it is applied to a defect, such as a surface, by overlapping the atomic densities and using the embedding function obtained for a given material from fitting to the universal binding energy relation, along with the pair repulsion term, to obtain the energy with a defect. Ferrante *et al.*[38] have applied these techniques to the adhesion across three different planes of Ni and for slip on two crystal planes of Ni. The results give the correct trends and shapes for the energy curves in both cases. The method has been applied successfully to a large number of defects in solids and has been extended to include alloys and covalently bonded solids. It is substantially simpler than the quantum-mechanical techniques described earlier.

4.2. Equivalent-Crystal Theory

We will now concentrate on the equivalent-crystal theory of Smith ad Banerjea[34,35] which, although at present not as versatile as EAM, gives quantitatively more accurate values for surface energies and surface relaxation. This technique also uses the universal binding energy relations. In this theory, the energy of a solid, with an arbitrary arrangement of atoms, is written as

$$E_{\text{solid}} = E_{\text{crystal}}(a) + \delta E(a) \tag{12}$$

Here, $E_{\text{crystal}}(a)$ is the energy of a crystal of the same material and having the same crystal structure as the ground-state crystal, but with a lattice constant a that may be different from the equilibrium (ground-state) lattice constant a_{m}; quantity $\delta E(a)$ is the difference between the energy of the actual solid with a defect and that of the crystal. Both quantities $E_{\text{crystal}}(a)$ and $\delta E(a)$ are necessarily functions of the lattice constant a of the crystal.

The difference in energy, $\delta E(a)$, can be expressed as a perturbation series, which usually needs to be carried only to the second order. One can find a value of the lattice constant, a_{e}, so that this perturbation series goes to zero:

$$\delta E(a_{\text{e}}) = 0 \tag{13}$$

in which case

$$E_{\text{solid}} = E_{\text{crystal}}(a_{\text{e}}) \tag{14}$$

Now the cohesive energy $E_{crystal}(a)$ of a crystal with lattice constant a can be simply evaluated using the universal binding energy relation (UBER)[27] [see Eqs. (7)–(9)], if one knows the three quantities ΔE, a_m, and l [see Eq. (9)]. These can be determined experimentally and have been tabulated for a number of materials.[27] As pointed out earlier, the UBER $E^*(a^*)$ has been shown to be very well approximated by the Rydberg function, given by Eq. (10), which can be evaluated trivially.

One other parameter is required in order to apply the ECT to metals. In evaluating the perturbative correction $\delta E(a)$, one needs to deal with integrals of the form

$$g(\mathbf{R}) = \int n_A(\mathbf{r})\upsilon(\mathbf{r} - \mathbf{R})d\mathbf{r} \tag{15}$$

where the atom in question is at the origin, **R** is the position vector of a neighboring atom, $n_A(\mathbf{r})$ is the electron density around an atom in the crystal, and $\upsilon(\mathbf{r})$ is the potential of an ion core. These integrals are assumed to be spherically symmetric, i.e., dependent only on the magnitude of **R** and independent of the direction of **R**, and may be parameterized as

$$g(R) = R^p e^{-\alpha R} \tag{16}$$

The exponent p is chosen most conveniently[34,35] to be $2n - 2$, where n is the principal quantum number of the highest, partially occupied s level in the atom. This follows from assuming that the only significant contribution to $n_A(\mathbf{r})$ comes from the highest, partially occupied s orbital, which has the largest range and thus the maximum overlap with neighboring ion-core potentials. The inverse decay length α is the fourth empirically determined parameter and has been computed for different metals by fitting to the experimental value of the vacancy formation energy E_{vac}. When considering integrals, such as that in Eq. (16), between other-than-nearest-neighbor atoms, it is usually sufficient to consider up to second-nearest neighbors only. One needs to include a multiplicative factor $e^{-R/\lambda}$ to account for the additional screening of the interactions with these atoms by the nearest-neighbor shell. The screening length λ is assumed to be proportional to the scaling length l introduced above and is determined by reference to a simple metal, e.g., sodium, for which the screening length is well known. So, with a knowledge of the four experimental quantities, ΔE, a_m, the equilibrium bulk modulus B or the scaling length l derived from it, and E_{vac} or the inverse length α derived from it, we can compute the total energy of an arbitrary assembly of atoms.

In practical applications of the ECT, one notes that in most situations involving localized defects or disorder, only a few of the atoms have their local environment changed from that in the equilibrium crystal. Further, because of screening, only the energy density in the vicinity of the localized defect will be changed from that in the equilibrium crystal. Hence it is convenient to associate a separate equivalent crystal, with lattice constant a_e, for each atom that has had its local environment altered. One then sums the energy changes associated with each such equivalent crystal in order to obtain the total defect energy.

It has already been mentioned that the integrals in Eq. (16) are assumed to be spherically symmetric. This is adequate for most situations, but in some cases, for instance, while computing surface relaxations in metals, it has been necessary to account for the anisotropy of these integrals. This has been done in a simple way by adding a "bond compression" energy for each nearest-neighbor bond that is shorter than its equilibrium length. This has the form[(35)]

$$E_{\text{bond}} = \sum_{n=1}^{N} \sum_{m=1}^{M_n} \frac{\Theta(a^*_{nm})}{L_{nm}} \Delta E\left[1 - E^*(a^*_{nm})\right] \tag{17}$$

Here N is the total number of atoms in the system, M_n is the number of nearest neighbors of the nth atom, $\Theta(x)$ is the Heaviside theta function, L_{nm} is the smaller of the two numbers M_m and M_n, and

$$a^*_{nm} = \left(\frac{d_{nm} - d_0}{d_0}\right)\left(\frac{a_{\mathrm{m}}}{l}\right) \tag{18}$$

where d_{nm} is the distance between nearest-neighbor atoms n and m, and d_0 is the equilibrium nearest-neighbor distance.

TABLE 3
Surface Energies (erg cm^{-2}) for Some Selected Metals

Element	Crystal face	(This work)	Theory (LDA)	(Other)	Experiment[a]
Al	(111)	920			
	poly				1169
	(100)	1280			
	(110)	1310	1090[b]		
Ni	(111)	2400		1450[c]	
	poly				2664
	(100)	3120	3050[d]	1580[c]	
	(110)	2980		1730[c]	
Cu	(111)	1830	2100[e]	1170[c]	
	poly				2016
	(100)	2380	2300[d]	1280[c]	
	(110)	2270		1400[c]	
Ag	(111)	1270		620[c]	
	poly			705[c]	1543
	(100)	1630	1650[d]		
	(110)	1540		770[c]	
Fe	(110)	1810			
	poly				2452
	(100)	3490	3100[d]	1693[f]	
W	(110)	3330			
	poly				4435
	(100)	5880	5100[g]	2926[f]	

[a]H. Wawra, *Z. Metallkd.* **66**, 395, 492 (1975).
[b]K. M. Ho and K. P. Bohnen, *Phys. Rev. B* **32**, 3446 (1985).
[c]S. M. Foiles *et al.*, *Phys. Rev. B* **33**, 7983 (1986).
[d]J. G. Gay *et al.*, *J. Vac. Sci. Technol. A* **2**, 931 (1984).
[e]J. A. Appelbaum and D. R. Hamann, *Solid State Commun.* **27**, 881 (1978).
[f]M. W. Finnis and J. E. Sinclair, *Phil. Mag. A* **50**, 45 (1984).
[g]C. L. Fu *et al.*, *Phys. Rev. B* **31**, 1168 (1985).

TABLE 4
Percent Changes in Interlayer Spacings due to Relaxation

Surface	$\Delta d_{n,n+1}$	Theory, ECM (%)	Theory, EAM[a] (%)	Experiment (%)	Technique
Al(111)	Δd_{12}	−3.6			
	Δd_{23}	1.8			
Al(100)	Δd_{12}	−5.0			
	Δd_{23}	2.1			
Al(110)	Δd_{12}	−10.1		−8.6 ± 0.8	LEED[b]
				−8.5 ± 1.0	LEED[c]
	Δd_{23}	4.8		5.0 ± 1.1	LEED[b]
				5.5 ± 1.1	LEED[c]
	Δd_{34}	−0.4		−1.6 ± 1.2	LEED[b]
				2.2 ± 1.3	LEED[c]
	Δd_{45}	0.0		0.1 ± 1.3	LEED[b]
				1.6 ± 1.6	LEED[c]
Ni(111)	Δd_{12}	−2.9	−0.05		
	Δd_{23}	1.4	0.00		
Ni(100)	Δd_{12}	−3.7	−0.002		
	Δd_{23}	1.7	−0.001		
Ni(110)	Δd_{12}	−6.8	−2.30	−8.7 ± 0.5	LEED[d]
				−4.8 ± 1.7	Ion scattering[e]
	Δd_{23}	3.2	0.08	3.0 ± 0.6	LEED[d]
				2.4 ± 1.2	Ion scattering[e]
Cu(111)	Δd_{12}	−2.8	−1.40	−0.7 ± 0.5	LEED[f]
	Δd_{23}	1.4	−0.05		
Cu(100)	Δd_{12}	−3.5	−1.4	−2.1 ± 1.7	LEED[g]
				−1.1 ± 0.4	LEED[h]
	Δd_{23}	1.6	−0.3	0.45 ± 1.7	LEED[g]
				1.7 ± 0.6	LEED[h]
Cu(110)	Δd_{12}	−6.5	−4.9	−8.5 ± 0.6	LEED[i]
				−5.3 ± 2.4	Ion scattering[i]
	Δd_{23}	2.7	0.2	2.3 ± 0.8	LEED[i]
				3.3 ± 1.5	Ion scattering[i]
Ag(111)	Δd_{12}	−2.3	−1.30		
	Δd_{23}	1.1	−0.04		
Ag(100)	Δd_{12}	−2.9	−1.90		
	Δd_{23}	1.4	−0.05		
Ag(110)	Δd_{12}	−5.4	−5.1	−5.7	LEED[h]
				−7.8 ± 2.5	Ion scattering[j]
	Δd_{23}	2.4	0.3	+2.2	LEED[h]
				4.3 ± 2.5	Ion scattering[j]

[a]S. M. Foiles *et al., Phys. Rev. B* **33**, 7983 (1986).
[b]J. N. Andersen *et al., J. Phys. C* **17**, 173 (1984).
[c]J. R. Noonan and H. L. Davis, *Phys. Rev. B* **29**, 4349 (1984).
[d]D. L. Adams *et al., J. Phys. C* **18**, 1753 (1985).
[e]R. Feidenhansl *et al., Surf. Sci.* **134**, 329 (1983).
[f]S. A. Lindgren *et al., Phys. Rev. B* **29**, 576 (1984).
[g]R. Mayer *et al., Phys. Rev. B* **35**, 3102 (1987).
[h]H. L. Davis and J. R. Noonan, *Surf. Sci.* **126**, 245 (1983).
[i]D. L. Adams *et al., Phys. Rev. Lett.* **49**, 669 (1982).
[j]Y. Kuk and L. C. Feldman, *Phys. Rev. B* **30**, 5811 (1984).

In order to test ECT, we present two sets of results in Tables 3 and 4, the surface energies for several planes of a number of transition and simple metals and surface relaxations for these metals. As can be seen, there is excellent agreement with the fully self-consistent quantum-mechanical calculations. We see also that the surface relaxations are well predicted. This is particularly impressive, since they represent very small energy changes and very small displacements. Thus we can conclude that ECT gives a very high degree of quantitative agreement with the much more difficult quantum-mechanical calculations; it is the difference between using a microcomputer and a Cray. It should be emphasized, however, that the fully quantum-mechanical results for model systems are absolutely essential for the development and verification of the semiempirical techniques.

Finally, this technique has been used to calculate the adhesive energy for a number of metals.[39] Similarly to Ferrante and Smith, the binding energy for adhesion of two semi-infinite slabs of the same metal in perfect registry across a given crystallographic plane has been computed as a function of the separation between the slabs. This has been done

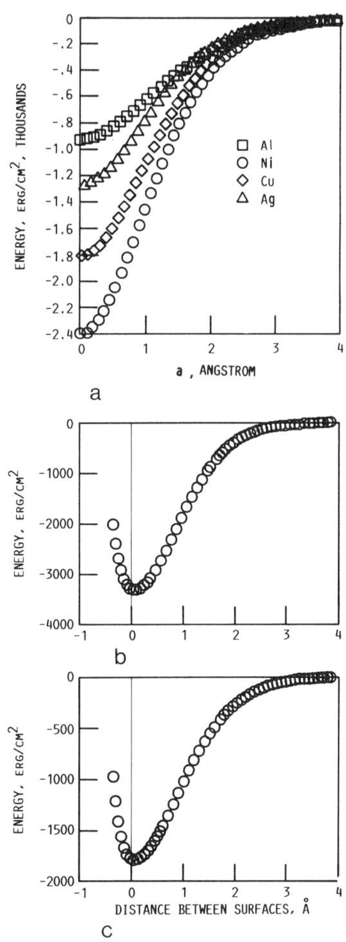

FIGURE 11. Rigid adhesive binding energy versus separation using ECT[33,34]: (a) for the (111) surface of Al, Ni, and Ag; (b) for the W(110) interface; (c) for the Fe(110) interface.

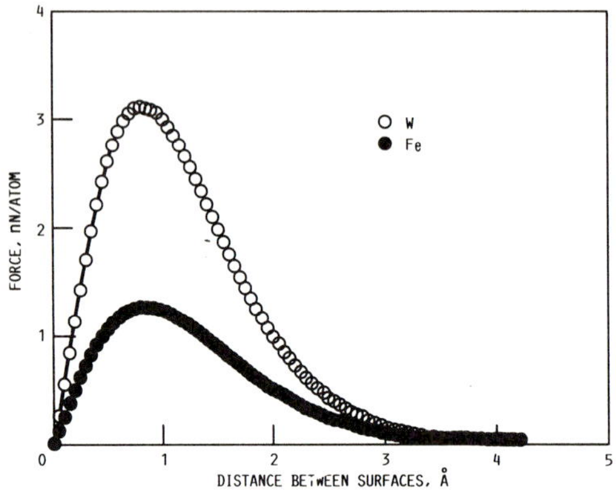

FIGURE 12. Rigid adhesive binding force versus separation for W(110) and Fe(110) surfaces.

for the lowest-index interfaces of each of four face-centered-cubic (fcc) metals—aluminum, nickel, copper, and silver—and for two body-centered-cubic (bcc) metals—iron and tungsten. In each case, the two semi-infinite slabs are separated rigidly with no surface relaxation or reconstruction being permitted. These metals, with the exception of tungsten, are known to show planar relaxation in their free surfaces. The tungsten (110) surface, which is known to reconstruct, has been included for comparison.

Values of the inverse decay length α have been given elsewhere.[35] For Cu, Al, Ag, and Ni, α is 29.44, 21.05, 33.36, and 30.15 in nanometers^{-1}, respectively. Representa-

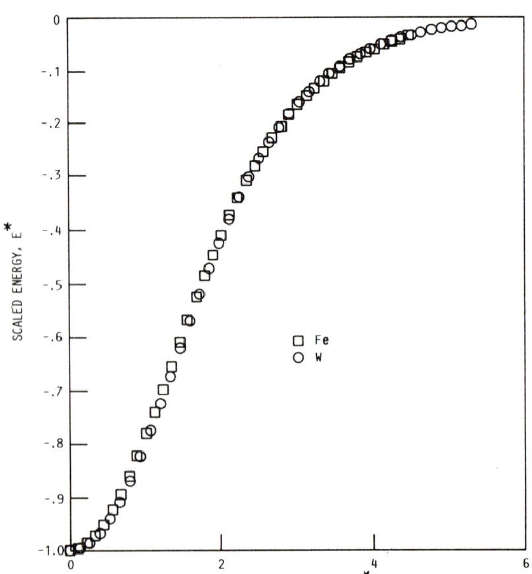

FIGURE 13. Scaled bcc (110) adhesive energies versus scaled separation.

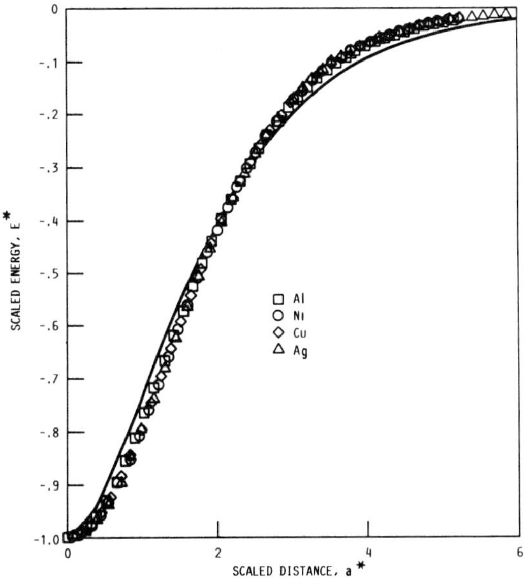

FIGURE 14. Scaled adhesive energies at a fcc (111) interface versus scaled separation with a fit to Eq. (10) (solid line).

tive results for the variation of the binding energy with separation for some of the metals are shown in Figure 11. Representative results for the variation of the force with separation are shown in Figure 12.

As with previous results in developing the UBER, these rigid-adhesion curves can be scaled, as in Eqs. (7)–(9), with a being the interslab separation, which is defined to be zero when the surface layers are at the equilibrium interplanar distance. Results for the scaling are shown in Figure 13. The *shape* of the rigid-adhesion curve is clearly seen to be universal, i.e., all the scaled curves fall on top of one another. For comparison, the Rydberg function is also plotted in Figure 14. It can be seen that the fit to the Rydberg function is not perfect. We feel this has to do with the way in which we have included the screening of interactions between atoms on opposite sides of the interface and is a topic of current development.

5. CONCLUDING REMARKS

Both fully quantum-mechanical calculations and semiempirical methods are now successfully being applied to interface and defect energetic problems. The new, semiempirical techniques are simple enough to be used by nonspecialists in quantum theory calculations. The accuracy of the prediction by the equivalent-crystal theory and the success of EAM gives hope that substantial progress will be made in the prediction of interface and defect energies. It is relatively straightforward to include relaxation and temperature effects in both the ECT and EAM through Monte Carlo and molecular dynamic techniques. Fully quantum-mechanical techniques will be useful for verification on a fixed configuration. It is now conceivable to attack real systems of interest in material science. Possibly these results will serve us and aid experimentalists in sorting out the many factors with which they must deal.

NOMENCLATURE

A	Cross-sectional area
a	Distance or lattice constant
a_e	Lattice constant of equivalent crystal
a_m	Equilibrium lattice constant
a_0	Equilibrium separation distance
$a(\mathbf{r})$	Localized electron-orbital wave function
a^*	Scaled separation distance
C_{ij}	Coefficients of electron orbitals
c	Speed of light in vacuum
d	Distance between nearest-neighbor atoms
d_0	Equilibrium nearest-neighbor distance
E	Energy
E^*	Scaled binding energy
E_{ad}	Adhesive binding energy
E^*_{ad}	Scaled adhesive binding energy
E_{bond}	Bond-compression energy
E_{crystal}	Energy of perfect crystal
E_{es}	Electrostatic energy
E_{ke}	Kinetic energy
E_{solid}	Energy of solid with defect
E_{vac}	Vacancy formation energy
E_{xc}	Exchange-correlation energy
e	Electronic charge
F	Embedding energy function
g	Perturbation integral
h	Planck's constant
$\hbar$	$h/2\pi$
$\mathbf{k}$	Wave vector
$\mathbf{k}_{\parallel}$	Wave vector parallel to an interface or surface
$\mathbf{k}_{\perp}$	Wave vector perpendicular to an interface or surface
L	Number of nearest-neighbor atoms
l	Scaling length
M	Total number of nearest-neighbor atoms
m_{e}	Mass of electron
N	Total number of atoms
n_{A}	Atomic electron density distribution
ΔV	Contact potential
$\mathbf{R}$	Position vector of an ion
$\mathbf{r}$	Position vector

T	Temperature, usually in Kelvin
V	Potential energy
V_{xc}	Exchange-correlation potential
Z	Atomic number
Z_0	Nuclear charge
α	Inverse decay length
β,γ,ν	Fitting constants
ΔE	Equilibrium binding energy
δE	Energy correction
ϵ	Energy eigenvalue or electron energy
$\Theta(x)$	Heaviside function
λ	Thomas–Fermi screening length
ρ	Electron density distribution
Φ	Electrostatic energy density
υ	Potential of ion core
Ψ	Electron wave function

REFERENCES

1. J. R. Smith and J. Ferrante, *Mater. Sci. Forum* **4,** 21 (1985); Panel Report on Interfacial Bonding and Adhesion, *Mater. Sci. Eng.* **83,** 169–234 (1986).
2. D. H. Buckley, *Surface Effects in Adhesion, Friction, Wear and Lubrication,* Elsevier, Amsterdam (1982).
3. M. D. Pashley, J. B. Pethica, and D. Tabor, *Wear* **100,** 7 (1984).
4. D. H. Buckley, NASA Tech. Note TN D-5822, National Aeronautics and Space Administration (1970).
5. J. Gerkema and A. R. Miedema, *Surf. Sci.* **124,** 353 (1983).
6. D. Tabor and R. H. S. Winterton, *Proc. R. Soc. London, Ser. A* **312,** 435 (1969).
7. J. N. Israelachvili, *Surface and Intermolecular Forces,* Academic Press, New York (1985).
8. G. Binnig, C. F. Quate, and Ch. Gerber, *Phys. Rev. Lett.* **56,** 930 (1986).
9. C. M. Mate, G. M. McClelland, R. Erlandsson, and S. Chiang, *Phys. Rev. Lett.* **59,** 1942 (1987).
10. D. M. Mattox and D. A. Rigney, *Mater. Sci. Eng.* **83,** 189 (1986).
11. N. W. Ashcroft and N. D. Mermin, *Solid State Physics,* Holt, Rinehart and Winston, New York (1976).
12. D. R. Clarke and D. Wolf, *Mater. Sci. Eng.* **83,** 197 (1986).
13. J. R. Smith and J. Ferrante, *Phys. Rev. B* **34,** 2238 (1986).
14. P. D. Bristowe and A. Brokman, *Scr. Metall.* **14,** 1129 (1980).
15. A. Yaniv, *Phys. Rev. B* **17,** 3904 (1978).
16. C. C. Pei, *Phys. Rev. B* **18,** 2583 (1978).
17. R. E. Thomson and D. J. Chadi, *Phys. Rev. B* **29,** 889 (1984).
18. P. Hohenberg and W. Kohn, *Phys. Rev.* **136,** B864 (1964).
19. W. Kohn and L. J. Sham, *Phys. Rev.* **140,** A1133 (1965).
20. N. D. Lang and W. Kohn, *Phys. Rev. B* **1,** 4555 (1970).
21. J. Ferrante and J. R. Smith, *Phys. Rev. B* **31,** 3427 (1985).
22. J. Ferrante and J. R. Smith, *Surf. Sci.* **38,** 77 (1973).
23. F. J. Arlinghaus, J. G. Gay, and J. R. Smith, *Phys. Rev. B* **21,** 2055 (1980).
24. J. R. Smith, J. G. Gay, and F. J. Arlinghaus, *Phys. Rev. B* **21,** 2201 (1980).
25. X. Y. Zhu, J. Hermanson, F. J. Arlinghaus, J. G. Gay, R. Richter, and J. R. Smith, *Phys. Rev.* **29,** 4426 (1984).
26. R. Richter, J. G. Gay, and J. R. Smith, *Phys. Rev. Lett.* **54,** 2704 (1985).
27. J. H. Rose, J. Ferrante, and J. R. Smith, *Phys. Rev. Lett.* **47,** 675 (1981).
28. P. Vinet, J. Ferrante, J. R. Smith, and J. H. Rose, *J. Phys. C* **19,** L467 (1986); P. Vinet, J. H. Rose, J. Ferrante, and J. R. Smith, *J. Phys. Cond. Matter* **1,** 1941 (1989).

29. J. H. Rose, J. R. Smith, F. Guinea, and J. Ferrante, *Phys. Rev. B* **29,** 2963 (1984).
30. D. P. Vincenzo, D. L. Alehand, M. Schluter, and J. W. Wilkins, *Phys. Rev. Lett.* **56,** 1925 (1986).
31. M. L. Payne, P. D. Bristowe, and J. D. Joannopoulos, *Phys. Rev. Lett.* **58,** 1348 (1987).
32. R. Car and M. Parrinello, *Phys. Rev. Lett.* **55,** 2471 (1985).
33. S. M. Foiles, M. I. Baskes, and M. S. Daw, *Phys. Rev. B* **33,** 7983 (1986).
34. J. R. Smith and A. Banerjea, *Phys. Rev. Lett.* **59,** 2451 (1987); J. R. Smith, T. Perry, and A. Banerjea, *Atomistic Simulation of Materials* (V. Vitek and D. Srolovitz, eds.), p. 279, Plenum, New York (1989).
35. J. R. Smith and A. Banerjea, *J. Vac. Sci. Technol. A* **6,** 812 (1988); *Phys. Rev. B* **37,** 10411 (1988).
36. J. K. Norskov and N. D. Lang, *Phys. Rev. B* **21,** 2131 (1980).
37. M. J. Stott and E. Zaremba, *Phys. Rev. B* **32,** 1564 (1980); *Can. J. Phys.* **60,** 1145 (1982); *Solid State Commun.* **32,** 1297 (1979).
38. J. Ferrante, G. H. Bozzolo, C. W. Finley, and A. Banerjea, *Mater. Res. Soc. Symp. Proc.* **119,** 3 (1988).
39. A. Banerjea, J. R. Smith, and J. Ferrante, *Mater. Res. Soc. Symp. Proc.* **140,** 89 (1989).

12

Hard–Soft Acid–Base (HSAB) Principle for Solid Adhesion and Surface Interactions

Lieng-Huang Lee

1. INTRODUCTION

The donor–acceptor interaction[1,2] and the acid–base interaction[3] have been reviewed. On many occasions, the two terms, though different, have been used interchangeably to describe the interactions involving the exchange of electrons between a donor and an acceptor. For polymer adhesion, Fowkes[4,5] and Bolger *et al.*[6] have pointed out the important role of the acid–base interaction in the formation of an adhesive bond.

For the acid–base interaction in solutions, in 1963, Pearson[7] proposed the hard–soft acid–base (HSAB) principle to describe some basic rules about the kinetics and equilibrium of the reaction. In this paper, we attempt to apply the HSAB principle to solid interactions with the aid of the frontier orbital method. We shall first describe the HSAB principle as it has been evolved in recent years[8–12] and then the band structures of solids. After we demonstrate the compatibility between the HSAB principle and the band structures in the solid state, we then illustrate with several examples of adhesion and surface interactions between metals and polymers. We are hoping that, in doing so, we can further understand the dynamics and energetics of surface interactions at the molecular level.

2. HSAB PRINCIPLE

2.1. HSAB Principle for Inorganic Reactions

In an inorganic reaction, the generalized Lewis acid–base interaction can be represented as

$$\underset{\text{(Acid)}}{A} \quad + \quad \underset{\text{(Base)}}{:B} \quad \rightleftharpoons \underset{\text{(A–B Complex)}}{A{:}B} \tag{1}$$

Lieng-Huang Lee • Webster Research Center, Xerox Corporation, Webster, New York 14580.

Pearson[7] has suggested that for the above interaction we classify acids and bases according to the absolute hardness, which will be defined later in Figure 1:

- A hard acid contains an acceptor atom of high positive charge and small size. It does not have easily excitable outer electrons and is not polarizable.
- A soft acid contains an acceptor atom of low positive charge and large size. It has several easily excitable outer electrons and is polarizable.
- A hard base contains a donor atom of low polarizability that is hard to reduce. It is associated with empty orbitals of high energy and hence inaccessible.
- A soft base contains a donor atom of high polarizability and low electronegativity. It is easily oxidized and associated with unoccupied low-lying orbitals.

There are two rules governing the acid–base interactions in solutions:

1. About equilibrium: Hard acids prefer to coordinate with hard bases, and soft acids to soft bases.
2. About kinetics: Hard acids react readily with hard bases, and soft acids with soft bases.

The above principle has been well tested in inorganic reactions.[13]

2.2. HSAB Principle for Organic Reactions and the Frontier Orbital Approach

For an organic reaction, the HSAB principle has been applied for the study of kinetics and equilibrium,[14] and the frontier orbital method has been used to illustrate the electrophilic and nucleophilic interactions[15] (Figure 2):

- A hard electrophile (or acid) has a high-energy LUMO (lowest unoccupied molecular orbital) and usually has a positive charge.
- A soft electrophile has a low-energy LUMO but does not necessarily have a positive charge.
- A hard nucleophile (or base) has a low-energy HOMO (highest occupied molecular orbital) and usually has a negative charge.
- A soft nucleophile has a high-energy HOMO but does not necessarily have a negative charge.

Furthermore, according to Koopmans' theorem,[16] the frontier orbital energies are given by

$$-E_{\text{HOMO}} = I \quad \text{and} \quad -E_{\text{LUMO}} = A \tag{2}$$

where I is the ionization potential (or energy) and A the electron affinity.

2.3. Perturbation Equation

Hudson and Klopman[17] have derived an equation to describe the effect of orbital perturbation of two molecules on chemical reactivity. Their equation for the interaction energy ΔE can be simplified by using the HOMO of a nucleophile and the LUMO of an electrophile:

$$\Delta E = -\underbrace{\frac{Q_{\text{Nu}}Q_{\text{El}}}{\epsilon R}}_{\text{The Coulombic term}} + \underbrace{\frac{2(C_{\text{Nu}}C_{\text{El}}\beta)^2}{E_{\text{HOMO}} - E_{\text{LUMO}}}}_{\text{The frontier orbital term}} \tag{3}$$

where Q_{Nu} and Q_{El} are the total charges for the nucleophile and electrophile, respectively; C_{Nu} and C_{El} are the coefficients of the atomic orbitals Nu and El, respectively, in the molecular orbital; β is the resonance integral; ϵ is the permittivity, and R is the distance between Nu and El. Quantity E_{HOMO} is E_{V}, the energy of the valence band edge, and E_{LUMO} is E_{C}, the energy of the conduction band edge.

On the basis of the above equation, the HSAB rules can be restated as follows:

- A hard–hard interaction (or reaction) is fast because of a large Coulombic attraction as described by the first term of Eq. (2).
- A soft–soft interaction (or reaction) is fast because of a large orbital interaction between the HOMO of the nucleophile and the LUMO of the electrophile as described by the second term of Eq. (2).

2.4. Electronegativity and Absolute Hardness

In the beginning when the HSAB principle was first introduced, the meaning of hardness was not easy to understand. Certainly, it is not the hardness that measures the resistance against deformation. Then, what is it? It was not until the last several years that the real definition of absolute hardness[8,9] received a quantitative and theoretical backing.[10–12]

First, let us go back one step to define absolute electronegativity, χ. According to Mulliken,[18,19] for atomic species, the electronegativity χ^{M} is the negative of chemical potential μ:

$$\chi^{\text{M}} = \tfrac{1}{2}(I + A) = -\mu \tag{4}$$

However, this relation is only applicable to a single atom, but not to a molecule. If the

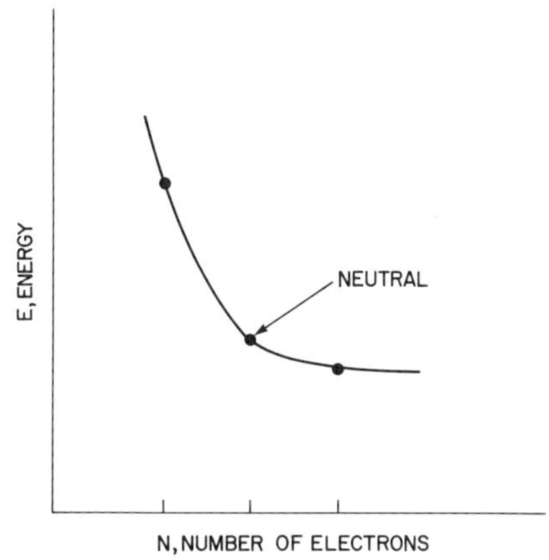

FIGURE 1. Plot of electronic energy vs. number of electrons for a fixed collection of nuclei [From R. G. Pearson, *J. Chem. Educ.* **64** (7), 563 (1987) (reproduced with permission).]

total electronic energy of an atom (or molecule) is plotted against the total number of electrons, a curve similar to that in Figure 1 is obtained. The first ionization potential for the species will be much larger than the electron affinity and much smaller than the second ionization potential. Assuming a differential curve, the electronegativity may be defined as

$$\chi^M = -(\partial E/\partial N)_Z = -\mu \tag{5}$$

where Z is the total number of nuclear charges in the atom.

The second derivative of the same is defined to be absolute (or chemical) hardness η given by

$$\eta = \frac{1}{2}(\partial^2 E/\partial N^2)_z \tag{6}$$

The corresponding operation definition is the corresponding finite-difference formula

$$\eta = \frac{1}{2}(I - A) \tag{7}$$

Hence, the absolute hardness η is the resistance of the chemical potential to change in the number of electrons. The opposite of η is the absolute softness, σ. Here we should point out that both μ and η are global properties at the ground state in the sense that they characterize the species as a whole.

According to the frontier orbital method,[20] the relationship of η with the energies of LUMO and HOMO is shown in Figure 2. Thus, from Figure 2,

$$\eta = \frac{1}{2}(E_{\text{HOMO}} - E_{\text{LUMO}}) \tag{8}$$

Here χ^M is the horizontal broken line in the middle of the energy gap. Hence the energy gap E_g is twice the absolute hardness η. This energy gap is an important link between chemistry and physics in the solid state.

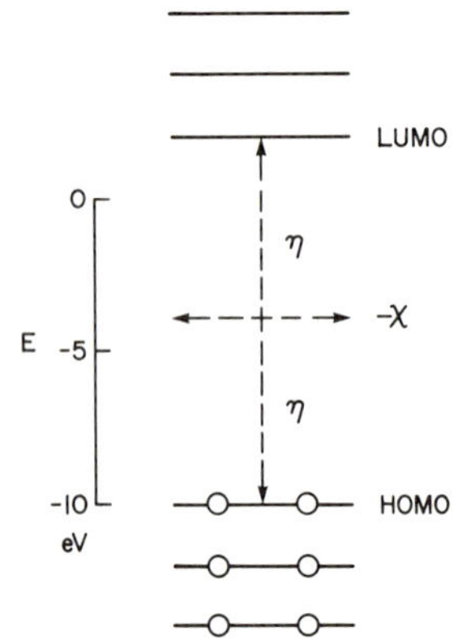

FIGURE 2. Orbital energy diagram for a molecule; HOMO, highest occupied MO; LUMO, lowest unoccupied MO. [From R. G. Pearson, *J. Chem. Educ.* **64** (7), 563 (1987) (reproduced with permission).]

3. EXTENSION OF THE HSAB PRINCIPLE TO SOLID INTERACTIONS

The applications of chemical hardness to crystals have been discussed by Shankar and Parr,[(21)] and those to metals and catalysis by Ugo[(22)] and by Yang and Parr.[(23)] However, in the literature, we have not found any formal application or extension of the HSAB principle to solid interactions. In this paper, we recently demonstrated that the extension of the HSAB principle to solid interactions is feasible in view of the electronic band structures of solids.[(24)] We shall discuss the physical meaning of the absolute hardness in terms of the average energy gap.

3.1. Electronic Band Structures of Solids

Solids can be classified as metals, semimetals, intrinsic semiconductors, and insulators. The band structures of solids[(25)] are illustrated in Figure 3. Monovalent metals,[(26)] such as Na, have a partially filled valence band, the lower half of which is occupied. The Fermi level is in the valence band but at the top of the occupied orbitals. Furthermore, there is still an energy gap between the valence band (occupied MO) and the conduction band (unoccupied MO). In some metals, such as the bivalent metals, the valence band is full but overlaps a higher unoccupied conduction band. In this case, the Fermi level is in the conduction band and the overlapped valence band. Thus, the electrons close to the Fermi level are still free to move as the extra bands supply the unoccupied states. In the latter case, there appears to be no minimum energy gap E_g^0, which is generally reported in the literature. However, it is not obvious that there should always be an average energy gap E_g^{Av}, which will be discussed later.

In addition to metals, there are semimetals,[(27)] such as graphite, whose valence band and conduction band can overlap. In general, their minimum energy gaps are very narrow. The third class of solids is the intrinsic semiconductor; its minimum energy gap E_g^0 is generally below 3 eV. Thus, thermal excitation alone can create an electron–hole pair to

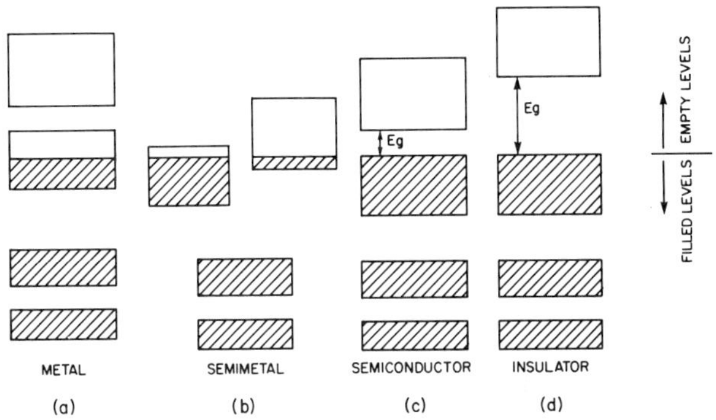

FIGURE 3. Band structures of solids: (a) metal, (b) semimetal, (c) semiconductor, (d) insulator. [From A. J. Epstein and J. S. Miller, *Sci. Am.* **241** (4), 52 (1979) (reproduced with permission).]

enhance conduction. The Fermi level of the intrinsic semiconductor[26] lies between the valence band (HOMO) and the conduction band (LUMO). Hence

$$E_F = \frac{1}{2}(E_{HOMO} + E_{LUMO}) = \frac{1}{2}(E_c + E_v) = -\chi \tag{9}$$

where E_F is the energy of the Fermi level. The fourth class of solids is the insulator, which includes most polymers. The minimum energy gap of an insulator is generally above 3eV. Therefore, thermal excitation alone cannot enhance the conduction of electricity.

3.2. Average Energy Gap and Absolute Hardness

The energy gaps generally reported in the literature are minimum energy gaps E_g^0 at room temperature.[28] On the other hand, the average energy gap E_g^{Av}[29,30] is defined as the difference in the energy between the conduction and valence bands:

$$E_g^{Av} = -(E_{HOMO} - E_{LUMO}) = -(E_v - E_c) = 2\eta \tag{10}$$

According to Phillips,[29] E_g^{Av} should contain both covalent (homopolar) component E_h and ionic (heteropolar) component C:

$$(E_g^{Av})^2 = (E_h)^2 + (C)^2 \tag{11}$$

For many $A^N B^{8-N}$ compounds, E_g^{Av} can be determined from dielectric measurements.

Moreover, the Phillips ionicity (or fractional ionic character) of a substance is defined by

$$f_i = \frac{C^2}{E_h^2 + C^2} = \frac{C^2}{(E_g^{Av})^2} \tag{12}$$

The E_g^{Av} and f_i values in Tables 1–3 are taken from papers by vanVechten.[31,32] Phillips and vanVechten[33] found that compounds with zinc blend and wurtzite structures have ionicities smaller than 0.785, while those with rock salt structures have ionicities greater than 0.785. We compare ionicities with calculated absolute hardness values in Tables 2–4.

3.3. Energy Gaps and Absolute Hardness Values for Semiconductors

3.3.1. Group IV Semiconductors

For comparison, both E_g^0 and E_g^{Av} for several diamond structure elements for Group IV semiconductors are listed in Table 1. In general, E_g^{Av} is always larger than E_g^0. From Table 1, the η values calculated from E_g^{Av} are rather close to those calculated from the A–I data; however, the agreement is still not good enough. Most of them, except carbon, have calculated η values around 2, or the η values around 3. Moreover, all four elements have ionicities of zero.

3.3.2. Group III_a–V_a Semiconductors

In Table 2, we compare ionicities and the calculated absolute hardness values for Group III_a–V_a semiconductors and SiC. These compounds are of the zinc blend structure.

TABLE 1
Minimum and Average Energy Gaps and Absolute Hardness Values for Group IV Atoms (eV)

	E_g^{0a}	E_g^{Ava}	η^b (calculated)	η^c
C	5.4	13.6	6.8	5.0
Si	1.1	4.8	2.4	3.4
Ge	0.7	4.3	2.2	3.4
α − Sn	0	3.1	1.6	3.1

[a] E_g^0 values[(28)] were cited by Burstein, Pinczuk, and Wallis.[(30)]
[b] η is calculated from E_g^{Av}.[(31,32)]
[c] η is the value obtained by Pearson.[(9)]

The chemical hardness values are below 3 eV for all compounds except SiC, AlN, and GaN. Thus, they are rather soft. In contrast, the ionicities are below 0.4 for all compounds except SiC, AlN, and GaN. SiC is low in f_i of 0.18.

3.3.3. Group II_b–VI_a Semiconductors

In Table 3, the ionicities and the calculated absolute hardness values for Group II_b–VI_a semiconductors are compared. In general, the ionicities are between 0.6 and 0.7 for zinc blend and wurtzite structures. They are more polar than the III_a–V_a compounds in Table 2. The chemical hardness values are higher than 3 eV for most of the compounds in

TABLE 2
Ionicities and Calculated Absolute Hardness Values (eV) for Group III_a–V_a Semiconductors and SiC[a]

Zinc blend	E_g^0	E_g^{Av}	η (calculated)	f_i
AlP	3.0	5.6	2.8	0.31
AlAs	2.16	5.2	2.6	0.27
AlSb	1.60	4.7	2.4	0.43
GaP	2.24	5.6	2.8	0.37
GaAs	1.35	5.2	2.6	0.31
GaSb	0.81	4.1	2.2	0.26
InP	1.27	5.2	2.6	0.42
InAs	0.36	4.6	2.3	0.36
InSb	0.18	3.7	2.8	0.32
SiC	3.1	9.2	4.6	0.18
AlN	3.8	11.0	5.5	0.45
GaN	3.25	10.7	5.4	0.50

[a] The E_g^0 values[(28)] were cited by Burstein, Pinczuk, and Wallis[(30)]; the E_g^{Av} and f_i values were obtained by vanVechten.[(31 and 32)]

TABLE 3
Ionicities and Calculated Absolute Hardness Values (eV) for Group II_b–VI_a Semiconductors[a]

Zinc blend	E_g^0	E_g^{Av}	η (calculated)	f_i
ZnS	3.54	7.8	3.9	0.63
ZnSe	2.58	7.1	3.6	0.63
ZnTe	2.26	5.8	2.9	0.61
CdTe	1.44	5.4	2.7	0.67
HgTe	0	(~2.5)	1.3	—
Wurtzite				
BeO	—	18.2	9.1	0.60
ZnO	3.7	11.8	5.9	0.62
CdS	2.4	7.1	3.6	0.69
CdSe	1.7	6.6	3.3	0.70
Rock salt				
CdO	2.5	9.0	4.5	0.71

[a]The E_g^0 values[28] were cited by Burstein, Pinczuk, and Wallis[30]; the E_g^{Av} and f_i values were obtained by vanVechten.[31 and 32]

Table 3 except the tellurides, thus these compounds are harder than III_a–V_a compounds in Table 2. However, there does not appear to be a direct correlation between ionicities and chemical hardness values. For example, BeO has a calculated η value of 9.1 but a f_i of 0.60. This is rather expected because absolute hardness should be related to the average energy gap, which is a function of both the covalent and ionic components.

3.3.4. Group IV_a–VI_a Semiconductors

Most of Group IV_a–VI_a semiconductors in Table 4 are actually narrow-gap semimetals. The absolute hardness values calculated from the average energy gaps are very low (1.2–2.0 eV). Thus, these compounds are very soft. However, the ionicities for these

TABLE 4
Ionicities and Calculated Absolute Hardness Values (eV) for Group IV_a–VI_a Semiconductors[a]

Rock salt	E_g^0	E_g^{Av}	η (calculated)	f_i
SnTe	0.2	2.33	1.2	0.61
PbS	0.41	4.0	2.0	0.83
PbSe	0.29	3.3	1.7	0.77
PbTe	0.32	2.55	1.3	0.74
Trigonal (distorted NaCl)				
GeTe	0.1	2.71	1.4	0.62

[a]The E_g^0 values[28] were cited by Burstein, Pinczuk, and Wallis[30]; the E_g^{Av} and f_i values were obtained by VanVechten.[31,32]

TABLE 5
Average Energy Gap and Hardness Values (eV) for Transition Metal Atoms

Metal	I^a	A^b	X^M	η^c	$E_g^{Av\,d}$ (calculated)
Ti	6.82	0.08	3.45	3.4	6.8
V	6.7	0.5	3.6	3.1	6.2
Cr	6.77	0.66	3.72	3.1	6.2
Mn	7.44	0	3.72	3.7	7.4
Fe	7.87	0.25	4.06	3.8	7.6
Co	7.8	0.7	4.3	3.6	7.2
Ni	7.64	1.15	4.4	3.3	6.6
Cu	7.73	1.23	4.48	3.3	6.6
Mo	7.10	0.75	3.9	3.2	6.4
Ru	7.40	1.5	4.5	3.0	6.0
Pd	8.34	0.56	4.45	3.9	7.8
Ag	7.58	1.30	4.44	3.1	6.2
Pt	9.0	2.1	5.6	3.5	7.0
Au	9.23	2.31	5.77	3.5	7.0

[a] C. E. Moore, *Natl. Stand. Ref. Data. Ser., Natl. Bur. Stand.* 34 (1970).
[b] H. Hotop and W. C. Lineberger, *J. Phys. Chem. Ref. Data* **14**, 731 (1985).
[c] These values are for isolated metal atoms [R. G. Pearson, *Inorg. Chem.* **27**, 734 (1988)].
[d] E_g^{Av} is calculated from η obtained from A–I values.

compounds are very high. For example, PbS has the highest ionicity of 0.83, and it is the best case showing that η and f_i are not directly related. However, with the combination of these two factors, it is easy to classify PbS as a very soft base. In fact, the entire group can be called very soft bases.

3.3.5. Transition Metal Atoms

We also attempt to calculate E_g^{Av} from η for several transition metal atoms (Table 5). The η values that fall between 3 and 4 eV indicate that most of these metals are rather "soft." Since there are not many E_g^{Av} values in the literature for comparison, we assume that they are reasonably good. We shall discuss the applications of the HSAB principle based on these η values.

3.4. Absolute Hardness of Solid

Now let us examine again the band structures of solids in Figure 3. In Figure 3a, the structure of metals with zero or low E_g values signifies that all bulk metals are "soft" as implied by A-I for the finite difference approximation. However, they do not appear to be as "soft" as the η values calculated for metal atoms from the A–I data (Table 5). Indeed, metals have been classified as amphoteric materials. Most of the metals are "soft" acids, and some of them "soft" bases. When two metals are brought together into close contact, one of them is assumed to be an acid, the other a base. In a paper by Cain *et al.*[(34)] the interaction at the Cu/Cr interface has been treated as an acid–base interaction. In this case, a soft base (Cu) and a soft acid (Cr) react preferentially. In fact, the Cr atom (η = 3.1 eV) is as soft as the Cu atom (η = 3.3 eV) (see Table 5).

Figures 3b and c show that both semimetals and semiconductors (Tables 1–4) should be rather "soft" because of relatively low E_g^0 and E_g^{Av} as well (most η values < 3.0 eV). Semiconductors, generally, react readily with metals even at ambient temperatures.[(35)] Thus, the interaction takes place as a "soft" acid (metal) with a "soft" base (semiconductor). Both chemical reaction and interdiffusion[(36)] jointly create a diffused "interphase" instead of an abrupt "interface."

In Figure 3d, it is noted that the band structure of an insulator contains a large gap. Thus, most insulators, regardless of organic or inorganic, are considered to be rather "hard" on the basis of the energy gap alone. Generally, they are harder than metals, semimetals, and semiconductors.

It is important to remember that the hardness of one material with respect to the other depends on its η value. The one with a lower η value is "softer" than that with a higher η, though they both can be "hard." Even in the same material, one functional group can be harder than the other. In this case, a new term, local hardness $\tilde{\eta}$,[(11)] is introduced to signify the relative hardness of each functional group.

4. APPLICATIONS OF THE HSAB PRINCIPLE TO ADHESION AND SURFACE INTERACTIONS BETWEEN METALS AND POLYMERS

The interactions between metals and polymers are good examples of the applications of the HSAB principle to solids. Since the metals discussed in this section are vacuum deposited, there are no oxide layers involved. When Cr is deposited on pyromellitic dianhydride–oxydianiline polyimide (PMDA–ODA PI), there appears to be some chemical bonding between Cr and PI. In this case, Cr is a "soft" acid, and PI is "hard" base. How can the bonding take place? According to Ho *et al.*[(37,38)] indeed the reaction does not proceed in the manner such as the Cr atom attacking one of the carbonyl groups (nucleophiles). What really happens is that the Cr atom delocalizes and forms a charge transfer (or acid–base) complex with the PMDA unit of the PI. In the frontier orbital terminology, the stabilization of the complex is achieved through the transfer of electronic charge from the metal d states of Cr (HOMO) to the LUMO of the π system of the PMDA unit of PI. In this manner, PI acts as a "soft" acid by accepting the electron and Cr (η = 3.1 eV) becomes a "soft" base by donating the electron. Other transition metals, such as Ti(η = 3.4 eV),[(39)] are also very reactive with PI. The general reaction presumably follows the same path of the acid–base interaction through a charge-transfer complex, which should involve a partial electron transfer.

One of the exceptions is Cu,[(40)] which does not interact with PI after the deposition and gives rise to a much weaker complex. Cain and Matienzo[(41)] have used the tight binding calculations of the extended Hückel type to find the relative acid strengths of Cu compounds in the following decreasing order: $CuF_2 > CuO > CuF \approx Cu_2O > Cu$. Indeed, the Cu ions react faster than Cu with functional polymers. The Cu^{2+} ion has a Fermi level lying below the top of the Cu 3d band, and some of the Cu 3d orbitals are unoccupied and able to accept electron pairs by acting as an acid. In the HSAB terminology, Cu becomes a "hard" acid in the form of Cu^{2+} ion because the hardness for Cu is 3.3 eV, and that for Cu^{2+}, 8.3 eV.

Generally, many transition metals react well with polymers, ceramics,[42] etc., partly because of the ease of oxidation and partly because of the availability of d-orbital electrons. Buckley and Brainard[43] have found that metals react with PTFE and PI during his pin-and-disk experiments under vacuum. With PTFE, the adhesive forces are three times the applied load. Now, we can also attribute these reactions to the HSAB interactions. In the literature, there should be numerous examples of the surface interactions that can be classified as the acid–base interaction. However, we are unable to discuss them all in this paper. One of the reasons is that the chemical hardness values for polymers have yet to be determined indirectly. We are hoping to publish the calculated values in the future.

It is important to point out that though chemical hardness values are rather similar for all metals, not all metals react in the same way or at the same rate with polymers. Since all metals are "soft" and most polymers are "hard," the metal–polymer reactions or adhesion, unlike the exceptional Cr–PI reaction mentioned in one of the preceding sections, are generally difficult to achieve. To circumvent the situation, most metals have to be transformed into harder counterparts, e.g., related oxides of higher chemical hardness values. This is exemplified in the cited case about the Cu^{2+} ions,[41] which are harder than the metal atoms, Cu. Consequently, the reactivities of the metal–polymer interaction are determined by the ease of the formation of oxides.[44] In theory, the more negative the H_f, the more reactive the interaction. For the same reason, Au and Ag are always nonreactive or inert toward polymers because these two metals are more electronegative than other metals. We shall explore, in detail, the metal–polymer adhesion in a separate paper on the chemistry and physics of solid adhesion.[45]

5. CONCLUSIONS

Our approach in this paper is somewhat speculative in correlating the average energy gap with absolute hardness. The concept of average energy gap has also been included in Shankar and Parr's paper[21] in their discussion about the structure stability diagram. Thus, we felt more comfortable about our discussions in this paper even without any direct experimental support.

We also discussed the relations between chemical hardness values and ionicities for semiconductors and semimetals. In general, these two parameters are not directly related. In a separate paper, we shall discuss that both μ and η are two important chemical parameters[21] which can be used to estimate the number of electrons transferred between systems A and B. For solids,[46] two physical properties, the work function and the average energy gap, can be used in place of μ and η.

The extension of the HSAB principle and the frontier orbital concept to solid interactions has been demonstrated. For the HSAB principle, the absolute hardness has been discussed along with the average energy gaps in solids. The mechanisms of interactions between different types of solids have been elaborated in light of the HSAB principle. Both the HSAB principle and the frontier orbital concept have been found to be beneficial in explaining the mechanisms for several metal–polymer adhesion and surface interactions. In practice, our new findings should be helpful to those in carrying out research in, or processing, thin-film microelectronic devices and thick-film multilayer structures.

ACKNOWLEDGMENTS

The author would like to thank Professor R. G. Pearson of the University of California at Santa Barbara for his valuable comments on the original manuscript in which we first applied the HSAB principle to solids.

NOMENCLATURE

A	Electron affinity
C	Ionic (heteropolar) component of the average energy gap
C_{El}	Coefficient of the atomic orbital of the electrophile
C_{Nu}	Coefficient of the atomic orbital of the nucleophile
ΔE	Interaction energy
E	Electronic energy
E_c	Energy of the conduction band edge
F_F	Energy of the Fermi level
E_g^0	Minimum energy gap
E_g^{Av}	Average energy gap
E_h	Covalent (homopolar) component of the average energy gap
E_{HOMO}	Energy of the highest occupied molecular orbital
E_{LUMO}	Energy of the lowest unoccupied molecular orbital
E_V	Energy of the valence band edge
f_i	Ionicity or partial ionic character
I	Ionization potential
N	Number of electrons for a fixed collection of nuclei
Q	Total electron charge
R	Distance between the electrophile and nucleophile
Z	Total number of nuclear charge in the atom
β	Resonance integral
ϵ	Permittivity
η	Absolute hardness
$\tilde{\eta}$	Local hardness
μ	Chemical potential
σ	Absolute softness
χ^M	Electronegativity (Mulliken)

REFERENCES

1. V. Gutmann, *The Donor–Acceptor Approach to Molecular Interactions,* Plenum Press, New York (1978).
2. B. V. Deryagin, N. A. Krotova, and V. P. Smilga, *Adhesion of Solids* (trans. R. K. Johnson), Plenum Press, New York (1978).
3. W. B. Jensen, The Lewis acid–base definitions: a status report, *Chem. Rev.* **78** (1), 1 (1978).
4. F. M. Fowkes, Donor–acceptor interactions at interfaces, *J. Adhes.* **4,** 155 (1972); also in: *Recent Advances in Adhesion* (L. H. Lee, ed.), p. 39, Gordon and Breach, New York and London (1973).
5. F. M. Fowkes, Role of acid–base interfacial bonding in adhesion. *J. Adhes. Sci. Technol.* **1** (1), 7 (1987).
6. J. C. Bolger and A. S. Michaels, Molecular structure and electrostatic interactions at polymer–solid

interfaces, in: *Interface Conversion* (P. Weiss and D. Cheevers, eds.), Chapter 1, Elsevier, New York (1969).
7. R. G. Pearson, Hard and soft acids and bases, *J. Am. Chem. Soc.* **85,** 3533 (1963).
8. R. G. Pearson, Recent advances in the concept of hard and soft acids and bases, *J. Chem. Educ.* **64** (7), 563 (1987).
9. R. G. Pearson, Absolute electronegativity and hardness: application to inorganic chemistry, *Inorg. Chem.* **27,** 734 (1988); also *J. Am. Chem. Soc.* **110,** 7684 (1988).
10. R. G. Parr and R. G. Pearson, Absolute hardness: Companion parameter to absolute electronegativity, *J. Am. Chem. Soc.* **105,** 7512 (1983).
11. M. Berkowitz, S. K. Ghosh, and R. G. Parr, On the concept of local hardness in chemistry, *J. Am. Chem. Soc.* **107,** 6811 (1985).
12. M. Berkowitz and R. G. Parr, Molecular hardness and softness, local hardness and softness, hardness and softness kernals, and relations among these quantities, *J. Chem. Phys.* **88** (4), 2554 (1988).
13. R. G. Pearson (ed.), *Hard and Soft Acids and Bases,* Dowden, Hutchinson and Ross, Stroudsburg, PA (1973).
14. R. G. Pearson and J. Songstad, Applications of the principle of hard and soft acids and bases to organic chemistry, *J. Am. Chem. Soc.* **89,** 1827 (1967).
15. N. S. Isaacs, *Physical Organic Chemistry,* Longman Scientific and Technical, with Wiley, New York (1987).
16. T. Koopmans, Ordering of wave functions and eigenenergies to the individual electrons of an atom, *Physica* **1,** 104 (1933).
17. R. F. Hudson and G. Klopman, *Tetrahedron Lett.* **12,** 1103 (1967); also Polyelectronic perturbation treatment of chemical reactivity, *Theor. Chim. Acta* **8,** 165 (1967).
18. R. S. Mulliken, Electronic structures of molecules XI. Electroaffinity, molecular orbitals and dipole moments, *J. Chem. Phys.* **3,** 573 (1935).
19. R. S. Mulliken, Electronic structures of molecules XII. Electroaffinity and molecular orbitals, polyatomic applications, *J. Chem. Phys.* **3,** 586 (1935).
20. K. Fukui, Role of frontier orbitals in chemical reactions, *Science* **218** (4574), 747 (1982).
21. S. Shankar and R. G. Parr, Electronegativity and hardness as coordinates in structure stability diagram, *Proc. Natl. Acad. Sci. USA* **82,** 264 (1985).
22. R. Ugo, General features of homogeneous catalysis with transition metals, *J. Soc. P. Ax. Edit.* Chim, Milano **51** (12), 1319 (1969).
23. W. Yang and R. G. Parr, Hardness, softness, and the Fukui function in the electronic theory of metals and catalysis, *Proc. Natl. Acad. Sci. USA* **82,** 6723 (1985).
24. L. H. Lee, Applications of the hard–soft acid–base (HSAB) principle to solid adhesion and surface interactions between metals and polymers, in: *New Trends in Physics and Physical Chemistry of Polymers* (L. H. Lee, ed.), p. 185, Plenum Press, New York (1989).
25. A. J. Epstein and J. S. Miller, Linear-chain conductors, *Sci. Am.* **241** (4) 52 (Oct. 1979).
26. F. Gutmann and L. E. Lyons, *Organic Semiconductor,* p. 17, Wiley, New York (1967).
27. D. L. Carter and R. T. Bate (eds.), *The Physics of Semimetals and Narrow-Gap Semiconductor,* Pergamon Press, Oxford (1971).
28. R. C. Weast (ed.), *Handbook of Chemistry and Physics,* p. E-108–113, CRC Press, Florida (1988–1989).
29. J. C. Phillips, Ionicity of the chemical bond in crystals, *Rev. Mod. Phys.* **42** (3), 317 (1970).
30. E. Burstein, A. Pinczuk, and R. F. Wallis, in: *The Physics of Semimetals and Narrow Gap Semiconductors* (D. L. Carter and R. T. Bate, eds.), p. 251, Pergamon Press, Oxford (1971).
31. J. A. vanVechten, Quantum dielectric theory of electronegativity in covalent systems. I. Electronic dielectric constant, *Phys. Rev.* **182,** 891 (1969).
32. J. A. vanVechten, Quantum dielectric theory of electronegativity in covalent systems. II. Ionization potentials and interband transition energies, *Phys. Rev.* **187,** 1007 (1969).
33. J. C. Phillips and J. A. vanVechten, Dielectric classification of crystal structures, ionization potentials, and band structures, *Phys. Rev. Lett.* **22,** 705 (1969).
34. S. R. Cain, L. J. Matienzo, and F. Emmi, Modeling of chemical interactions at Cr/Cu interfaces, *J. Phys. Chem. Solids* **50** (1), 87 (1989).
35. L. Braicovich, Adsorption and reaction of metals on elemental semiconductors, in: *The Chemical Physics of*

Solid Surfaces and Heterogeneons Catalysts (D. A. King and D. P. Woodruff, eds.), Chapter 6, Elsevier, Amsterdam (1988).

36. L. J. Brillson, The structure and properties of metal-semiconductor interfaces, *Surf. Sci. Rep.* **2,** 123 (1982).
37. P. S. Ho, B. D. Silverman, R. A. Haight, R. C. White, P. N. Sanda, and A. R. Rossi, Delocalized bonding at the metal-polymer interface, *IBM J. Res. Dev.* **32** (5), 658 (1988).
38. P. S. Ho, R. A. Haight, R. C. White, B. D. Silverman, and F. Faupel, Chemistry, microstructure and adhesion of metal–polymer interfaces, this volume, Chapter 14.
39. F. S. Ohuchi and S. C. Freilich, Metal polyimide interface: A titanium reaction mechanism, *J. Vac. Sci. Technol.* **A4** (3), 1039 (1986).
40. R. A. Haight, R. C. White, B. D. Silverman, and P. S. Ho, Complex formation and growth at the Cr- and Cu-polyimide interface, *J. Vac. Sci. Technol.* **A6** (4), 2188 (1988).
41. S. R. Cain and L. J. Matienzo, Acid–base properties of copper oxides and fluorides by means of band calculations: Relevance to metal–polymer adhesion, *J. Adhes. Sci. Technol.* **2** (5), 395 (1988).
42. D. H. Buckley, *Surface Effects in Adhesion, Friction, Wear and Lubrication,* Elsevier, Amsterdam (1981).
43. D. H. Buckley and W. A. Brainard, The atomic nature of polymer–metal interactions in adhesion, friction and wear, in: *Advances in Polymer Friction and Wear* (L. H. Lee, ed.), Vol. A, p. 315, Plenum Press, New York (1975).
44. D. M. Mattox, Thin film metalization of oxides in microelectronics, *Thin Solid Films,* **18,** 173 (1973).
45. L. H. Lee, The chemistry and physics of solid adhesion, this volume, Chapter 1.
46. L. H. Lee, Relevance of the density-functional theory to surface interactions and solid adhesion, presented to the Symposium on Acid–Base Interactions: Relevance to Adhesions Science and Technology, held at Lehigh University, Bethlehem, Pa., and to be published in the *J. Adhes. Sci. Technol.*

13

Interface Design for Thin Film Adhesion

John E. E. Baglin

1. INTRODUCTION

Technological demands for good adhesion performance of thin coatings on critical surfaces arise in many applications. Examples include reflective coatings on optical components, chrome plating for corrosion protection, metallic contacts and conducting patterns on semiconductor chips and their packaging substrates, enamel glazing, low-friction coatings, and magnetic thin films for data storage applications.

In all cases, the ideal bonding of coating to substrate should be physically strong and should not be weakened by the passage of time or the normal usage of the material. This means that the film/substrate interface must be chemically and thermodynamically stable within temperature ranges appropriate to the application, while retaining the bonding strength needed to withstand applied physical stresses. The requirement of chemical stability for the interface of materials A and B generally implies that A and B must have no bulk miscibility or reactivity, otherwise the interface would invite unlimited thermal interdiffusion of A and B. It is nevertheless possible to establish a stably bonded interface between such dissimilar materials, as discussed below.

The adhesion performance of a coating may depend upon the kind of usage in question. For example, the film may be required to survive repeated abrasion; it may be subjected to a direct peeling force; it may need to remain in place after multiple thermal cycles in which the interface is stressed by differential thermal expansion; or it must resist interface edge penetration by a corrosive fluid. Special considerations apply in the design of systems to suit each of these requirements. However, the interface failure mode common in all these cases is the initiation and propagation of a fracture in the interface region.[1] These are the phenomena to be inhibited in order to produce better adhesion, and with this in mind we refer to Figure 1 which displays schematically the development of an interface fracture, and indicates the principal factors which can influence the fracture process.[2,3] It is these factors which must all be considered in the design of coatings with good adhesion performance, whatever the intended service stress will be. These factors will be discussed in more detail below. Their quantitative importance to adhesion will then be discussed, and their effect on various adhesion characterization techniques. Section 4

John E. E. Baglin • IBM Almaden Research Center, San Jose, California, 95120-6099.

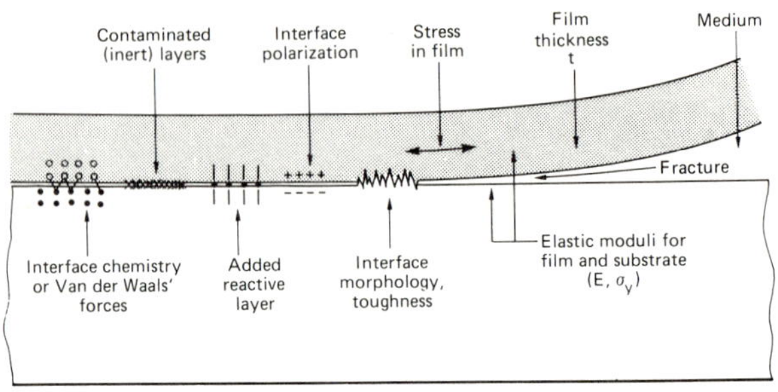

FIGURE 1. Schematic illustration of some factors affecting fracture at a thin film interface.

reviews new experimental work on the atomic-scale tailoring of interface structures for optimum adhesion, by means of ion beam processing. Finally, the summary in Section 5 suggests some areas offering good prospects for future research and study.

2. ORIGINS OF THIN FILM ADHESION

2.1. Interface Energy

A classical view of the energy of adhesion at a planar interface between materials A and B is illustrated in Figure 2. The energy of adhesion E_{ad} is defined as the energy gained per unit area by bringing materials A and B (surface free energies E_A and E_B, respectively) into contact to produce an interface free energy γ_{int}. Then

$$E_{ad} = E_A + E_B - \gamma_{int} \tag{1}$$

A large value of γ_{int} could proscribe adhesion ($E_{ad} < 0$), while a small value of γ_{int} can permit large E_{ad}. Measurements of γ_{int} for liquids on solids can be made directly from the contact angle of the liquid droplets on the wetted or nonwetted surface.(4) For solid coatings, γ_{int} may be inferred from more complex experiments, or it may sometimes be derived thermodynamically.

Factors expected to affect γ_{int} include the interface chemical bonding and its physical population, van der Waals forces and, in the case of metal–insulator interfaces, charge transfer and image charge bonding.(5) The latter two forces are weak, and would generally dominate only when electronic (chemical) bonding of interface atoms is not available. In the best situation, a stable configuration of interface atoms will experience electronic bonding to their neighboring atoms, which remain bonded in the film or substrate matrix.

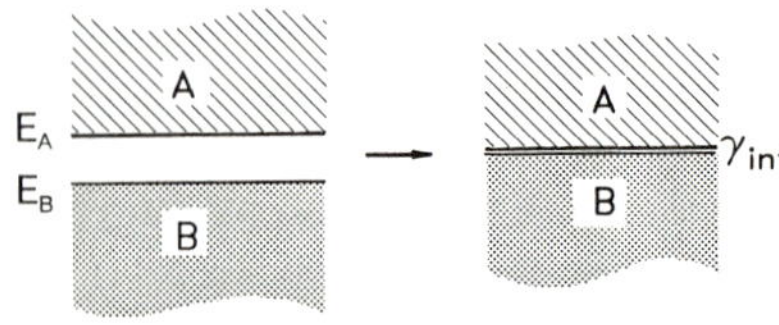

FIGURE 2. Free-energy change when materials A and B are joined.

Such bonds may be metallic, ionic, or covalent, and they should represent a continuous linkage across the interface (Figure 1).

Much insight into the physical bonding process has recently been provided by theoretical approaches to estimating E_{ad}. Ferrante and Smith[6,7] have made quantum-mechanical calculations of E_{ad} for intermetallic interfaces. Senbetu *et al.*[8] have used a density-functional formalism to obtain E_{ad} for metal–semiconductor interfaces.

2.1.1. Interface Phases

It has been proposed[3] that when interface layers of atoms display a stable bonding of the kind shown in Figure 1, the term "interface phase" would be appropriate to describe the condition of those interface layers. An example exists for copper films bonded to Al_2O_3, in which stable ternary Cu—O—Al bonding has been verified[9] at a sharp interface, even though no such ternary phase is known in the bulk. The apparent formation of a unique Cu—(Cu—O—Al)—Al_2O_3 configuration raises intriguing questions about how best to model or energetically characterize such an "interface phase." The ability to predict the existence of such phases in new systems of interest would greatly assist in planning practical interface tailoring with which to create stable adhesion; it would also help in recognition of systems where no stable interface phase exists for adhesion. At present, however, this general and important concept remains largely unexplored.

2.1.2. Contaminants

The interface energy depends heavily on the atomic species at the interface, and naturally is susceptible to radical change by submonolayer amounts of contaminant species. (For example, layers of native oxide on metals or silicon, or adsorbed OH layers on alumina can inhibit adhesion of deposited metal coatings.) Control or removal of such contaminant layers (e.g., by sputter etching *in situ* prior to vacuum deposition of the film) is critically important for process reproducibility in general.

It is worth noting, however, that in the right system, adsorbate species deliberately introduced might be used to provide interface chemical possibilities for adhesion which were otherwise unattainable.

2.2. *Interface Fracture*

Failure of film adhesion requires both the initiation and the propagation of a fracture in the interface region (Figure 1). Both are inhibited when strong electronic bonding produces a low value of γ_{int}. However, the failure thresholds are also affected greatly by a variety of other features, including the elastoplastic properties of the film and the interface roughness. To some extent, optimization of such mechanical properties can be used to enhance adhesion.

2.2.1. Delamination Threshold

As an illustration, we cite the work of Klokholm *et al.*[10,11] in which is evaluated the critical internal tensile stress S in a deposited film of thickness t and Young's modulus E, where the film will spontaneously delaminate from a rigid substrate for which the interface energy is γ_{int}. The Barenblatt fracture model[12] describes the condition for the onset of fracture when the stress energy stored in the film, $S^2t(1 - \nu)/E$, is large enough to initiate crack formation:

$$S^2t(1 - \nu)/E > \frac{1}{2} E_{ad}/(1 + \nu) \qquad (2)$$

where ν is the Poisson ratio (about 0.33).

Evidently, fracture is assisted by large values of S or t, or by small values of E. Externally applied energy (in the form of a peeling force, for example) can provide an energy term added to the stored stress energy, and thereby bring about delamination directly.

2.2.2. Crack Propagation

A uniform planar interface provides the simplest geometry for fracture propagation, requiring a constant input of energy at the fracture threshold. If, however, the crack encounters a rough interface region (as shown in Figure 1), its movement will be "pinned" by dispersion of energy at the asperities. Getting it to move again will require input energy (e.g., from a greater imposed peeling force) in excess of the planar fracture threshold. Thus, by roughening a substrate surface prior to film deposition, the adhesion performance can be improved for any given system. Such roughening can often be achieved by chemical etching or sputter etching of the substrate prior to film deposition. Its scale should presumably be of dimensions less than the film thickness.

2.2.3. Toughening

In favorable circumstances, interface fracture toughening may create an interface stronger than its surrounding materials and, depending on the nature of the applied stress, this could result in the delamination fracture being located within the substrate or film in the vicinity of the interface. One defense against this failure mode would lie in fracture toughening of a broad region on either side of the interface. An example of this effect is the work of Madakson *et al.*[13] in which ion implantation of Ti throughout a 500 Å region near a Cu/Al_2O_3 interface produced fine (100 Å) precipitates (believed to be $CuAlTiO_3$), whose presence would inhibit fracture development. A further example of the use of implantation is that of Hou and Stringer,[14] who implanted Y or Hf ($> 10^{16}$ ions/cm^2) at 400 keV into Ni–Cr prior to thermal oxidation, thereby greatly improving the adhesion of the grown oxide film.

3. *ADHESION PERFORMANCE*

3.1. *Dependence on Application*

As discussed above, the ability of a film to withstand various detaching assaults depends on a great variety of basic physical and chemical quantities which could be defined for the film and substrate, such as interface free energy, film thickness, Young's modulus, intrinsic stress in the film, and possibly even fracture toughness. However, the way in which these basic measurable quantities combine to affect "adhesion performance" depends entirely on the nature of the ordeal which the "performance" is intended to describe.[15] A simple illustration of this point is the contrast between a coating subjected to a peeling action normal to the interface, and a coating subjected to a scratch test, where interface shear will dominate, and the result will depend strongly on interface morphology. Although qualitatively it may always be said that better chemical bonding at the interface will always mean better adhesion performance, it cannot be assumed that the coating with adhesion optimized to resist peeling will automatically perform well in a scratch test.

In short, "adhesion" can only be quantified with reference to a specific type of test or usage. It is not in itself a basic physical property, such as the surface free energy or the film thickness, whose measurement can be accomplished by a variety of methods. A peel force measurement will not be helpful or relevant for an engineer who is concerned to prevent delamination caused by differential thermal expansion shear. While various tests can (and should) be made in order to derive the fundamental quantities such as γ_{int}, the long-held ambitions to identify a universal "adhesion" figure of merit must be abandoned, at least until theoretical models have succeeded in linking the various test criteria via the elastoplastic properties of the interface materials. Few such links exist today, and "adhesion performance" must therefore be defined and measured in the context of the application of interest. As an illustration of the differences in adhesion testing techniques and their diagnostic power, some of those techniques will be briefly reviewed in the next section.

3.2. Adhesion Measurements

3.2.1. Threshold Tests

Quick tests for thin film adhesion can be made with Scotch tape or a cotton bud (Q-tip). In the Scotch tape test, the adhesive tape is applied evenly and carefully to the film in question, and then pulled off in a direction normal to the substrate. Where film–substrate adhesion is poorer than that of the tape to the film, the film peels off. The Q-tip represents a rudimentary scratch test in which the cotton tip is rubbed back and forth on the film under finger pressure. The reproducibility of these tests is variable, and they provide only threshold information, where the threshold references, e.g., adhesion of tape on test surface, are themselves not well known, and subject to error due to surface impurities.

3.2.2. Peel Test

A typical peel test is illustrated schematically in Figure 3. In this case, the peel force P(g/mm) is measured while a strip of film a few mm wide is peeled at slow uniform speed, usually in a metallurgical test machine.[16] The test is quantitative, in that P represents a reproducible measure of adhesion performance for a given film–substrate system. This test also has the distinction of having been subjected to a detailed finite-element modeling for interface fracture and elastoplastic deformation of both film and

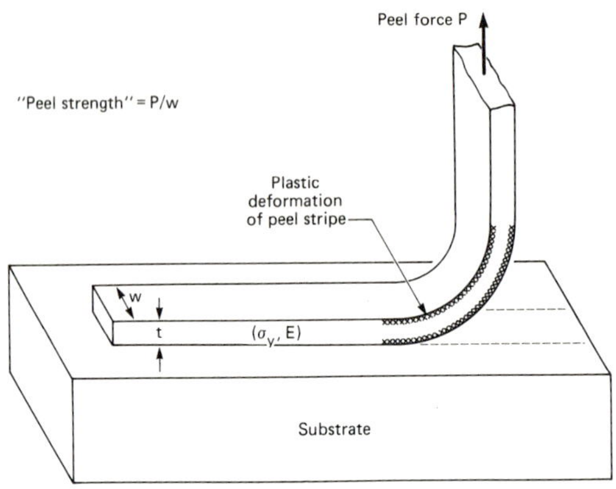

FIGURE 3. Peel test for adhesion.

substrate in the process. This model program was developed by K. S. Kim[17] for application to a general system whose elastic constants and dimensional parameters can be supplied. The result is a dimensionless plot relating P/γ_{int} to the mechanical properties of the film and substrate, as shown in Figure 4 for the case of Cu on Si.

The analysis serves to display the critical dependence of P, the peel force per mm, on both γ_{int} and other physical properties, a dependence unique to the normal peel test. It is interesting to note, for example, how the quantity P/γ may be increased by preparation of a film with low yield stress σ_y, or by reducing the actual thickness t of the coating.

In practice, this test is often implemented by overcoating the thin film ($\leq$ 10,000 Å) of interest with a stripe of metal some 10 μm thick, using a simple deposition mask. Practical problems may arise in obtaining reliable film/stripe adhesion, and in ensuring abrupt stripe definition to remove spurious edge effects in the measurement. However, the test is worth the effort, being intrinsically quantitatively understood and reproducible.

3.2.3. Scratch Test

In this test, a loaded diamond tip is drawn steadily across the surface of the test film, and the resulting groove is examined optically. At subcritical loads, a smooth groove is

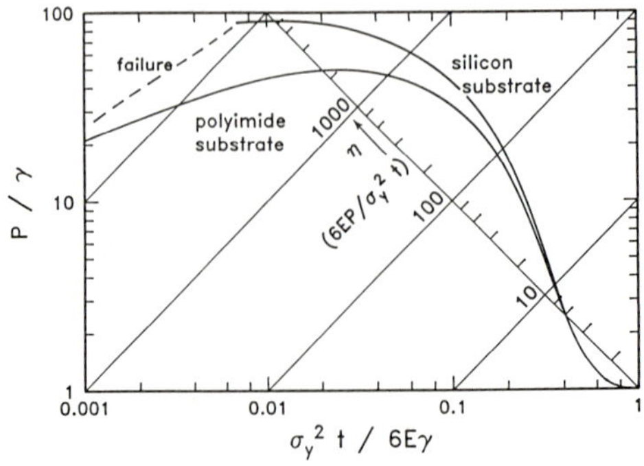

FIGURE 4. Universal Peel Diagram[17] calculated for Cu film on substrates of Si or polyimide.

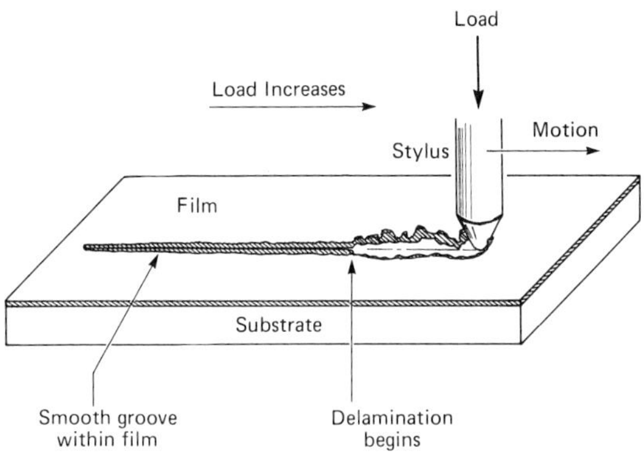

FIGURE 5. Scratch test for adhesion.

traced in a metal film, while at or above a critical load, ragged edges on the groove indicate that the film has been detached by the diamond (Figure 5). This critical load is the measure of adhesion yielded by the technique.[18] Instruments have now been developed in which a microindenter is fitted with acoustic sensors which will more sensitively identify the point at which delamination begins.

Since the scratching process primarily applies a shearing force at the interface, the delamination conditions can be expected to differ substantially from those of the peel test. An equivalent analysis of the fracture process is expected to show a different dependence on the mechanical properties and interface energy (a feature which supports the concept of testing with a technique related to intended application). However, the results should be equally reproducible, and in a similar way the effect of the relevant basic material properties should be identifiable, to practical advantage.

3.2.4. Pin Pull Test

A commonly used test for process control monitoring, this test poses large problems of reproducibility, and provides measurements not readily subjected to quantitative interpretation.

The test is illustrated in Figure 6 in one of several common configurations. A metal "pin" is either epoxied or brazed onto the surface of a matching "dot" of the film under test, and a force is applied normal to the test surface. The force is increased steadily until, at a critical value F, the pin and film detach from the substrate. The adhesion is then quoted as the quantity F/A, where A is the contact area of the pin.

With sufficient care, the ideal of force F applied precisely in the direction normal to the film, with no torque component, can be realized, especially when using aluminum pins which can yield to accommodate a small misalignment. However, it is not realistic to assume that the interface fracture is simultaneously initiated at all points of the pin contact. More probably, fracture will start at a local abnormality, perhaps at the pin edge; after that, lateral propagation of the crack will quickly release the pin. The threshold for initiating that crack will be the measured value of F; however, the argument for dividing that force by the area of contact A seems questionable. Furthermore, it appears that the test

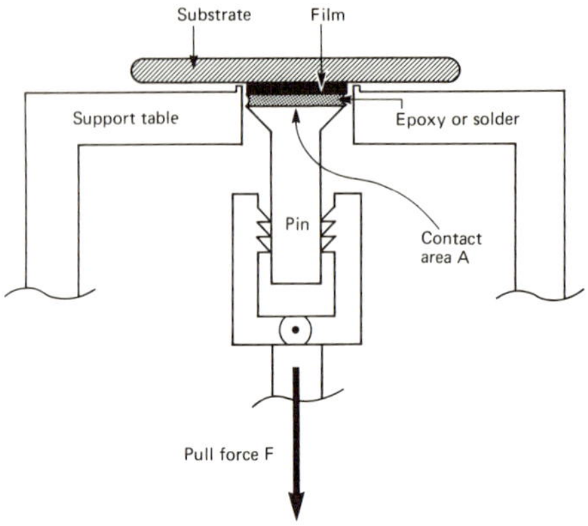

FIGURE 6. Pin pull test for adhesion.

must be unduly dependent on the perfection and uniformity of the likely sites for initial fracture at the perimeter of each test dot.

3.2.5. Other Tests

A variety of other techniques have been proposed for providing the normal force necessary to detach a film, some of which, such as the high speed centrifuge[19] or the pulsed loop current,[20] seem to avoid the complication of material deformation required by many tests. Recognizing, however, that all such tests ultimately must depend on the sequence of fracture initiation and propagation at the interface, it is not clear that they offer analytical or interpretive advantages over simpler peel or scratch tests. Since they do not directly simulate most service requirements for thin coatings, they may offer few practical advantages.

Microindenter tests for thicker coatings[21] appear to be workable when adhesion is low; however, quantitative interpretation will be very complex. Blister tests offer a sort of two-dimensional peel test[22] and require elaborate fixturing.

4. ADHESION ENHANCEMENT

4.1. Processing Strategies

Considering the variety of properties which determine the adhesion performance for a given film/substrate couple, it is clear that practical bonding can generally be optimized by suitable process design. Of primary concern is the development of interface chemical bonding where possible. This may be assisted by removal of contaminants from the substrate surface, or by addition of a reactive joining layer such as the Cr/CrO_x layer commonly used to attach Cu on SiO_2. The process may also be assisted by sputter-activation of substrate surface bonds prior to film deposition, or heating of the substrate during film deposition. It has long been recognized that adhesion is better on a surface previously roughened, generally by chemical etching. Such roughening serves to fracture-toughen the interface region, altering the geometry of the fracture process and providing an enhanced interface contact area.

4.2. Ion Beam Techniques

In recent years, it has become apparent that ion beam processing has much to offer in elegance, effectiveness, and economy for the improvement of thin film adhesion, addressing all the considerations above. Research results are encouraging, and significant advances have already been made. In this section, some of those findings will be reviewed together with the supporting information obtained about interface formation and structure. It is already evident that low-energy large-area Kaufman ion sources,[23] simple to incorporate in existing vacuum deposition equipment, will become a vital process tool for interface engineering in the future.

4.2.1. Ion Beam Stitching

When a beam of penetrating ions (100 keV to MeV energies) passes through a thin film/substrate interface (Figure 7), it has been found experimentally that improved adhesion almost always results. The implanted ions themselves play no part in the resulting interface structure, since in this case their energy is high enough to ensure that they come to rest deep inside the substrate. In their passage through the interface, however, these ions will produce dense trails of ionization and will sometimes produce direct collisional displacements of the matrix atoms. Evidently, the random disruption of electronic bonding and short-range atomic displacements thus produced at the interface can lead to the formation of some fresh, beneficial bonds to span the interface. Such a disturbance can also be supposed to disrupt any contaminant layer so that atoms from film and substrate may come into direct contact. This process has become known as "ion beam stitching," for obvious reasons.

It has also been found that heat treatment following irradiation will generally enhance the adhesion very considerably, leading to the concept of thermal relaxation of the disorder created by the energetic ions. Such relaxation would be expected to lead to the lowest

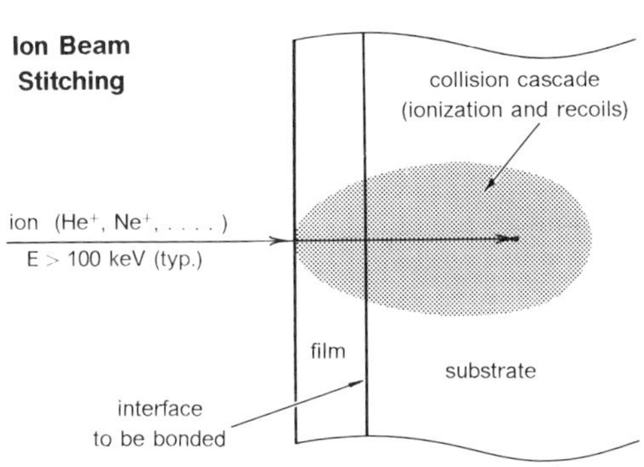

FIGURE 7. Ion beam "stitching."

TABLE 1

A Selection of Experimental Observations of Adhesion Enhancement by Ion Beam Stitching[a]

Film	Substrate	Beam	Threshold dose or range of doses (cm^{-2})	Adhesion test	Reference	Comment
Al	Soda glass	Ar 120 keV		Scratch	Collins *et al.*[24]	Measured adhesion vs. dose
	organic wash		~6E13	Pull		Increase × (50–100) max
	water wash		~1E15			
Au	Glass	S 12 MeV	1E12 to 1E15	Scotch Tape Scratch	Jacobson *et al.*[25]	
Cu	Al with native oxide	O 12 MeV	1E14 to 1E15	Scotch Tape Scratch		Cu–Al adhesion strong as deposited if no Al oxide
Au	Teflon	S 12 MeV	1E12	Scotch Tape		
Au	Si	Cl 20 MeV	1E15	Scotch Tape	Werner *et al.*[26] and Griffith *et al.*[27]	
	SiO_2	Cl 20 MeV	2E15			
	InP	Cl 20 MeV	5E14			
	Ferrite	Cl 20 MeV	3E13			
	Ferrite	F 5 MeV	2E15			
	W	Cl 20 MeV	2E14			
	CaF_2	Cl 20 MeV	2E14			
	Al_2O_3	Cl 20 MeV	5E15			
	GaAs	Cl 20 MeV	5E14			
	Teflon	He 1.5 MeV	1E13			
		H 1 MeV	1E14			
Cu	Alumina	He 200 keV Ne 280 keV	to 6E16	Peel	Baglin *et al.*[28,16]	Adhesion rises at ~5E15 cm^{-2}; saturates at ~2E16 cm^{-2}
	SiO_2 (suprasil)	He 200 keV Ne 280 keV		Peel		Poor adhesion; heating gives ×3
	Teflon	He 200 keV Ne 280 keV		Peel		Substrate damage by 5E15 leads to detachment

Cu	Cr with thin oxide	P 1 MeV	<2E15	Scotch Tape	Bøttiger *et al.*[29]	
		Ne 250 keV	<5E15	Peel		Strong bond
		He 200 keV	~2E16			Poor bond
		e 7 keV	5E17			Weak bond
Au	Teflon	γ 0.2 MeV to 3.0 MeV (1MeV mean)	5E6 rads	Pull	Sofield *et al.*[30]	20Å interface zone; XPS shows ternary complex
		e 240 eV to 3.5 keV	1E16			
Au	Mo with native oxide	Ni 1 MeV	>1E15	Scotch Tape	Pronko *et al.*[31]	No extensive interface mixing
Ag			>4E15			
Cu			>2E15			
Al			>2E15			
Cu	Al_2O_3 clean, water wash, ethanol wash	Ne 280 keV	to 2E16	Peel	Baglin[32]	Interface contaminant effects;
Au	Al_2O_3 clean, water wash, ethanol wash	Ne 280 keV	to 2E16	Peel		Interface contaminant effects
Fe	Soda-lime glass	H 100 keV	Fixed test dose 1E16	Scratch	Battaglin *et al.*[33-35]	Enhancement ×4; Diminished Fe-C bonding
	SiO_2	H 100 keV				Enhancement ×3
	SiO_2 on Si					Enhancement ×3

(*continued*)

TABLE 1 (Continued)

Film	Substrate	Beam	Threshold dose or range of doses (cm^{-2})	Adhesion test	Reference	Comment
Al	Glass					Enhancement ×5
	SiO_2		1E16			Enhancement ×2
Cu	Glass					Negligible effect
	SiO_2					Negligible effect
Au	Glass					No enhancement
	SiO_2					No enhancement
Cu	Mo	O 2 MeV	2E16	Scotch Tape and pin pull	Ingram and Pronko[36]	
		Si 3 MeV	5E15			
		Ni 3 MeV	1E15			
		Au 5 MeV	5E14			
Al	Si	hν 4-6 eV in air	1E14-1E16	Q-tip	Kellock *et al.*[37]	Adhesion enhanced Correlated with interface chemistry (XPS)
ZnS	SiO_2	e 1-3 keV	to 1E17	Scotch Tape Pull	Sofield and Trehan[38]	
		Cl 30 MeV	1E13-5E14	Scotch Tape		

Au	Si	Cl 20 MeV	2E13	Q-tip	Ingemarsson and Tombrello[39]	
	Al_2O_3		2E14	Scotch Tape		
	Fused silicon		1E15	Scotch Tape		
	SiO_2 grown		5E14	Scotch Tape		
Fe	Teflon	S 16 MeV	>1E12	Q-tip, Scotch Tape	Ingemarsson *et al.*[40]	Adhesion enhanced. CEMS shows chemical bonding
	PVC		5E12	Q-tip		Adhesion decreased with dose. Radiation damage?
Au	GaAs	O 100 keV 180 keV	2E13-2E15	Scotch Tape Pull	Bardin *et al.*[41,42]	Adhesion improved only when initial surface contaminated. Otherwise adhesion good as deposited
	GaAs water wash propanol	O 160/200 keV	5E15	Pull		
SiO_2 small particles	Fused silica	He 200 keV	1E14	Scotch Tape Q-tip	Musket *et al.*[43]	Adhesion enhanced
Si	Steel	Ar 120 keV	5E15	Pull Scratch	Ensinger *et al.*[44]	Enhancement > ×10

[a]For a more complete anthology, see References 2 and 3.

energy interface consistent with available chemistry and morphology, itself by definition an interface with enhanced energy of adhesion.

A selection of experimental observations of ion beam stitching is shown in Table 1. The qualitative success of the technique seems to apply to a wide range of materials which have no bulk chemical interactions (e.g., Cu on Al_2O_3) (and which therefore present the most acute problems for interface bonding). Despite the general success of this technique, it should be noted that not all systems responded well. Furthermore, much of the work reported was qualitative, based on a Scotch tape test or an abrasion test, both of which have thresholds at very weak levels of adhesion. Especially in the earlier work reported in Table 1, it is seldom clear whether substrate cleaning procedures would have permitted residual contaminant layers beneath the deposited films. In some more recent work, however, the improved adhesion is probably attributable to new chemical bonding at the interface. As in the case of bulk phases, it is reasonable to assume that such stable bonding will simply not be available in all systems.

"Ion beam mixing" is a term developed to describe the interdiffusion of materials across an interface subjected to energetic irradiation. The process has been used, for example, to achieve low-temperature growth of metal silicide layers, several thousand angstroms thick, starting from a metal film deposited on silicon. Its mechanism is largely one of radiation-assisted diffusion, where the necessary driving force is the heat of reaction of the film and the substrate.(45) No such driving force exists for the bulk penetration of most materials of Table 1, selected for their lack of reactivity. Consequently, the only interface mixing to be expected in these cases would be that of the few ballistically recoiling atoms within one or two monolayers of the interface. Adhesion can therefore not be considered a result of "ion beam mixing" over extended depths. This is in fact borne out by experiment, where only submonolayer transfer of species has been found at adherent $Cu-Al_2O_3$ interfaces after peeling.(16) Nevertheless, the same mechanism that produces ion beam mixing would clearly enable thermodynamically driven local migration of atoms (especially at elevated temperatures) to form an *interface phase,* provided such a stably bonded planar entity exists. In this sense, the term ion beam mixing is

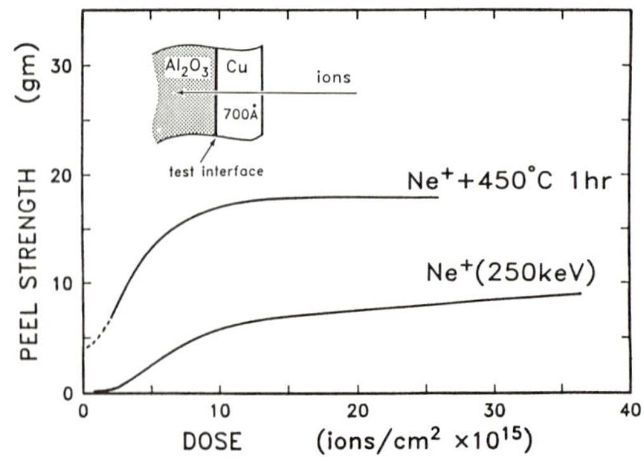

FIGURE 8. Ion beam stitching of Cu(700 Å) on Al_2O_3 (sapphire) substrates. Peel strength is displayed as a function of ion fluence, both before and after heat treatment.(16)

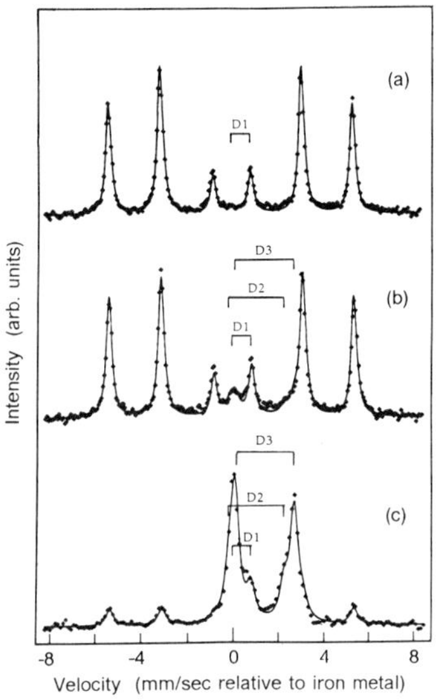

FIGURE 9. Conversion electron Mössbauer spectra from ^{57}Fe at the Fe–Teflon interface: (a) as prepared, (b) after irradiation with $10^{13}/cm^2$ S^{3+} (16 MeV), (c) after irradiation with $10^{14}/cm^2$ S^{3+}. Enhanced adhesion observed for (b) and (c) is correlated with the spectral components D1, D2, D3, believed to correspond to Fe—C (D1) and Fe—F (D3, D2) electronic bonding. (From Reference 46.)

probably correctly applied to the process of adhesion enhancement by interface rebonding.

Some of the mechanisms of ion beam stitching may be understood better from the following experimental examples.

1. Figure 8 displays the dependence of Cu–Al_2O_3 direct bonding upon fluence of 250 keV Ne^+ ions,[16] both before and after heat treatment. Stitching produces thermally stable bonding of 18 g/mm, while negligible attachment of the film would otherwise be obtained. The saturation at high doses implies that a dose of about one ion per interface atom does all that is possible to assist adhesion. At higher doses, a dynamic equilibrium interface structure is probably reached.

2. Figure 9 shows conversion electron Mössbauer spectra for ^{57}Fe deposited at the interface between an Fe film and a substrate of Teflon.[46] As deposited, the iron displays only a set of lines characteristic of α-Fe. However, after irradiation with 16 MeV S^{3+} ions

TABLE 2
Ion Beam Stitching to Overcome Contaminant Layers[a]

Substrate	Ion beam	Dose (ion/cm^2)	Peel strength (g)
Ni, no oxide	None	—	>200.0
Ni + native oxide	None	—	<0.1
Ni + native oxide	He^+, 200 keV	2×10^{14}	10.0–30.0
Ni + native oxide	He^+, 200 keV	2×10^{16}	30.0
Ni + native oxide	Ne^+, 200 keV	2×10^{16}	>200.0

[a] Peel strengths are reported for Pt (700 Å) deposited on Ni surfaces which in some cases had been oxidized.[47]

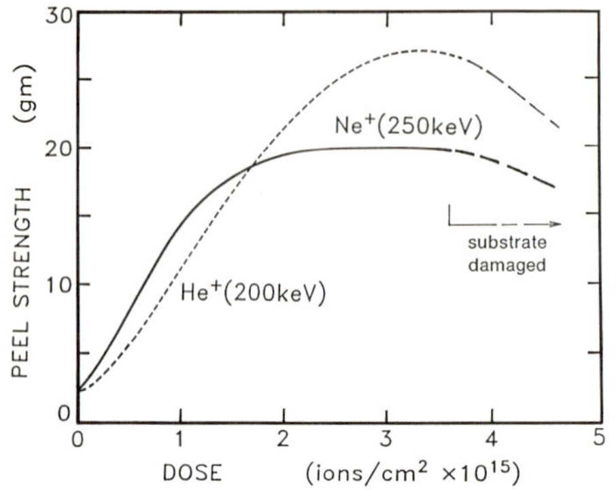

FIGURE 10. Ion beam stitching for Cu(700 Å) on Teflon. Peel strength is shown as a function of ion fluence for both He^+ and Ne^+ ions.[3]

($10^{14}/cm^2$), which substantially improved adhesion, the equivalent spectrum showed dominant new features identified with Fe—C bonding (D1) and Fe—F bonding (D2, D3). The observation implies a strong correlation between interface adhesion and the irradiation-induced formation of new interface chemical bonds.

3. Table 2 displays the peel strength observed for Pt films deposited on pure nickel deliberately coated with "contaminant" oxide layers.[47] The ability of stitching to overcome the oxide barrier and produce a strongly bonded interface is clearly seen. In similar tests, the adhesion of Cu to oxidized Cr was successful for oxide thicknesses up to 25 Å; for oxides thicker than 80 Å, adhesion was no longer achieved at doses of 10^{16} Ne^+/cm^2.

4. Figure 10 shows the adhesion improvement[2] for Cu deposited on Teflon and irradiated with Ne^+ ions at 250 keV or He^+ at 200 keV. Without irradiation, negligible adhesion could be found. At doses of only 2×10^{15} ions/cm², adhesion was raised to more than 20 g/mm. For large doses of ions, radiation damage weakened the polymer and adhesion declined; however, in practice, large doses would be unnecessary.

4.2.2. Pre-deposition Sputtering

By far the most successful and promising ion beam technique for adhesion enhancement has been that of pre-treatment of the substrate surface with a low-energy (500 eV) Ar^+ ion beam, *in situ* directly before vacuum vapor deposition of the film. Such a beam is expected to achieve the following things:

- remove unwanted contaminant layers at the substrate surface,
- randomly break terminating bond structures at the surface of a compound substrate, hopefully leaving active dangling bonds ready to receive the deposited film,
- alter the surface atomic abundances in a compound substrate, in order to present complex bonding sites of optimal composition to the arriving deposited atoms,
- in some instances, e.g., polymer substrates, produce a roughened surface whose structure will toughen the newly formed interface.

The first three expectations are illustrated in Figure 11, which shows schematically a normal surface of sapphire (Al_2O_3), terminated with oxygen atoms and coated with an

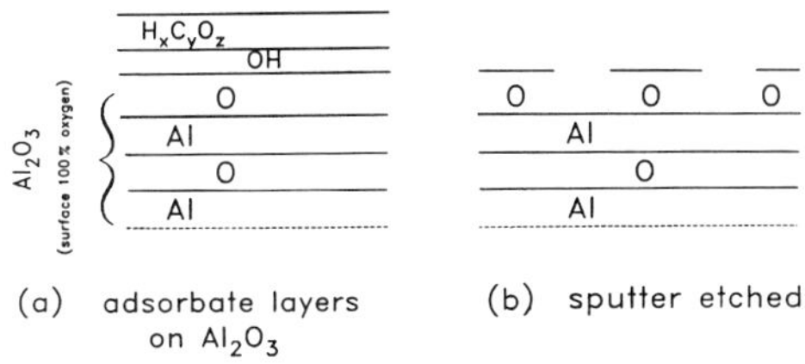

FIGURE 11. Schematic illustration of the anticipated changes in "surface" composition of a compound substrate (Al_2O_3) produced by low-energy ion sputtering. Layers of adsorbed contaminants are removed, and the terminating layer of oxygen atoms is depleted. Such processing *in situ* prior to film deposition offers some control for optimizing interface composition.

adsorbed OH layer resulting from a well-intentioned wash in clean water. Vapor-deposited Cu will not adhere well on either the hydrated or the oxygen terminated surfaces. Sputtering by the 500 eV Ar^+ beam from a Kaufman source will readily dispose of the OH layer and begin to erode the layer of oxygen atoms. At this point, some sites (with dangling bonds) will become available for arriving Cu atoms to lodge and bond in ternary Al—Cu—O configurations, provided that such configurations are stable. Continued sputtering will enrich the surface in Al up to some preferential-sputtering limit determined by, among other factors, the relative masses of Ar and substrate atom species, and the relative binding energies of Al and O in the sapphire surface.[48,49] At some point in the sputtering process, it is to be hoped that the average surface Al:O ratio will conform with that required for a stable ternary Al—O—Cu interface phase. By stopping sputtering at that moment, and promptly depositing Cu, the optimum receptive population of the substrate surface may be obtained. Post-deposition heat treatment should then enable the formation of the ordered interface phase, binding together the film and substrate.

In accord with these expectations, it was indeed found[9] that an optimum sputtering exposure existed, at which point the peel strength (following anneal) for Cu on sapphire was observed to be 120 g/mm, indicating a very strong and thermally stable direct bond (Figure 12). By repetition of the identical process up to submonolayer deposition of Cu in UHV, it was established by XPS that a new stable complex interface environment, probably involving Cu, O, and Al, could indeed be associated with the optimum adhesion. The corresponding spectra (Figure 13) clearly show a large Cu Auger peak forming, which is not consistent with simple binary phases of the Cu—O—Al system, and is believed to represent a ternary interface phase.

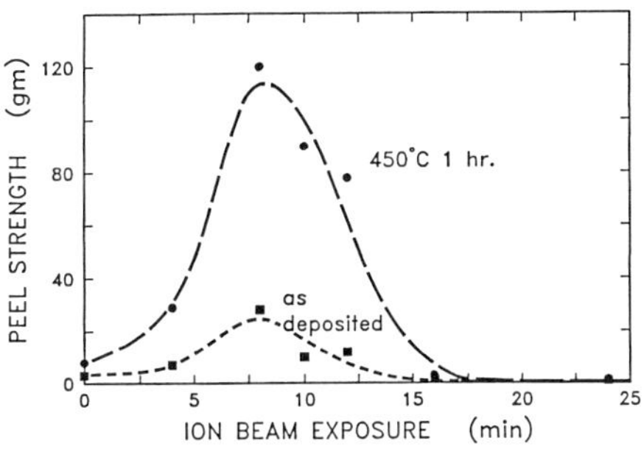

FIGURE 12. Adhesion of Cu on sapphire resulting from exposure of the substrate to 500 eV Ar^+ ions *in situ* before deposition of the Cu film. Subsequently, the sample was heated at 450 °C for 1 hour in a helium furnace. The existence of an optimal exposure to sputtering implies that a preferred interface composition was reached, corresponding perhaps to the Al:O ratio required for a stable ternary Al—Cu—O interface phase. (From Reference 9.)

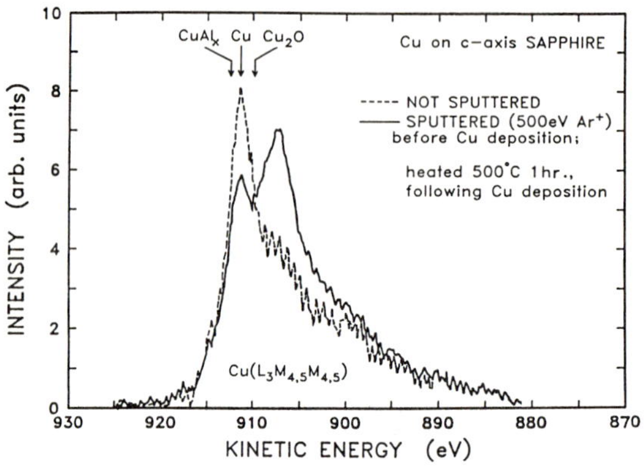

FIGURE 13. Electron spectra corresponding to $Cu(L_3M_{4,5}M_{4,5})$ Auger lines from Cu deposited thinly on *c*-axis sapphire and annealed for 1 hour at 450 °C. Dashed line: no pre-sputtering of substrate. Solid line: substrate exposed 6 min to 500 eV Ar^+ sputtering prior to deposition of copper. The new broad peak is believed to indicate a stable ternary Al—Cu—O interface environment. (From Reference 9.)

Similar success has been reported for the adhesion of Cu on Teflon.(50) Figure 14 displays the rise of peel strength with ion beam exposure time up to approximately 20 sec (at $50\mu A/cm^2$), after which the adhesion was strong and constant at 75 g/mm. Matching XPS studies displayed a new peak in the $Cu(2p_{3/2})$ spectrum believed to represent Cu—F bonding; however, Cu—C bonding was not evident. It is not easy to see why simple Cu—F bonding would produce interface adhesion, although it is a clear sign that the polymer surface was disturbed and Cu was indeed probably bonded at activated sites.

In addition, the Teflon surface was extensively roughened by the ion beam, raising the possibility that interface roughness was a key consideration in the successful adhesion. Probably both chemistry and roughness are in fact responsible.

A similar approach has been reported by Loh *et al.*(51) in which deposition of a metal film takes place simultaneously with 400 eV Ar^+ ion bombardment of the receiving surface. Termed "ion beam enhanced deposition," the process was qualitatively successful in bonding Au, Ag, or Cu to Teflon, where it seemed to be related to interface intermixing or roughening.

These examples serve to illustrate the power of low-energy ion beam techniques, especially for complex substrates. The broader application of this method to custom-tailored interface construction suggests intriguing possibilities for future development.

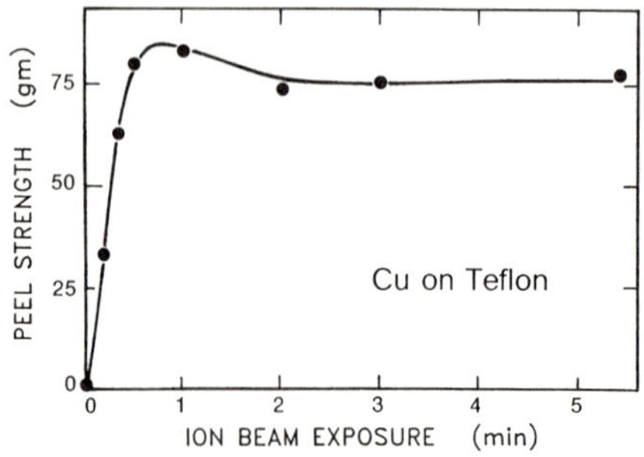

FIGURE 14. Adhesion of Cu deposited on Teflon immediately following exposure of the Teflon to 500 eV Ar^+ sputtering. (From Reference 50.)

5. SUMMARY

Recent advances in understanding the nature of thin film adhesion have highlighted the need to consider not only interface chemistry, but also the elastic behavior of joining materials, in order to create optimal bonding performance. New techniques now can be employed to modify and tailor not only the chemical structure, but also the morphology of interfaces in order to exploit the benefits of interface fracture toughening to the full.

An unresolved issue is the concept of the stable "interface phase," a joining layer or layers of atoms whose structure is thermodynamically stable when they are bounded, as they are, by the materials of film and substrate. It would be helpful to understand the stability criteria for such phases in order to predict the feasibility of tailoring direct adhesion in new systems of technological interest.

Interface tailoring by means of selective ion sputtering of substrates prior to film deposition appears to offer broad possibilities for future technology. It would seem to be a natural extension of the present applications of ion-assisted deposition, in which the composition of deposited compound phases can also be selectively engineered.

REFERENCES

1. M. D. Thouless, *Mater. Res. Soc. Symp. Proc.* **119,** 51 (1988).
2. J. E. E. Baglin, Ion beam effects on thin film adhesion, in: *Ion Beam Modification of Insulators* (P. Mazzoldi and G. W. Arnold, eds.), Chapter 15, Elsevier, Amsterdam (1987).
3. J. E. E. Baglin, Interface structure and thin film adhesion, in: *Ion Beam Deposition, Film Modification and Synthesis* (J. J. Cuomo, S. M. Rossnagel, and H. R. Kaufman, eds.), Noyes Publications, Park Ridge, NJ (1988).
4. L. E. Murr, *Interfacial Phenomena in Metals and Alloys,* Addison-Wesley, Reading, Mass. (1975).
5. A. M. Stoneham and P. W. Tasker, *J. Phys. C.* **18,** L543 (1985).
6. J. Ferrante and J. R. Smith, *Phys. Rev.* **B31,** 3427 (1985).
7. J. Ferrante, G. Bozzolo, C. W. Finley, and A. Banerjea, *Mater. Res. Soc. Symp. Proc.* **119,** 3 (1988).
8. L. Senbetu, J. G. Pronko, and T. T. Bardin, *Mater. Res. Symp. Proc.* **119,** 75 (1988).
9. J. E. E. Baglin, A. G. Schrott, R. D. Thompson, K. N. Tu, and A. Segmüller, *Nucl. Instrum. Methods* **B19/20,** 782 (1987).
10. E. Klokholm, IBM Internal Report No. RC9925, IBM—Research, Yorktown Heights, New York (April 1983).
11. J. J. Prescott, *Applied Elasticity,* p. 187, Dover, New York (1961).
12. G. I. Barenblatt, *Adv. Appl. Mech.* **7,** 55 (1962).
13. P. B. Madakson and J. E. E. Baglin, *Mater. Res. Soc. Symp. Proc.* **93,** 41 (1987).
14. P. Y. Hou and J. Stringer, *Mater. Res. Soc. Symp. Proc.* **119,** 205 (1988).
15. D. M. Mattox, in: *Adhesion Measurement of Thin Films, Thick Films and Bulk Coatings* (K. L. Mittal, ed.), NASTM Special Technical Publication 640, American Society for Testing and Materials, Philadelphia (1978).
16. J. E. E. Baglin and G. J. Clark, *Nucl. Instrum. Methods* **B7/8,** 881 (1985).
17. K. S. Kim, Elasto-Plastic Analysis of the Peel Test, University of Illinois Report No. UILU-ENG 85-6003 (March 1985); also *J. Eng. Mater. Technol.* **86-WA/EEP-3,** 10 (1986) and *Mater. Res. Soc. Symp. Proc.* **119,** 31 (1988).
18. J. Ahn, K. L. Mittal, and R. H. MacQueen, in Reference 15, p. 134.
19. J. W. Beams, J. Breazeale, and W. L. Bart, *Phys. Rev.* **100,** 1657 (1955).
20. S. Krongelb, in Reference 15, p. 107.
21. M. J. Matthewson, *Appl. Phys. Lett.* **49,** 1426 (1986).

22. J. A. Hinkley, *J. Adhes.* **16,** 115 (1983).
23. J. M. E. Harper, J. J. Cuomo, and H. R. Kaufman, *J. Vac. Sci. Technol.* **21,** 737 (1982).
24. L. E. Collins, J. G. Perkins, and P. T. Stroud, *Thin Solid Films* **4,** 41 (1969).
25. S. Jacobson, B. Jonsson, and B. Sundqvist, *Thin Solid Films* **107,** 89 (1983).
26. B. T. Werner, T. Vreeland, Jr., M. H. Mendenhall, Y. Qiu, and T. A. Tombrello, *Thin Solid Films* **104,** 163 (1983).
27. J. E. Griffith, Y. Qiu, and T. A. Tombrello, *Nucl. Instrum. Methods* **198,** 607 (1982).
28. J. E. E. Baglin, G. J. Clark, and J. Bøttiger, *Mater. Res. Soc. Symp. Proc.* **25,** 179 (1984).
29. J. Bøttiger, J. E. E. Baglin, V. Brusic, G. J. Clark, and D. Anfiteatro, *Mater. Res. Soc. Symp. Proc.* **25,** 203 (1984).
30. C. J. Sofield, C. J. Woods, C. Wild, J. C. Riviere, and L. S. Welch, *Mater. Res. Soc. Symp. Proc.* **25,** 197 (1984).
31. P. P. Pronko, A. W. McCormick, D. C. Ingram, A. K. Rai, J. A. Woolam, B. R. Appleton, and D. B. Poker, *Mater. Res. Soc. Symp. Proc.* **27,** 559 (1984).
32. J. E. E. Baglin, *Mater. Res. Soc. Symp. Proc.* **47,** 3 (1985).
33. G. Battaglin, M. Carbucicchio, R. Dal Maschio, F. Marchetti, P. Mazzoldi, and A. Valenti, XIV International Congress on Glass, New Delhi, India (March 1986).
34. G. Battaglin, P. Mazzoldi, and R. Dal Maschio, in: *Induced Defects in Insulators* (P. Mazzoldi, ed.), p. 235, Les Editions de Physique, Le Ulis Cedex (France) (1984).
35. M. Carbucicchio, A. Valenti, G. Battaglin, P. Mazzoldi, and R. Dal Maschio, *Radiat. Eff.* **98,** 21 (1986).
36. D. C. Ingram and P. P. Pronko, *Nucl. Instrum. Methods* **B13,** 462 (1986).
37. A. J. Kellock, J. Liesegang, G. L. Nyberg, and J. S. Williams, *Mater. Res. Soc. Symp. Proc.* **75,** 179 (1987).
38. C. J. Sofield and P. N. Trehan, *Radiat. Eff.* **98,** 35 (1986).
39. P. A. Ingemarsson and T. A. Tombrello, *Mater. Res. Soc. Symp. Proc.* **119,** 103 (1988).
40. P. A. Ingemarsson, T. Ericsson, A. Gustavsson-Seidel, G. Possnert, B. U. R. Sundqvist, and R. Wäppling, Proc. 12th International Conference of Hosei University on "Application of Ion Beams in Material Science," Tokyo (September 1987).
41. T. T. Bardin, J. G. Pronko, and D. K. Kinell, *Mater. Res. Soc. Symp. Proc.* **77,** 731 (1987).
42. T. T. Bardin, J. G. Pronko, L. Senbetu, and D. A. Kozak, *Mater. Res. Soc. Symp. Proc.* **119,** 147 (1988).
43. R. G. Musket, I. M. Thomas, and J. G. Wilder, *Appl. Phys. Lett.* **52,** 410 (1988).
44. W. Ensinger, M. Barth, and G. K. Wolf, *Nucl. Instrum. Methods* **B32,** 104 (1988).
45. W. L. Johnson, Y. T. Cheng, M. Van Rossum, and M.-A. Nicolet, *Nucl. Instrum. Methods* **B7/8,** 657 (1985).
46. P. A. Ingemarsson, T. Ericsson, R. Wäppling and G. Possnert, *J. Adhes. Sci. Technol.* **3,** 503 (1989).
47. D. K. Sood, Royal Melbourne Institute of Technology (MRC), Melbourne, Australia, private communication.
48. R. Kelly, in: *Physics and Chemistry of Solid Surfaces,* Vol. 5 (R. Vanselow and R. Howe, eds.), Chapter 7, p. 159, Springer-Verlag, Heidelberg (1984).
49. P. Varga and E. Taglauer, *J. Nucl. Mater.* **111/112,** 726 (1982).
50. Chin-An Chang, J. E. E. Baglin, A. G. Schrott, and K. C. Lin, *Appl. Phys. Lett.* **51,** 103 (1987).
51. Ih-Houng Loh, J. K. Hirvonen, J. R. Martin, P. Revesz, and C. Boyd, *Mater. Res. Soc. Symp. Proc.* **108,** 241 (1988).

14

Chemistry, Microstructure, and Adhesion of Metal–Polymer Interfaces

Paul S. Ho, Richard Haight, Robert C. White, B. D. Silverman, and F. Faupel

1. INTRODUCTION

In microelectronics, the large-scale integration of devices necessitates the use of multilayered metallization structures on the chip level and for packaging. Such structures are usually fabricated using alternate metal and insulating layers. This incorporates metal–insulator interfaces throughout the structure which, depending on the configuration and dimension of the devices, can have varying degree of complexities. These interfaces have to be designed with certain functional characteristics, particularly good chemical and adhesion properties. For this purpose, it is important to understand these basic properties of the metal–insulator interface.

Among the insulators, ceramics and polymers are the most commonly used. Comparing these two classes of materials, the ceramics have the advantage of high mechanical strength and thermal stability. For advanced packaging applications, the ceramic materials have certain undesirable characteristics, including high dielectric constant (about 6–8) and the difficulty in forming line patterns with small dimensions, e.g., less than about 10 microns. These factors restrict the speed and density of the packaging structure. In contrast, polymers generally have low dielectric constants of about 3 to 4 and can be processed by lithographic techniques to yield patterns with dimensions in the micron range, making them well suited for packaging applications.

The current trend in microelectronic packaging is to incorporate polymer layers into ceramic structures to make optimum use of the properties of these two classes of materials. An example is the metal–ceramic–polymer module shown in Figure 1, where PMDA-ODA (pyromellitic dianhydride-oxydianiline) polyimide, a high-temperature polymer, is incorporated on top of the ceramic substrate.[1] The use of the polyimide layer provides an additional wiring level which, with proper design, can enhance substantially

---

Paul S. Ho, Richard Haight, Robert C. White, B. D. Silverman, and F. Faupel • IBM Thomas J. Watson Research Center, Yorktown Heights, New York 10598. *Present address of Robert C. White*: Department of Electrical Engineering, Columbia University, New York, New York 10027. *Present address of F. Faupel*: Institut für Metall-physik der Universität Göttingen, D-3400 Göttingen, Federal Republic of Germany.

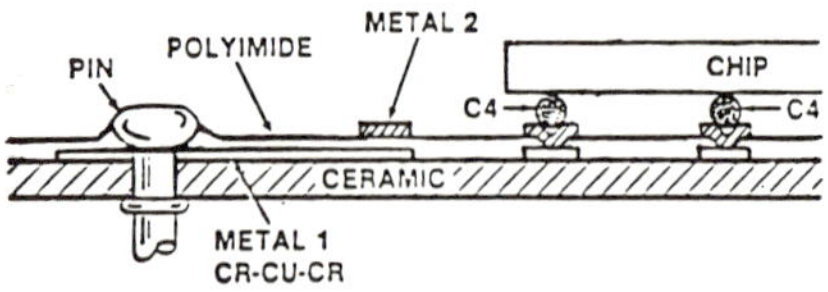

FIGURE 1. Schematic diagram of the structure of a metal–ceramic–polymer module used in microelectronic packaging.

the packaging performance by improving the electrical performance and wiring distribution. A similar approach has been used to incorporate several polyimide wiring layers into a multilayered ceramic structure for high-end packaging of mainframe computers.[2] One fundamental concern of such structures is the structural integrity, particularly when they are subject to heat treatment or during thermal cycling. The thermal process gives rise to thermal stresses because of the differential thermal expansions of metal, polymer, and ceramic substrate. Depending on the material combination, the stress can reach a very high level, exceeding the yield stress of the metal in some cases. To maintain the integrity of the structure under such circumstances, strong adhesion at the interface is needed.

The adhesion strength of an interface formed by bringing into contact two materials α and β can be defined by the work of adhesion. This is the work required to separate reversibly the interface to infinity per unit area, i.e.,

$$W_a = \gamma_\alpha + \gamma_\beta - \gamma_{\alpha\beta} \tag{1}$$

where W_a is the work of adhesion or the adhesion energy, γ_α and γ_β the surface tensions of α and β phases, respectively, while $\gamma_{\alpha\beta}$ is the interfacial tension between α and β. The parameter $\gamma_{\alpha\beta}$ is determined by two factors: the intrinsic characteristics of atomic or molecular bonding at the interface and the structural morphology of intermixing of the two materials.[3] The latter can be very important; for example, two polymers with relatively weak molecular bonding can form a strong polymer–polymer interface by promoting the interdiffusion, i.e., interpenetration of polymer segments, at the interface.

Depending on the combination of the materials forming the interface, the nature of the bonding can be significantly different. For example, it can vary from a strong and directional covalent bond in semiconductors to a weak van der Waals bond in molecular solids.[3] In this regard, the metal–polymer interface has distinct and interesting characteristics. While the metal is characterized by an ordered atomic structure with close packing density, the polymer has a disordered, loose, and interwound molecular structure. The bonding characteristics of such a structural combination are difficult to predict, *a priori*. In spite of its wide range of applications, the metal–polymer bonding has not been investigated until recently[4–7] (see also Chapter 1 by Lee).

This chapter reviews the results of the recent investigations on properties pertinent to the mechanical behavior of metal–polymer interfaces formed on PMDA-ODA polyimide. This polyimide is chosen in the discussion because it is commonly used in microelectronics and packaging, and the information about its interface is the most complete at this time. One key finding in the study of the metal–polyimide interface is its reactive nature which was manifested by material reactions at the interface, such as intermixing and cluster formation.[6,8] The extent of these reactions is largely determined by the nature of the interfacial chemistry and can affect significantly the structure and morphology of the interface. This material aspect of the interface and its correlation to chemistry are important for understanding the mechanical property of the interface, particularly for polymers designed for high-temperature applications.

The discussion first focuses on the studies on the characteristics of bonding between metal and the polyimide, and then on the interfacial structure and adhesion and deformation of the metal/polymer interface. The bonding characteristics have been investigated using surface spectroscopy techniques in combination with theoretical molecular orbital (MO) calculations.[6,7,9] The interfacial structure was examined by microanalytical techniques of transmission electron microscopy (TEM) and medium energy ion scattering (MEIS) in combination with Monte Carlo simulation. The adhesion and deformation characteristics were studied using the stretch deformation method. Due to the emerging interest and the increasing activities in studying metal–polymer interfaces, this chapter is not intended to be a comprehensive review, instead its discussion is focused on Cr and Cu, two metals with contrasting strong and weak interactions with the polyimide. The emphasis is placed on the current understanding of the microscopic nature of the metal–polyimide interface. For other polymer interfaces, one can refer to other review articles and conference proceedings.[10–13]

2. CHEMICAL BONDING AND ELECTRONIC STRUCTURE

2.1. Electronic Structure of Polyimide Surface

UV photoemission (UPS) and X-ray photoemission (XPS) spectroscopies were used to probe the electronic structure of the interface. In UPS, UV photons, which are generated by gas excitation or from a synchrotron source, with energy of about 100 eV or less are used to excite the valence electrons. The electrons with sufficient energy to escape from the surface can be detected and energy-analyzed by a spectrometer. The results give an energy distribution of the density of the valence states. Since the valence states are directly involved in the formation of the chemical bond, UPS seems to be ideally suited for probing the bonding characteristics. For metal–polymer interface studies, this is complicated by the fact that the polymer molecule contains various atomic elements. The energy levels of the valence states of these atoms or their complexes are close together and usually overlap in the energy range of the measurement. This makes it difficult to identify the specific complex or bonding site responsible for bond formation. In XPS, electron excitations are usually made by X-ray photons of energy exceeding 1000 eV. Thus, XPS measures the core states lying several hundred eV below the valence states. Unlike UPS, the core levels detected by XPS are well separated for the various atomic elements in the polymer molecule. Although those core states are not directly involved in the formation of chemical bonds, their energy levels and spectral features are affected due to bond formation. The atomic specificity of XPS is useful for determining the role of different atomic species in bond formation at the interface.

Both these spectroscopy techniques have very high surface sensitivity since they measure only those electrons within the escape distance of the surface, which is of the order of 30–40 Å. This restricts the probing depth of an interface to about the escape distance. As a result, most of the spectroscopy studies were carried out on interfaces with only few monoatomic layers of metal coverage. The surface sensitivity of these studies enables one to deduce the morphology of the metal coverage by measuring the intensity variation of the metal and the substrate as a function of metal coverage. The spectroscopy experiments were carried out during metal deposition on the polymer. By observing the changes of the spectral features during metal deposition, the spectroscopy techniques enable one to monitor the evolution of the valence and core states during the course of interface formation. Molecular orbital (MO) calculations were carried out to derive the electronic structure of various metal–polymer complexes. By combining the spectroscopy results with MO calculations, the nature of the chemical bonds at the interface can be deduced.

The experiments were carried out in a multitechnique ultrahigh vacuum chamber equipped with hemispherical electron spectrometer for UPS and XPS measurements. The vacuum chamber has incorporated facilities for ion sputtering, mass spectroscopy, gas exposure, and rapid sample introduction. In addition, it was connected to an evaporation chamber which was equipped with separate pumping and vacuum isolation using a gate valve. This arrangement allowed a clean *in situ* evaporation of metal while carrying out spectroscopy measurements without changing the sample position.

The samples were fabricated using Dupont 5878 polyimide. To insure a smooth substrate surface for the polyimide overlayer, a polished Si wafer was used with thin Al layers deposited on both sides. The polyimide layer of 100–300 Å was first spun onto the Al layer, then heated to 85 °C to drive off the n-methyl pyrrolidone solvent. This was followed by curing at 350 °C for 30 minutes in inert gas to complete the imidization reaction for formation of the polyimide molecular structure. This sample configuration provided a conducting Al layer under the polyimide for reducing electrical charging and a separate Al layer on the bottom for sample heating. All X-ray core level spectra were collected at room temperature in a grazing emission geometry (emission angle 10° with respect to the surface) in order to maximize surface sensitivity.

In order to use spectroscopy techniques to study the interface, the bonding characteristics of the polyimide surface were first investigated.[14] Since polymer surfaces are susceptible to damage under the common cleaning techniques, such as ion sputtering, considerable effort has been spent to establish a procedure to prepare a clean and intrinsic surface. For our samples, annealing was found to be most effective in obtaining a reproducible surface. As shown in Figure 2, the three as-prepared samples show subtle differences in surface chemistry which can be detected by UPS. For these UPS measurements, photons of 21.2 eV energy generated by a helium resonance lamp were used. Upon annealing at 360 °C for about $\frac{1}{2}$ h, the differences disappeared and the spectra converged to one typical of the polyimide surface, as represented by curve d in Figure 2. The nature of this annealing treatment was investigated by carrying out a series of annealing experiments on as-cured polyimide as a function of temperature. Evolution of water was detected and the corresponding changes in the spectral features showed that water desorption from the polyimide surface accounts for most of the changes observed during annealing.

To verify that the UPS spectrum in curve d of Figure 2 indeed represents the valence density of states (DOS) of polyimide, theoretical calculations of the molecular orbital structure of polyimide were carried out. Quantum chemical programs are readily available to carry out self-consistent-field molecular orbital (SCF MO) calculations (i.e., *ab initio*) of organic molecules of limited size. Such a program, GAUSSIAN-80,[15] was chosen to study the molecular orbitals of polyimide.[9,16]

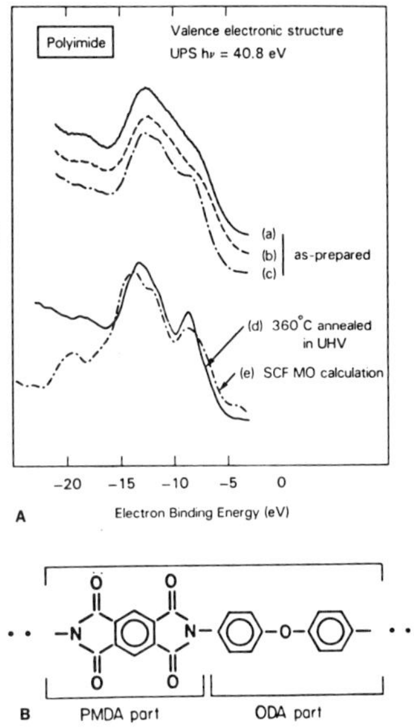

FIGURE 2. A. (a)–(c) UPS spectra from three different polyimide surfaces in the as-prepared state, (d) UPS spectrum after 360 °C *in-situ* annealing in UHV, (e) valence density of states from SCF MO calculations, broadened with 2.0 eV FWHM for comparison with experiment. B. Basic unit of the molecular structure of polyimide.

The fundamental unit (building block) of the polyimide chain is shown in Figure 2B. It consists of two parts: (1) a PMDA part which contains four C═O bonds, a benzene ring, and two N atoms; and (2) an ODA part which consists of two benzene rings connected through the ether oxygen. The calculations yield molecular orbitals whose energy distribution represents the DOS of the molecule, including both valence and core states. The theoretical results can be compared to the observed spectra by integrating the energy distribution of the DOS with a proper instrumental broadening of the spectrometer. The agreement between the measured (curve d) and calculated (curve e) valence DOS in Figure 2 is remarkably good. The main peaks near −10 and −14 eV are well reproduced by the calculation, and the shoulder near −13 eV also appears.

This agreement indicates that the 360 °C *in situ* annealing yields a UPS spectrum characteristic of intrinsic polyimide. It also permits us to identify the molecular orbital origin of the spectral peaks. The −10 eV peak consists of three types of carbon orbitals: those associated with the benzene rings, i.e., the C(2p) orbitals lying perpendicular to the molecule, the N(2p) lone-pair orbitals (also perpendicular to the molecule), and O(2p) lone-pair orbitals lying perpendicular to the molecule for the ODA part and along the molecular axis for the PMDA part. The peak near −14 eV is composed of the C—C σ bonds, C—O orbitals perpendicular to the molecular plane, and C—H π bonds. The peak structure between 17–20 eV is C(2s). In essence, the −14 eV peak represents valence states bounded together to form the backbone structure of the polymer, while the −10 eV peak contains the unbounded valence states, which are available to interact with the electronic states of the deposited metal atoms. As seen from subsequent discussions on interfaces, these valence states are primarily responsible for forming the chemical bonds at the interface.

Among the core spectra observed by XPS, the C(1s) spectrum provides the most useful information, so it is relied upon in probing the effect of metal interaction on the

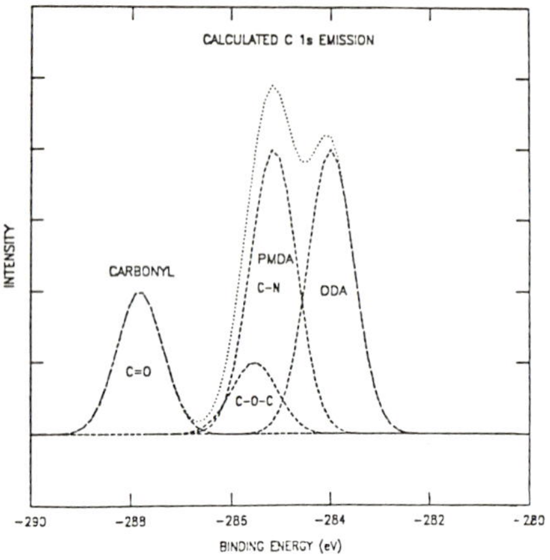

FIGURE 3. XPS C(1s) spectrum calculated for the clean PMDA–ODA surface. The deconvoluted contributions reflect the different chemical environment of the carbon atoms.

polymer chemistry. As shown in Figure 3, it consists of four peaks reflecting the different chemical environments of the carbon atoms in the PMDA–ODA polymer repeat unit (Figure 2B). These include one peak at −288 eV for the carbonyl carbon atoms, one at −285 eV for the aromatic carbons of the six-member rings in PMDA as well as carbon atoms bonded to nitrogen atoms, one at −286 eV for the carbon atoms bonded to the ether oxygen atom, and one at −284 eV for certain aromatic carbon atoms in the phenyl rings of ODA. The core level spectrum of oxygen exhibits only two peaks derived from the carbonyl and ether oxygens. The spectrum of nitrogen is simpler, consisting of only one peak associated with its single chemical environment in the repeat unit.

The electronic structure of the interface has been investigated for a number of metals, including Cu,[5,7,17,18] Ni,[5,19] Al,[20,21] Ti,[22,23] Cr,[5,18,24,25] Co,[26] Ag,[5] Au,[27] and Ge.[28] In the following two sections, results are discussed for Cr and Co to illustrate two contrasting interfaces of strong and weak bonding characteristics.

2.2. Chromium–Polymide Interface

The spectral series in Figure 4 illustrates the evolution of the C(1s) core states upon Cr deposition. The result indicates that for this strongly interacting metal, reaction occurs immediately upon deposition. With a coverage of only 0.12 Å Cr, the intensity of the carbonyl peak is reduced by about 50% and accompanied by a decrease in the intensity on the high binding energy side of the doublet associated with the PMDA unit. For further depositions of 0.2 Å and 1 Å coverages, the rapid reduction in the carbonyl and PMDA peak intensities continues. With 1 Å coverage, there is a discernible shift of the overall spectrum toward lower binding energy as well as a continued reduction in the emission intensities from the carbonyl and the PMDA fragment. This trend continues for further deposition of Cr. At coverages >2 Å we note a significant change in the C(1s) spectrum. The signal originating from the aromatic carbons in the ODA fragment (−284 eV) shifts by about 1.8 eV to lower binding energy, first exhibiting a shoulder and then a peak. This result can be explained as a consequence of continued overlayer formation where coverages above 2–3 Å result in Cr atoms located atop the phenyl rings of the ODA fragment

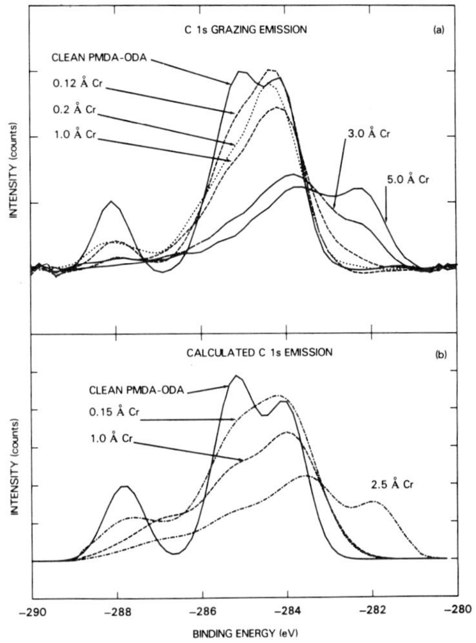

FIGURE 4. (a) C(1s) series with increasing Cr coverage. (b) Calculated C(1s) spectra with increasing Cr coverage. For 0.15 and 1.0 Å coverages, Cr atoms were distributed over the five- and six-member rings of PMDA. For the 2.5 Å coverage additional Cr has been placed atop the phenyl rings of ODA.

after saturating the interaction with the PMDA fragment. Such a matrixing effect results in interactions which give rise to the observed peaks.

Valence photoemission corroborates the studies described above. Figure 5 displays the changes in the valence spectra with increasing Cr coverage. Even with a coverage of only 0.24 Å a drastic decrease in the −10 eV peak can be observed, indicating the initial formation of chemical bonds between Cr and the unsaturated valence states of the PMDA

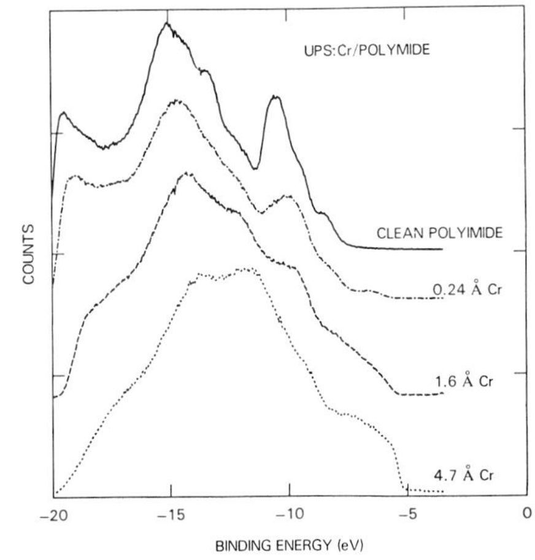

FIGURE 5. UPS spectra for increasing Cr coverages using 21.2 eV radiation.

fragment while the valence structure near −14 eV of the backbone bonding states is largely unaffected. Increasing coverage results in greater reductions of this peak with increasing contributions from emission from the d levels of the metal overlayer. A rigid shift in the spectra toward lower binding energy is also detected. Since Cr forms a charge transfer complex with the underlying polymer, a dipole can be expected to form which shifts the kinetic energy of the photoemitted electrons. This effect is observed in both core level as well as valence spectra with a shift toward lower binding energy of 1.2 eV saturating above 3–4 Å coverage.

The spectroscopy results were interpreted using electronic structures derived from molecular orbital calculations for various configurations of the Cr–polyimide complex.[6] Since chromium is at the beginning of the transition metal series and is essentially 3d electron poor, it is energetically favorable to be found at sites of high coordination. As an initial probe into the interaction of a single Cr atom with the polymer, calculations were performed for a number of different complexes involving one chromium atom interacting with the PMDA or the ODA unit. The most stable complex formed was the one with the Cr atom sitting above the central six-member ring, i.e., at the position of the highest coordination. Another stable complex with PMDA was found with Cr above the center of the five-member ring. The ODA part yields only one such site for complex formation with Cr, namely, the one above each of the two phenyl rings.

For the most stable Cr–PMDA complex, the lowest unoccupied a_2 π-molecular orbital (LUMO) interacts with the $3d_{xy}$ Cr orbital which is filled and lying in a plane parallel to that of the molecule.[9] These orbitals are shown at the top of Figure 6a. In the calculation, the distance of the Cr atom above the molecular plane was varied to determine the minimum energy configuration. The resultant bonding orbital is shown at the bottom

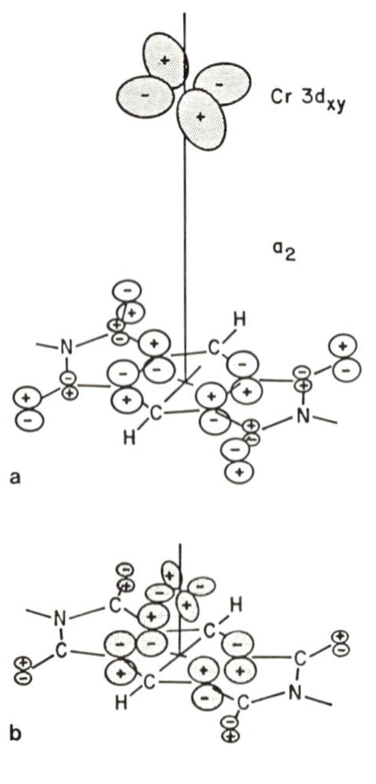

FIGURE 6. Lowest unoccupied level of the PMDA fragment and Cr $3d_{xy}$ orbital (occupied). The orbitals are shown at a large separation (noninteracting). (b) Occupied orbital of the complex that occurs as a result of the interaction of the unoccupied a_2 orbital and filled Cr $3d_{xy}$ orbital. For all orbitals, the opposite phase is indicated by + or − and the filled and empty orbitals are shaded and unshaded, respectively.

of Figure 6. In this complex, the charge transferred from the Cr atom resides mainly at four of the C atoms composing the central benzene ring; two of the C atoms on this ring receive no charge since these C atoms had essentially zero LUMO amplitude. There is also some small amount of charge transfer to the carbonyl C atoms. This transfer of charge shifts the carbonyl C(1s) levels to lower binding energy as well as causes an intensity reduction in the −288 eV peak. These changes in electronic states can explain the spectral features of the C(1s) observed with the initial coverage of Cr., i.e., a rapid reduction in the intensities and shift in energy of the carbonyl and PMDA peaks.

Similarly, when the chromium atom is placed above one of the five-member rings, there is charge transfer with a resultant shift of the C(1s) core levels to lower binding energy. Since the chromium atom is not equidistant from all of the carbonyl groups, the C(1s) carbonyl levels split into inequivalent pairs causing an overall broadening of the spectrum. Such an asymmetry in the charge transfer results in shifts of all the C(1s) levels associated with the carbon atoms on the central ring.

Results of the molecular orbital calculations can be evaluated by calculating the C(1s) spectra as a function of Cr coverage. Coverages are simulated with appropriate combinations of clean polyimide and Cr–PMDA and Cr–ODA complexes. As shown in Figure 4b, the calculated spectra provide a satisfactory account of the observed results, not only for the peak shifts but also for the overall spectral features.

A model of the interaction of Cr with PMDA–ODA can now be constructed. Cr initially bonds in highly coordinated sites on the PMDA fragment forming a Cr–polymer complex. Increased coverage results in saturation of such energetically attractive sites and forces Cr interaction with the phenyl rings of the ODA fragment. This occurs as the overlayer thickness approaches 1 monolayer. Central to the formation of the Cr–polymer complex is the charge transfer from the deposited Cr atom to the lowest unoccupied molecular orbital of the repeat unit of the monomer. The relative stability of these complexes depends on the site coordination. For Cr, the most stable complex is the one with the highest coordination on PMDA. The manner in which charge is transferred from Cr to the various atoms in the polymer repeat unit indicates that the chemical bond between Cr and polyimide is delocalized in nature. A similar mechanism of delocalized bonding can be applied to interpret the spectroscopic results of other metals (see the next section for Cu).

2.3. Copper–Polyimide Interface

With respect to the chemical reactivity, Cu represents a contrasting comparison with Cr because of its filled d-electron valence states. Core level studies[18,19] with increasing Cu coverage produced minimal changes in the observed spectra (Figure 7a). Even at coverages of up to 3 Å the carbonyl signal persists, being reduced only 40% relative to the clean spectrum. The intensity of the −285 eV peak decreases more rapidly than that at −284 eV, suggesting that Cu interacts primarily with the PMDA fragment. This is similar to Cr except that the reduction in peak intensity is significantly smaller at comparable coverages, indicating a weaker chemical interaction at the interface. Since Cu is more electron-rich than Cr, it is not expected to occupy high coordination sites. Such chemistry weakens the interaction with the underlying polymer and, consequently, incomplete coverage due to island formation is observed to occur.[6] A similar effect of chemical reac-

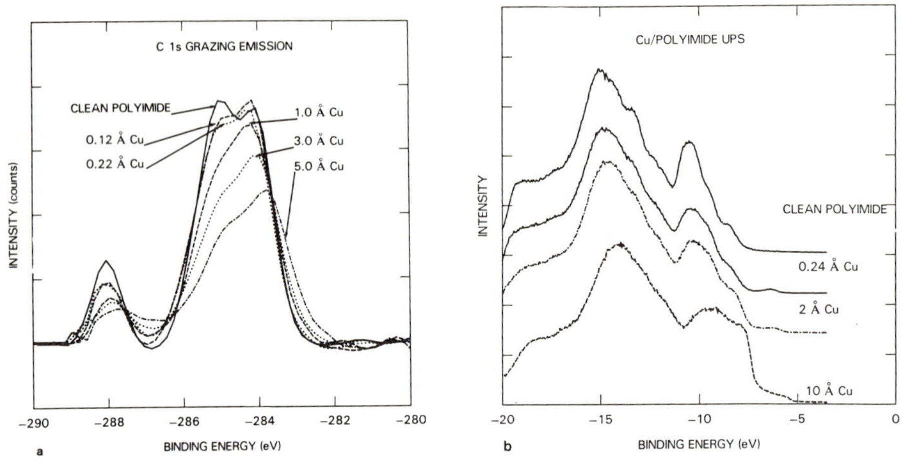

FIGURE 7. (a) C(1s) spectra for increasing Cu coverage. (b) UPS spectra for increasing Cu coverage.

tivity on the interfacial structure has been found to be important for all the metal–polyimide interfaces studied so far. This subject will be discussed further in Section 3.

These observations are consistent with the valence spectra of the Cu-covered polymer (Figure 7b). Although initial Cu interaction with PMDA was observed, as evidenced by the decrease of the −10 eV peak, this effect saturates. Even above 2 Å coverage the −10 eV peak is still observable, while at this coverage in the Cr case this peak had completely disappeared. This result indicates that patches of the polymer remained uncovered, consistent with island formation and weak Cu chemistry. Interestingly, a dipole shift is observed for this system although it is considerably smaller than in the Cr case with a shift to lower binding energy of only 0.4 eV at a coverage of 3 Å. Cu, then, presents a case with chemical reactivity significantly weaker than Cr when deposited onto the PMDA–ODA.

The Cu results can be interpreted in a similar manner as Cr. Although the relative energies of various complexes are difficult to determine accurately due to the weak interaction of Cu, the observed spectroscopic results are consistent with the formation of weak charge transfer complexes predominantly with the PMDA fragment. This shows that the concept of delocalized bonding described for Cr must also be applicable for Cu. Thus the nature of the metal–polymer bonding is different from the localized bonding observed for metal compounds, such as oxides and carbides.

In addition to Cu and Cr, the bonding characteristics to polyimide have been investigated for other 3d metals: Ni,[(5,19)] Ti,[(22,23)] and Co.[(26)] Together these studies reveal a consistent chemical trend in which the bonding strength and the chemical reactivity depend upon the number of the d electrons in the metal. The overall bonding characteristics are well described by the formation of metal–polymer complexes. On this basis, Ti is similar to Cr as it interacts strongly with the polyimide, while Ni is similar to Cu although its reactivity is somewhat higher. The trend in the chemical reactivity is also reflected in the interfacial morphology obtained with initial metal coverages. Again Ti is similar to Cr with uniform coverages, while Ni is close to Cu with island formation. Al, a metal with p valence states, has also been investigated.[20,21] Its bonding strength and chemical reactivity are intermediate between Cr and Cu.

3. INTERFACIAL MICROSTRUCTURE

Results of the spectroscopic studies on chemical bonding indicate that the chemical reactivity between the metal and polymer can significantly change the interfacial morphology, e.g., from a uniform coverage of Cr to an island formation of Cu. While those experiments discussed in Section 2 had been performed at room temperature, the effect of chemistry on interfacial structure is expected to be more pronounced at elevated temperatures. Such temperature dependence is relevant to a high-temperature polymer such as the polyimide. Additional experiments carried out at higher temperatures for Cu[6] and Al[20] have indeed confirmed the nature of thermal activation of the process. Since the interfacial tension is a strong function of the interfacial structure, this aspect of the material characteristics is important for understanding adhesion.

The microstructure and morphology of the interface have been investigated by cross-sectional TEM[29] and medium-energy ion scattering (MEIS).[8] For TEM studies, the samples have to be thinned to about 1000–2000 Å (for electron transparency) in a direction perpendicular to the interface. This required the development of an elaborate procedure for sample preparation, including cutting, glueing, mechanical polishing, and final ion milling.

Figure 8 shows cross-sectional micrographs of Cu/PI samples prepared at 293, 423, 523, and 573 degrees Kelvin. In all cases, copper is seen to migrate inside the polyimide films to form copper agglomerates that are nearly spherical in shape. Diffraction patterns obtained from the samples at the highest temperature indicate that each of the spheres is one single crystal of copper. One interesting aspect of these observations is that the spread in sizes of the spheres at any particular deposition temperature is very small. At high temperature, one sees a significant number of Cu aggregates resting at the PI/substrate interface. Apparently, the diffusion length of either the individual Cu atoms or of the agglomerates is sufficiently large so that the aggregates become pinned at this interface. At lower temperature, the diffusion length is shorter and the aggregates are imbedded nearer the surface of the polyimide film. The observed temperature dependence of Cu leads one to infer that Cu atoms have a high diffusivity in polyimide and, once diffusing inside, they aggregate because the binding energy between Cu atoms is higher than that of Cu to polyimide. These results are a consequence of the weak chemical interaction between Cu and the polyimide and are consistent with the spectroscopic studies.

Figure 9 shows the Cu peak measured by MEIS after deposition of increasing amounts of Cu at room temperature. The coverages have been indicated. Several observations can be made. First, the peak is much broader than the detector resolution (700 eV), even at the lowest coverage. This indicates immediately that the Cu is not forming a thin uniform film. It is either diffusing into the polyimide, or forming islands. Second, the peak shape hardly changes with increasing coverage. The dashed line through the upper curve is a magnified version of the ion scattering spectrum measured at the lowest coverage. This signifies that the depth distribution of the Cu does not change for coverages ranging from 1.11 to 8.35×10^{15} Cu atoms/cm^2. (One monolayer of Cu corresponds

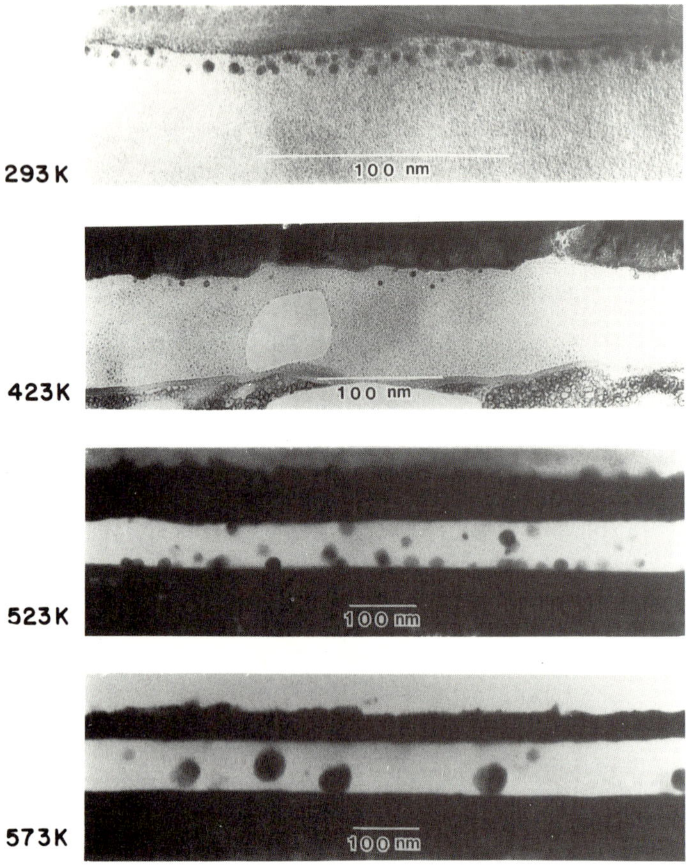

FIGURE 8. Cross-sectional TEM micrographs showing morphologies of Cu/polyimide interfaces formed at 293, 423, 523, and 573 K. Deposition rate is about one monolayer per minute.

to 1.93×10^{15} atoms/cm^2.) Third, the height of the Cu peak never reaches that expected for a continuous, pure Cu film. Even at the highest coverage studied, the surface of the polyimide has not been completely covered by Cu.

After the experiments at 300 K, Cu was evaporated on fresh samples at 423, 523, and 573 K. The results from MEIS measurements are shown in Figure 10. The number of Cu atoms/cm^2 evaporated on these samples is indicated for each temperature. From the drastic changes in the Cu peak shapes it is clear that Cu diffuses strongly into the polyimide. We note also that even though the coverages at higher temperatures are significantly larger than at lower temperatures, the observed signal intensities are much lower. This again is due to diffusion of Cu deep into the polyimide. The tails of scattering intensity at lower energy are a direct indication of this diffusion. However, the depth distributions observed do not resemble simple diffusion profiles. In particular, the occurrence of a maximum concentration below the surface indicates that simple diffusion is not taking place. Instead, the spectra indicate the existence of Cu clusters buried in the

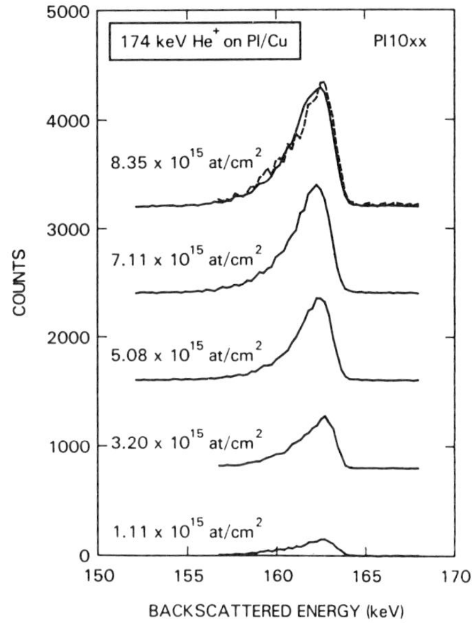

FIGURE 9. Ion scattering spectra from Cu evaporated on polyimide at room temperature. The total coverage increases from bottom to top as indicated. The dashed line in the upper curve is a magnified version of the lowest spectrum, showing that the peak shape does not change with increasing coverage.

polyimide. Calculations have been carried out to derive ion scattering spectra for spherical particles buried in a homogeneous medium. Qualitatively, the computation supported the explanation that with increasing deposition temperature, the Cu diffuses to greater depth, forming larger clusters. Thus the ion scattering results are consistent with TEM observations.

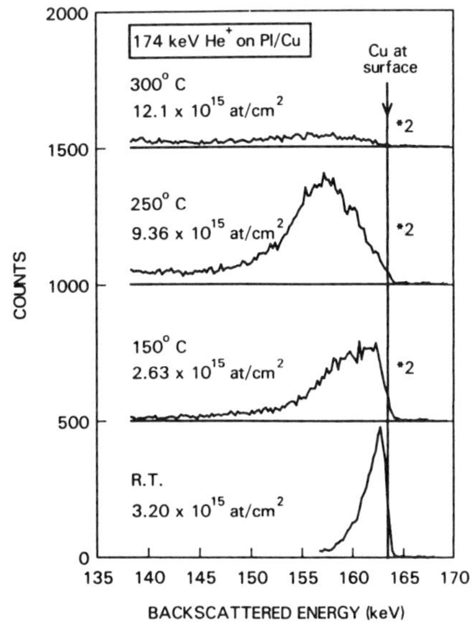

FIGURE 10. Cu ion scattering spectra measured after evaporation of Cu at various temperatures. Each spectrum was taken on a fresh sample. Note that the scale is enlarged by a factor of 2 in the upper three curves. Absolute coverages were determined in a separate 2 MeV ion backscattering setup.

To elucidate the role of chemical interactions that are expected to be important in the formation of the metal–polymer interface, related Monte Carlo calculations on a model metal–polymer interface have also been carried out.[29] These calculations were performed on a two-phase system, simulating the metal–polymer interface. The polymer was idealized by a set of interaction sites with the metal. For the case of Cr, such sites may well be the high coordination sites associated with the PMDA–ODA polymer.[9] The interaction sites were allowed to move about, consistent with the energetics of the model. The calculations have been performed on a two-dimensional square grid and the number of interaction sites that one introduces provides a measure of the openness of the polymeric structure. Metal atoms are "deposited" at the top surface and diffuse into the polymer according to the standard Monte Carlo algorithm. All single metal atom jumps and single polymer interaction site jumps are considered as a function of the interaction energies and temperature. Nearest-neighbor and next-nearest-neighbor interactions have been assumed between the polymer interaction sites and between the metal atoms. Only a nearest-neighbor interaction has been assumed between the metal atoms and interaction sites.

Figure 11 presents results of the Monte Carlo simulation for the Cu/PI interface. Figure 11a shows the distribution of Cu atoms just after the deposition of six monolayers of Cu. One notes that the Cu agglomerates are near the surface with, however, a number of free single Cu atoms distributed throughout the bulk. The other two figures, Figures 11b and 11c, show that migration and agglomeration can occur as a result of annealing even after the deposition has been terminated. The extent of migration and agglomeration depends on the annealing temperature and time. The observation that a significant number of agglomerates appear to become pinned at the PI/substrate interface cannot be re-

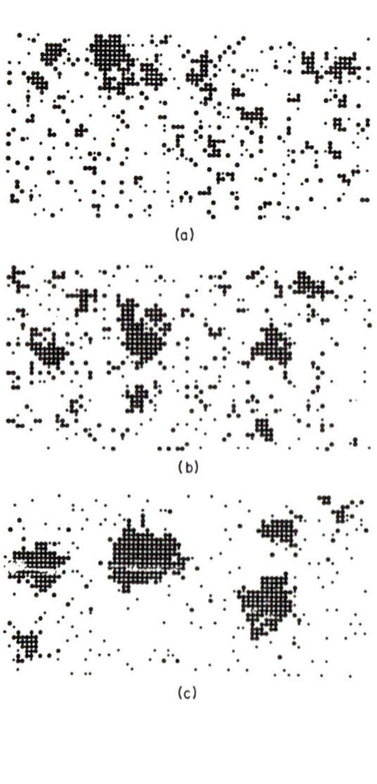

FIGURE 11. Results of Monte Carlo simulation for the Cu/polyimide interface: (a) as-deposited and (b) and (c) showing migration and agglomeration of Cu atoms after deposition as a function of time.

produced by the simulation, since no interaction that would yield such pinning has been incorporated into the model. The morphology for deposition at lower temperature has also been simulated. It has been found that the reduced thermal activation yields complexes that are closer to the PI surface, consistent with TEM observation in Figure 8.

Formation of the metal–polymer interface is also a function of the deposition rate of Cu. The micrographs shown in Figure 8 were obtained from samples prepared with extremely slow evaporation rates (about one monolayer per minute). When this rate is increased the details at the interface change significantly. For evaporation at room temperature, a slight increase in the evaporation rate to a few monolayers per minute causes the Cu to form a uniform film at the surface on the polyimide instead of penetrating into the film. This is shown in Figure 12a. Annealing of the sample after the rapid room-temperature evaporation will not result in the formation of Cu clusters into the polyimide but yields island formation on top of the polyimide (Figure 12b). This is apparently the regime for which single metal atoms deposited at the surface have a good chance of interacting with other metal atoms at the surface before they can diffuse into the bulk. When the evaporation rate is increased greatly (a few monolayers per second), even the high-temperature evaporation does not produce Cu clusters inside the polyimide film.

Figures 13a and b show the microstructures obtained when Cr is evaporated at slow rates, onto hot and cold polyimide, respectively. These micrographs show clearly that there is no formation of Cr precipitates inside polyimide and intermixing due to Cr

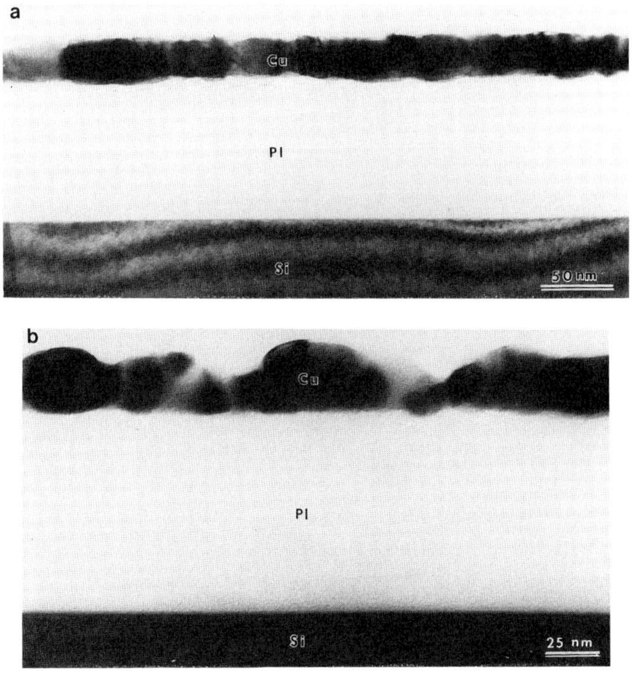

FIGURE 12. Cross-sectional TEM micrographs showing morphologies of Cu–polyimide interfaces formed with a deposition rate of several monolayers per minute: (a) as-deposited and (b) upon annealing at 300 °C.

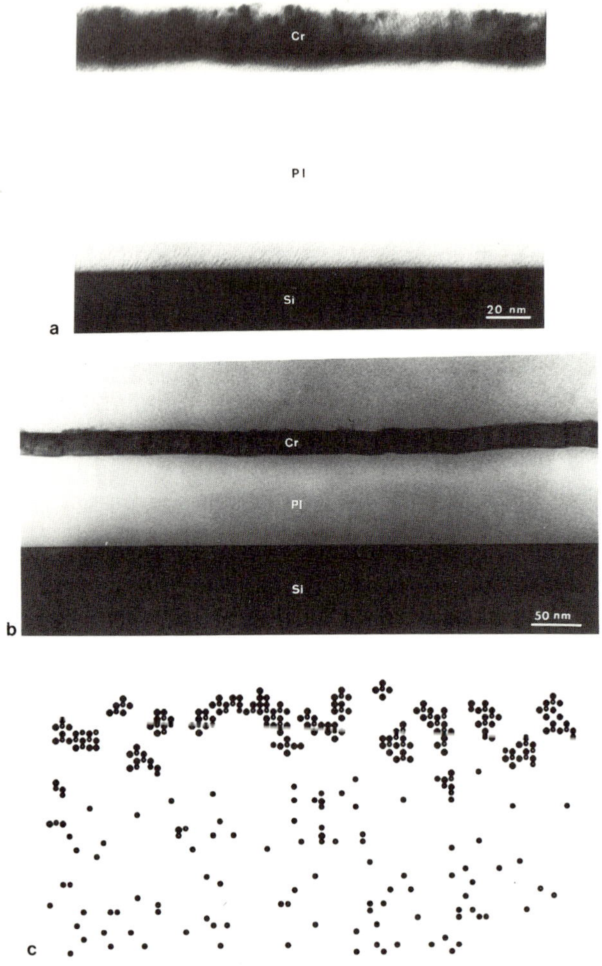

FIGURE 13. Cross-sectional TEM micrographs showing morphologies of Cr/polyimide interfaces formed with deposition temperatures at (a) 25 °C and (b) 300 °C. (c) Result of Monte Carlo simulation for a strongly interacting interface.

penetration. The intermixing, if present, is limited to the resolution of the microscope, i.e., about 10 Å.

Figure 13c shows the result of the Monte Carlo simulation for a strongly interacting metal. One sees that most of the metal atoms get hung up at the surface and are unable to penetrate the film. Spherical aggregates are not formed at the surface, since the interaction with the polymer has been assumed to be stronger than interaction between metal atoms. Even though the interaction is chosen to be stronger than that used in the Cu/PI interface simulation, it is still no greater than several tenths of electron volts.

The composition distribution measured by MEIS on TEM samples[8,29] confirmed the intermixing of Cu into polyimide and the formation of Cu clusters, while no penetration was detected for Cr. The extent of the intermixing of Cu was again found to be strongly dependent on the deposition temperature.

Observations on other interfaces of Al[20] and Ni[19] reveal morphological features intermediate between Cr and Cu. Al is closer to Cr in that it tends to form uniform film on the polyimide in most of the cases studied, even though with slow evaporation rate (a few monolayers per minute) and at temperatures of about 300 °C Al atoms have been found to agglomerate to form surface islands.[20] In comparison, Ni is similar to Cu as it tends to diffuse into the polyimide to form Ni clusters, particularly under the condition of a low evaporation rate and a high deposition temperature.

The observed effects of evaporation rate and temperature on the interfacial structure reveal the kinetic nature of the formation process of the metal–polymer interface. For example, in the case of Cu, the high binding energy between Cu atoms favors the formation of Cu clusters in the polyimide or surface islands. However, the actual morphology of the interface depends on the rate of Cu diffusing into polyimide relative to that along the surface and the outcome is controlled by the deposition rate and temperature. For Cu, its bonding to the polyimide is weak, so one can promote the indiffusion by a high deposition temperature and a low rate. Under such conditions, an interface will be formed with extensive intermixing and cluster formation of Cu in the polyimide. For other metals, such kinetic effects on interfacial structure will be different, depending on the bonding energy between the metal and the polyimide. For strongly bonded metals, such as Cr, the rate of diffusion into the polyimide is very low so that, in most of the cases studied, only a uniform interface without island formation was obtained. Such an interplay of the energetics of chemical bonding and the kinetics of the formation process plays an important role in controlling the structure and morphology of the interface. It also explains the overall correlation between the interfacial structure and the chemical reactivity observed for the metals studied so far.

The observed variation in the interfacial structure and morphology can influence significantly the mechanical property of the interface, such as the adhesion strength. For example, the results of the Cu–polyimide interface suggest that the strength of this weakly bonded interface can be improved by promoting the extent of the intermixing at the interface.

4. ADHESION AND DEFORMATION

Considering the basic question of measuring the adhesion energy, the combination of metal and polymer provides an interesting, but challenging, problem. The adhesion energy can be defined as the minimum energy per unit area required to separate two surfaces in contact[3] [see Eq. (1)]. In practice, the separation requires mechanical deformation of the two materials forming the contact. The polymer, being viscoelastic, will sustain a large deformation in the separation process. Consequently, a substantial portion of the energy input will be consumed in deforming the polymer. This, plus the energy for deforming the metal, is significantly larger than the adhesion energy, thus making an accurate determination of the adhesion energy difficult. This difficulty affects most of the techniques used so far, including the peeling and blister tests. The problem is exemplified by the peeling test where a recent analysis by Kim *et al.*[30] shows that to extract the adhesion energy requires detailed calculations based on information about the deformation behavior of the test structure.

To evaluate the mechanical performance of a multilayered structure, adhesion is not the only crucial parameter; the deformation characteristics and failure mode are important to consider as well. For these purposes, we have developed a stretch-deformation method for direct measurement of the adhesion energy.[31] By a combination with optical and electron microscopies, this method can be used also to observe the process leading to delamination.

Instead of measuring the force for separating the interface as in the previous tests, this method determines the energy required to stretch the structure to a complete delamination by measuring the force and elongation simultaneously. The configuration of the test sample is shown in Figure 14, consisting of a polymer substrate clamped at both ends and a metal overlayer. When the substrate is stretched, energy is transferred through the interface to deform the metal overlayer. In the process, the energy input is partitioned into three parts for deforming the substrate, the overlayer, and the interface. The portion for deforming the substrate can be subtracted from results obtained in a separate stretch test of the substrate. To separate the other two parts is less straightforward, as it depends on the material combination and the delamination characteristics of the sample. For this purpose, the test is supplemented with optical and electron microscopies whereby the various stages of sample deformation can be identified. These microscopy observations are useful not only for adhesion measurements, but also for detailed observations of the microstructure change and the failure mode of the interface.

This method has been applied to study a number of bilayered structures formed by depositing metal layers on PMDA–ODA polyimide substrates, including Cu, Cr, and Cu/Cr. Again, we will discuss the results of Cu and Cr. These two materials are interesting because their contrasting ductile and brittle characteristics provide a good example with which to probe the effect of metal property on the delamination behavior.

The sample substrates were prepared by spinning PMDA–ODA polyamic acid solution (DuPont 5878) on oxidized Si wafers. Prior to metal evaporation, the substrates were processed through a complete curing cycle in nitrogen, then baked in vacuum at 150 °C to

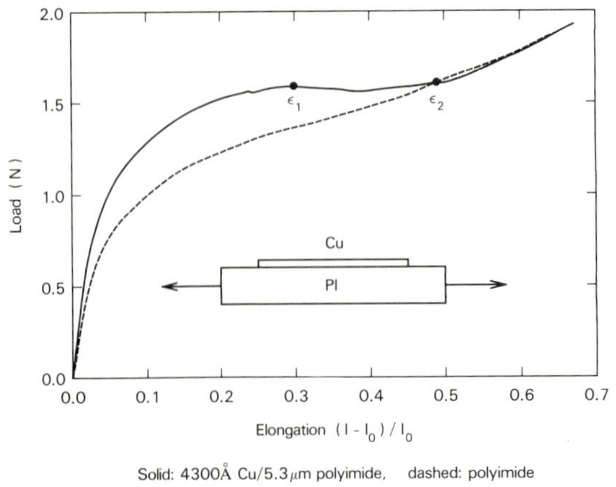

FIGURE 14. Load vs. elongation curves for Cu/polyimide and polyimide structures. Insert shows the sample geometry for a stretch-deformation test.

remove water and chemical residues. The metal overlayers were e-gun evaporated at room temperature through a metal mask to define the sample geometry. Stripes of 2 mm width were cut and peeled off from the wafer to form the test samples. For comparison, peel test samples were prepared together with the stretch deformation samples on the same substrate.

To record the load vs. elongation curves, a computer-controlled tester was constructed. The tester employed a stepping motor with 1 μm resolution to produce the stretch deformation and a piezoelectric transducer to measure the force experienced by the sample. The sample stage was mounted on an optical microscope equipped with a video camera for *in situ* recording of sample deformation.

The load vs. elongation curve obtained for a 4300 Å Cu/5.3 μm polyimide structure is shown in Figure 14, together with that for the polyimide substrate. While the pure polyimide deforms under stretching in a typical viscoelastic manner, the addition of a thin Cu overlayer does not seem to change, at a first glance, the overall deformation behavior significantly. Nevertheless, the presence of the Cu layer, in spite of only about 10% of the polyimide thickness, requires approximately 30% additional load for the same deformation. The additional amount of energy is consumed for deforming the Cu layer and the interface, including a portion for interfacial delamination, i.e., the adhesion energy. To extract the adhesion energy, the onset as well as the end point of delamination have to be determined. For this purpose, optical microscopy was used. To examine the progression of the deformation process, it is useful to look at the difference of the deformation curves. As shown in Figure 15, the differential deformation curve (solid line) exhibits an initial range of about 2% of elastic deformation of the Cu overlayer. Then Cu yields, following by plastic deformation up to about 20% elongation, as indicated by a plateau in the deformation curve. In this range, no cracks are observed; instead, scanning electron microscopy (SEM) reveals slip bands emerging to the surface and deformation twins are observed by transmission electron microscopy (TEM) in the bulk of the Cu film. After 20% elongation, cracks start to form in the Cu film with a preferred orientation perpendicular to the tensile stress. At about 30% strain (ϵ_1 in Figure 14), the film starts to delaminate at the edges of cracks, as shown in the SEM micrograph of Figure 16a, and often cracks start at

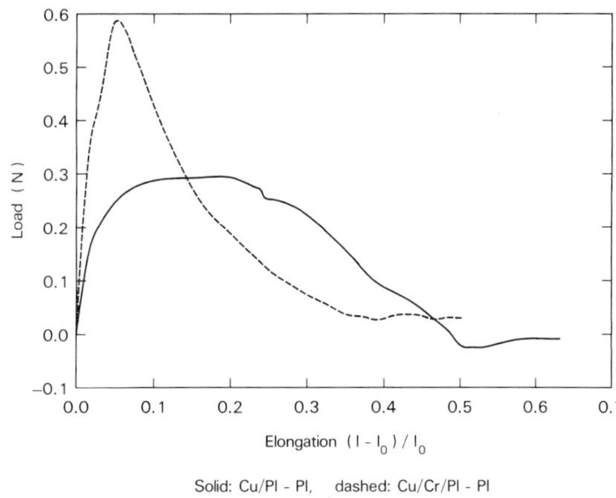

FIGURE 15. Differential load vs. elongation curves showing the difference in load between metal/polyimide and polyimide structures for metal layers of Cu (solid line) and Cu/Cr (dashed line).

several points at the same time. After a complete delamination (ϵ_2 in Figure 14), the deformation behavior converges to that of the substrate, as expected.

Based on the observed deformation behavior, one can determine the adhesion energy according to

$$W_a = \frac{1}{b_0} \int_{\epsilon_1}^{\epsilon_2} [F(\epsilon) - F_s(\epsilon)] d\epsilon \qquad (2)$$

where b_0 is the original film width and F the applied load, the subscript s denoting the substrate. If Eq. (2) is applied to the result of Cu/polyimide, one obtains $W_a = 9.6 \pm 0.5$ J/m^2, corresponding to a peel strength of 0.98 ± 0.5 g/mm. Kim *et al.*(30) obtained a value of 0.6 g/mm from analysis of peel test results. Considering the experimental error margins for measuring such small adhesion energies and the assumptions involved, the agreement is satisfactory.

From Figure 14, one can see that the integral in Eq. (2) for determining W_a is only a small fraction, about 4% of the total energy input for a complete delamination of the bilayered sample. This demonstrates the nature of the problem for determining the adhesion energy between two highly deformable materials. Because a very large fraction of the input energy is consumed to deform the materials forming the interface, it is difficult to extract the small fraction of the adhesion energy without making a direct measurement as in the stretch-deformation test. Even for the stretch-deformation method, the measured adhesion energy is only an upper limit since a small contribution from plastic deformation of the materials cannot be excluded.

A finite-element analysis(32) has been carried out to deduce the stress distribution in the layer structure, which reveals a high shear stress concentrating at the sample edges during stretch deformation. Similar stress concentration at the crack edges is expected and will drive the interface to delaminate between cracks. Thus, crack formation following by delamination between cracks is an effective process for relieving the deformation energy. This process was observed by SEM in the Cu/polyimide structure (Figure 16a) and can be inferred also from the fast decrease in the load of the differential curve starting from the beginning of crack formation. Crack formation usually requires a certain critical energy in the metal layer, thus the overall deformation behavior depends on the thickness combination and deformation characteristics of the metal/polymer structure. To explore this effect, we have investigated the Cr and Cu/Cr polyimide structures.

Typical brittle behavior was observed in a 5000 Å thick Cr film deposited on polyimide at 150 °C. The film had an average grain size of about 100 Å. After deposition, a high stress was generated in the Cr/polyimide structure which was sufficient to cause film fracture. At this stage, an irregular crack network can readily be observed by SEM or TEM. Upon further straining of the polyimide, the crack density does not change significantly; instead a preferential straining of the polyimide is observed. The morphology of the structure after 12% straining is shown in the TEM micrograph of Figure 16b. Here the elongation of the polyimide between Cr islands in the straining direction almost completely accounts for the overall elongation. At this point, some delamination becomes visible.

Because of the early crack formation, the adhesion energy of Cr was not measured on the Cr/polyimide sample. For that purpose, a structure prepared by incorporating a 800 Å

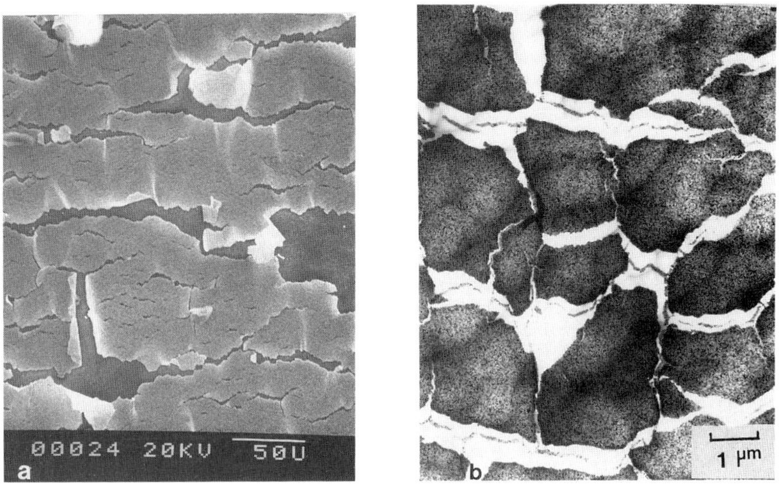

FIGURE 16. (a) SEM micrograph showing the delaminating of CU on polyimide starting at crack edges and (b) TEM micrograph showing crack morphology in a Cr/polyimide structure after 12% elongation.

interfacial Cr layer into the Cu/polyimide test sample was used. The load vs. elongation curve for this structure is plotted in Figure 17 while the differential curve is shown in Figure 15 (dashed line). It is clear that the deformation behavior of the sample is significantly altered by the presence of the Cr interfacial layer. Indeed, not only the initial slope increases about a factor of two; the onset of crack formation and the subsequent delamination have been shifted to a substantially lower range of about 5% elongation. These changes in the delamination process cannot be accounted for by the difference in the elastic properties of Cr vs. Cu (Young's moduli for Cr and Cu are 27.9 and 13×10^{11} dyn/cm^2, respectively[33]), and have to be attributed to the different deformation characteristics of the Cr/polyimide combination.

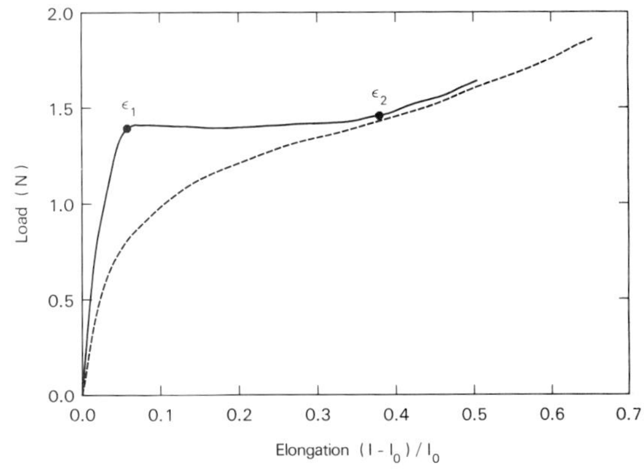

FIGURE 17. Load vs. elongation curves for Cu/Cr polyimide and polyimide structures. The thicknesses of Cu and polyimide layers are the same as the sample in Figure 14 and the Cr layer is 800 Å thick.

The differential deformation curves in Figure 15 demonstrate clearly the effect due to the interfacial Cr layer. Compared with the Cu/polyimide structure, the Cu/Cr/polyimide structure exhibits no plateau region of plastic deformation; instead crack formation and delamination commence at about 5% of elongation and continue to a complete delamination at about 40%. This suggests that the onset of the delamination process is controlled by crack formation in the Cr layer. This is consistent with the observation by optical microscopy which reveals about the same crack density in Cu and Cr. The Cu layer plays a role also since it absorbs a large portion of the deformation energy, so its deformation characteristics can influence the progression of the delamination process. Additional experiments employing different Cu/Cr thickness ratios show that the extent of transition from ductile to brittle behavior can be controlled by the Cu/Cr thickness ratio.[34]

The adhesion energy for the Cr/polyimide interface determined according to Eq. (2) from the Cu/Cr polyimide results is 33 ± 8 J/m^2, corresponding to 3.4 ± 0.8 g/mm. This compares favorably with the value of 4.3 g/mm obtained by the peel test. Comparison of the adhesion energies of the Cu and Cr interfaces shows that the Cr/polyimide interface is significantly stronger. Since the microstructure of the Cr interface is sharp and with a minimal amount of atomic mixing, its high adhesion must be attributed to the strong chemical bonding at the interface. This is a clear demonstration that chemical bonding at the interface plays an important role in determining the adhesion strength of the interface.

5. SUMMARY

In this chapter, we have reviewed the results of recent studies on the chemistry, microstructure, and adhesion of the metal–polyimide interface. Information on these aspects of material properties are important for understanding the mechanical behavior of metal–polymer interfaces. A basic approach has been used to study the metal–polyimide interface under well-controlled conditions using a combination of analytical techniques. These include the use of photoemission spectroscopy to monitor the valence and core states, TEM and MEIS to observe the morphology and microstructure. These experimental investigations have been complemented by theoretical studies, including molecular orbital calculations to deduce the electronic states of the metal–polymer complex and Monte Carlo simulation to derive the interfacial morphology. Such an approach has yielded detailed microscopic information on the chemical and material characteristics of the metal–polyimide interface. The material reactions, such as intermixing and cluster formation, reveal a fundamental nature of this class of interfaces. The interfacial reactivity and its correlation to chemistry are important for understanding the mechanical behavior of the interface, particularly for high-temperature polymers, such as the polyimide.

Among the metals studied, the chemical bond between Cu and the polyimide is the weakest, so Cu atoms diffuse readily into polyimide to form clusters. In contrast, Cr bonds strongly to polyimide, giving rise to a high-strength interface with uniform morphology. The difference in the bonding characteristics is reflected in the adhesion strength and deformation behavior of these interfaces.

A stretch deformation method has been developed for quantitative study of adhesion and deformation of metal/polymer layered structures. This method provides a direct measurement of the interfacial adhesion energy without resorting to complex stress analysis to extract the energy, as in other tests. This has been applied to Cu and Cu/Cr layered structure and the adhesion energies obtained are consistent with peel tests. The delamination process was found to occur generally in the sequence of elastic and plastic deformation, crack formation, and interfacial delamination. The detailed characteristics depends largely on the combination of metal and polymer. This was demonstrated by the incorporation of a Cr interfacial layer, which alters the ductile deformation behavior of a Cu/polyimide structure to a brittle behavior.

In summary, the metal–polyimide interface exhibits interesting chemical and mechanical properties. Many of the observed characteristics seem to be general for other polymers as well, particularly for those with high-temperature applications. The success of using metal–polymer interfaces for electronic packaging applications may well depend on the extent of fundamental understanding of such interfacial properties.

ACKNOWLEDGMENTS

The authors wish to thank many co-workers in the IBM T. J. Watson Research Center, in particular P. O. Hahn, J. W. Bartha, G. W. Rubloff, F. K. LeGoues, A. Rossi, P. Sanda, S. T. Chen, C. Feger, and C. H. Yang. During the course of this work, the authors have also benefited from discussions with many colleagues from the IBM Almaden Research Center, the System Technology Division of Endicott, New York, and the General Technology Division of Fishkill, New York.

REFERENCES

1. D. J. Bendz, R. W. Gedney, and J. Rasile, *IBM J. Res. Dev.* **26,** 278 (1982).
2. T. Watari and H. Murano, *Proceeding of the 35th Electronic Components Conference,* p. 193, IEEE (1985).
3. For a general discussion on adhesion and bonding, see Chapter 1 of this book; see also J. J. Gilman, *J. Appl. Phys.* **31,** 2208 (1960).
4. J. M. Burkstrand, *J. Vac. Sci. Technol.* **16,** 363 (1979); *J. Appl. Phys.* **52,** 4795 (1981).
5. N. J. Chou and C. H. Tang, *J. Vac. Sci. Technol.* **A2,** 751 (1984).
6. P. S. Ho, P. O. Hahn, J. W. Bartha, G. W. Rubloff, F. K. LeGoues, and B. D. Silverman, *J. Vac. Sci. Technol.* **A3,** 739 (1985).
7. P. O. Hahn, G. W. Rubloff, J. W. Bartha, F. LeGoues, and P. S. Ho, *Mat. Res. Soc. Symp. Proc.* **40,** 251 (1985).
8. R. Tromp, F. K. LeGoues, and P. S. Ho, *J. Vac. Sci. Technol.* **A3,** 782 (1985).
9. A. R. Rossi, P. N. Sanda, B. D. Silverman, and P. S. Ho, *Organometallics* **6,** 580 (1987).
10. C. B. Duke, *Int. J. Quantum Chem.* **13,** 267 (1979).
11. T. J. Fabish, *CRC Critical Reviews in Solid State and Materials Science,* p. 383, Chemical Rubber, Cleveland (1979).
12. E. D. Feit and C. W. Williams (eds.), *Polymer Materials for Electronic Applications,* Vol. 184, Symposium of the American Chemical Society, Washington (1982).
13. E. Geiss, K. N. Tu, and D. R. Uhlmann (eds.), *Proceedings on Electronics Packaging and Materials Science,* Materials Research Society (1985).
14. P. O. Hahn, G. W. Rubloff, and P. S. Ho, *J. Vac. Sci. Technol.* **A2,** 756 (1984).
15. J. S. Binkley, R. A. Whiteside, R. Koishnan, R. Seeger, D. J. DeFrees, H. B. Schlegel, S. Topiol, L. R. Kahn, and J. A. Pople, *Quantum Chemistry Program Exchange (QCPE)* **10,** 416 (1981).
16. B. D. Silverman, P. N. Sanda, P. S. Ho, and A. R. Rossi, *J. Polym. Sci., Polym. Chem. Ed.* **23,** 2857 (1985).

17. R. C. White, R. A. Haight, B. D. Silverman, and P. S. Ho, *Appl. Phys. Lett.* **51,** 481 (1987).
18. R. A. Haight, R. C. White, B. D. Silverman, and P. S. Ho, *J. Vac. Sci. Technol.* **A6,** 2188 (1988).
19. J. W. Bartha, P. O. Hahn, G. W. Rubloff, and P. S. Ho (to appear).
20. J. W. Bartha, P. O. Hahn, F. K. LeGoues, and P. S. Ho, *J. Vac. Sci. Technol.* **A3,** 1390 (1985).
21. Lj. Atanasoska, H. M. Meyer III, S. G. Anderson, and J. H. Weaver, *J. Vac. Sci. Technol.* **A5,** 3325 (1987).
22. F. S. Ohuchi and S. C. Freilich, *J. Vac. Sci. Technol.* **A4,** 1039 (1986).
23. J. L. Jordan, P. N. Sanda, J. F. Morar, C. A. Kovac, F. J. Himpsel, and R. A. Pollak, *J. Vac. Sci. Technol.* **A4,** 1046 (1986).
24. J. L. Jordan, C. A. Kovac, J. F. Morar, and R. A. Pollak, *Phys. Rev.* **B36,** 1369 (1987).
25. M. J. Goldberg, J. G. Clabes, A. Viehbeck, and C. A. Kovac, *Mat. Res. Soc. Symp. Proc.* **108,** 225 (1988).
26. S. G. Anderson, H. M. Meyer III, and J. H. Weaver, *J. Vac. Sci. Technol.* **A6,** 2205 (1988).
27. H. M. Meyer III, S. G. Anderson, Lj. Atanasoska, and J. H. Weaver, *J. Vac. Sci. Technol.* **A6,** 30 (1988).
28. Lj. Atanasoska, H. M. Meyer III, S. G. Anderson, and J. H. Weaver, *J. Vac. Sci. Technol.* **A6,** 2175 (1988).
29. F. K. LeGoues, B. D. Silverman, and P. S. Ho, *J. Vac. Sci. Technol.* **A6,** 2200 (1988).
30. K. S. Kim and J. Kim, *Trans. ASME, J. Eng. Mater. Technol.* **110,** 266 (1988).
31. P. S. Ho and F. Faupel, *Appl. Phys. Lett.* **53,** 1602 (1988).
32. S. T. Chen, private communication.
33. E. A. Brandes (ed.), *Smithells Metals Reference Book,* Butterworths, London (1983).
34. F. Faupel, C. H. Yang, S. T. Chen, and P. S. Ho (to appear).

15

Aspects of Bioadhesion

Robert E. Baier and Anne E. Meyer

1. INTRODUCTION

The joining of substances in wet, salty, biochemically active environments is perforce an event of—at least—water displacement from interfaces. When living cells or tissues are to be intimately mated, or that mating prevented, consideration must be especially given to interaction of the hydrated peripheries of the particles/materials in question.[1] If neither, or only one, of the potentially adherent members easily sheds its bound water layer, prompt practical underwater adhesion is weak to negligible, although it might later become quite secure as excess water is expressed from the joint. Natural mucosal surfaces that resist bioadhesion to most other substances, living or synthetic, provide excellent examples. Witness the inherent resistance to biofouling of the inside of the cheek, the outside of the porpoise and killer whale,[2] and the uncompromised endothelial lining of blood vessels,[3] for instance.

When stable bonds are formed (however slowly) between originally hydrated surfaces, it is often through the intermediary of H-bonding of pendant hydroxylated amino acid side chains or oligosaccharide groups extending from polypeptide backbones.[4] Some glycoproteins and extracellular proteoglycans adsorb to solid surfaces in this manner, in preference to their continued solvation by water. Even when the approaching surfaces can rapidly dispatch their boundary water, however, strong adhesion is not assured. This is especially the case with intrinsically low surface energy, hydrophobic solids like Teflon and paraffin.[5] Conversely, the fact that a surface is intrinsically hydrophilic is no guarantee of resistance to the formation of tenacious bonds. Witness the essentially irreversible soiling that eventually destroys the clarity of hydrogel ("soft") contact lenses.

Main determinants of the strength of adhesive interactions in the biological environment are the spontaneous deposition and subsequent degree of conformational change (denaturation) of biomacromolecules that precede any close approach of even microscopic objects.[6] This is particularly true in the interaction of biological systems with synthetic or engineering materials, where the rapid, spontaneous deposition of glycoproteinaceous "interface conversion layers" is the first noted event in the process of dental plaque

---

Robert E. Baier and Anne E. Meyer • Industry/University Center for Biosurfaces, State University of New York at Buffalo, Buffalo, New York 14214.

formation,[7] thrombus formation,[8] biofouling of ship bottoms,[9] and tissue bonding of prosthetic appliances.[10] Obviously, one must be sensitive to the varying definition of *biocompatibility* in these diverse systems, where lack of adhesion is the preferred result for artificial hearts,[11] blood vessels,[12] and food pasteurizers,[13] while strong adhesion is desired for artificial hips,[14] dental implants,[15] and trickling filter sewage-treatment surfaces.

Often, of course, the events observed have been the opposite of those desired, such as the formation of dental plaque on the naturally hydrophilic surfaces of teeth[16] and the lack of tissue integration with hydrophobic, percutaneous catheters (leading to infection, inflammation, and adverse results in both cases).[17] Accepting the ubiquity of protein deposition as the first discernible event in the interaction of most materials in biological fluids (saliva, blood, seawater, etc.),[18] as noted on surfaces varying from net negative to neutral to positive, rough or smooth, hydrophilic or hydrophobic, one must inquire about measurable surface properties that predict the tenacity of subsequent adhesive bonds with particulate matter, living or dead. The required dehydration of the interfacial zone is accomplished—usually—by entropically driven deposition of proteins.[19] Surprisingly, standard concepts of wetting, spreading, and surface energy as applied in nonliving systems allow significant correlation with bioadhesive phenomena as well.[20] In essence, the degree of protein adhesion, as the first "priming" event, reflects the interaction potential of the dehydrated surface as dictated by its outermost atomic constitution.[21] Simple contact angle measurements in air are often sufficient to predict the likely adhesive response to various natural and synthetic surfaces under water.[22]

There is an apparent anomaly in this approach. Water-wettable (hydrophilic) surfaces, usually taken to be of high surface-free energy as in the case of ceramics, metals, glasses, and metal oxides, can be—in fact—resistant to spreading of organic matter (oleophobic) and, therefore, operationally of low energy. Thus, hydrophilicity *cannot* be correlated unambiguously with high surface energy, and it is a serious error to calculate thermodynamic surface energy terms from (usually unstable) contact angle measurements with aqueous fluids alone.[23] Rather, the empirical value of critical surface tension, determined with noninteracting liquids according to the method popularized by Zisman,[21] correlates best with practical adhesive phenomena observed in living systems. For example, the critical surface tensions of the oral mucosa, of the insides of blood vessels, and of the outsides of dolphins all fall in the nonadhesive zone between 20 and 30 dyn/cm (erg/cm^2)[24]; tooth surfaces, shipbottom paints, pasteurizer-grade stainless steel, and most dental restorative materials display critical surface tensions above 30 dyn/cm, where stronger adhesion is both predicted and measured.[25] Surfaces rich in polar moieties, especially carboxyl groups, seem capable of bonding most strongly to biomass in the living state, given that these surfaces can first be dehydrated.

In concert, then, with earlier chapters of this volume, it can be seen that adhesion in living systems follows similar rules. Dehydration of the interacting surfaces is the additional requirement that allows close enough approach of the potential adherends to allow *nonspecific* (London–van der Waals) attraction via atomic forces long associated with adhesive phenomena in nonaqueous systems. The dehydration event is almost always spontaneously effected by the selective deposition of (interestingly, often the most hydrated) proteins from the solution phase adjacent to solid nonphysiologic materials. The retention of this "primer coat" is a function of both the surface energetics of the substratum and any specific secondary interactions (such as chemical bonding, lectin–receptor binding, antigen–antibody recognition, etc.) that might subsequently occur. Where the interaction potential is low, as at apolar, low-surface-energy materials like methyl-group-dominated medical-grade silicones, the detachment of both the deposited protein layers and secondarily attached biomass is quite facile, requiring only modest shear stresses.[26] Conversely, strong biological bonds and tissue integration with implanted prostheses can be encouraged by treating the materials to display high surface energies with numerous polar binding sites.[27] Functioning, substitute blood vessels and improved, retentive dental implants have already resulted from implementation of these concepts. This chapter presents exemplary aspects of a few of these developments, to inform new practitioners of the art of bioadhesion and to promote its acceptance into the broader, general science of adhesion.

2. ANALYTICAL METHODS MOST APPLICABLE TO BIOADHESION STUDIES

Only a few techniques have been proven reliable for the rapid, nondestructive instrumental analysis of biological surfaces. Some, including those suitable for human skin in its natural state or modified by therapeutic or cosmetic preparations, are generally available and in routine use.[28,29] One of these methods, employing the sensitive spectroscopic process known as multiple attenuated internal reflection,[30] can produce readily interpretable signatures of the infrared or UV/visible absorbing groups within the few molecular layers of matter defining the true biological interface with exterior phases. These spectral records provide both qualitative and semiquantitative information about the actual compositions, rates of accumulation, or modifications of important boundary layers.

Exemplary data characterizing natural "skins" in the human oral cavity, and analyses of their chemistries, thicknesses, and probable influences on subsequent bacterial adhesion have been published.[7,31] Scanning electron microscopy of "acquired pellicles" on clean test objects immersed in the fluid/air environment of the human mouth revealed that, for periods of typically 2 hours, the naturally deposited films are universally of glycoproteinaceous composition and mostly (but not completely) free of bacterial colonization.

As already described in detail,[32–36] a half dozen more simultaneous or sequential analytical techniques are also applicable to probing the structure and composition of such biological deposits without requiring their manipulation or removal from the substrata of interest. These methods include, besides the acquisition of absorption spectra by the internal reflection technique, (1) the determination of film thicknesses and refractive indexes by ellipsometry, (2) the measurement of surface electrical states by recording contact potentials, (3) the inference of wettabilities, strengths of adhesion, operational surface free energies and critical surface tensions by direct measurements of liquid/solid/gas contact angles, (4) the morphologic inspection of such layers by direct scanning electron microscopy at useful magnifications of up to 20,000 times without requiring overcoating the samples with sputtered metallic conducting films, (5) the simultaneous analysis by energy-dispersive X-ray techniques of the presence and relative abundances and locations of all elements of atomic number higher than sodium, and (6) the identification of actual crystalline forms present, by way of glancing angle X-ray diffraction. Newer analytical techniques, such as X-ray photoelectron spectroscopy (XPS, or ESCA), are taking their place alongside these other methods in the elucidation of interfacial mechanisms.

3. EXPERIMENTAL REQUIREMENTS

One important lesson from years of research into adhesive strengths and properties of biological deposits is this: experiments must always be done in a manner allowing both knowledge and control of the exposure geometries, flow fields, shear forces, and boundary hydrodynamics during all the deposition and/or detachment processes investigated. It is not acceptable in bioadhesion research to terminate a biological exposure (for the purpose of noting the accumulating microfouling films, for instance) by "rinsing" a sample in an unspecified or nonquantifiable manner. Neither is it acceptable to equate the forces needed to detach or distract a biological deposit from a solid substratum with the thermodynamic strength of adhesion of the attached elements. Cell and film separations from dissimilar phases are not simply the reverse of the first adhesive events. Rather, the failures of bioadhesive joints are usually mixtures of the cohesive and adhesive types. Use of cylindrical tube geometries and flat plate geometries with known, controlled flow rates and shear fields has been most useful in overcoming problems in this regard. Special "flow cells" can be employed, for example, during both the first exposure of test surfaces to biological media and in the final rinsing steps required to remove loosely held debris, co-adsorbed interfering salts, or entrained materials.[(37)]

While controlling the flows and hydrodynamic stresses in such test devices, it is somewhat more complicated—but generally valuable—to incorporate various strain gages, transducers, temperature sensors, heaters, electrodes and the like into the flow circuits. These allow one to monitor the dynamic events accompanying biological film deposition and adhesion of fouling elements. The technique of streaming potential measurement, for instance, allows direct calculation of the important boundary phase parameter of the plane of shear called the zeta potential.[(38,39)] It is also possible to apply "voltage clamping" circuitry so that the primary events of biofilm deposition and secondary events of microbial attachment, polymer exudation, and mineralization can be observed on substrata controlled electronically at any desired surface potential/charge condition.[(38,39)]

4. BIOADHESION OF MICROBES

Using flow cells and controlled geometries as described, a most revealing study was completed of the early events of microbiological fouling of model heat exchange surfaces in contact with the warm subtropical waters of the Gulf of Mexico.[(40)] The Gulf of Mexico findings were compelling in their demonstration of rapid deposition of humic-like conditioning films on heat exchange surfaces; these adsorbed films in turn became colonized within 24 to 72 hours by pioneer bacteria of both flagellated and unflagellated types. Within 144 hours of exposure to the flowing Gulf of Mexico water, deposits of polymer exudates were revealed around the pioneer microorganisms by using a special adaptation of scanning electron microscopy, which eliminates the need for obscuring electroconductive layers of sputtered metals. At the same exposure time, deposition of calcareous matter, usually associated with diatoms and other algal forms, was noted. Patches of extreme microbiological diversity, with ecological succession already in progress, were also observed on test surfaces in the period between 72 and 144 hours. Figures 1–6 illustrate this sequence of events.

It has been exciting to discover that microbial organisms, even those specialized for effective colonization of "foreign surfaces," can be almost quantitatively excluded from successful adhesion to test substrata[41] that exhibit surface properties in the bioabhesive, or adhesion-resistant "biocompatible," zone identified a decade earlier from studies dealing with mammalian cells, notably those in blood.[42,43] One surface-energy-conversion coating, formed by covalent binding of polydimethylsiloxane to glass, mineral, or metallic surfaces, is excellent in this regard, exposing arrays of closely packed methyl groups to the colonizing environment. Such methyl groups exhibit a critical surface tension or apparent surface free energy of about 22 dyn/cm, in the minimally bioadhesive zone. This statement does not mean that such surfaces are not rapidly colonized by microbes, but rather that their *retention* to such substrata is relatively easily overcome when mechanical forces are applied.

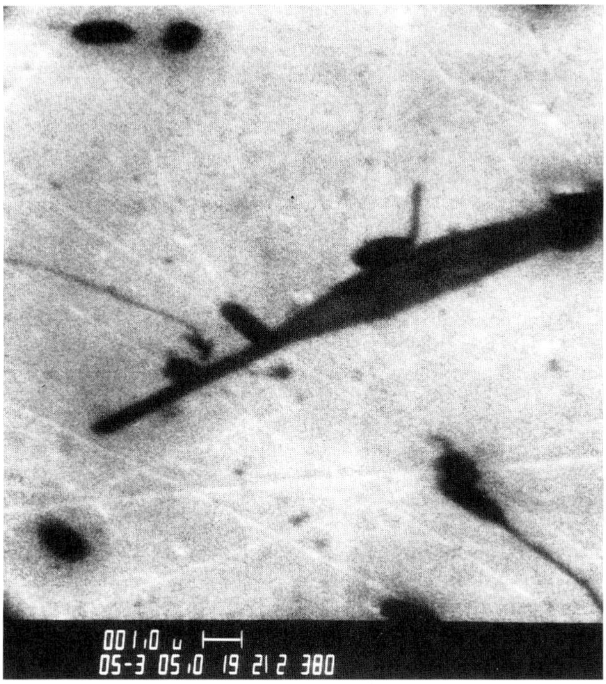

FIGURE 1. After 72 hours exposure to flowing seawater from the Gulf of Mexico, the smooth metallic test plate has acquired a proteinaceous "conditioning" film (overall gray background) and some pioneer bacteria. Because the metallic substratum is conductive, overcoating the sample for scanning electron microscopic analysis is not necessary. Original magnification: 5000×; note micrometer scale below photograph.

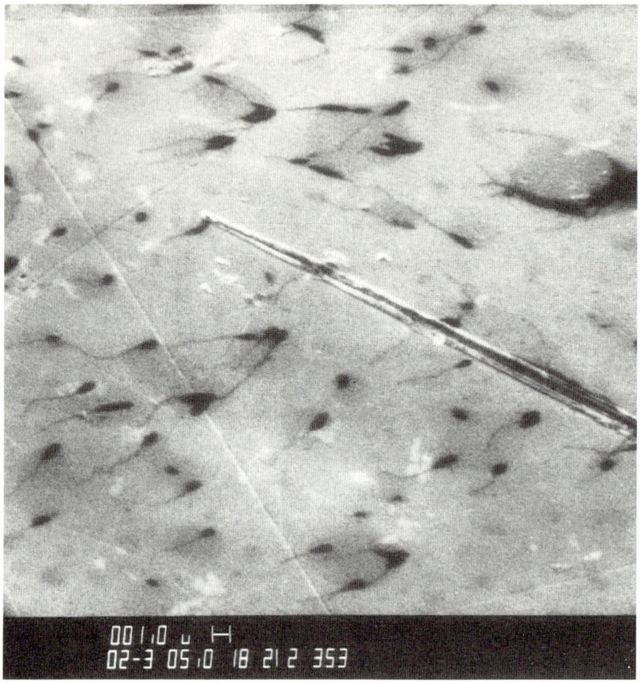

FIGURE 2. Within 144 hours, microscopic analysis of metallic test plates exposed to flowing Gulf of Mexico seawater reveals additional bacteria, bacterial exudates (note darker patches around the bacteria), and some mineralization (white patches). Original magnification: 2000×; note micrometer scale.

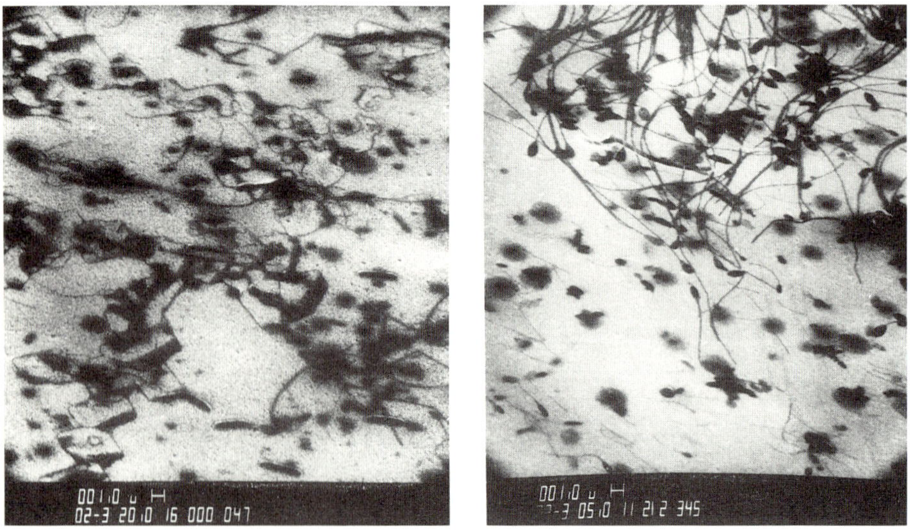

FIGURE 3. These photos (original magnification: 2000×) show two different views of metallic test plates exposed to flowing Gulf of Mexico seawater for 9 days. By this time, the substrata are heavily populated by bacteria. Note micrometer scale below photograph.

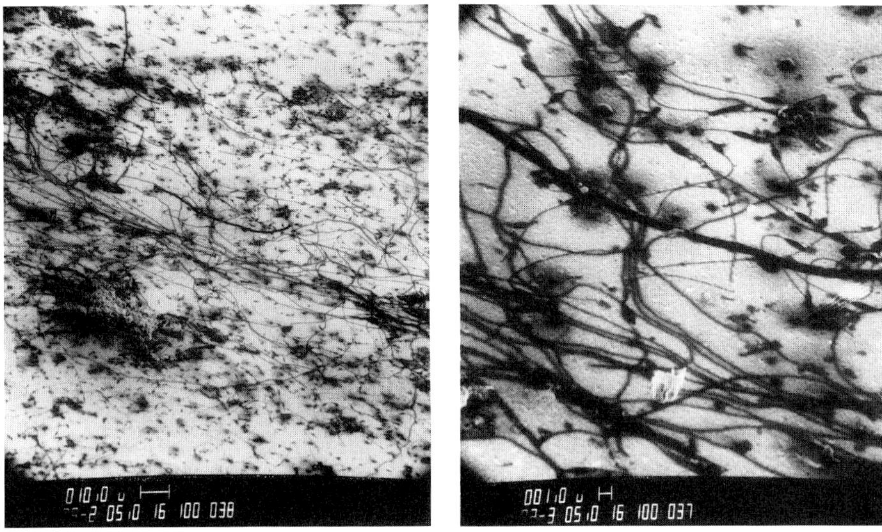

FIGURE 4. Electron photomicrographs of one area of a metallic text plate exposed to Gulf of Mexico seawater for 12 days. Original magnifications: 500× (left) and 2000× (right).

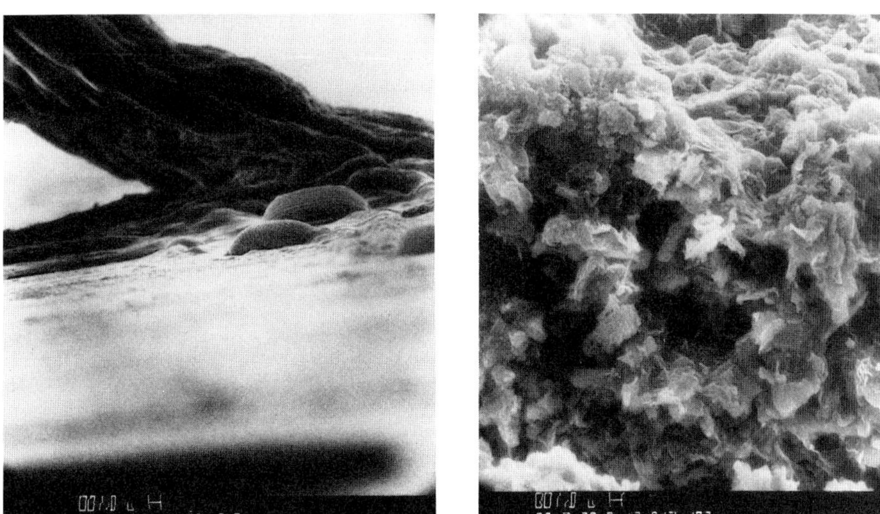

FIGURE 5. After the formation of the conditioning film and initial colonization by microorganisms, the fouling film can take many forms depending upon the exposure conditions and the organisms present. On the left (original magnification: 2000×), we observe how a particular species of algae [*Enteromorpha intestinalis* (L.) Link] attaches to and grows from a substratum; also note the diatoms in the foreground. On the right (original magnification: 2000×), we see the complex and very thick mineralized film that formed on a glass plate after 3 months exposure to flowing seawater; closer inspection of this film revealed alternating layers of proteinaceous and mineralized material. (Electron microscopist: M. S. Fornalik.)

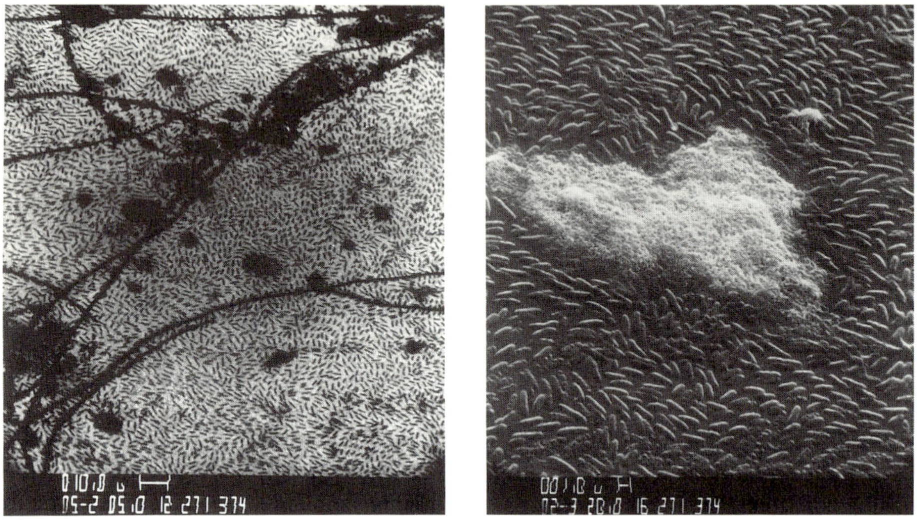

FIGURE 6. Similar biofilms form at all natural interfaces. These photos display the extensive proteinaceous and bacterial films formed at air/liquid interfaces. On the left is shown (original magnification: 500×) the film formed from fresh river water as the water was kept stagnant for one week. The film is seen in stark black and white contrast because it was transferred from the water surface to a metallic substratum of sufficient conductivity to allow this type of image formation without requiring further metallic overcoating. Each small rod is a bacterial cell; the larger masses are algal forms and bacterial colonies. On the right (original magnification: 2000×), the film from the same water sample was collected on a clean glass plate. Since the glass substratum is not conductive, it required overcoating with evaporated metal before imaging in the scanning electron microscope; thus, much of the contrast and detail observed in the prior photos is lost. (Electron microscopist: M. S. Fornalik.)

The early colonizing film of microorganisms is common on test substrata of every quality so far considered and in every water condition so far tested.[44] Materials evaluated to date include a variety of the inert and toxic metals and metal alloys, and a large range of organic coatings useful for adjusting the critical surface tensions of metallic support plates. Water quality tested, so far, has ranged from that of untreated natural ocean water, to that same water modified by intermittent chlorination or by electrolytic enrichment in copper and aluminum ions. Both such latter treatments have been correlated with at least partial suppression of the biofouling film growth in past trials.[45]

Beneath the first layer of pioneer colonizing bacteria is, in any natural system and without exception, a spontaneously deposited protein-dominated "conditioning" film. This interface conversion layer is first contributed from the macromolecular constituents of the flowing water itself. Many such components are available within the ubiquitous 5 to 10 parts per million background concentration of "humic" or "gelbstoffe" substances in all natural coastal waters where fouling is most intense.

After pioneer (usually short rod-like) bacteria attach and secrete their slimy exudates into their zones of influence, a second population of attaching organisms enriches the fouling layer on top of the pre-adsorbed conditioning film. Continuing deposition of macromolecules is often noted. This second wave of attaching species is often dominated by prosthecate microorganisms that can utilize the scarce nutrients in natural ocean waters by enhancing their surface-to-volume ratios. These prosthecate microorganisms are oligotrophic, living in nutrient-poor conditions. Their preferential attachment to and growth upon the highly concentrated, nutrient-rich film at solid/liquid boundaries seems to represent an escape from oligotrophy not completely consistent with their presumed resistance to growth in media of higher organic content. When the substratum colonized by these organisms is free of leachable toxic atoms, and when the water is similarly free of poisons that can interfere with bacterial metabolism and growth, the attached species secrete mucilagenous polymers around themselves. Filamentous growths then proliferate over the entire surface, making it an excellent filter medium for collection of other oceanic organisms and debris, as well as an excellent culture medium for the growth and multiplication of both the attached and entrapped species. As noted with dental plaques of longstanding in the oral cavity, or the thrombotic deposits of long duration in artificial hearts, significant mineralization (usually calcification) can and does occur with time. Mineralization is enhanced in oceanic environments by the binding of many inherently mineral-rich diatomaceous species having silicaceous or calcareous skeletons.

It is both interesting and important to note that the strength of biological adhesion in these natural circumstances can be controlled by adjustments of the critical surface tension of the substrata. This control technique is now under scrutiny for the purpose of mitigating biofouling of power plant heat exchange equipment, without the use of toxic substances such as chlorine.[9,44,46,47]

5. BIOADHESION OF MACROBIOTA

Of great concern to the operators of ships is the macrofouling community composed of barnacles, tube worms, encrusting bryozoa, and bushy hydroid growths that impede the efficient movement of water over originally smooth ship hulls. Even when killed by various toxic agents or parasites, their remaining, strongly cemented, empty shells cause enormous increases in drag coefficients and excessive consumption of fuel. Barnacle attachment begins with a barely visible, soft-bodied larval form called the cyprid. Cyprids explore for potential settlement and growth sites using functions for which a miniature adhesive tester might be designed. The critical dependence of the success of the cyprid colonization process on the previous presence of bound protein films, at least, and perhaps bacterial overlying forms as well, has been highlighted in the work of Crisp and Meadows.[48]

The correlation developed from early studies, given as a plot of the degree of biological fouling versus the critical surface tension of the material,[49] shows the same sharp minimum strength of biological adhesion as was noted with similar data from studies in blood and tissue.[50]

6. BIOADHESION FROM BLOOD

In the development of artificial hearts, major continuing problems are the unwanted adhesion of thrombotic and coagulum masses, as well as scale-forming minerals from the blood stream itself.[51] Materials selection is severely limited to those elastomers that can withstand the 10^8 or more flexing cycles necessary during the forecast period of service of such devices.[11]

During even short-term implantation of artificial heart components, and especially of flexing chambers, it is immediately obvious that the deposition of unwanted blood-borne materials is unavoidable.[3] Adjustments in flow dynamics, velocity profiles, and other mechanical design features minimize—but have not yet eliminated—problems with these deposits. Implantation times of weeks are now routine in many research centers as the devices continue to improve. However, serious further consequences of surface deposition of both organic and inorganic matter, predominantly on the flexing surfaces of the elastomeric heart chambers, remain a major problem and might be considered the limiting parameter for further progress in development of this important artificial internal organ.

The preferred materials of construction for artificial hearts, some substitute blood vessels, and left ventricular assist devices are hardy elastomers. Polyether types of polyurethanes can be fabricated for this purpose to exhibit surface enrichment of "biocompatible" methyl groups, provided by direct surface alloying or admixture of polydimethylsiloxane in the original polymer blend.[52] These surfaces are usually easiest, among hydrophobic polymers, to keep smooth and free of entrapped air prior to their first exposure to blood. Even in the face of such stringent precautions, the best synthetic materials still acquire spontaneously deposited "conditioning" films from blood (usually dominated by the plasma protein fibrinogen) prior to any cellular adhesion. On well-tailored surfaces, after this "conditioning" film is in place, arriving blood platelets do attach, but typically maintain their discoid shapes without disgorging their granules and/or clotting factors. On inappropriate surfaces, platelets go through a distress syndrome called "viscous metamorphosis" that triggers gross thrombosis and frank activation of the coagulation cycle. Segmented polymorphonuclear leukocytes, the abundant phagocytic white cells that usually are excluded from the walls of flow devices by fluid dynamics factors, then secondarily stream to the wall and colonize the platelet- and platelet-debris-strewn foreign surface. This is usually without any beneficial effect on improperly "tailored" surfaces, where attaching white cells, thus, contribute directly to the growing adherent cellular mass and further accelerate the thrombotic episode. The culmination of these events usually is gross occlusion of the prosthetic devices.

An exception is when prosthetic materials having critical surface tensions between 20 and 30 dyn/cm are used. This choice does not prevent the white cells from streaming aggressively toward platelet- and protein-film-coated substrata, but upon their arrival, these cells seem able to function as a "clean-up squad." They emit fibrinolytic enzymes that weaken the attachment of the original fibrinogen films to the more passive lower-energy substrata, so that interfacial shear forces can detach the entire bound mass. A long-term dynamic equilibrium with the fresh flowing blood ensues. A plasma protein film acts as a "passivation" layer, eliminating the strangeness of the interface between prosthetic material and the blood. Continued mechanical function is preserved indefinitely.

The same strategy has been successful for tailoring the critical surface tensions of biomedical organ substitutes to exhibit desired degrees of biological adhesion. General approval was granted in 1979 by the U.S. Food and Drug Administration for peripheral vascular reconstruction using surface-prepared and stabilized human umbilical cord veins. The human umbilical cord typically is delivered in lengths of 2 to 3 feet, containing blood vessels of both arterial and venous anatomies surrounded by an interesting ground substance called Wharton's gel.[53] The natural pavements of endothelial cells of the blood vessels are lost during their manual dissection from fresh or frozen umbilical cords, exposing the highly platelet-adhesive subendothelial lining. This "sticky" but otherwise physiologically tolerable layer, along with its subjacent basement membrane and highly elastic internal lamina, is converted to a bioabhesive material by using, as a primary quality-control criterion, the critical surface tension parameter. Processing conditions have been developed to maintain this value in the mid-20 dyn/cm for the finally tanned prostheses. The international experience with these prostheses includes numerous successful human implantations, saving life and limb from advanced atherosclerosis. This is achieved with long-term maintenance, up to a decade in some ongoing cases, of the integrity of the internal elastic lamina and preservation of nonadhesive surface qualities.

7. BIOADHESION IN HARD AND SOFT TISSUES

Implants buried in tissues must become firmly invested by the host tissue in order to provide the immobile, firm, infection-free performance that is usually required. The bonding integrity of the tissue phase immediately adjacent to the implant must be nearly perfect to prevent bacterial seepage, colonization, and resulting inflammatory responses. These conditions must be achieved and maintained from environments that are often bloody, nonsterile, and contaminant-rich during device placement.

Even though implants into hard and soft tissues have been used clinically for decades, the results obtained have been ambiguous at best. Many failures have been noted, mainly as a result of loss of tissue bonding, but also as a result of plaque and debris accumulation and very poor healing around protruding posts at epithelial junctions. It is not uncommon for implants to be extruded completely from their placement sites within only a few weeks. Trials in animals, and applications in humans, have often proceeded too rapidly, with an unfortunate absence of both necessary and sufficient knowledge of the surface properties of the materials.

Knowledge of the zones of surface energy that can favor or minimize adhesion in biological environments now makes it possible to design and develop improved prosthetic materials for a number of important applications. For example, basic knowledge developed by surface analytical methods introduced earlier revealed the intrinsic surface properties of natural dental enamel.[7] Data for the same surfaces, after various treatments (for example, with citric acid or phosphoric acid etching solutions), opened the way to obtaining long-term retention of a whole generation of pit and fissure sealant materials and dental cements that have entered general dental and orthodontic practice.[54–56] Glow discharge (plasma) treatments[57] of biomedical devices of even such physically large dimensions as artificial hips can be quite beneficial in improving their receptivity to "bone cements" and other adhesives used to bind them securely within the prepared femoral shafts.

Evaluation of the tissue response to surface-treated biomedical implants showed dramatically increased cellularity when glow-discharge-treated or relatively high-surface-energy test specimens were implanted. Tenacious tissue bonding to higher surface energy—but not lower energy—implants was noted during these evaluations, through an aggressive "healing" response involving recruitment of fibroblast/fibrocyte cells to the vicinity of the implant surface and their organization into adhesive capsular tissue.

The cellular response observed within the fascial tissue that reorganized against low-surface-energy implants comprised only thin, but dense, scar-like zones. The remaining deeper tissue included only a few elongated nuclei from responding fibroblasts, enmeshed within a loose connective tissue matrix. Scanning electron micrographs of the interfacial layer of tissue illustrated the weakness of the cell-poor subadjacent tissue and its ease of separation and fragmentation during the specimen handling procedure.

Shifting the point of observation from the tissue side of the interface to the low-surface-energy implants themselves, scanning electron micrographs showed that even modest mechanical forces could disrupt the tissue/implant bond quite rapidly. Bond failure was predominantly of the interfacial type rather than by cohesive failure in either of the bulk phases. Scanning electron microscopic views of textured, low-energy substrata made it further apparent that clean separation of the host tissue from low-surface-energy foreign bodies can occur even when surface textural variations are present.

"Interfacial failure" of the type discussed here does not mean that the base, initial implant is actually re-exposed by mechanical failure of the tissue bond to its face. Rather, the failure is in the integrity of the glycoproteinaceous "glue line," preventing cellular layers beyond it from obtaining permanent liaison with the original substratum. Further examination of the tissue faces adjacent to low-surface-energy implants showed that rounded fibroblast cells dominated the surface zone of the host tissue. The spheroidal nature of the cellular population corresponded with the low cell density and poor intercellular bonding obtained in this zone. Indeed, the tissue layers peeled like the layers of an onion, one or two from the other, in thin sheets.[15] This further illustrates the poor mechanical integrity of the capsule with which low-surface-energy implants become invested during their implantation periods.

As already noted, when the identical bulk materials have their surface energies elevated by a treatment such as exposure to a gas plasma, almost no scar-like amorphous separating layer is present at the implant boundary. High magnification views of the tissue interface region, as mechanically separated from a high-surface-energy implant, reveal no scar-like, fibrous separating layers. Rather, the cells fragment in a manner consistent with cohesive failure within the cells themselves. Confirming this interpretation, scanning electron micrographs illustrate the surface condition of the originally high-surface-energy implants to be dominated by membrane fragments and internal cellular debris left behind by cohesive failure in the tissue mass itself (after removal of the implant).

Thus, modification of the surface energy state of equally smooth implants can lead to circumstances as divergent as no practical adhesion at all versus the strongest practical adhesion of which the system is capable (that is, the cohesive strength of the weaker of the two bound phases—in this case, the biological tissue itself). The higher energy surfaces induce not only abundant cellularity, but morphologic features of cell flattening, spreading, and tight integration within a fibrous matrix. The same cells from the same hosts over the same time period go on to differentiate to strikingly different populations, depending upon the surface energies of the substrata against which they reside. This observation can lead to numerous practical applications in diverse fields of biotechnology. The question remains for future work to determine if graded responses can be selected, as desired for varying biomedical device or bioreactor requirements. Can one obtain selective cell retention and modest adhesion strength or must it be an all-or-none phenomenon?

8. RESEARCH PRIORITIES FOR FUTURE BIOADHESION STUDIES

Many examples of biological adhesion share striking similarities. For instance, the microbial fouling processes that occur in the oral cavity compare well with events in cooling water structures and processing equipment for various dairy products.

An unfulfilled need exists for improved understanding on a basic level of the deposition and binding of macromolecules and of microorganisms at solid and semisolid (hydrogel, tissue) surfaces. Detailed knowledge of these fundamental processes is required to inform our attempts to control bioadhesion on a general basis. Specific areas for study include the effects of different surface properties (texture, charge, chemistry, energy) on retention of macromolecules, seeking evidence for selective binding of particular components or subfractions, determining the actual degree of coverage of the original surfaces, assessing the orientation of the adsorbed molecular entities, and defining the modifications to the original surface properties resulting from the adsorption events. Further, it is important to learn more about the specific bacteria or cells first colonizing materials, particularly noting responses to molecules concentrated at the interfaces. The resultant tendency for attached bacteria to produce polymeric exudates must be examined carefully. These exudates both permanently bind the microorganisms to the materials in question and engulf the growing, metabolizing units in a protective "coat of slime."

It is extremely important to apply the most modern analytical methods to the problems of identifying the nature of any exopolymers produced, addressing as well their mode of production and changes with time and conditions. The reactions between these exuded polymeric products and the first adsorbed, usually glycoproteinaceous, films that provide direct binding of cells, microbes, or other organisms to the starting surfaces must be ascertained. Improved measures of the actual strengths of adhesion between the exopolymers and the films coating solid surfaces of differing physical chemical states must be developed. The practical focus of these efforts should be on the identification of any "weak links" in the chain of cellular colonization, growth, and binding that may be subject to intended or inadvertent interruption.

Other aspects of bioadhesive events are of equally urgent import, specifically as they influence the events occurring at the initial attachment interfaces. It is recognized that diffusion and other transport processes in the bulk biological phases may limit the rates of attachment of biomass to different surfaces. Further, it is known that the compositions, adhesive and cohesive strengths, and densities of the attached cellular layers, as well as the geometry and hydrodynamic features of specific locales, can heavily influence the conditions of biofilm formation. These same factors influence the removal of biodeposits from the interfacial zones. Based upon the "zones of biocompatibility" critical-surface-tension criteria reviewed here, one important lesson for microfouling limitation is that complete prevention of initial film formation and early bacterial adhesion is not necessary: appropriate adjustments of material surface properties can affect—indeed, enhance or limit—the rate and reliability of re-entrainment of original deposits into the adjacent aqueous phases.

Another strong recommendation is the selection of uniform testing methods for initial material properties, film adsorption, and cellular attachment, so that direct comparisons of different materials in different biological circumstances can be made.[58] What seems to be required are standardized units that can be mounted in numerous biological milieus, exposing test panels to measurable flow rates, shear forces, nutrient conditions, and so on.

The purpose, of course, is to provide biofouling indices with comparable meanings under different conditions, at different times, and in different sites. For example, availability and regular use of inexpensive, hygienic, standardized test units could provide a continually expanding data base concerning the effects of material surface properties, diets, and prophylactic treatments on rates of dental plaque formation under a variety of physical, chemical, and biological conditions.

More effective means of cleaning solid surfaces also are required, since it is obvious that no current techniques are routinely able to completely remove cellular and adsorbed film layers closest to the material surfaces without threatening the material's integrity. When incomplete cleaning is accepted, subsequent buildup of secondary biofouling deposits is inevitable and can occur at rates much more rapid than those initially observed on truly clean (or beneficially coated) substrata. Thus, research directed at more complete removal of adherent biomass and its precursor films, by both mechanical and chemical cleaning techniques, is essential.

9. SUMMARY AND CONCLUSIONS

Events of bioadhesion are critical in both medical environments (blood, tissue, etc.) and engineering environments (food handling, ocean engineering, biotechnology applications, etc.). Different circumstances or environments require different degrees or extremes of adhesion. Generally, in the blood environment, in food handling equipment, and in the oceanic environment, *prevention* of adhesion is sought; adhesion *promotion* needs tend to dominate in tissue environments and in "biotechnology" applications. There is initial colonization (macromolecular and/or cellular) of all engineering systems exposed to biological media. Relative *strengths of adhesion* are related to initial surface tensions of the solid substrata, and reflect the *integrity* of the first "conditioning" films, which are usually dominated by proteins (even in the case of biomineral deposits). Fluid flow effects (such as shear stress) also play a critical role in the formation of biological fouling films, often controlling the rate of film component deposition and re-entrainment.

The identities and two- and three-dimensional structures of conditioning films in a variety of biological environs must be determined to better understand how man-made materials can be optimized for the desired (bioadhesion prevention or promotion) interfacial consequences. This can be accomplished by application of a sequence of nondestructive surface analytical techniques to biological films as diverse as human skin *in situ* and the "skin" on one's teeth. Plaque formation in the human oral cavity follows a path similar to that of fouling of heat exchangers: "conditioning" film deposition, microorganism attachment, exopolymer production, and mineralization occur in both circumstances. Preferred surface energy ranges can favor or inhibit permanent biological adhesion. Useful trends are seen in recent research results from medical device trials including those for artificial hearts, dental implants, artificial hips, and substitute blood vessels. Observations of calcification of the blood-contacting surfaces of flexing elastomeric heart-assist sacs, and improvements in tissue adhesion to glow-discharge-treated metallic implants are examples of the new findings.

Finally, research priorities in the field of bioadhesion surfaces must be brought into focus, emphasizing the need for additional attention to the effects of different surfaces on macromolecular adsorption, to the properties of the various cells colonizing immersed surfaces, to the nature of the exopolymers anchoring cells to surfaces, to reactions between exopolymers and surface "conditioning" films, to transport processes in the bulk aqueous phases, to the geometry and hydrodynamics of specific systems, and to improved techniques for removal of biological films.

10. POSTSCRIPT: UNIVERSALITY OF THE PRINCIPLE OF BIOADHESION

It is both gratifying and impressive to witness the continuing accumulation of examples wherein the minimum "strength" of biological adhesion is repeatedly associated with the same defined range of surface energies, or critical surface tensions, for solid substrata in extremely diverse biological circumstances. For example, Grinnell has demonstrated the minimum binding strength (minimum resistance to detachment, more properly) for cells in tissue culture to be associated with materials known from other measurements to have critical surface tensions between 20 and 30 dyn/cm.(59,60) Work in our laboratory confirms, with substrata ranging in surface free energy from about 10 to 50 dyn/cm, that the spread areas and associated degrees of distortion of human blood platelets, freshly obtained from plasmapheresis, are minimized for surfaces in the zone between 20 and 30 dyn/cm.(61) Even the shear rate or shear force induction of hemolytic damage to circulating red blood cells has been shown to be minimized when the critical surface tensions of the plates of the shearing devices were adjusted to the mid-20 dyn/cm range.(62)

Indeed, one of the most compelling proofs of, at least, the bioengineering utility of surface energy modification to minimize biological adhesion has been in the successful, now long-term, implantation of substitute blood vessels and heart valves and, more recently, total artificial hearts and left ventricular assist devices, that do not accumulate blood clots or thrombotic masses during their pulsatile contact with unmodified blood. Since surfaces contacting blood have been scrutinized for more than two decades, the instruction we gain from examining the usual and unusual sequelae of blood cell adhesion to nonphysiologic surfaces provides valuable lessons in the general features of biological adhesion in other, less well-studied, circumstances.(63,64) Thus, it is seen that deliberate modification of the surface properties of immersed solid materials, specifically to low-surface-energy ranges between 20 and 30 dyn/cm, prevents successful interface conversion by the spontaneously depositing glycoproteinaceous macromolecules. This induction of a poor "primer coat" inhibits the practical, long-term adhesion of biological cells, even those cells specialized in surface colonization.

Having accepted and met the challenge of establishing "biocompatibility," in the sense of preventing biological adhesion in some cases and promoting that adhesion in others, with prosthetic devices, a good bit of the mystique associated with the placement of new synthetic materials in biological environments has been overcome. The stage is set, therefore, for the even greater challenge of developing processes for surface conversion of a wider diversity of engineering materials and of the architecturally superior materials of natural origin that might be taken from animal or human donors for use in life- and limb-saving applications in unrelated human recipients.

ACKNOWLEDGMENTS

Many of the aspects of bioadhesion reviewed in the foregoing pages were investigated cooperatively with colleagues at numerous locations, all of whom are sincerely thanked for the privilege of their association. Special appreciation is due Dr. V. DePalma, Dr. C. Akers, Dr. D. Goupil, Ms. R. Zontek, Mr. E. Gasiecki, and Mr. R. King, all of whom contributed significantly to some of the projects. Dr. J. Natiella, Dr. M. Meenaghan, Dr. H. Flynn, Dr. L. Weiss, Dr. G. Nancollas, Dr. C. Mouton, Dr. S. Sorensen, and Dr. J. Lippes, all senior colleagues at the State University of New York at Buffalo, provided leadership in many of the program areas, as well as shared critical experimental resources. Dr. W. Pierce and Mr. H. Donachy of Hershey Medical Center, Pennsylvania State University, are especially thanked for allowing the authors to study the artificial heart devices from their laboratory. Dr. P.-O Glantz of the University of Lund was instrumental in initiating and carrying out the studies in the oral cavity, and Dr. K. C. Marshall of the University of New South Wales generously contributed expertise in microbial fouling. The authors' primary colleagues in studies of natural and artificial blood vessels have been Dr. W. Abbott of Harvard Medical School and Dr. H. Dardik of the New Jersey College of Medicine and Dentistry. Financial support from the Office of Naval Research, the National Institutes of Health, the Department of Energy, and several corporations made this work possible.

REFERENCES

1. L. Weiss, *The Cell Periphery, Metastasis and Other Contact Phenomena,* Wiley, New York (1967).
2. R. E. Baier, H. Gucinski, M. A. Meenaghan, J. Wirth, and P.-O. Glantz, Biophysical studies of mucosal surfaces, in: *Oral Interfacial Reactions of Bone, Soft Tissue & Saliva* (P.-O. Glantz, S. A. Leach, T. Ericson, eds.), pp. 83–95, IRL Press Limited, Oxford, England (1984).
3. R. E. Baier, The organization of blood components near interfaces, *Ann. N.Y. Acad. Sci.* **283,** 7–36 (1977).
4. J. H. Waite, Adhesion in byssally attached bivalves, *Biol. Rev.* **58,** 209–231 (1983).
5. G. A. Young and D. J. Crisp, Marine animals and adhesion, in: *Adhesion-6* (K. W. Allen, ed.), pp. 279–313, Applied Science Publishers, Barking, Essex (1982).
6. R. E. Baier, Initial events in microbial film formation, in: *Marine Biodeterioration: An Interdisciplinary Study* (J. D. Costlow and R. C. Tipper, eds.), pp. 57–62, Naval Institute Press, Annapolis, MD (1984).
7. R. E. Baier, Occurrence, nature, and extent of cohesive and adhesive forces in dental integuments, in: *Surface Chemistry and Dental Integuments* (A. Lasslo and R. P. Quintana, eds.), Chapter 5, pp. 337–391, C. C. Thomas, Publisher, Springfield, IL (1973).
8. R. E. Baier and R. C. Dutton, Initial events in the interaction of blood with a foreign surface, *J. Biomed. Mater. Res.* **3,** 191–206 (1969).
9. R. E. Baier, Influence of the initial surface condition of materials on bioadhesion, in: *Proceedings of the Third International Congress on Marine Corrosion and Fouling,* pp. 633–639, Northwestern University Press, Evanston, IL (1973).
10. R. E. Baier, J. R. Natiella, A. E. Meyer, and J. M. Carter, Importance of implant surface preparation for biomaterials with different intrinsic properties, in: *Tissue Integration in Oral & Maxillo-Facial Reconstruction* (D. van Steenberghe, ed.), pp. 13–40, Excerpta Medica, Amsterdam (1985).
11. J. W. Boretos, W. S. Pierce, R. E. Baier, A. F. Leroy, and H. J. Donachy, Surface and bulk characteristics of a polyether urethane for artificial hearts, *J. Biomed. Mater. Res.* **9,** 327–340 (1975).
12. R. E. Baier, C. K. Akers, J. R. Natiella, M. A. Meenaghan, and J. Wirth, Physiochemical properties of stabilized umbilical vein, *Vasc. Surg.* **14,** 145–157 (1980).

13. R. E. Baier, Modification of surfaces to reduce fouling and/or improve cleaning, in: *Fundamentals and Applications of Surface Phenomena Associated with Fouling and Cleaning in Food Processing* (B. Hallstrom, D. B. Lund, and Ch. Tragardh, eds.), pp. 168–189, Division of Food Engineering, Lund University, Alnarp (Sweden) (1980).
14. R. E. Baier, Conditioning surfaces to suit the biomedical environment: recent progress, *J. Biomech. Eng.* **104,** 257–271 (1982).
15. R. E. Baier, A. E. Meyer, J. R. Natiella, and J. M. Carter, Surface properties determine bioadhesive outcomes: methods and results, *J. Biomed. Mater. Res.* **18,** 337–355 (1984).
16. P.-O. Glantz and R. E. Baier, Recent studies on nonspecific aspects of intraoral adhesion, *J. Adhes.* **20,** 227–244 (1986).
17. C. W. Hall and J. J. Ghidoni, Indirect percutaneus cannula construction, *J. Surg. Res.* **25,** 122–128 (1978).
18. R. E. Baier, *Applied Chemistry at Protein Interfaces,* Advances in Chemistry Series, **145,** American Chemical Society, Washington, D.C. (1975).
19. W. Norde and J. Lyklema, Thermodynamics of protein adsorption, *J. Colloid Interface Sci.* **71,** 350–366 (1979).
20. R. E. Baier, E. G. Shafrin, and W. A. Zisman, Adhesion: mechanisms that assist or impede it, *Science* **162,** 1360–1368 (1968).
21. W. A. Zisman, Relation of the equilibrium contact angle to liquid and solid constitution, *Advances in Chemistry Series* **43,** 1–51, American Chemical Society, Washington, D.C. (1964).
22. R. E. Baier, Comments on cell adhesion to biomaterial surfaces: conflicts and concerns, *J. Biomed. Mater. Res.* **16,** 173–175 (1982).
23. R. J. Good, M. K. Chaudhury, and C. J. van Oss, Theory of adhesive forces across interfaces: (2) interfacial hydrogen bonds as acid–base phenomena and as factors enhancing adhesion, in: *Fundamentals of Adhesion* (L.-H. Lee, ed.), Chapter 4, this volume.
24. R. E. Baier and A. E. Meyer, Biosurface chemistry for fun and profit, *Chemtech.* **16,** 178–185 (1986).
25. R. E. Baier, Modification of surfaces to meet bioadhesive design goals: a review, *J. Adhes.* **20,** 171–186 (1986).
26. R. E. Baier and A. E. Meyer, Surface energetics and biological adhesion, in: *Physicochemical Aspects of Polymer Surfaces,* Vol. 2 (K. L. Mittal, ed.), pp. 895–909, Plenum Publ. Corp., New York (1983).
27. R. E. Baier and A. E. Meyer, Implant surface preparation, *Int. J. Oral Maxillofacial Implants* **3,** 9–20 (1988).
28. R. E. Baier, Noninvasive, rapid characterization of human skin chemistry *in situ, J. Soc. Cosmet. Chem.* **29,** 283–306 (1978).
29. R. E. Baier, Surface chemistry in epidermal repair, in: *Epidermal Wound Healing* (H. I. Maibach and D. T. Rovee, eds.), pp. 27–48, Year Book Medical Publishers, Chicago (1972).
30. N. J. Harrick, *Internal Reflection Spectroscopy,* Interscience, New York (1967).
31. R. E. Baier and P.-O. Glantz, Characterization of oral *in vivo* films on different types of solid surfaces, *Acta Odontol. Scand.* **36,** 289–301 (1978).
32. R. E. Baier and W. A. Zisman, Wettability and multiple attenuated internal reflection infrared spectroscopy of solvent-cast thin films of polyamides, *Macromolecules* **3,** 462–468 (1970).
33. R. E. Baier and G. I. Loeb, Multiple parameters characterizing interfacial films of a protein analogue, polymethylglutamate, in: *Polymer Characterization: Interdisciplinary Approaches* (C. D. Craver, ed.), pp. 79–96, Plenum Press, New York (1971).
34. R. E. Baier, Blood compatibility of synthetic polymers: Perspective and problems, in: *Polymers in Medicine and Surgery* (R. Kronenthal, Z. Oser, and E. Martin, eds.), pp. 139–159, Plenum Press, New York (1975).
35. R. E. Baier, Physical chemistry of the vascular interface: composition, texture, and adhesive quality, in: *Vascular Grafts* (P. N. Sawyer and M. J. Kaplitt, eds.), pp. 76–107, Appleton-Century-Crofts, New York (1978).
36. R. E. Baier and W. M. Abbott, Comparative biophysical properties of the flow surfaces of contemporary vascular grafts, in: *Graft Materials in Vascular Surgery* (H. Dardik, ed.), pp. 79–103, Symposia Specialists, Miami, FL (1978).
37. V. A. DePalma and R. E. Baier, Flow cell and method for continuously monitoring deposits on flow surfaces, U.S. Patent #4,175,233 (Nov. 20, 1979).
38. V. A. DePalma and R. E. Baier, Microfouling of metallic and coated metallic flow surfaces in model heat

exchange cells, Proceedings of the Ocean Thermal Energy Conversion (OTEC) Biofouling and Corrosion Symposium, pp. 89–106, United States Department of Energy (1978).
39. V. A. DePalma, Apparatus for zeta potential measurement of rectangular flow cells exposed to flowing protein solutions, *Rev. Sci. Instrum.* **51,** 1390–1395 (1980).
40. V. A. DePalma, D. W. Goupil, and C. W. Akers, Field demonstration of rapid microfouling in model heat exchangers: Gulf of Mexico, November 1978, Proceedings of the 6th OTEC Conference, U.S. Department of Energy, 12.14-1-9 (1979).
41. R. E. Baier, Adhesion to different types of biosurfaces, in: *Dental Plaque and Surface Interactions in the Oral Cavity* (S. A. Leach, ed.), pp. 31–47, Information Retrieval Inc., Arlington, VA (1980).
42. R. E. Baier and V. A. DePalma, The relation of the internal surface of grafts to thrombosis, in: *Management of Arterial Occlusive Disease* (W. A. Dale, ed.), pp. 147–163, Year Book Medical Publishers, Chicago (1971).
43. R. E. Baier, The role of surface energy in thrombogenesis, *Bull. N.Y. Acad. Med.* **48,** 257–272 (1972).
44. A. E. Meyer, R. E. Baier, and R. W. King, Initial fouling of nontoxic coatings in fresh, brackish, and sea water, *Can. J. Chem. Eng.* **66,** 55–62 (1988).
45. R. E. Baier, V. A. DePalma, A. E. Meyer, R. W. King, and M. S. Fornalik, Control of heat exchange surface microfouling by material and process variations, in: *Fouling in Heat Exchange Equipment* (J. M. Chenoweth and M. Impagliazzo, eds.), HTD- Vol. 17, pp. 97–103, The American Society of Mechanical Engineers, New York (1981).
46. S. C. Dexter, Influence of substrate wettability on the formation of bacterial slime films on solid surfaces immersed in natural sea water, in: *Proceedings of the Fourth International Congress on Marine Corrosion and Fouling,* pp. 137–144, Cent de Rech et D'Etud Oceanograph., Boulogne, France (1977).
47. S. C. Dexter, J. D. Sullivan, J. Williams III, and S. W. Watson, Influence of substrate wettability on the attachment of marine bacteria to various surfaces, *Appl. Microbiol.* **30,** 298–308 (1975).
48. D. J. Crisp and P. S. Meadows, Adsorbed layers: the stimulus to settlement in barnacles, *Proc. R. Soc. London, Ser. B* **158,** 364–387 (1963).
49. D. W. Goupil, V. A. DePalma, and R. E. Baier, Physical/chemical characteristics of the macromolecular conditioning film in biological fouling, in: *Proceedings of the Fifth International Congress on Marine Corrosion and Fouling,* pp. 401–410, Graficas Orbe, Madrid, Spain (1980).
50. R. E. Baier, Substrata influences on the adhesion of microorganisms and their resultant new surface properties, in: *Adsorption of Microorganisms to Surfaces* (G. Bitton and K. C. Marshall, eds.), pp. 59–104, Wiley-Interscience, New York (1980).
51. W. S. Pierce, H. J. Donachy, N. Rosenberg, and R. E. Baier, Calcification inside artificial hearts: inhibition by warfarin-sodium, *Science* **208,** 601–603 (1980).
52. W. S. Pierce and H. J. Donachy, Polymer alloy blood compatible surface, United States Patent #4,312,920 (January 26, 1982).
53. R. E. Baier, Advanced biomaterials development from 'natural products', *J. Biomater. Appl.* **2,** 615–626 (1988).
54. M. G. Buonocore, Bonding to hard dental tissues, in: *Adhesion in Biological Systems* (R. S. Manly, ed.), Chapter 15, pp. 225–254, Academic Press, New York (1970).
55. A. J. Gwinnett and A. Matsui, A study of enamel adhesives. The physical relationship between enamel and adhesive, *Arch. Oral Biol.* **12,** 1615–1620 (1967).
56. A. J. Gwinnett and A. Matsui, The scientific basis of the sealant procedure, *J. Prevent. Dent.* **3,** 15–28 (1976).
57. R. E. Baier and V. A. DePalma, Electrodeless glow discharge cleaning and activation of high-energy substrates to insure their freedom from organic contamination and their receptivity for adhesives and coatings, Calspan Corporation, Report No. 176, 17 pp., Calspan Corporation, Advanced Technology Center, Buffalo, NY 14225 (1970).
58. R. E. Baier, Selected methods of investigation for blood-contact surfaces, *Ann. N.Y. Acad. Sci.* **516,** 65–77 (1987).
59. F. Grinnell, M. Milam, and P. Srere, Adhesion of cells to surfaces of diverse chemical composition and inhibition of adhesion by sulfhydryl binding reagents, *Arch. Biochem. Biophys.* **153,** 193–198 (1972).
60. F. Grinnell, Cellular adhesiveness and extracellular substrata, *Int. Rev. Cytol.* **53,** 65–144 (1978).
61. R. E. Baier, V. A. DePalma, D. W. Goupil, and E. Cohen, Human platelet spreading on substrata of known surface chemistry, *J. Biomed. Mater. Res.* **19,** 1157–1167 (1985).

62. M. C. Williams, Shear-induced hemolysis: materials and blood composition, Report No. N01-HB-3-2952, 58 pp., National Technical Information Service, Springfield, VA (1975).
63. R. E. Baier, Surface chemical factors presaging bioadhesive events, *Ann. N.Y. Acad. Sci.* **416,** 34–57 (1983).
64. R. E. Baier, Surface properties influencing biological adhesion, in: *Adhesion in Biological Systems* (R. S. Manly, ed.), Chapter 2, pp. 15–48, Academic Press, New York (1970).

About the Contributors

Lieng-Huang Lee (editor) is a Senior Scientist of Xerox Corporation. He received his B.S. degree in Chemistry from Xiamen (Amoy) University Xiamen, China, and his M.S. and Ph.D. degrees in Chemistry from Case Institute of Technology, Cleveland, Ohio. He is the editor of 12 books on adhesion, friction, and wear of polymers. In addition, Dr. Lee has published over 75 technical papers and holds twenty-nine U.S. patents. He is a member of the Board of Editors for the *Journal of Adhesion*. In 1976, he was the Chairman of the Division of Organic Coatings and Plastics Chemistry of the American Chemical Society. In 1983, he was chosen as one of the distinguished scholars by the National Academy of Sciences and the Chinese Academy of Sciences, and was awarded a Visiting Professorship in Chemistry by Xiamen University in 1986, and an honorary professorship by the Chinese Academy of Sciences in 1988. Dr. Lee is listed in several biographical directories, including *World's Who's Who in Science*.

John E. E. Baglin is a research manager in the Physical Science Department at the IBM Almaden Research Center. His current interests include ion and electron beam interactions in solids, thin film interactions, and the basic phenomena of thin film adhesion. His publications number more than 120, addressing topics in thin film physics, laser annealing, materials characterization, and nuclear physics. He received the Ph.D. in Physics from the University of Melbourne in 1963, and held appointments at Iowa State University/Ames Laboratory and at Yale University prior to joining the IBM Research Division in 1972. Dr. Baglin is a Counsellor and Past President of the Materials Research Society, and serves on the Materials Education Council.

Robert E. Baier is Research Professor of Biophysical Sciences and Co-Director of the National Science Foundation Industry/University Cooperative Research Center for Biosurfaces at the State University of New York at Buffalo. He received his Ph.D. in 1966 from the State University of New York at Buffalo in Biophysics and is a licensed Professional Engineer in the states of New York and Ohio. From 1968 through 1984, he was a Staff Scientist of the Advanced Technology Center of Arvin/Calspan (formerly the Cornell Aeronautical Laboratory of Cornell University) in Buffalo, New York and was awarded the Society for Biomaterials Clemson Award in 1983 for his contributions to basic research in the biomaterials field. From 1985 to July 1989, Dr. Baier was Director of the New York State Center for Advanced Technology in Health-care Instruments and Devices, leaving that post to devote more time to the Center for Biosurfaces.

Amitava Banerjea received his early Physics education from Delhi University, India, and his M.Tech. from the Indian Institute of Science. In 1983 he was awarded his Ph.D. in theoretical solid-state physics at Case Western Reserve University. His research interests have included the statistical mechanics of one-dimensional systems, the magnetic properties of light actinide compounds, and the energetics of solids. At present he spends part of his time in studying high-temperature superconductivity and the rest in studying various materials phenomena.

F. Brochard-Wyart studied at ENSET (Ecole Normale Supérieure de l'Enseignement Technique) in 1964 and gained her degree in Solid State Physics at the University of Orsay. She became a theorist in the Liquid Crystal Group (Orsay) and obtained her Ph.D. degree in 1974. In 1977, she moved to the Condensed Matter Group (Collège de France) and worked on the dynamics of polymers. In 1985, Dr. Brochard-Wyart was appointed Professor of Chemical Physics at Université P. et M. Curie and is now a member of the "Structure and Reactivity at Interfaces" Laboratory at this university, working mainly on wetting, dewetting, and phase transitions in monolayers.

Manoj K. Chaudhury graduated with a Ph.D. degree in Chemical Engineering from the State University of New York at Buffalo in 1984. He carried out his doctoral research under Professor R. J. Good on topics related to contact angle and wettability. Currently, Dr. Chaudhury is employed at Dow Corning Corp., Silicone Research Division. His research interests are in surface chemistry and adhesion. Dr. Chaudhury is the recipient of the V. K. LaMer Award in Surface Chemistry and the H. Doan Industrial Scientist Award. In 1988, he did research work in the Whitesides Laboratory at Harvard University.

P. G. de Gennes studied at Ecole Normale Supérieuere between 1951 and 1955. He has been Professor of Physics at Collège de France since 1971 and, since 1976, Director of Ecole de Physique et Chimie. Dr. de Gennes has published over 300 technical papers and three books: *Superconducting Metals and Alloys* (Benjamin, 1964), *The Physics of Liquid Crystals* (Oxford, 1974), and *Scaling Concepts in Polymer Physics* (Cornell, 1979). He has won many prizes and, in 1988, was the recipient of the Award in Polymer Chemistry of the American Chemical Society. Dr. de Gennes is a member of the French Academy of Sciences, the Dutch Academy of Sciences, the American Academy of Arts and Sciences, and the Royal Society. Dr. de Gennes won the 1990 Wolf Prize in Physics.

F. Faupel is presently employed at the Institut für Metall-physik of the University of Göttingen, West Germany. His research interests are in the field of material transport and thermodynamics of amorphous and crystalline alloys, high-temperature superconductors, and polymers. He received his Ph.D. in physics from the University of Göttingen in 1985. During 1985–1987 Dr. Faupel carried out research at University of Göttingen. In 1987 and 1988 he worked at the IBM Thomas J. Watson Research Center at Yorktown Heights, New York on metal–polymer interfaces and thin film reactions.

John Ferrante received his B.S. in Physics from Case Institute of Technology and his Ph.D. in Theoretical Solid-State Physics from Case Western Reserve University. He is currently a senior scientist in the Office of the Chief Scientist at the NASA Lewis Research Center. His early research was in experimental surface physics with applications in electric propulsion and thermionics. Later, Dr. Ferrante contributed to the application of surface science to tribology. Most recently, he has concentrated on solid-state theory applied to surfaces and defect energetics in solids, with particular emphasis on interfacial adhesion.

Robert J. Good received his B.A. in Chemistry from Amherst College in 1942, his M.S. from the University of California, and his Ph.D. from the University of Michigan. He taught from 1956 to 1959 at the University of Cincinnati, and, since 1964, has been Professor of Materials Science at the State University of New York at Buffalo. In 1957, with one of his first students (Louis A. Girifalco) and with Dr. Gerard Kraus, he proposed the "geometric mean" relation for the energy of adhesion between two phases across an interface and the energies of cohesion of the separate phase. Recent interests include hydrogen bonding across interfaces, contact angles of liquids on solids, surface thermodynamics, microemulsions, the role of viscoelasticity and elongational viscosity in practical adhesion, and the theory of passivity. He received the Kendall Award of the ACS in 1976 and the Schoellkopf Award of the Western New York Section of the ACS in 1979. He is the author or coauthor of 120 papers and one patent, and is coeditor of three books.

W. Gutowski obtained his M.Sc. in Mechanical Engineering in 1971 and his Ph.D. in Materials Science in 1975 from the Technical University of Stettin, Poland. In 1981, he immigrated to

Australia. In 1983, he joined the CSIRO (Commonwealth Scientific and Research Organization) Division of Building Research and is a Group Leader in the Adhesives and Adhesion Group. Dr. Gutowski has published over 30 papers and holds 29 patents.

Richard Haight obtained his Ph.D. in Physics from the State University of New York at Albany (1983). He held a postdoctoral position (1983–1985) at AT&T Bell Laboratories, where he developed a laser-based photoemission source for picosecond studies of transiently excited materials. Since 1985 he has been a research staff member at the IBM Watson Research Center.

Dan A. Hays obtained a doctorate in physics from Rutgers University in 1966. Following a postdoctoral position at the University of Pittsburgh, he joined Xerox Corporation in 1968 where he is a principal scientist in the Webster Research Center near Rochester, New York. His research interests include contact electrification, charged particle adhesion, and xerographic development systems.

Paul S. Ho is Manager of the Interface Science Group in the Physical Sciences Department at the Thomas J. Watson Research Center of IBM. He obtained a B.S. degree from National Cheng Kung University, Taiwan and a Ph.D. degree from Rensselaer Polytechnic Institute. Dr. Ho joined IBM in 1972 from Cornell University, where he was Associate Professor in the Department of Materials Science and Engineering. His current research is in the areas of thin films and interfaces in multilayered structures for VLSI applications. He has authored or coauthored many technical patents and published extensively in the area of thin films and interfaces. Dr. Ho is a Fellow of the American Physical Society.

Hatsuo Ishida received both his B.S. degree in Chemical Engineering and his M.S. degree in Industrial Chemistry from Doshisha University, Kyoto, Japan, and his Ph.D. degree in Macromolecular Science from Case Western Reserve University, Cleveland, Ohio. He is currently Professor of Macromolecular Science at Case Western Reserve University. Dr. Ishida is also Director of the C. R. Newpher Polymer Composite Laboratory and a newly established Organic Ferromagnetic Laboratory. He has pioneered in the molecular characterization of composite interfaces. Since 1983, he has been General Chairman of the International Conference on Composite Interfaces. Dr. Ishida is also active in the adhesion and corrosion protection of coating systems. He is on the advisory board of the *Journal of Adhesion*, and has over 150 papers and four edited books to his credit.

Anne E. Meyer is a Lecturer with the Faculty of Engineering and Applied Sciences at the State University of New York at Buffalo and is Program Manager of the National Science Foundation Industry/University Cooperative Research Center for Biosurfaces. She was Associate Director of the University's Surface Science Center (1985–1989), the precursor organization to the Center for Biosurfaces, and was Associate Director of the New York State Center for Advanced Technology in Health-Care Instruments and Devices for the 1988/89 program year. Ms. Meyer was head of the Surface Science Section of the Advanced Technology Center of Arvin/Calspan Corporation (Buffalo, New York) from 1983 to 1986; she joined the company as an analytical chemist in 1974.

James D. Miller is a senior chemist with Amoco Performance Products, Inc. and works in the areas of carbon fibers and advanced composites. He received his B.S. from the University of Michigan and M.S. and Ph.D degrees in Macromolecular Science from Case Western Reserve University. Research interests include fiber surface chemistry and surface spectroscopies.

Edwin P. Plueddemann is a Scientist Emeritus of research at Dow Corning Corporation; he obtained his Ph.D. degree from Ohio State University in 1942. He is the author of a book, *Silane Coupling Agents* (1982), and about 80 technical articles. Dr. Plueddemann also holds 80 U.S. patents. He won several awards, including the American Chemical Society Award for Creative Invention (1984) and the Adhesion Society Award for Excellence in Adhesion Science (1988).

B. D. Silverman is a Research Staff Member at the IBM Thomas J. Watson Research Center, Yorktown Heights, New York. Dr. Silverman received his Ph.D in Theoretical Solid-State Physics at Rutgers University in 1959 and, in 1969, joined the IBM Research Laboratory in San Jose, California, where he managed a group responsible for developing materials for optical computer memories. He was transferred to Yorktown Heights in 1973, managing a group involved with the physics of conducting organic materials. Dr. Silverman has worked in such divergent areas as ferroelectronics, ultrasonics, quantum electronics, optical computer memories, organic metals, chemical carcinogenesis, and polymers. He is a fellow of the American Physical Society.

John R. Smith earned his B.S. degree at Toledo University and his Ph.D. degree in Physics at Ohio State University. He served in the U.S. Army at NASA Lewis Research Center and then did postdoctoral work at the University of California, San Diego, before joining the General Motors Research Laboratories, where he is now a principal research scientist. He is a fellow of the American Physical Society. Research interests are in the theory of solid surfaces, electronic and magnetic properties, chemisorption, superlattices, adhesion, metal-contact electronic structure, and defects and universal features of bonding in solids.

Carel J. van Oss studied in the Netherlands and France (Ph.D. 1955, Sorbonne, Paris; Physical Biochemistry). In 1968, Dr. van Oss joined the Department of Microbiology, SUNY at Buffalo. Currently he is a Professor of Microbiology and Adjunct Professor of Chemical Engineering at the State University of New York at Buffalo. He is Head of the Laboratory of Immunochemistry of the Department of Microbiology and a member of the Ernest Witebsky Center for Immunology at SUNY at Buffalo. Dr. van Oss is the author of 235 papers and numerous abstracts, and coauthor or coeditor of five books; he holds eight patents. His current research involves studies of polar and apolar interactions in macroscopic systems, cell separation methods, the nature of the interaction between antigens and antibodies, and physicochemical and immunochemical cell surface characterization.

Robert C. White received his Ph.D. in Chemical Physics from the University of Utah. He worked at the IBM Thomas J. Watson Research Center in the Physical Sciences Department until January 1988, when he joined Columbia. Dr. White has worked on quantum chemistry of strained organic molecules, gas-phase photochemistry of radicals from organics, temperature-dependent angle-resolved photoemission, and the chemistry at metal–polymer interfaces. Currently his efforts include investigation of polymer surface and interface processes and control for microelectronics packaging.

Richard P. Wool is on the faculty of the Department of Materials Science and Engineering at the University of Illinois, Urbana. He received his Ph.D. from the University of Utah in 1974. His primary research interests include studies of the dynamics, structure, and strength of polymer interfaces with application to many practical systems. Experimental, theoretical, and computer simulation techniques are used interactively to study the interrelations between dynamics, structure, and strength of interfaces.

Author Index

Subject Index